A Philosophy Reader from the Circle of Miskawayh

This volume presents the first complete edition of Oxford, MS Marsh 539, a hitherto unpublished philosophy reader compiled anonymously in the eastern Islamic world in the eleventh century. The compilation consists of texts on metaphysics, physiology and ethics, providing excerpts from Arabic versions of Greek philosophical works (Aristotle, Plotinus, Galen) and works by Arabic authors (Qusṭā ibn Lūqā, Fārābi, Miskawayh). It preserves fragments of Greek–Arabic translations lost today, including Galen's *On My Own Opinions*, the *Summa Alexandrinorum*, and Themistius on Aristotle's *Book Lambda*. The philosophy reader provides a unique insight into philosophical activity of the place and time of the well-known philosopher Miskawayh, showing us which works had entered the mainstream and were considered necessary to know. Elvira Wakelnig's edition includes a new facing-page English translation and a rich commentary which identifies the source texts and examines the historical and philosophical context of each passage.

ELVIRA WAKELNIG works on the transmission of Greek philosophy into Arabic and on Arabic philosophy. She has held research positions at the Universities of Bochum, Warwick and Vienna and is the author of *Feder, Tafel, Mensch* (2006).

A Philosophy Reader from the Circle of Miskawayh

Edited and translated by ELVIRA WAKELNIG

CAMBRIDGE
UNIVERSITY PRESS

CAMBRIDGE
UNIVERSITY PRESS

University Printing House, Cambridge CB2 8BS, United Kingdom

Published in the United States of America by Cambridge University Press, New York

Cambridge University Press is part of the University of Cambridge.

It furthers the University's mission by disseminating knowledge in the pursuit of education, learning and research at the highest international levels of excellence.

www.cambridge.org
Information on this title: www.cambridge.org/9780521193610

First published 2014

Printed in the United Kingdom by TJ International Ltd. Padstow Cornwall

A catalogue record for this publication is available from the British Library

ISBN 978-0-521-19361-0 Hardback

Contents

Preface

The present volume would never have arisen without Rotraud Hansberger and Simon Swain, who submitted a project proposal to the Arts and Humanities Research Council (AHRC) to edit, translate and study the Bodleian manuscript Marsh 539 at the Department of Classics & Ancient History, University of Warwick. To them and to the AHRC for funding the project are owed my first and foremost thanks.

By the time the funding was granted, Rotraud Hansberger had already been whisked off to a Junior Research Fellowship at King's College Cambridge, and I found myself in the lucky position of being asked to undertake the project in her place.

In what followed I was given the opportunity to spend three amazing years in Oxford and Warwick working on a most interesting manuscript and enjoying the company of many colleagues and friends, some of them in London as well. To all the people who made these three years such an outstanding experience I want to express my sincerest thanks.

There are a number of people whom I would like to thank in person. First of all, Simon Swain, who, apart from playing a major role in putting this project together, took a vivid interest in its progress throughout, gave invaluable advice and read the entire draft – and certain bits and pieces more than once – in order to discuss the Arabic and especially to help with the establishment of the English translation.

Rüdiger Arnzen also went through the text, translation, commentary and introduction and made very helpful comments and suggestions for different readings of the Arabic and the improvement of the translation, and saved me from many a mistake.

Harry Munt is the third person who read my entire manuscript and made countless invaluable suggestions for the improvement of the English text.

Emilie Savage-Smith encouraged me with her constant interest in the project and provided a lot of advice, especially – but not only – for the physiological passages. She also kindly let me use her microfilms of the Arabic version of Galen's *De usu partium*.

Fritz Zimmermann lent me a further microfilm of that particular text and helped immensely with my understanding of the Platonic, Pseudo-

Platonic and Pseudo-Aristotelian passages (25–9) and with deciphering some difficulties in the Plotinian passages – we once spent a very pleasant afternoon at St Cross trying to figure out how some manuscript readings could be altered so that they would make sense. We also tackled a number of problems in the Arabic Reading Class at the Warburg Institute organised by Charles Burnett and Peter Adamson, the participants of which studied parts of my text and came up with a number of helpful suggestions.

All Souls college was kind enough to host a seminar funded by the AHRC which Simon Swain and I organised, on philosophical compendia, thus focusing on the genre in which the Marsh manuscript is written.

In addition I had the opportunity to present part of the text and translation, as well as a number of my results, in the 'Research in Progress' Seminar at the Oriental Institute of the University of Oxford, organised by Geert van Gelder. I want to thank, in particular, him and Nefeli Papoutsakis for many valuable comments.

Oliver Overwien, Marwan Rashed and Philipp Vallat were kind enough to provide detailed answers to some particular questions. I am very grateful to Peter Starr for obtaining a manuscript reproduction for me in Turkey. Meryem Sebti and Uwe Vagelpohl generously shared with me some manuscript reproductions they had acquired. Musab Hayatli helped with solving a number of tricky Arabic puzzles and suggested some alternative readings. Vicky Bell took a particular interest in finding the appropriate title for the volume and we spent a pleasant evening at the Angel exploring different alternatives, none of which – unfortunately – made it to the title page in the end. Over a cup – or rather several cups – of tea, Stefan Hagel solved many of the problems which I had with displaying the edition and translation, which were, needless to say, caused by my incompetence and not by his wonderful CTE programme.

Furthermore, I would like to thank the Bodleian Library for unrestricted access to their amazing treasures and for the permission to reproduce samples of the manuscript; the Oriental Institute Library for full access; the Süleymaniye Library in Istanbul for allowing me to browse through their scanned manuscript holdings; and St Cross for accepting me as a member of Common Room.

Finally I should like to thank the Cambridge University Press readers for their valuable comments and for convincing the Press to offer me a contract. I am grateful to Hilary Gaskin for commissioning the present volume and to Leigh Mueller, Tom O'Reilly and Anna Lowe for all their efforts.

Vienna, April 2011

Postscriptum

As the present volume had still not been published by the end of October 2013 and the problems with the Arabic script proved insurmountable, Stefan Hagel saved the day and my book by layouting the entire manuscript brilliantly in his *Classical Text Editor* at very short notice. To him, I owe the deepest gratitude and the publication of this volume.

I also want to take this opportunity to thank Christophe Erismann for his continuous encouragement and for his invaluable scientific advice.

Vienna, November 2013

Introduction

I Marsh 539: a philosophy reader

The present volume contains the first complete edition of the Arabic text preserved in the Oxford manuscript, Bodleian, Marsh 539. The manuscript was first mentioned in Rosenthal's 1940 article 'On the Knowledge of Plato's Philosophy in the Islamic World' because of its excerpts from the Platonic *Leges* and the three proofs for the immortality of the soul. In a further three-part article, entitled 'Aš-Šayḫ al-Yûnânî and the Arabic Plotinus Source', Rosenthal identified excerpts in the Marsh manuscript which are attributed to the 'Greek Sage' as an important witness to the Arabic Plotinus Source. Since then, the Greek Sage has attracted the attention of a number of scholars, but until recently no attempt was undertaken to study the Marsh manuscript in its entirety. The first step towards such a study was taken by Cottrell in her article 'L'Anonyme d'Oxford (Bodleian Or. Marsh 539): bibliothèque ou commentaire?'. The present volume hopes to be a second step in the same direction as it puts at the disposal of the scholarly community the complete Arabic text of the Marsh manuscript together with an English translation and a commentary which primarily aims at providing some context for the broad variety of texts and topics gathered in the manuscript. This should provide a basis for further research on the Marsh manuscript in its entirety as well as on the Arabic Plotinus, since some of the Marsh manuscript's material of the Greek Sage has so far been only available in Lewis' English translation, published in the Henry-Schwyzer edition of the *Enneades*.

The Marsh manuscript is incomplete at the beginning and the end, and therefore offers no indication of its title, author or composition date. The question arises as to how much textual loss the manuscript has suffered. Since no other manuscript copy of the text has surfaced so far, any information must be gained from an analysis of the text itself.

The text is a compilation of philosophical material taken from various Greek and Arabic authorities quoted either by name or anonymously. The

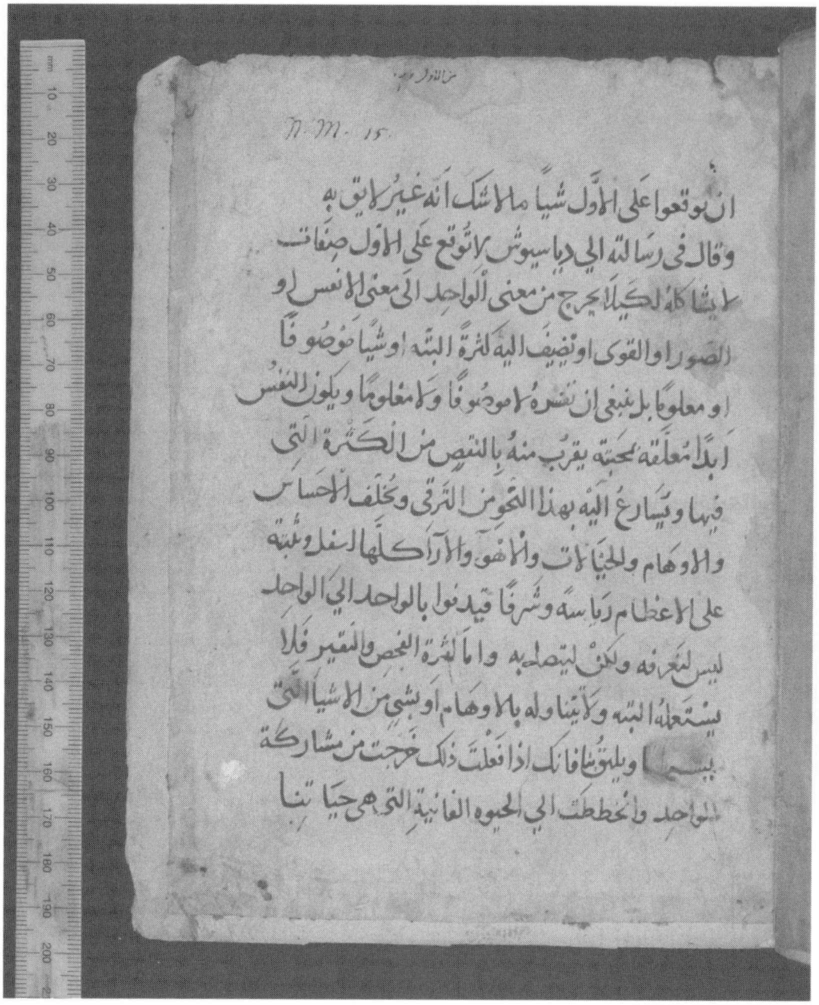

ان يوقعوا على الاول شيا مما لا شك انه غير لايق به
وقال في رسالته الي ديانسيوس لا توقع على الاول صفات
لا يشاكله لكيلا يخرج من معنى الواحد الى معنى الانس او
الصور او القوى او نضيف اليه لثرة البتة او شيا موصوفا
او معلوما بل ينبغي ان نثبته لا موصوفا ولا معلوما ويكون النفس
ابدا متعلقة بحبته يقرب منه بالنقص من الكثرة التي
فيها ويسارع اليه بهذا النحو من الترقي وتخلف الحواس
والاوهام والخيالات والاهوا والاراء كلها اسفل وتنبه
على الاعظام رياسه وشرفا فيدنوا بالواحد الي الواحد
ليس ليعرفه ولكن ليتصل به واما لثرة الفحص والتفتيش فلا
يستعمله البته ولا يتناوله بالاوهام او بشي من الاشيا التي
يبسمعها وليلوثنا فانك اذا فعلت ذلك خرجت من مشاركة
الواحد وانحططت الي الحيوه الغانية التي هي حياتنا

MS Marsh 539, fol. 5a (first page of the text).

exact sources are indicated only in a few cases.[1] The subjects dealt with are metaphysics, psychology, natural sciences in the guise of physiology, ethics and politics, in other words most of the Late Antique branches of philosophy. Thus the compilation may be described as a philosophy reader

[1] Explicitly mentioned are Aristotle's *Metaphysica* in passages (7) and (9), *Physica* (70), *Ethica* (202) and the Ps.-Aristotelian *R. al-Siyāsa al-ʿāmmīya* (18), Plato's *Phaedo* (27-8) and *Leges* (203). Less clear are the references to a *Risāla ilā Diyānusiyūs* (2) and Proclus' *Minor Elements* (14), and the references to 'a discourse' of Plato (25) and Aristotle (26) and to Aristotle 'somewhere else' (207) are completely vague. However, even as far as the more exact references are concerned, only that to the *R. al-Siyāsa al-ʿāmmīya* is unambiguous, as will be seen in the discussion of the passages.

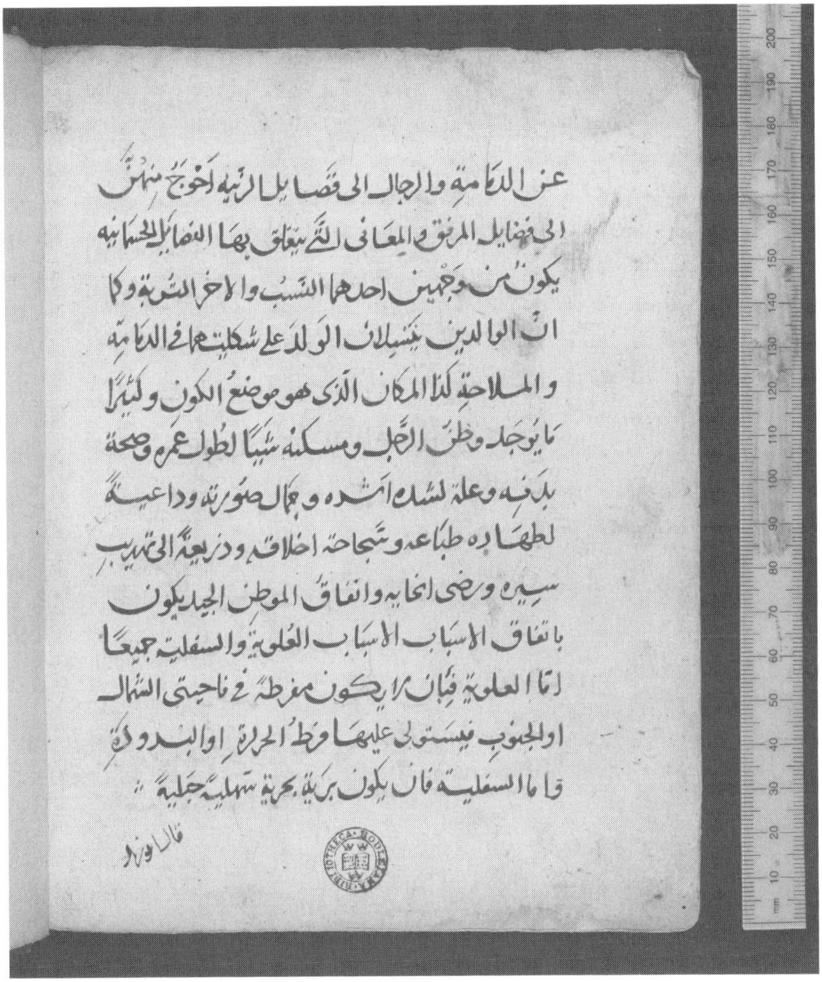

MS Marsh 539, fol. 133b (last page of the text).

and this description provides an appropriate term with which to refer to the text from now on, namely as the *Philosophy Reader* (*PR*).[2] However, what purpose the *PR* served for its readers is hard to tell with any certainty. It may have been used as a manual or textbook for studies or as an anthology of the common philosophical knowledge of its time.[3]

[2] Zimmermann ('Origins') refers to the text as 'the Oxford Anthology' and Cottrell ('L'Anonyme') as 'l'anonyme d'Oxford'.

[3] I hope to publish a thorough study of the literary genre of the *PR* in the near future.

II The author: a student of Miskawayh?

A closer look at the material contained in the *PR* gives some clues about the unknown compiler and places him within the immediate or extended circle of the tenth-century philosopher from Ray in Khorasan, Miskawayh (around 936–1030).

The first indication of a link between the compiler of the *PR* and Miskawayh may be gained from the selection of Arabic authors the compiler quotes by name. They are, in order of appearance in the text, al-Fārābī in passage (46), Abū ʿAlī Miskawayh (84, 91,[4] 94, 210), Qusṭā ibn Lūqā (145), Abū Zayd (182) and Abū l-Ḥasan al-ʿĀmirī (213). Two observations are obvious, namely that Miskawayh is the sole authority cited by name more than once,[5] and that the other quoted authorities are either his predecessors or contemporaries. This holds true also for the Arabic authors cited anonymously in the *PR*, i.e. Ḥunayn ibn Isḥāq in passages (115) and (119), ʿAlī ibn Rabban al-Ṭabarī (132–8, 187, 192, 194) and the *Ikhwān al-Ṣafāʾ* (142 and maybe 167). Thus it is possible to establish a *terminus ante quem non* for the *PR*, which may also be the *terminus ad quem*, at Miskawayh's lifetime.

The mention of Abū Zayd, i.e. al-Balkhī, and Abū l-Ḥasan al-ʿĀmirī are particularly interesting as they belong to the same philosophical tradition and cultural milieu as Miskawayh. Like him they stand in the legacy of al-Kindī and spent their lives in the east of the Islamic world with various degrees of contact with the different courts there. Early in his life al-ʿĀmirī studied with Abū Zayd, who is credited with having brought the Kindian tradition to the east,[6] and later on he was at the court of Abū l-Faḍl ibn al-ʿAmīd (d. 970), the vizier of the Buyid Emir Rukn al-Dawla (reg. 932–76). It was the very same court where Miskawayh served as a librarian for seven years. Although al-Tawḥīdī criticises Miskawayh for not having profited from the presence of al-ʿĀmirī during the time they both spent under the same patronage, they certainly knew each other person-

[4] Here the text reads only Abū ʿAlī for short.

[5] Abū Zayd, i.e. Abū Zayd al-Balkhī was probably cited for a second time in the final part of the *PR* which is not preserved in the Marsh manuscript due to the textual loss at the end of the manuscript. *Qāla Abū Zayd* is the catchword given in the margin of the last existing folio-page and thus a clear indication that the following quotation was meant to be from him. Qusṭā ibn Lūqā is quoted a second time anonymously in passage (112), and the passages following upon passage (213) may all be derived from al-ʿĀmirī.

[6] On which see Adamson, 'Kindian Tradition'.

ally.[7] This may explain why the longest continuous excerpt in the *PR*, covering nine and a half folios, is quoted from al-ʿĀmirī.[8]

The second feature which links the compiler of the *PR* to Miskawayh is that one-fifth of the material in the Marsh manuscript, i.e. about twenty-five folios, is excerpted from three different writings of his. It is remarkable that all passages explicitly ascribed to him by name derive from one single work, his *Fawz al-aṣghar* (*Minor Triumph*). Yet the quoted version is not the edited standard text, but an alternative version attested only by the excerpts in the *PR* and the *K. al-Ḥikma* (*Book of Wisdom*).[9] The question of which version is the earlier is impossible to answer with the little evidence known today. It may, however, be recalled that the standard version was composed to answer three questions on God, the soul and prophecy, which were addressed to Miskawayh by a certain ruler, and that it is not only known as *al-Fawz al-aṣghar* (*Minor Triumph*), but also as *al-Jawāb fī l-masāʾil al-thalāth* (*Answer to Three Questions*) or similar.[10] Could it thus be the case that Miskawayh composed the treatise to answer these three questions using material he had previously written down in the *Fawz*, i.e. the alternative version partly preserved in the *PR*, and that, due to its rigid tripartite structure, the new treatise was such a success that it replaced the original *Fawz*? It seems that even the compiler of the *PR* already had difficulties finding a complete copy of the alternative *Fawz*, as gaps left in the Marsh manuscript probably indicate.[11] The two other writings by Miskawayh used in the *PR* are *Tahdhīb al-akhlāq* (*Refinement of Character*) and *Tartīb al-saʿādāt* (*Succession of the Kinds of Happiness*), from which mainly quotations ascribed to Aristotle are taken up in passages (201) and (206–7). The compiler of the *PR* repeats these ascriptions and may thus have felt no need to refer to Miskawayh as well. He applies the same approach when he quotes Hippocrates, Aristotle or Plato in the words of ʿAlī ibn Rabban al-Ṭabarī in passages (132, 136), (133) or (187, 192, 194), respectively. How-

[7] Al-ʿĀmirī's arrival at Ibn al-ʿAmīd's court is reported in Miskawayh's *Tajārib al-umam*, II, p. 277, and in al-Tawḥīdī's *Akhlāq al-wazīrayn*, pp. 344–5. Al-Tawḥīdī's criticism of Miskawayh is found in the *K. al-Imtāʿ*, I, p. 36. For an account of the persons involved, see Kraemer, *Humanism*, pp. 207–59.

[8] That is, if the quotation introduced by *qāla Abū l-Ḥasan al-ʿĀmirī* in passage (213) does continue until the end of passage (222). There the Marsh manuscript breaks off and the catchword *qāla Abū Zayd* is the first indication that the compiler of the *PR* stops quoting al-ʿĀmirī, since the text of passages (213–22) is not interrupted once by a single *wa-qāla* or similar phrase.

[9] On the *K. al-Ḥikma* see below, pp. 42–3.

[10] See Wakelnig, 'New Version', p. 85, n. 8.

[11] On these gaps, see *ibid.*, p. 89 and n. 22 and below, pp. 54–6.

ever, some excerpts from the *Tahdhīb* cited in passages (109–11) and (185)[12] which are not ascribed to anyone in particular but seem to be Miskawayh's own phrasing are also not marked as such in the *PR*. Whatever the reason for that may be, Miskawayh is the second most important source used in the *PR*, second only to the excerpts in the name of the Greek Sage, which take up about thirty-two folios.

The Greek Sage is, in turn, another feature which ties the compiler of the *PR* to Miskawayh. This enigmatic appellation, which must denote Plotinus, seems to be used first by Miskawayh in his *Ḥikma al-khālida* (*Eternal Wisdom*) to refer to a student of Diogenes and possessor of wisdom. There the Greek Sage is also said to have presented his wisdom in well-known works which were so easily available that Miskawayh did not feel any need to quote samples from them.[13] This statement from *al-Ḥikma al-khālida* was repeated in the entry on Diogenes in the *Ṣiwān al-ḥikma* (*Cabinet of Wisdom*) as the surviving *Muntakhab* (*Selection*) of the *Ṣiwān* attests. Yet, as further attested by the *Muntakhab*, the author of the *Ṣiwān* fortunately added a separate entry for the Greek Sage in which he presented a selection of the Sage's sayings. These sayings are most probably derived from the Sage's well-known works mentioned by Miskawayh, since the author of the *Ṣiwān* must have known Miskawayh personally, as his entry on the latter suggests, and seems to have had access to the same books.[14]

Apart from Miskawayh, and the *Ṣiwān*-tradition which appears to

[12] Passage (185) is, in the *PR*, introduced with a consecutive *wa-qāla* which seems to refer to the previously quoted Galen. A further particular case is passage (200), which the *PR* ascribes to Plato although the beginning is parallel to a passage in Miskawayh's *Tahdhīb* where he quotes from an allegedly Aristotelian treatise *Faḍāʾil al-nafs*, and the second part is parallel to the *Ethica Nicomachea*.

[13] Miskawayh, *al-Ḥikma al-khālida*, 216.16–18:
وديوجانس هذا صاحب الشيخ اليونانيّ ومعلّمه . والشيخ اليونانيّ هو صاحب الحكمة التي ظهرت منه في كتبه المعروفة به ، وليس هذا موضع ذكرها . فمن أحبّ أن يطالعها فليقرأها من تلك الكتب فإنها موجودة .
For a discussion and translation of the passage see Zimmermann, 'Origins', p. 209.

[14] See al-Qāḍī, '*Ṣiwān*', p. 113, where she quotes from the entry on Miskawayh in the *Muntakhab* (ed. Dunlop, line 3418) the following: 'As for what *I heard from him* [i.e. Miskawayh] of his ways and means, and what *I saw* of his good behaviour and his pure virtues ...' (tr. al-Qāḍī; translator's emphasis). How much material by the Greek Sage (*GS*) the *Ṣiwān* contained is uncertain, as only two abridgements of the text, the *Muntakhab* and the *Mukhtaṣar*, survive. However, it must have been more comprehensive than the selection which the *Muntakhab* preserves, since Shahrastānī's *K. al-Milal* and Shahrazūrī's *Rawḍat al-afrāḥ* quote their *GS*-material from the *Ṣiwān*, but have more of it than the *Muntakhab*. Unfortunately, there is no overlap between the surviving *GS*-material of the *Ṣiwān*-tradition and the *PR*, yet both have some overlaps with the *Theology of Aristotle* (*ThA*), the other major Plotinus Source in Arabic. On this see Zimmermann, 'Origins', pp. 208–9, and Rosenthal, 'Aš-Šayḫ', I, pp. 469–70, and II, pp. 384–440.

depend on the former, the Greek Sage is also known to Abū l-Ḥasan Aḥmad b. Muḥammad al-Ṭabarī who cites him as an authority on the subject of generation and procreation in his *Muʿālajāt al-Buqrāṭīya*, unfortunately without quoting any of the Sage's own words. Abū l-Ḥasan was the court physician of the Buyid Rukn al-Dawla and al-Tawḥīdī reports that he presented the vizier al-Ṣāḥib ibn ʿAbbād with al-ʿĀmirī's works.[15] His *Muʿālajāt al-Buqrāṭīya* further provide the only known evidence of an Arabic translation of Proclus' *De malo*, apart from Miskawayh's short reference to it in his *Fawz*.[16] All this links Abū l-Ḥasan neatly to the philosophical circle in the east of the Islamic world in the tenth and eleventh centuries, in which Miskawayh must have played a major role.

Finally, it should be mentioned that much of the Greek material contained in the *PR* overlaps with Miskawayh's own philosophical discussions, and in some instances the *PR* seems to present Miskawayh's source texts before giving the latter's own account of a doctrine.[17] This makes one wonder whether the compiler copied Miskawayh's personal notes, in which he may have jotted down excerpts from his source texts as well as from his own works. Be that as it may, I hope to have shown that it is highly plausible that the compiler of the *PR* was part of Miskawayh's circle, and maybe his student or a student of one of his students.[18]

15 See al-Tawḥīdī, *Akhlāq al-wazīrayn*, pp. 115–16. On Abū l-Ḥasan, see Ibn Abī Uṣaybiʿa, *ʿUyūn*, I, p. 321, and Hasnawi, 'Élève', pp. 43–4.

16 Abū l-Ḥasan's mention, which may somehow go back to the Greek text (cf. the reconstructed text by Boese in Procli Diadochi *Tria opuscula*, 52.16–18), reads (see MS Marsh 158, fol. 18a13–14; cf. Sezgin's facsimile edn, 25.16–18):
ولبرقلس مقالة بيّن فيها طبيعة الخير والشرّ وبرهن على أنّهما متضادان متقابلان وأنّهما قوّتان وأنّ لكلّ قوّة منهما أفعال وتأثيرات .
whereas Miskawayh only refers to a book which Proclus devoted to it, i.e. the clarification of the good and the bad (*Fawz*, 82.13–83.1).

17 Cf., e.g., the material of the *De anima*-Paraphrase in passage (79) with passage (84) excerpted from the *Fawz*, and, further, passage (92) quoted from the *DA*-paraphrase with passages (94–5) taken from the *Fawz*.

18 It is difficult to name any particular person who may have been the compiler of the *PR*. A possible candidate could be Abū l-Qāsim al-Kātib, also known as Ghulām al-ʿĀmirī, whom al-Qāḍī ('*Ṣiwān*', p. 118) suggests as a likely author of the *Ṣiwān*. This possibility would be corroborated by the fact that the Marsh manuscript had been ascribed to a certain Abū l-Qāsim after it had come into Marsh's possession, were it not that this ascription was probably based on a misinformed statement on the flyleaf (on which below, p. 52). However, there is no apparent overlap between the *Ṣiwān*, at least as attested in its two surviving summaries, and the *PR*, which could be expected if they originated from the same author.

III The contents: a fusion of the Late Antique curriculum and Neoplatonism?

The text of the *PR* as contained in the Marsh manuscript starts midsentence on the first remaining folio of the original manuscript, which is numbered as folio 5.[19] It is impossible to say when the foliation was done and whether it may give any indication of the extent of the textual loss the manuscript suffered at the beginning. If the foliation originates from a time when the manuscript was still complete, not much text would be lost, especially if the first pages contained a separate title-page and maybe some kind of table of contents.

The *PR* in its present shape is a continuous text without any divisions, sections or chapters. The only structuring elements are the introduction of *qāla* 'he says, he has said', often in combination with the name of the quoted authority or a simple consecutive *wa-qāla*, the insertion of a blank space between consecutive passages, and the beginning of a new line or three dots to mark the end of an excerpt or a topic.[20] My division of the *PR* into 222 passages follows these indications wherever possible, but neglects them where they oppose the logical sequence of the text or where a comparison with sources or parallel texts suggests another division.

The first topic dealt with in the *PR* is God, followed by a lengthy account of the soul which also includes some discussion of intellect and nature; next is a thorough exposition of the human body with special attention to sense perception, followed by a presentation of the good, happiness, the division of the sciences and the virtues. The last extant passage on physical virtues was obviously not intended to be the final one as the catchword in the margin of the last remaining folio of the original manuscript clearly shows. How much text is missing at the end is even harder to guess than the extent of the textual loss at the beginning. Given the sequence of subjects dealt with in the *PR*, the discussion of God was most probably the first topic, but whether the discussion of virtues was meant to be the last one is impossible to judge.

An assessment of the contents of the *PR* would become even more difficult if the suggestion by Zimmermann that the text contained in the Marsh manuscript may actually be just an abridgement were correct.[21]

[19] Whether the flyleaf numbered fol. 4 belongs to the *PR* as well is difficult to tell (see below, p. 52).

[20] On which see also Zimmermann, 'Origins', pp. 209 and 217.

[21] *Ibid.*, p. 239, n. 387: 'One wonders whether the original [of the Oxford anthology] may

However, this suggestion could only be ascertained or refuted by the discovery of hitherto unknown material.

The broad outline of subjects dealt with in the *PR* in its surviving form recalls the Neoplatonic hierarchy of being which descends from the One via intellect, soul and nature to the human body. The corresponding upward motion, the reversion of the beings to the One, may be expressed by broaching the subjects of happiness and ethics.[22] An attempt at merging this Neoplatonic approach with the Late Antique division of sciences might be behind Miskawayh's discussion of the branches of knowledge as attested in the *PR*, passage (211). Contrary to the accepted tradition, Miskawayh starts his division of theoretical philosophy with the highest science which is free, i.e. metaphysics, and then lists the ancillary sciences, i.e. mathematics and natural sciences. Had he replaced mathematics with psychology as al-Kindī had done, his division would have been applicable to, at least, the beginning of the *PR*.[23] However, the structure of his *Fawz al-aṣghar* may be compared to the *PR*, as it, especially in the standard edition, addresses the topics of God, the soul and prophecy, and the discussion of prophecy starts off with a detailed discussion of man and the human body in general.

Turning to a more thorough examination of the separate sections, the arrangement of the individual passages seems less coherent. Topics are dealt with, dropped and taken up again. Repetition is commonplace. The sections further vary in character. Whereas the first section on God must have been compiled from a great number of different sources – unfortunately still for the most part unidentified – and may be described as doxographic, the sections on the soul and on the functions of the human body are derived from fewer sources (the Greek Sage, the *DA*-paraphrase, Miskawayh's *Fawz* or Galen's *De usu partium* (*UP*), and Nemesius). They seem, therefore, to be more unified and are most probably always consistent with the compiler's own views.

not have been a little less erratic in its indication of sources. It is possible that the Oxford copy is but an abridgment.'

22 Rüdiger Arnzen (personal communication) has suggested this interpretation of the structure of the *PR* to me and ventured the guess that the missing textual portion at the end of the *PR* may have dealt with eschatology and the like.

23 On al-Kindī on mathematics and psychology, see Endreß, 'Mathematics', pp. 129–30.

IV The individual passages: a detailed table of contents

The following detailed table of contents of the passages contained in the *PR* provides the number which I have given to each passage, the name of the quoted authority when indicated, a title with which I summarise the contents of the passage, and its source where identified:

God

(1) On an incorrect description of God
(2) (Plato?) in his *Letter to Dionysius* on the indescribability and oneness of God
(3) On how the contemplation of God guarantees His providence
(4) On the ascent to the heavens
(5) Hermes on God's imperceptibility
(6) Pythagoras on the four divine and the four natural numbers
(7) Aristotle in the *Metaphysics* on God being one, eternal, not a body and unmoved (Aristotle Source)
(8) On the definition of time, on the motions of the heavens, and on the one universal principle (Aristotle Source)
(9) Aristotle in the *Metaphysics* on the substance above the heavens (Aristotle Source)
(10) On the principles of motion (Aristotle Source with a tr. of Aristotle, *Physica*, 198a35–198b2)
(11) On linear and circular motion (Aristotle Source)
(12) On finite and infinite powers and their corresponding motions (Aristotle Source)
(13) On the necessary existence of a substance above all substances and a universal principle (Aristotle Source)
(14) Proclus in the *Minor Elements* on the First Cause as the giver of all good and being above everything else
(15) On God's imperceptibility and being called by the noblest names known
(16) On the veneration for and beliefs in God (Proclus' commentary on the *Dicta aurea*?)
(17) Themistius on God as the universal principle and being omniscient (Paraphrase of Book *Lambda*)
(18) Aristotle in the *Letter on Gouverning in General* to Alexander on God (*R. al-Siyāsa al-ʿāmmīya*)

(19) On God's creation of the first principles *ex nihilo* and His indescribability (*R. al-Siyāsa al-ʿāmmīya*)
(20) The Stoics on God's corporeality
(21) Parmenides on the Ultimate Cause being one, simple and above everything else
(22) On how neither imagination nor intellect grasps the Ultimate Cause
(23) Pythagoras on intellect and the First
(24) Heraclitus on the First being not a body, but indescribable
(25) Plato in one of his discourses on God's creation and the soul's descent into this world, and on the chances and responsibilities man has due to his soul
(26) Aristotle in one of his discourses on God being the Good, on the relation of intellect and soul to the Good and to each other, and on the descent of the soul
(27–8) Socrates in the *Phaedo* on the afterlife (unknown tr. of Plato, *Phaedo*, 63b6–8 and 63c2–4; 69c3–7)
(29) Aristotle on the four worlds and the position of the soul in them; Plato on the spirit
(30) The Greek Sage on the Unmoved First Agent and the emanation of intellect (*GS*)
(31) On the One's creation of multiple things and on how everything is moved by its desire for Him (*GS*)
(32) On the Unmoved First Agent and on the emanation of intellect and everything else (*GS*)
(33) On the uniqueness and oneness of the First Creator and on all things being in the intellect (*GS*)
(34) On light, sun and moon as similes for the First, second and third (*GS*)
(35) On the True One being with all things (*GS*)
(36) On God being above everything (*GS*)
(37–8) On how the intellect acquires the forms from the First (*GS*)
(39) On the First Creator not being a form and on intellect not being a word (*GS*)
(40) On the First One being without any multiplicity and on the enfolding effects, i.e. intellect, soul, heavenly and earthly bodies (*GS*)

The soul

(70) Aristotle in the *Physics* on motion and rest and on moving 'primarily' (Aristotle Source with a tr. of *Physica*, 192b21–3)
(71–4) Hippocrates, Galen and others on the definition of nature
(75) On the soul's descent into the sensible world (*GS*)
(76) On the localisation of the tripartite soul in the brain, heart and liver (*GS*)
(77) Plato on the soul being inside and outside this world (*GS*)
(78) Aristotle on the soul being the perfection of the natural body (*DA*-paraphrase)
(79) On the soul being a simple substance and why it is better described as perfection than as form (*DA*-paraphrase)
(80–3) On the soul not being a body (*DA*-paraphrase)
(84–6) Abū ʿAlī Miskawayh on the soul being simple and neither a body nor an accident (Miskawayh, *Fawz*)
(87) Plato's proof that the soul is self-moving (Miskawayh, *Fawz*)
(88–9) Plato on two kinds of life and (the philosophers) on two kinds of death (Miskawayh, *Fawz*)
(90) Aristotle proving that the soul does not perish upon its separation from the body (*DA*-paraphrase)
(91) Abū ʿAlī on the three Platonic proofs for the immortality of the soul and Proclus' exposition (Miskawayh, *Fawz*)
(92) Aristotle on how the soul knows the simple and composite things (*DA*-paraphrase)
(93) On the intellect's and the soul's relation to the body (*DA*-paraphrase)
(94) Abū ʿAlī Miskawayh on how the soul perceives the simple and composite things (Miskawayh, *Fawz*)
(95) On the soul being one, not having parts and not perceiving through different powers (Miskawayh, *Fawz*)
(96) The Greek Sage's arguments for the incorporeality of the soul (*GS*)
(97) Plato on the soul not being evil (*GS*?)
(98) On the embodied soul being an excellence
(99–100) On nature being the soul's tool and like a potter
(101) The soul's increase in knowledge as a proof for its immortality
(102) Pythagoras on the divine element in man and his travel after death (*Waṣāyā Fīthāghūrus al-dhahabīya*)
(103) Zeno on the soul's relation to the body
(104) Plato on the rational soul and its immortality

The human body

(141) Hippocrates on the development of the embryo (Hippocrates Source)
(142) The wise men on how the planets influence embryonic development (*Ikhwān al-Ṣafāʾ*)
(143–4) On procreation and the recreation of body parts (Galen, *UP*)
(145) Qusṭā ibn Lūqā on the spirit (Qusṭā ibn Lūqā)
(146) Galen on the spirit and on the brain, heart, liver (Galen, *De propriis placitis*)
(147) Aristotle on the soul's perception of sensibles (*DA*-paraphrase)
(148–54) On vision, hearing, smell, taste and touch (*DA*-paraphrase)
(155–6) On excessive sense perception and on the lack of sense perception in plants (*DA*-paraphrase)
(157–9) On the senses, their organs and the corresponding elements (Nemesius and some unidentified *DA*-Source)

Lacuna in the Marsh manuscript, probably meant to be filled with:

(160) On the elements which correspond to the five senses, on excessive sense perception, and on the different degrees of refinement of the senses (Miskawayh, *Fawz*)
(161) On the common sense (Miskawayh, *Fawz*)
(162) On imagination, sense perception, intellect and discernment
(163–5) The Greek Sage on vision (*GS*)
(166–7) Aristotle and others on vision and hearing (unidentified Aristotle Source and *Ikhwān al-Ṣafāʾ*?)
(168–74) Aristotle on hearing
(175) On echoes
(176–7) Plato and Aristotle on smelling
(178–80) On the definition of light and whether it is a body or not (*DA*-paraphrase)
(181) The Greek Sage on light being a quality (*GS*)
(182) Abū Zayd on colour
(183) The Sage on the augmentative power, appetite and anger being in the liver (*GS*)

Lacuna in the Marsh manuscript

(184) Galen on passionate love
(185) On shame, generosity, nobility, piety, liberality and embarrassment (Miskawayh, *Tahdhīb*)
(186) On laughter and weeping
(187) Plato on laughter, weeping and joy (ʿAlī ibn Rabban)
(188) On the innate disposition
(189) On smallpox

(190) On the reason for the tremor and paleness in complexion of someone frightened
(191) Six kinds of fear (Nemesius)
(192) Plato on the symptoms of fear (ʿAlī ibn Rabban)
(193) Apollonius on the symptoms of passionate love
(194) Plato on hair (ʿAlī ibn Rabban)
(195–6) Hippocrates on hair (Hippocrates Source)
(197) Galen on the relation between climate and physique (Galen, *De temperamentis*, as contained in a gloss in the *K. fī l-Firāsa*)
(198) On what the flesh, the velocity of movements and the voice indicate (*K. fī l-Firāsa*)
(199) On properties and accidents of the body and the example of laughter

Ethics

(200) Plato on the pure good (Miskawayh, *Tahdhīb* and Aristotle, *al-Akhlāq*, or *Ikhtiṣār al-Iskandarānīyīn*, 1094a3, 1097a19–21)
(201) Aristotle on the noble, praiseworthy, potential and beneficial goods (Miskawayh, *Tahdhīb*)
(202) (Aristotle) in the *Ethics* on the desired good, on the two- and threefold division of the good, and on happiness (*Ikhtiṣār al-Iskandarānīyīn*)
(203) Plato in the *Laws* on the divine and human goods (unknown tr. of Plato, *Leges*, 631b-c)
(204) The Greek Sage on the Pure Good (*GS*)
(205) Aristotle on the definition of happiness
(206) Aristotle on the five parts of happiness and how Pythagoras, Socrates and Plato define happiness as exclusively belonging to the soul (Miskawayh, *Tahdhīb*)
(207) Aristotle elsewhere on the three genera of happiness (Miskawayh, *Tartīb al-saʿādāt*)
(208) On the happy man (*Ikhtiṣār al-Iskandarānīyīn*?)
(209) On the definition of excellence
(210) Abū ʿAlī Miskawayh on the three genera of happiness (Miskawayh, *Fawz*)
(211–12) On the division of theoretical and practical wisdom and on the prophets being the physicians of the soul (Miskawayh, *Fawz*)
(213) Abū l-Ḥasan al-ʿĀmirī on knowledge (al-ʿĀmirī, *al-Itmām*?)
(214) On physical and psychic virtues (al-ʿĀmirī, *al-Itmām*?)
(215–6) On knowledge (al-ʿĀmirī, *al-Itmām*?)

(217) On generosity (al-ᶜĀmirī, *al-Itmām*?)
(218–20) On leadership (al-ᶜĀmirī, *al-Itmām*?)
(221) On the three parts of the soul and the four cardinal virtues (al-ᶜĀmirī, *al-Itmām*?)
(222) On physical virtues (al-ᶜĀmirī, *al-Itmām*?)

V The sources of the *PR*: an overview of what was read in the tenth and eleventh centuries in the Islamic East

As the table of contents shows, about twenty sources of the *PR* can be established with great certainty so far. A number of them have not survived in their entirety, so the *PR* preserves fragments of these otherwise lost texts. These fragments are particularly valuable as the compiler of the *PR* is very thorough when quoting his sources and hardly ever tampers with them. Even if he shortens a text, he does not do so by summarising it in his own words, rather he simply omits sentences and resumes the text where it seems beneficial to his purpose. His most drastic textual intervention is rearranging the order of his source's sentences.

In what follows I present, in alphabetical order of authors, a short overview of the sources and their relation to the *PR*.

Alexander of Aphrodisas, *De intellectu – Fī l-ᶜAql ᶜalā raʾy Arisṭūṭālīs*

Text (Greek): ed. Sharples in Alexander, *Mantissa*, pp. 42–52; tr. Sharples in Alexander, *Supplement*, pp. 24–44; **(Arabic):** ed. Badawī in *Commentaires*, pp. 31–42; ed. Finnegan in 'Texte arabe du ΠΕΡΙ ΝΟΥ', pp. 181–99.[24]

The Greek *De intellectu* comes second in a collection of twenty-six treatises supplementing Alexander's *De anima* and known as *Mantissa*. The unity of *De intellectu* and Alexander's authorship of the entire treatise or parts of it are disputed.[25]

It is not clear whether the entire *Mantissa* was translated into Arabic, yet individual treatises certainly were. The most influential among them

[24] For manuscripts overlooked in these editions, see Geoffroy, 'Tradition', p. 196, n. 20.
[25] See Sharples, *Supplement*, p. 24.

was *De intellectu* which was translated by Isḥāq ibn Ḥunayn under the title *Fī l-ʿAql ʿalā raʾy Arisṭūṭālīs* (*On the Intellect according to Aristotle's Doctrine*).[26]

The *PR*'s passage (45) must ultimately go back to the Arabic Alexander, although it is not certain whether the compiler had direct access to *De intellectu* or used some kind of summary or commentary. In the latter case, his source may have been al-Fārābī's commentary on either Alexander's *De anima* or *De intellectu*, from which he most probably excerpted passage (46).[27]

(The 'Alexandrians'), *Ikhtiṣār al-Iskandarānīyīn – Summa Alexandrinorum*

Text (of the extant Arabic fragments of Books I, VI, VII and the additional book VII): ed. Dunlop in 'Taimur Pasha', pp. 260–63; ed. Badawī in *al-Akhlāq*, pp. 439–45; **(quotations of Books I–V, VIII–X):** Mubashshir, *Mukhtār al-Ḥikam*, ed. Badawī, pp. 209–14;[28] **(of the Arabic-Latin tr. and its reworking 'Redactio Patavina'):** ed. Marchesi in *L'Etica*, pp. XLI–LXXXVI (repr. in Badawī, *al-Akhlāq*, pp. XLI–LXXXVI, i.e. pp. 446–91); ed. Fowler in 'Manuscript Admont', pp. 195–252; **(of a Hebrew abridgement of Books I–VI in Shem Tov ibn Falaquera's *Shelemut ha-maʿasim*):** ed. Jospe in *Torah and Sophia*, pp. 417–38.

Ikhtiṣār al-Iskandarānīyīn (*Summary of the Alexandrians*), the Arabic version of the *Summa Alexandrinorum* of Aristotle's *Ethica Nicomachea*, is a text in eleven books no longer extant in its entirety. Most recently Ullmann has challenged the hitherto accepted assumption that the *Summa* was composed in Greek by Nicolaus, whose summaries of Aristotle's works were translated into Syriac and Arabic.[29] In a detailed study Ullman has first

[26] See Gätje, *Psychologie*, pp. 70–1.

[27] Philippe Vallat (personal communication) mentioned to me the possibility of the existence of these two commentaries lost today.

[28] For a list of exact correspondences between Mubashshir and the Greek *Ethica Nicomachea* and the Latin translation of the *Summa*, ed. Marchesi, see Dunlop, 'Introduction', pp. 63–4 or his 'Summa', p. 254; for a comprehensive comparison quoting the passages in question see Ullmann, *Nikomachische Ethik 2*, pp. 82–121.

[29] For a comprehensive presentation of this former assumption see Dunlop, 'Introduction', pp. 62–79. The identity of this Nicolaus with Nicolaus of Damascus (first century BC) has, however, been debated. Whereas Drossaart Lulofs (*Nicolaus*, pp. 1–15) and Dunlop ('Introduction', p. 76) assume it to be correct, Fazzo ('Nicolas', pp. 110–25) convincingly challenges it and suggests, based on the Arabic evidence and especially Ibn Buṭlān, that there may well have existed a Nicolaus of Laodicea in the fourth century AD who composed not only summaries of individual Aristotelian works but also *De philosophia Aristotelis*. The only extant Syriac manuscript of excerpts of *De philosophia Aristotelis* ascribed to Nicolaus lists a 'Study of Habits' among his other works in the preface. Although this preface is, according to Drossaart Lulofs (*Nicolaus*, pp. 24, 95–6),

shown that the Fez unicum of the Arabic translation of the *Ethica Nicomachea* is a fusion of two separate translations, an older one made by Usṭāth (Books V–X and the additional, non-Aristotelian Book VII) and a younger one by Isḥāq ibn Ḥunayn (Books I–IV). In a second step he has demonstrated that the *Summa* is most probably based on this fused Arabic version of the *EN* and must thus originate in Arabic.[30] A *terminus post quem non* for its composition is provided by Miskawayh's *Tahdhīb* which is the first Arabic work known to quote extensively from the *Summa*.[31]

The complete Arabic text of the *Summa* seems lost, but some abbreviated fragments (of Books I, VI, the additional Book VII and Book VII) survive in a Cairo manuscript. Additionally there are the quotations in Mubashshir's *Mukhtār al-Ḥikam* (of Books I-V and VIII-X) which, as Dunlop shows, remain in the sequence of the *Summa* and are, judging by a comparison with the Arabic-Latin translation, sometimes closer to the original text of the *Summa* than the fragments.[32] The Arabic-Latin translation was undertaken by Hermannus Alemannus and survives in a possibly shortened version. There also exists a reworking of the translation termed *Redactio Patavina* and completed in the second half of the thirteenth century by Engelbert of Admont. Both texts seem to be, at times, shorter than the Arabic model.[33] There is further a Hebrew abridgement of Books I–VI in Shem Tov ibn Falaquera's *Shelemut ha-maʿasim* (*The Perfection of Actions*).

The *PR* quotes the *Summa* in passage (202), yet explicitly ascribes the quotation to the '[*Nicomachean*] *Ethics*' which makes one wonder whether the compiler was aware of the fact that he did not have the complete Aris-

late, inauthentic and hardly corresponds to the preserved text, Fazzo ('Nicolas', p. 114) entertains the possibility that this title may refer to the *Summa*. The ascription of the *Summa* to the Alexandrians instead of Nicolaus could have been explained in analogy to the case of Nicolaus' summary of the Ps.-Aristotelian *De plantis* which seems to have reached Shemtov Falaquera with an ascription to the Alexandrians instead of the correct one to Nicolaus (on which, see Drossaart Lulofs and Poortman, *Nicolaus*, pp. 348–52).

30 See Ullmann, *Nikomachische Ethik 2*, pp. 15–122. Dunlop's argument for Ibn Zurʿa being the translator of an alleged Greek original of the *Summa* (presented in his 'Introduction', pp. 68–71, and 'Summa', pp. 261–3) is dealt with by Ullmann, *Nikomachische Ethik 2*, pp. 75–6.

31 See Dunlop, 'Summa', p. 255, and 'Introduction', pp. 29–31; and Harvey, 'Greek Library', pp. 499–501.

32 For a description of these fragments, which can be shown at times to abbreviate and at times to elaborate the Arabic translation of the *Summa*, and of the quotations in Mubashshir, see Dunlop, 'Introduction', pp. 79–85.

33 On the Latin versions, see Zonta, 'Aristote', p. 198. See also the example which Dunlop ('Taimur Pasha', p. 254) presents, in which the Latin translation is shorter than the text preserved in the fragments, although this could also be due to an addition to the Arabic *Summa* in the fragments.

totelian work in front of him, but rather an abridgement. It is, of course, possible that he had access to both works as Dunlop assumes for Miskawayh.[34] However, this assumption has been questioned by Zurayk and Harvey,[35] and if not even Miskawayh had access to the complete Aristotelian work it seems unlikely that a compiler standing in his tradition would. Passage (202) is put together from different statements taken from the first chapter of the *Summa*, and has some parallels to the quotations in Mubashshir and the Latin translation, but also material that is not attested anywhere else. Part of the passage is also repeated in a slightly different version in passage (208), which may in fact not derive from the *Summa* directly, but, for example, from Miskawayh quoting the *Summa*, as is the case with the Aristotelian quotations in passages (206–7).

Passage (209) may also derive from the *Summa*, but since the compiler of the *PR* does not give any ascription and since there is no parallel in any of the remaining witnesses to the *Summa*, it is impossible to be certain. In any case there remains no passage of the *PR* which would necessitate the assumption that the compiler had access to the Aristotelian text of the *EN*.

ʿAlī ibn Rabban al-Ṭabarī, *Firdaws al-ḥikma* and its *Mukhtaṣar*

Text: ed. al-Ṣiddīqī; ed. al-Jundī;[36] **(of the *Mukhtaṣar*):** Istanbul, MS Aya Sofya 4857, fols. 5b-57b.

ʿAlī ibn Sahl Rabban al-Ṭabarī's *Firdaws al-ḥikma* (*Paradise of Wisdom*) in 7 parts (*nawʿ*), 30 sections (*maqāla*) and 360 chapters (*bāb*) is an Arabic compendium of foremost medical knowledge, but also includes material on natural philosophy, physiology, psychology, meteorology, astronomy and astrology, and magic. According to its author it was, after a long period of preparatory work, finally completed in 850 and, later on, translated into Syriac by ʿAlī ibn Rabban himself.[37] The main medical sources are Hippocrates and Galen,[38] whereas for natural philosophy Aristotle is widely used,

[34] See Dunlop, 'Introduction', p. 31.

[35] See Harvey, 'Greek Library', pp. 499–500.

[36] For a list of partial (German) translations see Ullmann, *Medizin*, p. 120, n. 4.

[37] See *Firdaws*, p. 2 and p. 8. On this and for further information on ʿAlī ibn Rabban, see Meyerhof, 'ʿAlī ibn Rabban', pp. 43–56, 59–68; and Ullmann, *Medizin*, pp. 119–20. ʿAlī ibn Rabban was a Persian Christian by birth – his father's title 'Rabban' being a Syriac honorary title ('our master'), not a reference to a Jewish rabbi, as often assumed – but later on converted to Islam and even composed two polemics against the Christians.

[38] For further details, see Ullmann, *Medizin*, p. 121; also Meyerhof, 'ʿAlī ibn Rabban', p. 56,

namely his *De caelo*, *Meteorologica*, *De generatione et corruptione*, the *K. al-Ḥayawān*[39] and the Arabic *DA*-paraphrase.[40] Of particular interest is the account of Indian medicine, which remains unique in Arabic medical literature.[41]

The *Firdaws* was very popular and often quoted in the Arabic-Islamic tradition, especially in medical literature, but also in philosophical compilations.[42] Due to its length it also seems a perfect candidate for summary reworkings, and, indeed, there is at least one *Mukhtaṣar Firdaws al-ḥikma* of 34 chapters preserved in an Istanbul manuscript. This contains most of the first part on general philosophical principles (the elements, change, generation and corruption, action and being acted upon, etc.), most of the second section of the second part (on the soul and the senses), most of the fifth part on the objects of sense perception, and most of the third section of the seventh part on spheres and the celestial bodies. The text of the chapters taken up is hardly ever changed; in some cases the final passages are missing.[43] The character of this *Mukhtaṣar* strips the *Firdaws* of any

who mentions that ʿAlī quotes his contemporaries Ibn Māsawayh and Ḥunayn, and wonders whether Ḥunayn dedicated his *K. al-ʿAshr Maqālāt fī l-ʿayn* to ʿAlī.

39 The nineteen books or chapters (*maqālāt*) of the Arabic *K. al-Ḥayawān* contain, first in chs. 1–10 the *Historia Animalium* (Arisṭūṭālīs, *Ṭibāʿ*, ed. Badawī); next, in chs. 11–14, *De partibus animalium* (*Arabic Version*, ed. Kruk); and last, in chs. 15–19, *De generatione animalium* (Aristotle, *GA*, ed. Brugman and Drossaart Lulofs). In the medieval Arabic world there seems to have been no awareness that the *K. al-Ḥayawān* is, in fact, a combination of three independent Aristotelian treatises. Thus Ibn Abī Uṣaybiʿa added the title *K. Aʿḍāʾ al-ḥayawān* from Ptolemy's list of Aristotelian writings to his own in which he had already included the *K. al-Ḥayawān* (see *Arabic Version*, ed. Kruk, p. 17).
I would like to suggest that ʿAlī ibn Rabban may have had access to an early Syriac translation of Nicolaus, *De philosophia Aristotelis* (on which see Fazzo, 'Nicolas', p. 103, n. 8) as ʿAlī shares some common topics with al-Baladī who, in turn, refers to Nicolaus' *De philosophia Aristotelis* as his source, on which see commentary to passage (134).

40 See Arnzen, *Aristoteles'* DA, p. 72. Meyerhof ('ʿAlī ibn Rabban', p. 62) and Ullmann (*Medizin*, p. 122) remark that ʿAlī ibn Rabban does not yet quote the works of Greek authors in the Arabic translations of Ḥunayn's school, but paraphrases the Syriac translations. This verdict cannot be extended to the *DA*-paraphrase.

41 Ullmann, *Medizin*, p. 122; see also Meyerhof, 'ʿAlī ibn Rabban', pp. 63–5.

42 For the medical field see Meyerhof, 'ʿAlī ibn Rabban', pp. 61–2. Two philosophical compilations which quote the *Firdaws* are the *K. al-Ḥikma* (on which below, pp. 42–3) and a certain *Kitāb fīhi Ārāʾ al-ḥukamāʾ fī l-ṭabīʿīyāt wa-fīhi al-ārāʾ wa-l-kalimāt al-rūḥānīya li-l-mutaqaddimīn*. The latter is described by Daiber ('Doxographie', pp. 4980–1) who does not refer to the *Firdaws*. However, the quotation in the name of Pythagoras which he mentions actually derives from ʿAlī ibn Rabban.

43 I give a preliminary overview of the chapters contained in the *Mukthaṣar* as preserved in MS Aya Sofya 4857 in correspondence to the text of the *Firdaws* edited by al-Ṣiddīqī: chs. 1–2 ≈ 9.5–13.2; chs. 3–9 ≈ 14.3–27.11; ch. 10 ≈ 58.12–14, p. 59; ch. 11 ≈ 48.1–50.2; chs. 12–20 ≈ 60.1–82.19 (with 75.5–20 considerably shortened on fol. 33a7–12; with 76.6–77.6 and 80.5–81.4 missing); chs. 21–24 ≈ 356.2–365.3; ch. 25 ≈ 368.19–369.22; chs.

medical importance and transforms it merely into a philosophical introduction. Although this description sounds as if the *Mukhtaṣar* may well have been the source of the *PR* where it anonymously quotes ʿAlī ibn Rabban, this cannot be the case, as the *PR*, in passages (132–8)[44] and (194) uses material from the first section of the second part not contained in the abridgement. In fact, only one and a half passages which the *PR* quotes from the *Firdaws*, i.e. the second half of passage (187) and passage (192) occur also in the *Mukhtaṣar*. The *PR* must thus quote from either the complete *Firdaws* or a different abridgement.

It is also interesting to observe that the *Firdaws* and the *PR* share a common source, namely the Arabic version of the *DA*-paraphrase. There are even passages, i.e. (78), (90) and (92), where the *PR* draws on the same material of the paraphrase as ʿAlī ibn Rabban. However, in these cases the compiler of the *PR* was obviously not tempted to quote from the *Firdaws*, but kept to the original source.

al-ʿĀmirī, *Kitāb al-Itmām li-faḍāʾil al-anām*

It may seem surprising to suggest the *K. al-Itmām li-faḍāʾil al-anām* (*The Perfecting of the Virtues of Man*) of al-ʿĀmirī, which no longer exists and to which no fragments quoted in later sources are attributed, as a possible source of the *PR*. There are, however, some indications which make it likely that the text excerpted in passages (213–22) derives from exactly this work.[45] First, there is the general agreement between the title *The Perfecting of the Virtues of Man* as preserved in a list of al-ʿĀmirī's writings which he himself provides in the *K. al-Amad* (ed. Rowson, pp. 52–5) and the contents of the excerpt in the *PR*. The latter contains a division of virtues into physical and psychic and a lengthy presentation of the four cardinal virtues. Second, the beginning of passage (213) is very similar to a maxim presented in al-ʿĀmirī's surviving *K. al-Iʿlām bi-manāqib al-Islām* (*An Exposition of the Merits of Islam*) where the reader is referred to the *K. al-Itmām* for a detailed discussion of the maxim and what it entails.

26–31 ≈ 541.3–556.23 (with 549.17–550.4 and pp. 552–3 missing); ch. 32 ≈ 88.17–89.11; ch. 33 ≈ 94.6–96.24; ch. 34 ≈ 522.3–523.15. No note has been taken of the marginal glosses occurring on 13 folios.

44 These passages occur also in the *K. al-Ḥikma* which may suggest that, in the case that both compilations used not the complete *Firdaws* but some kind of abridgement similar to the *Mukhtaṣar* of the Aya Sofya 4857, they shared a common source.

45 On the assumption that the entire text of passages (213–22) derives from one single author and source, see above, pp. 4–5. If only the first part of it were by al-ʿĀmirī, the suggestion that the *K. al-Itmām* was its source would apply to it alone.

(Aristotle), *De anima*-paraphrase

Text: ed. Arnzen with German tr. in *Aristoteles' De anima*, pp. 180–351; Istanbul, MS Aya Sofya 4156, fols. 54a-106a.[46]

The anonymous Arabic paraphrase of Aristotle's *De anima* must go back to a lost Greek model, but also draws on the Arabic Plotinus and Proclus Sources.[47] Like the latter, it is not a translation in the strict sense of the word, but rather a free reworking which must originally have been much longer than that preserved today.[48]

The Greek model's main sources are the *DA*-commentaries of Philoponus and Stephanus, which provide a *terminus post quem* at the turning of the sixth to the seventh century and thus date the paraphrase to the last period of Alexandrian philosophical activity.[49]

The Arabic translation, in turn, can be dated before 850, the year of the completion of ʿAlī ibn Rabban al-Ṭabarī's *Firdaws*, which uses the *DA*-paraphrase. A possible reference to the paraphrase in the *Fihrist* may be seen in the mention there of a *talkhīṣ* ('abridgement') of the *DA* by the Alexandrians which was summarised by Ibn al-Biṭrīq.[50]

The structure of the Arabic *DA*-paraphrase includes, apart from the actual body of text, a preface and an *incipit* which recalls the works of the Kindī-circle.[51] The paraphrase survives in at least two versions which both seem to be less comprehensive than the original. The version which is preserved in the Madrid manuscript Escorial ²649 (fols. 203b-26a) and edited by Arnzen attests, in itself, to two different recensions.[52] The second ver-

[46] For a description of the MS, see Sebti, 'Copie', pp. 401–3.

[47] Arnzen (*Aristoteles' DA*, pp. 104–7) points out that the Greek paraphrase seems to have been used directly or indirectly by the Byzantine philosopher Sophonias (thirteenth/fourteenth century). The Arabic rendering shows signs of having used an Arabic Plotinus Source which was probably still arranged in the order of the Greek *Enneades* (*ibid.*, p. 121) and an Arabic version of Proclus' *Elementatio theologica* which had not yet been divided into the corpus of *Proclus Arabus* and the *Liber de causis*, and which contained more material than both of them (*ibid.*, pp. 124–9). However, Arnzen further suggests that there was mutual influence, i.e. that the *DA*-paraphrase also influenced the Arabic Plotinus and Proclus Sources, which would be evidence for the scholarly collaboration taking place in the circle of al-Kindī (*ibid.*, pp. 123–4).

[48] *Ibid.*, pp. 82–3.

[49] *Ibid.*, pp. 94–5 and 104.

[50] *Ibid.*, pp. 75 and 145.

[51] The preface and the *incipit* were not written by the translator, but added at a later stage, although at a stage when the body of the text was still more complete than today, as the preface covers parts of the Aristotelian text no longer extant in the actual paraphrase; see *ibid.*, pp. 21–33; and Sebti, 'Copie', pp. 403–10.

[52] One of these recensions is further attested in a Persian translation of the first half of the thirteenth century by the philosopher Bābā Afḍaladdīn Kāshānī; see Arnzen, *Aris-*

sion entitled *Kitāb fīhi Ṣifat al-nafs, ta'līf al-ḥakīm al-fāḍil Arisṭāṭālīs* (*Book on the Description of the Soul, composed by the virtuous sage Aristotle*, fol. 54a) has recently been discovered in the Istanbul manuscript Aya Sofya 4156 (fols. 54a-106a) by Sebti and has not yet been extensively studied. The Aya Sofya copy, which must date to the 1320s is rather corrupt, containing many omissions, mistakes and several misplaced folios. Sebti suggests, mainly on terminological grounds, that the Aya Sofya version is closer to the original than the edited version.[53] Unfortunately, it does not preserve any significant amount of material not present in the edited version.

The *PR* quotes the *DA*-paraphrase extensively in passages (78–83), (90), (92–3), (147–56) and (178–80),[54] but it is impossible to ascertain from which version. The wording is often closer to the edited version, but hardly ever exactly the same. Yet, at times, it is closer to the Aya Sofya verison or even different from both of them.

Aristotle Source, hitherto unknown

Evidence: *K. al-Ḥaraka*; *Mukhtār min Kalām al-ḥukamā'*;[55] *PR* passages (7–13) and (70).

The textual material presented in passages (7–13) of the *PR* may derive from one or several hitherto unknown Arabic Aristotle Sources. The independent existence of this text or texts is attested by partial parallels to the *PR* passages in the *K. al-Ḥaraka* and the *Mukhtār min Kalām al-ḥukamā' al-arbaʿa al-akābir* (*Selection from the Sayings of the Four Major Philosophers*), which are not possible sources of the *PR*. The material derives most probably from some translation with commentary, as the exact reproduction of *Physica* 192b21–3 and 198a35–b2 from a translation different from Isḥāq ibn Ḥunayn's suggests.[56] Further Aristotelian passages covered, although

toteles' DA, pp. 5–18. The text of the Escorial manuscript starts, after the *basmala* and mentioning the prophet and his family, without indication of any title with the words: *hādhā mā dhakara al-faylasūf fī Kitāb al-Nafs*.

53 Sebti, 'Copie', pp. 402 and 410–13.

54 Passages (159), (168) and (170–4) may quote from a no longer extant part of the *DA*-paraphrase.

55 The passage in question is found neither in the *Muntakhab* nor the *Mukhtaṣar Ṣiwān al-ḥikma* and thus most probably derives from a source other than the *Mukhtār*'s *Ṣiwān* Source. However, the possibility remains that it was present in the original *Ṣiwān*, but not taken up in any of its surviving summaries.

56 It may be the *vetus* translation since the *K. al-Ḥaraka*, when quoting *Physica* 198a35–b2, refers to Aristotle's work by the title *Samʿ al-Kiyān*. This is, as Kraus ('Ibn Muqaffaʿ', p. 7, n. 2) shows, the title used only in earlier times.

not in the form of translation, are taken from *Physica*, Books III–IV and VIII, maybe *De caelo*, Book I, chs. 5–6, and *Metaphysica*, Book *Lambda*, chs. 6–10. Part of the material has a distinct Neoplatonic ring.

Ps.-Aristotle, *Physiognomonica – Kitāb fī l-Firāsa* or *Kitāb al-Firāsa*

Text: ed. Ghersetti with Italian tr.

The translation of the Ps.-Aristotelian *Kitāb fī l-Firāsa* (*Book on Physiognomy*) from the Greek was carried out by Ḥunayn ibn Isḥāq, as he himself informs us in his first gloss to the Arabic text. Incorporated within the translation there are in total fourteen glosses to the Ps.-Aristotelian text introduced by '*qāla* Ḥunayn' which have either doctrinal or terminological and philological character. In the doctrinal glosses Ḥunayn quotes verbatim from the works of Hippocrates and Galen (*Epidemiae* VI, Galen's commentary on it, and Galen's *De temperamentis* II) to prove some Ps.-Aristotelian statements in the *K. fī l-Firāsa* wrong. Ibn al-Nadīm's doubts about the authorship of Aristotle may well be based on these critical glosses, as Grignaschi suggests. His further suggestion that Ibn Abī Uṣaybiʿa's reference to a commentary (*sharḥ*) on the *Physiognomonica* by Ḥunayn is to these glosses seems less certain.[57]

The compiler of the *PR* quotes from the body of the Ps.-Aristotelian text in passages (113) and (198) as well as from one of the glosses in passage (197). In the latter case he follows Ḥunayn and correctly attributes the quotation to Galen, but then seems to overlook the end of the gloss and introduces the following quotation of Ps.-Aristotle with a continuative '*wa-qāla*' (198).

Ps.-Aristotle, *Risālat al-Siyāsa al-ʿāmmīya*

Text: ed. Maróth in *Correspondence*, pp. 25–84; ed. Grignaschi in 'Siyâsatu-l-ʿâmmiyya', pp. 97–197.

The *Risālat al-Siyāsa al-ʿāmmīya* (*Letter on Gouverning in General*) is part of the epistolary novel or cycle *Rasāʾil Arisṭāṭālīs ilā l-Iskandar* (*Aristotle's Letters to Alexander*), which is based on a Greek model, maybe translated via Syriac into Arabic and reworked by Sālim Abū l-ʿAlāʾ, the secretary of Hishām ibn

[57] See *Fī l-Firāsa*, ed. Ghersetti, pp. xvii–xviii and xxii; and Grignaschi, 'Physiognomie', pp. 285 and 288–90.

ᶜAbd al-Malik who reigned in the first half of the eighth century.[58] The cycle consists of fourteen letters, ten narrative pieces and two gnomologies which are allegedly addressed to Alexander by Aristotle. Maróth who has edited the entire cycle assumes that the letters were written by the Greek compiler in the sixth century, whereas the other pieces were composed by various unknown authors in earlier times.[59] The *R. al-Siyāsa* is presented as the eighth letter and is, according to Grignaschi, the core, 'la pièce maîtresse', of the Arabic version of the entire cycle.[60] Maróth suggests that it originally existed as a self-contained treatise which was then inserted by the Greek compiler into the cycle as letter eight. To account for the diversity in language of the Arabic version, he further suggests that it was independently translated into Arabic at an early date, but taken into consideration when the entire cycle was translated later.

Passages (18) and (19) of the *PR* must ultimately go back to the *R. al-Siyāsa*, yet it cannot be known whether the compiler of the *PR* himself shortened the excerpted material so elegantly and put it into a new context, or whether he drew on an already existing reworking of it. That the *R. al-Siyāsa* was popular and often reworked is suggested by the fact that Mubashshir's *Mukhtār al-Ḥikam* also quotes part of the same material as the *PR*.[61]

al-Fārābī, see Alexander

Galen, *De propriis placitis – Fī mā yaᶜtaqiduhu raʾyan*

Text (Greek): ed. Boudon-Millet and Pietrobelli with French tr. in 'Galien', pp. 172–213; **(partial Greek text and Arabic-Latin tr.):** ed. Nutton with tr. in *Galen*, pp. 54–125.

Allegedly the last work Galen wrote, *De propriis placitis* (*On my own Opinions*) addresses philosophical topics such as the createdness of the world and the essence of the soul, and divides them according to the level of cer-

[58] For this and the following see Gutas, 'Graeco-Arabic Epistolary "Novels"', pp. 59–66.

[59] See his edition and reconstruction of the composition process of the cycle in *Correspondence*, esp. pp. 97–8; see also Grignaschi, 'Siyâsatu-l-ᶜâmmiyya', his 'Rasāʾil' and 'Roman' on the entire novel and on the *R. al-Siyāsa* in particular.

[60] 'Rasāʾil', p. 8.

[61] On the popularity of the entire novel, see Maróth, *Correspondence*, pp. 95–7.

tainty which can be attained in their knowledge (certain, probable or no knowledge).[62]

In his *Risāla fī Dhikr mā turjima min kutub Jālīnūs* (*Epistle about the Translation of Galenic Works*) Ḥunayn lists a Syriac and two Arabic translations of the work entitled *Fī mā yaʿtaqiduhu raʾyan* (*On What he Believes as an Opinion*), one by Thābit ibn Qurra and a second by ʿĪsā ibn Yaḥyā, one of Ḥunayn's students.[63] Unfortunately none of them is known to survive and the many references to *De propriis placitis* in Arabic writers, almost all of them to Galen's agnostic approach to the question of the eternity of the world and the essence of the soul, are hardly ever literal.[64]

There exists, however, an Arabic-Latin translation and a lengthy Hebrew quotation covering ch. 2 and the beginning of ch. 3 in Ibn Falaquera's *Sefer ha-Mebaqqesh*, which must ultimately go back to an Arabic text.[65]

By comparison with the recently rediscovered Greek text and the Arabic-Latin translation, it becomes apparent that the *PR*, in passage (146), quotes from *De propriis placitis*.[66] The quotation is composed of five phrases drawn from different places in the Greek text which may be an indication that the compiler of the *PR* used not the original Arabic translation, but rather some reworking. On the other hand, he proves in other passages that he is capable of shortening and rearranging his sources considerably, but intelligibly – for example, in passages (18) and (131) – which he may well have done in case of *De propriis placitis*.

[62] For the Arabic text, see edn Bergsträsser in *Ḥunain*, pp. 27–8, entry 49. The text explicitly states that Ḥunayn had already corrected part of Ḥubaysh's translation, but was at the time correcting the rest of it. An addition by a later hand notes that he then translated Book XVII. This seems to indicate that Ḥunayn provided a new translation for the entire Book XVII, which may explain why the Paris manuscript preserves two different translations of that very book (see Savage-Smith, *Galen on Nerves*, p. 8).

[63] See passage (143) below.

[64] See the Greek text p. 172; and Nutton's discussion in *Galen*, pp. 47–8.

[65] See the later recension (A) of the *Risāla*, entry 113 (ed. Bergsträsser in *Ḥunain*, 46.22–47.4):
كتابه فيما يعتقده رأياً . هذا الكتاب أيضاً مقالة واحدة يصف فيها ما عُلم وما لم يُعلم ، وقد ترجمه أيّوب إلى السريانيّة وترجمتُه | إلى السريانيّة لإسحق ابني وترجمه إلى العربيّة ثابت بن قرّة لمحمد بن موسى وترجمه عيسى بن يحيى إلى العربيّة وقابل به إسحق الأصل وأصلحته [وأصلحته Bergsträsser in a 'nachträgliche Verbesserung', p. 53 وأصلحه : MS] لعبد الله بن إسحق .
The earlier recension (B) does not mention a second Arabic translation by ʿĪsā ibn Yaḥyā (Bergsträsser, 'Neue Materialien', p. 32, table) and neither do Ibn al-Nadīm, al-Qifṭī and Ibn Abī Uṣaybiʿa.

[66] On the references to the Galenic work see Nutton, *Galen*, p. 40; and Zonta, *Interprete*, pp. 15–16.

Galen, *De usu partium* (*UP*) – *Kitāb fī Manāfiʿ al-aʿḍāʾ*

Text (Greek): ed. Helmreich in *Claudii Galeni opera omnia*, vols. III–IV; ed. Kühn; and tr. May in *Galen On the Usefulness*; **(Arabic):** ed. and tr. Savage-Smith in *Galen On Nerves* (partial edn of Book XVI); Paris, Bibliothèque Nationale de France, MS Arabe 2853 (complete copy; Book XVII even in two different translations; dated 1283); Manchester, John Rylands Library, MS Arabic 809 (complete copy; seventeenth century); Bethesda, Maryland, National Library of Medicine, MS A 30.1 (contains only middle of Book I – middle of Book XV; part of Books V and VI occur twice; seventeenth century); Madrid, Escorial, MS Arab. 850 (contains only Books IX–XVII; dated 1145).

In his *De usu partium* Galen explains in seventeen books how the body operates and why the way it does is the best possible way arranged for by a benevolent Creator and a providential nature. As for the Greek-Arabic translation, Ḥunayn states in his *Risāla fī Dhikr mā turjima min kutub Jālīnūs* that he translated *De usu partium* into Syriac, because the previous translation by Sergius [of Reshʿayna] was bad, but he also corrected the Arabic translation done by his nephew Ḥubaysh.[67] This may be the reason why existing Arabic manuscripts claim that the translation entitled *K. fī Manāfiʿ al-aʿḍāʾ* (*On the Uses of the Body Parts*) is by Ḥunayn. The Arabic translation remains more or less faithful to the Greek original, although, for example, the mention of nature and its providence is replaced by reference to the Creator.[68]

The compiler of the *PR* quotes the *K. fī Manāfiʿ al-aʿḍāʾ* in passages (118), (120), (123), (125), (127–8), (131), (140), (143) and (144). He mostly reproduces his model faithfully, yet in passages (118), (131) and (143) he shortens and rearranges the Galenic original heavily. Since the *K. fī Manāfiʿ al-aʿḍāʾ* is, except for the partial edition of Book XVI by Savage-Smith, still unedited, I present the Arabic version of the Galenic text underlying these *PR* passages in appendix 1.[69]

The *K. fī Manāfiʿ al-aʿḍāʾ* seems to have been popular among philosophers, who may have felt particularly attracted to its argumentation from design. In his *Tartīb al-saʿādāt* (103.18), Miskawayh, for example, refers his readers to it to learn about the immense wisdom displayed in the well-arranged human body.

[67] See Nutton, *Galen*, pp. 22–33; and Zonta, *Interprete*, pp. 102–8.

[68] See Wakelnig, 'Fragments Galen'.

[69] For an edition of all the passages of *UP* cited in the PR, see Wakelnig, 'Fragments', appendix, pp. 230–8.

Hippocrates Source, hitherto unknown

Evidence: al-Majūsī, *Kitāb Kāmil al-Ṣināʿa al-ṭibbīya*; *PR* passages (139), (141), (195–6).

Four passages which the *PR* ascribes explicitly to Hippocrates have word-for-word parallels in al-Majūsī's *Kitāb Kāmil al-Ṣināʿa al-ṭibbīya* (*Complete Book of the Medical Art*). There, only one section has a definite ascription, which reads 'Hippocrates and Galen agree that ...'. The absent ascriptions in the other instances make al-Majūsī an unlikely source of the *PR*, although the *K. Kāmil al-Ṣināʿa* probably preceded it by about 100 years, and al-Majūsī's material in question is far more elaborate than the parallel passages in the *PR*. Thus both writings may draw on a common source of Hippocratic material which has most probably been reworked and maybe enhanced by Galenic texts.[70]

Ḥunayn ibn Isḥāq, *al-Masāʾil fī l-Ṭibb li-l-mutaʿallimīn* and alternative recensions

Text (of the *Masāʾil*): ed. Abū Rayyān, ʿArab and Mūsā;[71] **(of the Arabic-Latin version):** ed. Maurach in 'Johannicius', pp. 151–74; tr. Withington in 'Galenic System', pp. 386–96;[72] **(of the *Masāʾil ʿalā Ṭarīq al-taqsīm wa-l-tashjīr*):** Berlin, MS 6258 (= Sprenger 1885), fols. 1a-18b; Oxford, Bodleian, MS Huntington 602, fols. 48b-71a.

Ḥunayn's introductory work to Galenic medicine exists in at least two different Arabic versions.[73] The *Masāʾil fī l-Ṭibb li-l-mutaʿallimīn* (*Questions on Medicine for Students*) are the long recension, which presents the material in question-and-answer form and enjoyed wide popularity in the Islamic world. The *Masāʾil ʿalā Ṭarīq al-taqsīm wa-l-tashjīr* (*Questions in Form of Divisions and Tables*) arranges part of the material of the *Masāʾil fī l-Ṭibb* in tabular form. The assumption of the existence of a third Arabic recension, entitled *Kitāb al-Mudkhal fī l-ṭibb* (*Introduction to Medicine*) and in the struc-

[70] Weisser (*Zeugung*, p. 33), talking about Hippocratic, Galenic and Aristotelian texts on procreation and embryonic development, points out that there is always the possibility of a transmission through Byzantine or early Islamic compendia or other reworkings, which can only rarely be defined any further.

[71] The translation of the *Masāʾil* by Ghalioungui was unavailable to me.

[72] For manuscripts of the *K. al-Mudkhal fī l-ṭibb*, see Ullmann, *Medizin*, p. 117, n. 10. For a discussion of Maurach's edition and one further edition, see Weisser, 'Isagoge', pp. 229–30. Based on the use of different manuscripts, Withington's translation is slightly shorter than Maurach's Latin edition.

[73] On which see Ullmann, *Medizin*, pp. 117–18.

ture of a text-book, was based on the Latin version, known as the *Isagoge Johannitii ad Tegni Galeni*. However, according to Jacquart and Troupeau, the *Masāʾil fī l-Ṭibb* and the *Kitāb al-Mudkhal fī l-ṭibb* seem to correspond in the extant Arabic manuscripts, and the Latin version may be explained as a reworked translation of the *Masāʾil fī l-Ṭibb*.[74] The Latin *Isagoge* was particularly well attested and influential in the Middle Ages.[75]

Passages (115) and (119) of the *PR* make use of material presented in the *Masāʾil fī l-Ṭibb* and, to a lesser extent, also in the *Masāʾil ʿalā Ṭarīq al-taqsīm*. However, as the wording is quite different, the compiler may actually quote from a reworking or a summary of Ḥunayn.

Ikhwān al-Ṣafāʾ, *Rasāʾil* or *Kitāb Ikhwān al-Ṣafāʾ*

Text: edn Beirut.

The well-known encyclopaedia of philosophy and sciences was compiled in the tenth century by the so-called 'Brethren of Purity' (*Ikhwān al-Ṣafāʾ*), allegedly a group of intellectuals in Baghdad who have kept their aura of mystery until today.[76] Although the modern editions are all entitled *Rasāʾil* in reference to the about 50 or so epistles of which the work consists, most of the MSS actually give the title as *Kitāb Ikhwān al-Ṣafāʾ*.[77]

The *PR* quotes extensively from the 25th epistle, *Fī Masqaṭ al-nuṭfa* (*On Insemination*) in passage (142), which is ascribed to the 'wise men' (*al-ḥukamāʾ*). This does not necessarily prove that our compiler had access to the entire encyclopaedia, as individual treatises are known to have circulated separately.

The *PR* may further quote from *Fī Nawādir al-falāsifa fī l-mūsīqā* (*Rare Sayings of the Philosophers on Music*) preserved in the fifth epistle *Fī l-Mūsīqā* (*On Music*) of the *Ikhwān al-Ṣafāʾ*. Their *Fī Nawādir al-falāsifa fī l-mūsīqā* is partly based on the section on music in Ḥunayn ibn Isḥāq's *Nawādir al-falāsifa* and partly on yet unidentified sources.[78] Since the material quoted in the *PR*,

[74] Jacquart, 'À l'aube', p. 211 and n. 6.

[75] Weisser ('Isagoge', pp. 230–5) deems it probable that the material behind Ḥunayn's versions derives from a Late Antique or early Byzantine Greek background, as Temkin, Gracia and Vidal have suggested, but disproves the possibility that the Latin version could have been translated directly from the Greek as the latter two suggested.

[76] On the question of authorship, see Stern, 'Authorship' and 'New information'.

[77] There are different divisions of the encyclopaedia which count either 51, 52 or 53 *rasāʾil*. For the title of the work, see Diwald, *Arabische Philosophie*, pp. 16–17.

[78] Ḥunayn's *Nawādir al-falāsifa* is no longer extant, but fragments of it, among them the philosophers' aphorisms on music, have survived in al-Anṣārī's *Ādāb al-falāsifa* and its Arabic-Hebrew translation (see Zakeri, *Persian Wisdom*, pp. 59–61, and 'Ādāb al-falāsifa',

passage (167), has no overlap with Ḥunayn, it may derive either directly from the *Ikhwān al-Ṣafāʾ* or from their source.[79]

Miskawayh, alternative recension of his *Fawz al-aṣghar*

Textual witnesses: *K. al-Ḥikma*, Istanbul, MS Esad Efendi 1933, fols. 56b1–78b15; *PR*, passages (47–9), (84–9), (91), (94–5), (160–1), (210–12).

The alternative recension of Miskawayh's *Fawz al-aṣghar* (*Minor Triumph*) is not known to have survived as an independent work, but only in the form of lengthy fragments excerpted from it for two compilations, namely the *PR* and the *Kitāb al-Ḥikma* preserved in the Istanbul manuscript Esad Efendi 1933. The excerpts in the two compilations correspond word for word with each other, whereas they deviate in varying degrees from the parallel passages in the standard version of the *Fawz*. Also, the two compilations are independent of each other, since, on the one hand, the material in the *K. al-Ḥikma* is far more comprehensive than the excerpts in the *PR*,[80] and, on the other, the *PR* attributes its quotations correctly to Miskawayh who is not named in the *K. al-Ḥikma* at all. The *PR* further preserves one additional passage not found in the *K. al-Ḥikma*.[81] All this is strong evidence for the existence of an alternative recension of the *Fawz* in the name of Miskawayh which must have been the common source of the *PR* and the *K. al-Ḥikma*.[82]

pp. 185–7, where he states that Ḥunayn and his son 'for the time being ... have equal claim to the authorship of "Aphorisms of philosophers on music"' and 'can be reckoned only among the many contributors to' al-Anṣārī's compilation; English translation of the section on music from the Hebrew by Werner and Sonne, 'Philosophy and Theory of Music'). On Ḥunayn's and *Ikhwān al-Ṣafāʾ*'s section on music, see Rosenthal, 'Graeco-Arabic Works', pp. 261–5.

79 Therefore the *PR* has also no overlap with the section entitled *Fī Nawādir al-falāsifa* in al-Kindī's treatise *Fī Ajzāʾ khabarīya fī l-mūsīqā* (*On the Informative Parts of Music*), which Rosenthal has shown to be parallel to Ḥunayn's *Nawādir* ('Graeco-Arabic Works', pp. 262–3).

80 There are two regards in which the *K. al-Ḥikma* is more comprehensive. First, it contains about twice as much *Fawz* material as the *PR*; and second, it takes its excerpts from a complete copy of the alternative *Fawz* version, whereas the *PR* relies on an incomplete copy as deliberately blank lacunae in the text indicate (on which, see above, p. 5, and below, pp. 54–6). The second regard is not strictly proof of the independence of the two texts, as the *PR* could simply have drawn on an incomplete copy of the *K. al-Ḥikma*.

81 The *PR* explicitly ascribes four of its six fragments from the alternative *Fawz* to Abū ʿAlī Miskawayh or Abū ʿAlī (see Wakelnig, 'New Version', p. 89); the additional material is found in passage (85), on which, see *ibid.*, pp. 93–4.

82 On the reasons why Miskawayh may have had to compose two different versions of the *Fawz*, and which of them may have been the earlier, see above, p. 5.

Miskawayh, *Tahdhīb al-akhlāq*

Text: ed. Zurayk; and tr. Zurayk in *Refinement*.

Miskawayh's *Tahdhīb al-akhlāq* (*Refinement of Character*), which is also known by the title *Ṭahāra* or *Ṭahārat al-nafs* (*Purification of the Soul*), is an ethical treatise in six discourses which discuss the soul and its faculties, human perfection, the good and happiness, justice, love and friendship, and the soul's health.[83] It draws on a number of Greek sources among which Aristotle and Plato are particularly prominent, but which also include Bryson,[84] Galen, Porphyry and Themistius.

The relation between the *Tahdhīb* and the *PR* is not completely apparent. Passages (109–11) and (185) may derive from the *Tahdhīb*, but present the text in a highly restructured arrangement which does not seem to fit with the usual quotation technique of the compiler of the *PR*. Passage (185) even has some additional sentences not found in the *Tahdhīb*. Even more surprising is passage (200), the first sentence of which parallels the *Tahdhīb*'s quotation of the Pseudo-Aristotelian treatise *Fī Faḍāʾil al-nafs*, whereas the second sentence parallels the *EN*, yet the *PR* ascribes the entire passage to Plato. More accurately quoted are passages (201) and (206), which are ascribed to Aristotle in both works and thus seem to be excerpted by Miskawayh himself from some other source. All this makes one wonder whether the compiler of the *PR* excerpts from the *Tahdhīb* or some other writing by Miskawayh no longer extant, as is the case with the *Fawz*.

Miskawayh, *Tartīb al-saʿādāt wa-manāzil al-ʿulūm*

Text: ed. Awjabī in *Ganjīna*, pp. 99–127.

The full title *Tartīb al-saʿādāt wa-manāzil al-ʿulūm* (*Succession of the Kinds of Happiness and Steps of the Branches of Knowledge*) is given by Miskawayh in his *Tahdhīb*, whereas the manuscripts read *K. al-Saʿāda* (*Book on Happiness*), *K. Tartīb al-saʿāda* (*Book on the Succession of Happiness*) or *al-Risāla al-musiʿda* (*Epistle which brings happiness*).[85] As indicated by the full title, the *Tartīb* dis-

[83] See Zurayk in the introduction to his tr. of Miskawayh, *Refinement*, p. xviii. One late manuscript gives an enhanced title *Tahdhīb al-akhlāq wa-taṭhīr al-aʿrāq*, and in another manuscript the last discourse is divided into two. The sixth discourse contains a section on the fear of death, which also circulated as a separate treatise entitled *R. fī l-Khawf min al-mawt* or something similar.

[84] On the Arabic text used by Miskawayh, see Swain, *Family*.

[85] See Gutas, 'Paul', p. 231, n. 1, where he describes the older editions and some manu-

cusses two main subjects, i.e. happiness and the division of the sciences according to Aristotle and his writings. Gutas has shown that part of the second section is strikingly similar to al-Fārābī's *Iḥṣāʾ al-ʿulūm* (*Enumeration of the Sciences*) and suggested that the common source for both, Miskawayh and al-Fārābī, is 'an introductory work composed by Paul the Persian (in Pehlevî?) on the philosophy of Aristotle, modelled on the late Alexandrian prolegomena of Aristotle'.[86]

In passage (207), the *PR* quotes a *Tartīb* section ascribed to Aristotle. There is further a resemblance in contents and structure between the *Tartīb* and a section in the alternative recension of the *Fawz* quoted in the *PR*, passage (211), and the *K. al-Ḥikma*. As the section is not paralleled in the standard *Fawz* version, its resemblance to the *Tartīb* corroborates the assumption that it belongs to the alternative *Fawz* and Miskawayh's authorship. Part of this assumed *Fawz* section is also similar to al-Fārābī's *Iḥṣāʾ*, so that it may draw on the same Paul the Persian Source as the *Tartīb*.

Nemesius, *De natura hominis* – *Kitāb al-Abwāb fī ṭabīʿat al-insān*

Text (Greek): ed. Morani; tr. Sharples and van der Eijk in *On the Nature of Man*; **(Arabic):** ed. Haji-Athanasiou in 'Le traité de Némésius'.[87]

De natura hominis was composed by the bishop Nemesius of Emesa at the end of the fourth century. It covers the topics of man's position in the universe, body and soul, physiology, voluntary action, fate and providence, and shares similarities with some slightly earlier works by Gregory of Nyssa which may be one reason why *De natura hominis* was eventually attributed to the latter in the Greek, Syriac, Armenian and Arabic tradition from the ninth century onwards.[88]

There is evidence for at least three independent Arabic translations or partial translations:[89] first, the excerpts from the first 30 chapters pre-

scripts as well; and Awjabī, p. 99, whose edition is based on an Iranian manuscript not mentioned by Gutas.

86 Gutas, 'Paul', p. 244.

87 The edition is part of a Parisian doctoral thesis in two volumes of which I had access only to the second volume containing the edition. The edition is based on the Damascus MS, while deviating readings of the more original version preserved in three other MSS are given in the apparatus. As the *PR* apparently uses the more original version, I have retrieved its readings from the apparatus and used them in my comparison of the two texts.

88 See Samir, 'Versions', pp. 100–1.

89 See *ibid.*, pp. 140–2.

served in Ps.-Apollonius, *Kitāb Sirr al-khalīqa* (*Book of the Secret of Creation*), which can be dated to the beginning of the ninth century;[90] second, a fragment of a translation summarising the first chapter contained in a Sinai manuscript.[91] It is the only Arabic witness of *De natura hominis* which ascribes the work correctly to Nemesius, whereas all others give Gregory of Nyssa as the author. Samir dates this translation to the ninth century. Third, there is a complete translation entitled *Kitāb al-Abwāb fī ṭabīʿat al-insān* (*Book of the Chapters on the Human Nature*), which survives in two recensions: the more original one is preserved in three Coptic manuscripts and its later reworking is contained in the Damascus manuscript Ẓāhirīya 4871.[92] The original translation is quoted by al-Kindī in his *Risāla fī Ḥudūd al-ashyāʾ wa-rusūmihā* (*Treatise on the Definitions and Descriptions of Things*) which thus provides a *terminus ante quem* for the translation. The manuscripts ascribe it to Isḥāq ibn Ḥunayn, whereas the early fourteenth-century encyclopaedist Shams al-Riyāsa Abū l-Barakāt ibn Kabar ascribes it to Ḥunayn. Samir is inclined to the latter ascription for chronological reasons and because the style of the translation resembles, as he claims, Ḥunayn's in trying to render the sense of the Greek in good Arabic using glosses and clarifications.[93]

The *PR*, in passages (108), (121–2), (124), (126), (129–30), (157–8) and (191) quotes at length from the complete translation in its more original recension, yet without ever referring to the author other than by 'He says'.[94] As many of the quoted passages occur in the context of the exposition of Galen's physiology, the reader of the *PR* often gets the impression

[90] See Weisser's edition of Ps.-Apollonius, *Buch über das Geheimnis der Schöpfung*, pp. 537–632.

[91] Samir ('Versions', p. 111) also refers to another independent fragment which is an adaptation of ch. 18 in the Oxford MS, Huntington 240, fols. 114b-115b.

[92] For a description of the MSS, see Samir, 'Versions', pp. 110–11.

[93] See *ibid.*, pp. 142–6.

[94] Whether the compiler of the *PR* simply did not know who the author was or whether he did not want to quote a Christian, i.e. either Nemesius or, more likely, the alleged author Gregory of Nyssa, must remain doubtful. The position of the compiler towards other Christian authors is ambiguous, as he refers by name to Qusṭā ibn Lūqā in passage (145), but not to Ḥunayn ibn Isḥāq in passages (115) and (119). The case of ʿAlī ibn Rabban is hard to assess; he was a Christian with a Jewish-sounding element in his name and, although he converted to Islam at the end of his life, the compiler of the *PR* need not necessarily have known that. On the other hand, in almost all the passages of the *Firdaws* used in the *PR* ʿAlī ibn Rabban quotes other authorities by name. So the compiler of the *PR* may not have felt obliged to give two authorities for the same quotation, just as he only gives the name of Aristotle when he actually quotes Miskawayh in passages (201) and (206–7).

of reading Galen's own words. This is, however, not far off the mark, as Nemesius makes abundant use of Galen's works in his *De natura hominis*.

Plato Source, hitherto unknown

Evidence: *PR*, passages (27–8) and (203).

The *PR* attests to one or two genuine Plato Sources by the more or less accurate translations of two *Phaedo* quotations (the first at 63b6–8, 63c2–4; the second, 69c3–7) and one quotation from the *Leges* (631b-c). As to the *Phaedo*, the *PR* may ultimately share a common source with al-Bīrūnī, who quotes from the Platonic dialogue in his *Taḥqīq mā li-l-Hind* (*Investigation of what is [said about] India*). However, the *Leges* source of the *PR* seems to be unrelated to al-Fārābī's *Jawāmiᶜ* (*Synopsis*) of Plato's work.

(Plotinus), *Dicta Sapientis Graeci – The Sayings of the Greek Sage (GS)*

Text (of the *PR* passages): ed. and tr. Rosenthal in 'Aš-Šayḫ', I, pp. 474–92, II, pp. 370–83; ed. Badawī in *Aflūṭīn*, pp. 184–94; ed. D'Ancona with Italian tr. in *Plotino*, pp. 239–46 (*Enn*. IV, 8); tr. of all *GS* passages in the *PR* by Lewis in 'Plotiniana'; for one further possible *GS* passage with edn and tr., see Zimmermann, 'Origins', pp. 219–21;

(of the passages in the *Ṣiwān al-ḥikma* tradition): ed. and tr. Rosenthal in 'Aš-Šayḫ', II, pp. 384–400; ed. Badawī in *Aflūṭīn*, pp. 195–8; see also *Muntakhab*, ed. Dunlop, lines 1122–70; and ed. Badawī, 172.13–176.3; Shahrastānī, *al-Milal wa-l-niḥal*, ed. al-ᶜAbd, 456.5–460.14;

(of the *R. fī Bayān ᶜĀlamay al-rūḥānī wa-l-jismānī*): ed. and tr. Rosenthal in 'Aš-Šayḫ', III, pp. 42–65.

The figure of the 'Greek Sage' (*al-Shaykh al-Yūnānī*) still remains one of the puzzles of the Arabic transmission of the Greek Neoplatonist Plotinus.[95] There must have existed a text in the name of the Greek Sage, termed by Zimmermann *GS, which contained an unknown number of extracts from the Arabic adaptation of Plotinus' *Enneades* IV–VI and shared a common source, namely the Arabic Plotinus Source, with the *ThA*. Other than the latter, the influence of the material in the name of the Greek Sage must have been rather limited and even the mention of the Greek Sage seems

[95] See Rosenthal, 'Aš-Šayḫ', I, pp. 461–73; Zimmermann, 'Origins', pp. 113 and 208–14; Aouad, 'Théologie', pp. 574–80; and D'Ancona, 'Greek Sage', pp. 159–70.

restricted to people linked, in one way or another, to Miskawayh.[96] The passage mentioning the Greek Sage in *al-Ḥikma al-khālida* is, as Zimmermann points out, cited and enhanced by the quotation of sayings of the Sage in the *Ṣiwān al-ḥikma*, as the surviving *Muntakhab* of the *Ṣiwān* attests. Some of the *Ṣiwān* material of the *GS* was then taken over by Shahrastānī into his *al-Milal wa-l-niḥal* (*Book of Religions and Sects*) and by Shahrazūrī into his *Rawḍat al-afrāḥ* (*Garden of Rejoicings*).[97] Independent of the *Ṣiwān* tradition, the *PR* is the most extensive witness of the *GS* and quotes the material either in the name of the Greek Sage, the Sage or Plato or without attribution in passages (30–44), (50–69), (75–7), (96–97), (163–5), (181), (183) and (204). However, the *GS* material in all these texts has some overlap with the *ThA*. This corroborates the assumption that the Arabic Plotinus Source must have been a common source of the *GS* and the *ThA*.

One step further away from the original *GS* than the texts so far mentioned is a curious little treatise entitled *R. li-l-Shaykh al-Yūnānī fī Bayān ʿĀlamay al-rūḥānī wa-l-jismānī* (*Treatise by the Greek Sage Explaining the Two Worlds, That of the Spirit, and That of the Body*). The unknown author sets out to rearrange and clarify the doctrines of the Greek Sage which are, as he claims, 'very confused and contradictory' (tr. Rosenthal, p. 43). The result is a presentation of and commentary on Plotinian thought.

The Greek Sage is further mentioned, yet unfortunately without any of his genuine words quoted, by Miskawayh's contemporary, the physician Abū l-Ḥasan al-Ṭabarī, in his *Muʿālajāt al-Buqrāṭīya*, as an authority on generation and procreation; by the unknown author of the Ps.-Fārābian *Sharḥ Risālat Zaynūn al-kabīr al-yūnānī* (*Commentary on the Treatise by the Elder Zeno, the Greek*) in his introductory remark; and by al-Bayhaqī in his entry on al-Fārābī quoting the title of a (Ps.-)Fārābian work entitled *Commentary on the Virtues of Zeno and [on?] a Book* Timaeus *by the Greek Sage*, which may or may not refer to the commentary just mentioned.[98]

[96] See above, pp. 6–7
[97] See Rosenthal, 'Aš-Šayḫ', I, pp. 469–70.
[98] On the *Sharḥ R. Zaynūn*, see Zimmermann, 'Origins', pp. 213–14, who translates the sentence in question as follows: 'I have come across writings by the elder Zeno, pupil of Aristotle, and by the Greek Shaykh, presented (*shrḥ*: expounded) by the Christians with some omissions and additions. Now I present them separately, as they ought to be presented, beginning with a treatise by the elder Zeno, the Greek.' The reference in al-Bayhaqī which reads *Sharḥ Faḍāʾil Zaynūn* [*Zaynūn*, Cottrell : *al-Zaytūn*, al-ʿAjam] *wa li-l-Shaykh al-Yūnānī Kitāb Tīmāwus* was discovered by Cottrell in the edition of the *Tatimmat Ṣiwān al-ḥikma* by al-ʿAjam, p. 41; see her 'L'Anonyme', p. 420.

Ps.-Pythagoras' *Dicta aurea* and the commentaries ascribed to Iamblichus and Proclus

Text of the Golden Verses (Greek): ed. Diehl and Young in *Theognis*; repr. with tr. in Thom, *Verses,* pp. 94–9; **(Arabic):** al-Anṣārī, *Ādāb al-falāsifa*, ed. Badawī, pp. 116–19; Miskawayh, *al-Ḥikma al-Khālida*, ed. Badawī, pp. 225–8; ed. Cheikho, Eddé and Malouf in *Traités inédits*, pp. 60–3; *Muntakhab Ṣiwān al-ḥikma*, ed. Dunlop, lines 499–549; and ed. Badawī, pp. 119–23; lemmata in Iamblichus' commentary, ed. Daiber with German tr. (see below).[99]

Texts of the Arabic commentaries (Iamblichus): ed. Daiber with German tr. in Iamblichus, *Pythagorica*; **(Proclus):** ed. and tr. Linley Ibn aṭ-Ṭayyib, *Proclus' Commentary.*

The Pythagorean *Dicta aurea*, the *Golden Verses*, first composed in the Hellenistic period, were translated as *Waṣāyā Fithāghūrus al-maʿrūfa bi-l-dhahabīya* into Arabic in the ninth century, possibly by an older contemporary of Ḥunayn ibn Isḥāq, and from then on widely known and quoted.[100] Miskawayh incorporated them into his *Ḥikma al-khālida* and al-Anṣārī in his *Ādāb al-falāsifa* (*Aphorisms of the Philosophers*), which may or may not indicate that they were part of Ḥunayn's *Nawādir al-falāsifa* (*Anecdotes of the Philosophers*).[101] They were also part of the *Ṣiwān al-ḥikma* as the extant *Muntakhab* attests.[102] According to Daiber, the Greek version which was translated into Arabic also contained Iamblichus' commentary on the *Dicta aurea*. This commentary is lost in Greek, but is still extant in Arabic, although less well attested than the *Dicta aurea* as such.[103]

[99] There is also Ullmann's edition of the *Golden Verses* in his doctoral thesis, *Griechische Spruchdichtung im Arabischen*, Tübingen, 1959, pp. 32–50, which I have not been able to consult.

[100] For the dating of the composition, see Thom, *Verses*, pp. 57–8, who suggests around 350–300 BC; for the Arabic translation, see Daiber, 'Doxographie', p. 4985; and for the different headings under which it is preserved, see Iamblichus, *Pythagorica*, p. 38, critical apparatus.

[101] They are also part of the Istanbul MS Koprülü I, MS 1608, fols. 29b-31a, on which see Zakeri, *Persian Wisdom*, pp. 59–73 and esp. p. 71.

[102] Rosenthal ('Documents', p. 111) shows that the *Ṣiwān* version is closer to the Greek than Miskawayh's, although he assumes that both must derive from Ḥunayn's florilegium. The *Dicta aurea* also occur linked to the *Mukhtār min Kalām al-ḥukamāʾ* (see Gutas, *Literature*, pp. 56–7). The *Ṣiwān* version further contains an explicit reference to Iamblichus' commentary in the title, although no material of the commentary is taken up (ed. Dunlop, line 499):

وهذا مجموع من كتاب ايامبليخس لتبيين وصايا فيثاغورس المعروفة بالذهبيّة .

The version of the *Dicta aurea* in the *Ādāb al-falāsifa* (119.1–3) is the only one that quotes a short excerpt from Iamblichus' commentary (see Daiber in Iamblichus, *Pythagorica*, p. 16).

[103] See Daiber, 'Doxographie', p. 4985, and in Iamblichus, *Pythagorica*, pp. 15–16. The attri-

Independent of this transmission is an extant Arabic summary of another Neoplatonic commentary on the *Dicta aurea* ascribed to Proclus. No such Proclean commentary survives in Greek, but Westerink has discovered a possible reference to it in the tenth-century Byzantine lexicon *Suda*, which derives its information from the sixth-century biographer Hesychius of Miletus. The *Suda* lists a commentary on the *Dicta aurea* among the works of Proclus Procleius (probably end of the fourth century AD), not the better-known Proclus Diadochus.[104] The possibility that a commentary by Proclus Procleius survives in an Arabic summary is corroborated by the hitherto rather puzzling fact that Ibn Buṭlān, according to Yāqūt, claims that Proclus was from Latakia, the ancient Laodicea.[105] This claim must surely refer to Proclus Procleius who is said to have been from Laodicea in Syria (*Suda*, Π 2472) and has thus left his trace in Arabic. Whether Ibn Buṭlān was conscious of the existence of two different Greek philosophers by the name of Proclus remains uncertain. More certain, however, is that Ibn al-Nadīm, the author of the famous book catalogue *al-Fihrist* (completed 987/8), only knows of one Proclus, namely Proclus Diadochus, whom he lists in his catalogue. To him Ibn al-Nadīm ascribes a commentary on the *Dicta aurea* in Syriac of which a mere three pages had been translated into Arabic by Thābit.[106]

The surviving Arabic summary of the commentary on the *Dicta aurea* by, most probably, Proclus Procleius is the work of Ibn al-Ṭayyib which bears the title *Istithmār li-Maqālat Fīthāghūrus al-maʿrūfa bi-l-dhahabīya, tafsīr Buruqlus* (*Extracting the Fruit from Pythagoras' Treatise known as The Golden, Proclus' Commentary*) at the beginning and the slightly varied title *Thamarat*

bution to Iamblichus is debated, especially as Iamblichus' comments on the *Dicta aurea* in his *Protrepticus* differ from the Arabic commentary. Daiber explains this fact by a development of Iamblichus' thought, whereas Thom suggests that a later Neoplatonist influenced by Iamblichus may have been the author of the commentary (see his *Verses*, pp. 17–21).

104 See Westerink, 'Proclus commentateur', pp. 74–8. The entry in the *Suda* reads (Π 2472): ⟨Πρόκλος,⟩ ὁ Προκλήϊος χρηματίσας, Θεμεσίωνος, Λαοδικείας τῆς Συρίας, ἱεροφάντης. ἔγραψε Θεολογίαν, Εἰς τὸν παρ' Ἡσιόδῳ τῆς Πανδώρας μῦθον, Εἰς τὰ χρυσᾶ ἔπη, Εἰς τὴν Νικομάχου Εἰσαγωγὴν τὴν ἀριθμητικήν· καὶ ἄλλα τινὰ γεωμετρικά.

105 The claim is made in Ibn Buṭlān's account of his visit to Latakia in 1054 which is discussed by Dunlop ('Introduction', pp. 74–6). The text of Yāqūt (*Geographisches Wörterbuch*, IV, 339.18–19) further calls Proclus the author of the *Arguments for the Eternity of the World*, which may indicate confusion on the part of either Ibn Buṭlān or a later transmitter or Yāqūt who wanted to display his familiarity with Proclus and thus added this mismatched reference to the *Arguments*.

106 Ibn al-Nadīm, *Fihrist* (252.16–8) in the section on Proclus:
كتاب تفسير وصايا فيثاغورس الذهبيّة نحو مائة ورقة ويوجد سريانيّ عمله لابنته وكان ثابت نقل
منه ثلثة أوراق ثم توفّى ولم يتمّه .

Maqālat Fīthāghūrus al-dhahabīya bi-tafsīr Buruqlus (*The Fruit from Pythagoras' Golden Treatise with Proclus' Commentary*) at the end.[107] It is not clear whether Ibn al-Ṭayyib summarised his *Istithmār* on the basis of a Syriac version or of an unattested Arabic translation. It is further unclear how much he changed his source. Whereas Linley allows for the possibility that Ibn al-Ṭayyib's 'work is the product of his own inspiration at least in part', Daiber points out that on other occasions Ibn al-Ṭayyib seems to reproduce his sources faithfully.[108] As for contents, it is apparent that the *Istithmār* covers the entire *Dicta aurea* starting with a discussion of the first verse of the standard edition and ending with the last. However, as the correspondences between the commentary and the verses are often not very obvious, it is possible that some material is omitted within the commentary.

Passage (16) of the *PR* is strikingly parallel to Ibn al-Ṭayyib's *Istithmār* in contents, but completely different in wording. The simplest explanation for this parallel is to assume a common source for both texts, namely an Arabic translation of Proclus' commentary. Thus the *PR* probably preserves a fairly accurate picture of the translation, as our compiler hardly ever tampers with his sources. Passage (16) may further derive from the same source as passages (14) and (15), since passage (14) is ascribed to Proclus' *Usṭūkhūsīs al-ṣughrā*, whereas passages (15) and (16) are introduced by a continuative 'he says'. In this case the Proclean commentary may have been known by the title *Usṭūkhūsīs al-ṣughrā* (*Minor Elements*) in Arabic.

Passage (102) of the *PR*, however, quotes the standard Arabic version of the *Dicta aurea*, not Proclus' commentary.

Passage (6), although quoted in the name of Pythagoras, may or may not be linked to the *Dicta aurea*.

[107] Rosenthal ('Symbolism', p. 274) remarks that the term *istithmār* or *thimār* was Ibn al-Ṭayyib's 'favorite word for the numerous brief summaries of the contents of Greek works prepared by him. They stressed what he considered the most important points made in them.

[108] See Linley in Ibn al-Ṭayyib, *Proclus' Commentary*, p. viii, and Daiber, 'Review Linley', p. 135. There Daiber refers to a collection of Arabic texts excerpted by Ibn al-Ṭayyib which is entitled *K. al-Nukat wa-l-Thimār al-ṭibbīya wa-l-falsafīya* to strengthen his claim. However, this collection has not been properly studied yet. For a list of the treatises excerpted in it, see Daiber, 'aṭ-Ṭayyib', p. 630, n. 8, and Ferrari, *Kategorienkommentar*, p. 28.

Qusṭā ibn Lūqā, *Risāla fī l-Farq bayn al-rūḥ wa-l-nafs*

Text: ed. Troupeau; ed. Gabrieli in 'Risālah', pp. 627–55; ed. Cheikho in 'Risāla', pp. 97–109; ed. Ülken in *Ibn Sina*, pp. 81–94; tr. Livingston in 'Qusṭā', pp. 63–74.

The Christian Qusṭā ibn Lūqā al-Baʿlabakkī (820–912) was a contemporary of Ḥunayn, a physician, philosopher and translator like him, and spent his last years in Armenia. His *Kitāb* or *Risāla fī l-Farq bayn al-rūḥ wa-l-nafs* (*Book* or *Treatise on the Difference between the Spirit and the Soul*) based on the medical tradition of Galen was highly influential in the Arabic as well as the Latin world.[109] The Arabic text has been edited three times, each time based on a different manuscript. The fourth forthcoming edition by Troupeau combines all hitherto available manuscript evidence, which often varies considerably.[110]

The *PR* quotes from the *R. fī l-Farq* anonymously in passage (112), but refers to Qusṭā ibn Lūqā for the excerpt presented in passage (145).

Themistius' Paraphrase of Book *Lambda* of Aristotle's *Metaphysica*

Text (Arabic, partial): of the complete translation (chs. 1–2), ed. Badawī in *Aristū*, pp. 329–33; of the abridgement (chs. 6–9), ed. Badawī in *Aristū*, pp. 12–21; and partial tr. Pines in 'Themistius' Commentary', pp. 180–5 (corresponds to Badawī, pp. 16–21); **(Hebrew)**: ed. Landauer in *Commentaria Aristotelem Graeca* V, 5; and partial tr. Pines in 'Themistius' Commentary', pp. 185–6 (corresponds to Landauer, p. 30); French tr. of the existent Arabic fragments and the Hebrew tr., Brague in Thémistius, *Paraphrase*, pp. 47–126.

Themistius' paraphrase of Book *Lambda* of Aristotle's *Metaphysica* is lost in Greek and only partly extant in Arabic, in which, at least, a complete translation and an abridgement must have existed. Extant is the beginning (ch. 1 and beginning of ch. 2) of the once complete translation entitled *Maqālat al-Lām, sharḥ Thāmisṭiyūs, tarjamat Isḥāq ibn Ḥunayn* (*Book Lam[bda], Themistius' Commentary, translated by Isḥāq ibn Ḥunayn*). It was used by Ibn Rushd in his long commentary on the *Metaphyisca* and translated into Hebrew. This Arabic-Hebrew translation has survived. Further extant is part of the second half (chs. 6–9) of an Arabic abridgement by the title *Min Sharḥ Thāmisṭiyūs li-ḥarf al-lām* (*From Themistius' Commentary on the Letter*

[109] The aforementioned philosophical compilation *K. Ārāʾ al-ḥukamāʾ fī l-ṭabīʿiyāt* quotes the *R. fī l-Farq* (see Daiber, 'Doxographie', p. 4981, n. 44).
[110] On all the manuscripts, see Livingston, 'Qusṭā', p. 62.

Lam[bda]). Ibn Sīnā consulted it when writing his commentary on Book *Lambda*, chs. 6–10, in the *K. al-Inṣāf* (*Book of Impartial Judgement*).[111]

The *Fihrist* attests to an Arabic translation by Abū Bishr Mattā, whereas the fragment of the complete version and the Arabic-Hebrew translation attribute the translation to Isḥāq ibn Ḥunayn. A possibility to account for these facts is to assume that Mattā was the author of the abridgement, either by translating an unknown Greek abridgement or, more likely, by abridging Ḥunayn's translation.[112]

The quotation in the *PR*, passage (17) is, as far as can be judged on the incomplete textual basis, parallel to the abridgement and the Hebrew translation. It can thus be best explained as deriving from the one complete Arabic translation from which both the abridgement and the Arabic-Hebrew translation stem.

VI Some parallel and similar texts to the *PR*: Greek legacy in Arabic

In addition to the sources there are a number of texts which are either parallel or strikingly similar to the *PR* and may thus indicate a common source which they share with it. Since some of these texts have either a complicated history of transmission, are not particularly well known, or have not yet been edited, I present, in alphabetical order of authors, a short overview of them, in which I also outline their relation to the *PR*.

Ps.-ʿĀmirī, *Kitāb al-Saʿāda wa-l-isʿād*

Text: facsimile of Minovī's transcript of the only existing MS in Minovi, *al-ʿĀmirī's as-saʿādah*; ed. ʿAṭīya in *al-Fikr*.

The *Kitāb al-Saʿāda wa-l-isʿād* (*Book on Seeking and Causing Happiness*) is a compilation of ethical and political material taken from Greek, Arabic and

[111] An edition of Ibn Sīnā's commentary can be found in Badawī, *Arisṭū*, pp. 22–33. Pines discusses it in 'Themistius' Commentary', p. 178, n. 3 and pp. 191–3.

[112] See Badawī, *Arisṭū*, pp. (16)-(17); Pines, 'Themistius' commentary', pp. 177–8, n. 3, who states that the possibility of the abridgement being done from the complete Arabic translation is much more probable than the alternative assumption of two different Greek recensions which may have been translated into Arabic by the same person, but he stresses that 'If this [first] hypothesis is accepted, it should be recognized that the abridgement denotes considerable skill'; and Bertolacci and Reismann in the introduction to their edition of Thābit ibn Qurra, *Maqāla*, pp. 721–2.

Persian sources and arranged according to subjects. The compiler calls himself Abū l-Ḥasan ibn Abī Dharr, but could not be identified by the first two scholars to discuss the *K. al-Saʿāda*, Kurd ʿAli and Arberry.[113] The latter, however, established that no source later than the first half of the tenth century was quoted in the compilation and thus suggested this as a *terminus post quem*. The first editor of the *K. al-Saʿāda*, Minovi, identified the compiler with Abū l-Ḥasan Muḥammad ibn Yūsuf al-ʿĀmirī. He did not give any argument for this identification except a suggested emendation of the name Abū l-Ḥusayn Yūsuf ibn Abī Dharr at the beginning of the *K. al-Saʿāda* to Abū l-Ḥasan Muḥammad ibn Yūsuf Abī Dharr, and the incorrect claim that al-ʿĀmirī's *K. al-Fuṣūl fī l-Maʿālim al-ilāhīya* gives its author's name as Abū l-Ḥasan Muḥammad ibn Abī Dharr Yūsuf al-ʿĀmirī, when in fact, the single manuscript copy of the treatise does not include the element 'Ibn Abī Dharr'.[114] Nevertheless Minovi's improbable identification has been accepted widely ever since.[115]

As to its form and structure, the *K. al-Saʿāda* is, although less broad in scope, similar to the *PR* and shares with the latter at least a couple of common sources, namely Themistius' paraphrase on Aristotle's Book *Lambda* and *Ikhtiṣār al-Iskandarānīyīn*.

Anonymous, *Kitāb al-Ḥikma*

Text: Istanbul, Süleymaniye Kütüphanesi, MS Esad Efendi 1933, fols. 56b-109b.

The *Kitāb al-Ḥikma* (*Book of Wisdom*) is the title given by a later hand to an anonymous and apparently untitled compilation preserved in the Istanbul miscellany Esad Efendi 1933. The *K. al-Ḥikma* mainly deals with three subjects, i.e. metaphysics, meteorology and physiology, drawing on a variety of sources not specified by the compiler.[116] The main source for metaphys-

[113] However, Arkoun ('La conquête', p. 151) and Ghurāb (in his edition of al-ʿĀmirī's *Iʿlām*, p. 13) imply that Arberry has identified the author of the *K. al-Saʿāda* with al-ʿĀmirī, whereas he ('Arabic treatise', p. 22) has, in fact, only 'concluded that Abu l-Ḥasan was a Persian having access to a library in which those [Greek and Sassanian] sources, in Arabic translation, were still available. Such a library could certainly have been found in Bagdad; but the book leaves an impression of having been written for the edification of an independent Persian ruler, whom the author was perhaps ambitious to serve as vizier.' Yet Arberry later accepted Minovi's identification ('Javidhan', p. 146, n. 1).

[114] See his 'Az Khazāʾin', p. 59.

[115] For example by the editor of four genuine works of al-ʿĀmirī, Khalīfāt (see *Rasāʾil*, pp. 104–24, where he discusses the question at length), and the second editor of the *K. al-Saʿāda*, ʿAṭīya (*al-Fikr*, pp. 82–5).

[116] For a more detailed discussion see Wakelnig, 'New Version', pp. 86–8.

ics is Miskawayh's alternative *Fawz*, for meteorology Aristotle's *Meteorologica* in the version of Yaḥyā ibn al-Biṭrīq, and for physiology Ps.-Apollonius' *K. Sirr al-khalīqa*. A fourth source, used also in the section on meteorology, but mainly in the section on physiology, is ʿAlī ibn Rabban's *Firdaws*.[117] Interestingly, all the physiological excerpts from the *Firdaws* are also taken up in the *PR*, which may suggest that the two compilations do not draw on the complete text of the *Firdaws*, but some kind of summary. The *K. al-Ḥikma* and the *PR* further have Miskawayh's alternative *Fawz* recension in common, and the *K. al-Ḥikma* may also draw on a Plotinus Source similar to the one of the *PR*.[118] All this makes one wonder whether the two compilations share a pool of source texts.

Anonymous, *Risāla fī Ārāʾ al-ḥukamāʾ al-yūnānīyīn*

Text (partial): ed. Badawī in *Aflāṭūn*, pp. 306–30.

The *Risāla fī Ārāʾ al-ḥukamāʾ al-yūnānīyīn* (*Treatise of the Opinions of the Greek Wise Men*) is, as the title suggests, a doxographical compilation of Greek wisdom, preserved in the Teheran manuscript Kitābkhāna-i Markazī 2103. It has, as far as I am aware, not attracted any scholarly attention except for the partial edition of the Platonic statements by Badawī in his volume *Aflāṭūn* or *Platon en Pays d'Islam*.[119] To judge by this partial edition, the *R. fī Ārāʾ al-ḥukamāʾ* is arranged according to topics (for example the Creator, virtues, medicine, definitions, intellect), for each of which a number of different Greek authorities are quoted by name, with their doctrines. The *R. fī Ārāʾ al-ḥukamāʾ* is thus a doxography *par excellence*. It has similar contents to the *PR* and has three parallels to it, in passages (4) and (104–5) which must derive from a common source or sources.

Ps.-Aristotle, *Kitāb al-Ḥaraka*

Text: Ankara, Üniversitesi Dil ve Tarih-Coğrafya Fakültesi Kütüphanesi, MS Ismail Saib I 1696, fols. 93a-108b; Istanbul, Süleymaniye Kütüphanesi, MS Hacı Mahmud 5683, fols. 120b-140a.[120]

[117] The meteorological excerpt is taken from the *Firdaws*, 27.4–11; for the physiological excerpts, see their parallels in the *PR*, passages (132–8).

[118] See passage (77).

[119] At the end of 2013, I have gained access to a reproduction of the manuscript, for which I want to express my gratitude to Gerhard Endreß. I now work on a study of the entire text.

[120] At the time when Pines ('Hitherto Unknown Arabic Extracts', p. 287) first discussed

The *Kitāb al-Ḥaraka* (*Book on Motion*) attributed to Aristotle preserves philosophical material deriving ultimately from Aristotle's *Physica*, *Metaphysica*, *Meteorologica*, *De caelo* and *De generatione et corruptione*, and Proclus' *Elementatio theologica* and *Elementatio physica*.[121] The complete title is given in two slightly different versions at the beginning and end of the treatise, namely first as *K. al-Ḥaraka min al-thaqīl li-Arisṭāṭālīs* (*Book on Motion due to Heaviness by Aristotle*), then as *K. al-Ḥaraka bi-l-burhān li-Arisṭāṭālīs* (*Book on Motion with Demonstration by Aristotle).* The source texts the compiler used seem to have not been the known Arabic translations of the Greek works in question, but re-workings which, in some cases, may have included commentary material. Further peculiarities of the compilation are the repetition of the same Greek material in different Arabic versions and the apparent lack of a final redaction.[122] The present state of the *K. al-Ḥaraka* as preserved in two Turkish miscellanies suggests that it is a later, maybe twelfth-century compilation of earlier material, which judging from the vocabulary and the terminology used must partly go back to the days of al-Kindī and most probably to his circle.[123]

The *K. al-Ḥaraka* and the *PR* share a common Aristotle Source from which they quote a lengthy continuous section covering the material of Appendix 2 and passages (10–13). As both texts seem, at times, to cite a more comprehensive version of their source than the other, the possibility that either the *K. al-Ḥaraka* is the source of the *PR* or vice versa can be ruled out.

Ibn al-Ṭayyib preserving an Arabic version of the Alexandrian *Prolegomena*

Text: ed. Gyekye in *Ibn al-Ṭayyib's Commentary*, pp. 9–27; tr. Dunlop in 'Existence', pp. 79–93.

this manuscript, the folios were unnumbered except for the last one as fol. 140. In the reproduction of the manuscript at my disposal there exists a foliation, which is at first at odds with Pines' numbering as according to it, the *K. al-Ḥaraka* starts on fol. 120b. Yet, as it has a fol. 127 and a fol. 127m the two numberings are in agreement from fol. 128 onwards.

121 For a discussion of the two MSS which preserve the *K. al-Ḥaraka*, one of which is a copy of the other, and for a first description of the treatise, see Taylor, 'Neoplatonic Texts', pp. 251–64. I am currently preparing an edition and study of the *K. al-Ḥaraka*.

122 E.g. prop. 20 of Proclus' *Elementatio theologica* is taken up twice.

123 As the four other writings contained in the miscellanies can be dated no later than the twelfth century, it is reasonable to assume that this is also the *terminus post quem non* for the *K. al-Ḥaraka*.

Ibn al-Ṭayyib's commentary on Porphyry's *Eisagoge* starts off with a text which is, as Gutas shows, closely related to the Greek text of David's *Prolegomena*, his introduction to philosophy, which preceded the *Eisagoge* in the Alexandrian curriculum.[124] Gutas therefore concludes that 'In all probability it is an Arabic translation of an abridgment of David or an abridgment of still another introduction based on David, later incorporated by Ibn al-Ṭayyib as a preface to his commentary on Porphyry's Εἰσαγωγή'; but, as he points out, 'This whole area requires further study.'[125]

Ibn al-Ṭayyib marks the end of his *Prolegomena* section with the phrase, 'So let us break off the discussion in this doctrine, at the point where the introduction breaks off, which the Alexandrians prefaced to consideration of Porphyry's *Eisagoge*' (tr. Dunlop, p. 93).

Mubashshir ibn Fātik, *Mukhtār al-Ḥikam wa-maḥāsin al-kalim*

Text: ed. Badawī.

Mukhtār al-Ḥikam wa-maḥāsin al-kalim (*The Choicest Maxims and Best Sayings*) is the only work known to be extant by the scholar Abū l-Wafāʾ al-Mubashshir ibn Fātik. It consists of sections dealing individually with Greek authors such as Hermes, Plato, Aristotle and Galen, presenting biographical information and quotations of their words. Mubashshir himself

[124] Ibn al-Ṭayyib's commentary was, due to a misplaced fly-leaf, falsely ascribed to al-Fārābī (see Stern, 'Ibn al-Ṭayyib', pp. 419–20). In his translation of the introductory part, Dunlop ('Existence', p. 78) states that 'it stands in near relation to the prologue of Ammonius, son of Hermeias (fifth century A. D.) to Porphyry's *Eisagoge*'. Yet he continues, 'in view of the presence of additions in our text which can be paralleled exactly from Syriac works ... it seems reasonably certain that the direct Greek original was not Ammonius. Not improbably the present text presents a lost Greek introduction to philosophy, based on Ammonius, written by John Philoponus (sixth century A. D.) and placed by him in front of his commentary on the *Eisagoge*.' However, Gutas (*Literature*, pp. 225–6, n. 3) presents a table of correspondences between David's *Prolegomena* and the Arabic version.

On the 'Introduction to Philosophy' in the late Alexandrian *Prolegomena* in the tradition of Ammonius, see Westerink, 'Commentators', pp. 344–5; Hein, *Definition*, pp. 33 and 54–130. There is further evidence of the Alexandrian *Prolegomena* in Arabic, e.g. in a fragment by Qusṭā ibn Lūqā (ed. Daiber with German tr. and commentary in 'Qosṭa').

[125] See Gutas (*Literature*, pp. 225–6). However, Stern (*Isaac Israeli*, p. 16, n. 1) concludes, based on the fact that Ibn al-Ṭayyib discusses only the first question of David's *Prolegomena*, but omits the following two, that 'it seems that the commentary on which Ibn al-Ṭayyib relies (no doubt through the intermediacy of his masters, more expecially [*sic*] Ibn al-Khammār, who was an important commentator) was a representative of the Olympiodorian school, independent of both Elias and David'.

indicates in his work that it was written in 1048/9.[126] As for his sources, Rosenthal observes that the *Ādāb al-falāsifa* seem to be of minor importance and that Mubashshir probably draws on the same source or sources as the *Ṣiwān al-ḥikma* tradition.[127] He must also share – at least some sources – with the *PR*: passage (19), which quotes from the *R. al-Siyāsa*, is paralleled partly in the *Mukhtār al-Ḥikam* as are passage (184), which gives Galen's alleged account of love, and passage (202) which draws on the *Ikhtiṣār al-Iskandarānīyīn* of the *EN*. Passages (103–4) are, further, similar to material on Plato and Zeno used by Mubashshir.

Ps.-Plato, *Multaqaṭāt Aflāṭūn al-ilāhī*

Text: ed. Badawī in *Aflāṭūn*, pp. 246–92.

This Ps.-Platonic text entitled *Multaqaṭāt Aflāṭūn al-ilāhī* (*Selected Readings from [the Works of] the Divine Plato*) is formally structured like a gnomologium, yet it covers a variety of mainly philosophical subjects (for example epistemology, metaphysics, physics and medicine) among which ethics is not prevalent.[128] It is divided into roughly 300 sections most of which are introduced by 'He says' or similar expressions. For a third of them, parallels in the Arabic gnomological literature can be found. As its three main sources Arnzen establishes the Arabic version of Galen's paraphrase of the Platonic *Timaeus*, the *Taqwīm al-siyāsa al-mulūkīya* (*The Correct Establishing of Royal Gouverning*), which is another Ps.-Platonic gnomologium, and a third unspecified corpus of philosophical, scientific and gnomological excerpts. The *Multaqaṭāt Aflāṭūn* were probably compiled in the late eleventh or early twelfth century.

The *PR* has no exact parallels to the *Multaqaṭāt Aflāṭūn*, but the two compilations share some similar psychological and ethical material.[129]

Ps.-Plato, *Risāla fī l-Radd ʿalā man qāla inna al-insān talāshā wa-fanā baʿda mawtihī*

Text (longer recension): ed. Türker with Turkish tr. in 'Traité', pp. 57–63; partial edn Cottrell with French tr. in 'L'Anonyme', pp. 435–6; **(shorter recension):** ed. Badawī in *Aflāṭūn*, pp. 337–9.

[126] See Rosenthal, 'al-Mubashshir', p. 133.

[127] *Ibid.*, pp. 135–6. Rosenthal presents these observations with due caution, as neither the original version of the *Ādāb* nor the *Ṣiwān* are known to be extant.

[128] See for this and the following, Arnzen, 'Contents', pp. 13–16 and 18–21.

[129] See passages (116) and (214).

The *Risāla fī l-Radd ʿalā man qāla inna al-insān talāshā wa-fanā baʿda mawtihī* (*Refutation of Someone Who Says that Man Corrupts and Vanishes after his Death*) is a short philosophical compilation existing in two recensions of either three or nine sections which derive from different, partly unidentified sources.[130] The manuscript tradition ascribes it to either Plato or al-Fārābī. However, Rosenthal seems to believe that the little treatise is of rather recent date.[131]

The first section, which supplies the title, explains, in the alleged words of Plato, why man cannot be said to perish completely after death. The second and third sections in the name of 'the wise man' (*al-ḥakīm*) correspond to the Arabic version of Proclus' *Elementatio theologica*, props. 142 and 123, as preserved in the *Kitāb al-Īḍāḥ fī l-khayr al-maḥḍ* (*Book of the Exposition of the Pure Good*), the *Liber de causis*, chs. 23 and 5. Sections (4–9) are only included in the longer recension which has so far emerged in only one manuscript.[132] The fourth section, which is introduced by the words 'The wise man (*al-ḥakīm*) says in his Letter to Wisāsūs', is the one parallel to passage (2) of the *PR*. Sections (5) and (6) are introduced by 'He says', which may or may not indicate that they derive from the same source as section (4). As they deal with the first principle and cause, Türker assumes that they may be taken from the Arabic Plotinus Source. The seventh section presents an alleged Platonic quotation from the *Leges*, the eighth an excerpt in the name of Abū l-Faḍl, i.e. the Buyid vizier Ibn al-ʿAmīd, and the ninth section is ascribed to Abū Yazīd which may be a misspelling of Abū Zayd, i.e. al-Balkhī.[133]

The *R. fī l-Radd* cannot be the source of the *PR* since the *PR*'s version of the passage it shares with the *R. fī l-Radd* is more elaborate. It also seems

[130] The subdivision into nine sections and their description follows Türker's study of the *Risāla* in 'Traité', pp. 30–47. Cottrell further divides Türker's first part into two ('L'Anonyme', p. 432). The first three sections which constitute the entire shorter recension, amount to about half of the text of the longer recension.

[131] See 'Aš-Šayḫ', I, p. 471.

[132] Of the seven manuscripts currently known to preserve the treatise, only Istanbul, Üniversite Kütüphanesi, MS 1458, fols. 105b-106b (in Hindu-Arabic numerals and ink), or fols. 103b-104b (European foliation in pencil) contains the longer version. For the MSS, see Cottrell, 'L'Anonyme', p. 430; Rosenthal, 'Aš-Šayḫ', I, p. 471, n. 2 (on a MS Rampur); and Daiber, 'Findings', p. 37, no. 127 (on two MSS Lucknow).

[133] Section (8) has parallels to an extant letter to ʿAḍud al-Dawla by Abū l-Faḍl ibn al-ʿAmīd, in whose library Miskawayh worked for seven years (ed. Daiber with German tr. as the third letter in his collection of six letters of Ibn al-ʿAmīd, *Naturwissenschaft*, pp. 44–9 and commentary, pp. 132–6; cf. also Cottrell, 'L'Anonyme', p. 438, n. 71). For the possible emendation of Abū Yazīd, see Türker, 'Traité', pp. 46–7. Al-Balkhī was the teacher of Miskawayh's contemporary al-ʿĀmirī. Both teacher and student are quoted in the *PR*.

unlikely that the *PR* is the source of the *R. fī l-Radd*, because the latter introduces the parallel passage differently, namely by ascribing it to the 'wise man', which could refer to either Plato or Aristotle. Thus both compilations must ultimately go back to a common source, although most probably via some intermediaries. This is particularly likely as the two texts seem to derive from the same cultural and intellectual milieu and to draw on similar material.

(Plotinus), *Risāla fī l-ʿIlm al-ilāhī*

Text: ed. Badawī in *Aflūṭīn*, pp. 165–83; ed. Kraus with partial French tr. in Anawati, 'Néoplatonisme', pp. 182–221; tr. Lewis in 'Plotiniana'.

The *Risāla fī l-ʿIlm al-ilāhī* (*Epistle of Divine Science*) is ascribed to al-Fārābī, but, according to Zimmermann, this ascription is secondary, along with the title.[134] The treatise is in fact another textual witness of the once existing Arabic Plotinus Source from which it draws its partial rendering of *Enn.* V 9, V 3, V 4 and V 5. The topics covered are mainly intellect and the First Cause. The length of the *R. fī l-ʿIlm* is about that of one of the longest chapters in the *ThA*. However, it has no overlap with either the *ThA* or the *GS* and is not quoted in the *PR*.

(Plotinus), *Uthūlūjiyā Arisṭāṭālīs – Theologia Aristotelis (ThA)*

Text: ed. Badawī in *Aflūṭīn*, pp. 1–164; ed. Dieterici in *Uthūlūjiyā*; German tr. Dieterici in *Theologie*; tr. Lewis in *Re-examination* and 'Plotiniana'.

The *Uthūlūjiyā Arisṭāṭālīs* (*Theology of Aristotle*) is so far the most important witness to the existence of an Arabic version of Plotinus' *Enneades*.[135] It exists in two recensions, the so-called *Short Version* or *Vulgata*, which is extant in a number of Arabic manuscripts, and the *Long Version*, of which Judaeo-Arabic fragments have survived as well as an Arabic-Latin translation and excerpts from a Hebrew translation. Whereas the *Short Version* consists of a preface, a long list of headings,[136] and a main body of text

[134] See Zimmermann, 'Origins', p. 214. For a discussion of the *R. fī l-ʿIlm*, see *ibid.*, pp. 214–17; Aouad, 'Théologie', pp. 571–4; Adamson, *Arabic Plotinus*, p. 7.

[135] See, e.g., Aouad, 'Théologie', pp. 544–70; Zimmermann, 'Origins'; Adamson, *Arabic Plotinus*; D'Ancona, 'Pseudo-*Theology*'.

[136] They allegedly announce the topics discussed in the *ThA*, but in fact correspond only

divided into ten chapters, the *Long Version* contains only a short section of the preface and fourteen chapters of text.[137] Although it is widely assumed that the *Long Version* draws on the *Short* one and introduces additional material (for example the doctrine of the Divine Word), other hypotheses have been brought forward as well.[138] Be that as it may, in the context of the *PR* only the *Short Version* is of importance. Thus, in what follows, any reference to the *ThA* is to this recension.

The Plotinian material preserved in the *ThA* derives from *Enn.* IV–VI alone, as is the case with all other fragments of the Arabic Plotinus.[139] The material is a free paraphrase of the Greek text rather than a translation in the strict sense of the word – a 'translation-cum-paraphrase of Plotinus' as Adamson calls it – and arranged in a completely new order.[140] As clearly stated in the preface of the *ThA*, the work is a product of the Kindī-circle, namely Aristotle's *Theology*, expounded by Porphyry, translated by ᶜAbd al-Masīḥ ibn ᶜAbdallāh ibn Nāᶜima al-Ḥimṣī and edited (*aṣlaḥa*) by al-Kindī himself.[141]

Since the *ThA* and the quotations in the name of the Greek Sage most probably derive from the same Arabic Plotinus Source, it is not surprising that the *ThA* shares some parallel passages with the *PR*, i.e. passages (31), (51), (54), (57), (66) and (96).

Proclus, *Corpus Proclus Arabus (PA)* and *Kitāb al-Īḍāḥ fī l-khayr al-maḥḍ – Liber de causis* (*Ldc*)

Text (*PA*): ed. Endreß with German tr. in *Proclus Arabus*; **(additions to *PA*):** ed. and tr. Zimmermann in 'Proclus Arabus'; **(*Ldc*):** ed. Bardenhewer with German tr. in *Pseudo-aristotelische Schrift*; ed. Badawī in *al-Aflāṭūnīya*, pp. 3–33; ed. Taylor

to *Enn.* IV 4, 1–34, the Arabic text of which is scattered over the *ThA* and *GS* (see Zimmermann, 'Origins', p. 126).

[137] Lewis (*Re-examination*, p. iii) assumes that, as, e.g., Books X and XI of the Latin 'begin where the Arabic has a change of topic but no indication of a new section', the fourteen chapters are the original division which may, following Steinschneider's suggestion, be an 'imitation of the fourteen Books of the *Metaphysics*'.

[138] See Aouad, 'Théologie', pp. 568–9.

[139] The latest fragments found so far were detected by Hansberger in the *K. al-Ḥiss wa-l-maḥsūs* (see her 'Plotinus Arabus').

[140] Adamson, *Arabic Plotinus*, p. 6. Whether this order is arbitrary and caused by 'a physical catastrophe which befell' the *ThA* (i.e. Zimmermann's big-bang theory, for which, see his 'Origins', p. 130), or deliberate, is still debated. For the latter position, see D'Ancona, 'Pseudo-*Theology*', esp. pp. 85 and 95–112.

[141] See, e.g., Zimmermann, 'Origins', pp. 118–22; Aouad, 'Théologie', pp. 546–8.

in *Liber de Causis*, pp. 136–277; **(*Ldc* II):** ed. Thillet and ʿUḍayma with French tr. in 'Proclus Arabe'.

The Arabic transmission of Proclus' *Elementatio theologica* still poses many questions. An entire Arabic translation of the Greek work is not known to survive, but one most probably existed formerly. Extant are two traditions which present the Proclean material in varying degrees of reworking. First, there is the corpus known as *Proclus Arabus* which consists of twenty propositions of the *Elementatio*, and which has been edited, translated and studied by Endreß. Zimmermann later added two more propositions to the corpus. Second, there is the *Ldc*-tradition. It is attested by at least two treatises bearing witness to a translation and reworking of either the entire or part of Proclus' *Elementatio*, most probably undertaken in the circle of al-Kindī.[142] The *Kitāb al-Īḍāḥ fī l-khayr al-maḥḍ* (*Exposition of the Pure Good*) is an Arabic reworking of 34 Proclean propositions which has become particularly well known in the Medieval Latin translation entitled *Liber de causis*. In recent years a further Arabic treatise similar to the *K. al-Īḍāḥ* has surfaced and, as it is untitled in the single known manuscript, has been termed *Liber de causis II*. This treatise renders 30 propositions of Proclus' *Elementatio*, 25 of which also occur in the *Liber de causis*, though often phrased differently.

The *PR* has no verbatim parallel to the *PA* or *Ldc* material, but shares many of their doctrinal peculiarities since it stands in the same tradition.[143]

Ps.-Sijistānī, *Ṣiwān al-ḥikma*, its abridgements and the *Mukhtār min Kalām al-ḥukamāʾ al-arbaʿa al-akābir*

Text (*Muntakhab Ṣiwān al-ḥikma*): ed. Dunlop; ed. Badawī; **(*Mukhtaṣar Ṣiwān*):** ed. Kartanegara; **(*Mukhtār min Kalām al-ḥukamāʾ*)**: ed. and tr. Gutas in *Literature*, pp. 62–213.

The *Ṣiwān al-ḥikma* (*Cabinet of Wisdom*) is a collection of biographical and doxographical material by Greek and Arabic philosophers, most probably

142 See, e.g., D'Ancona, *Recherches*, and, on the role of al-Kindī, *ibid.*, esp. pp. 155–94; D'Ancona and Taylor, 'Liber de causis'. A further, probably independent, witness of the reworking of Proclus' *Elementatio* as *Ldc* is al-ʿĀmirī's *K. al-Fuṣūl fī l-Maʿālim al-ilāhīya*, a paraphrase of 25 Proclean propositions (on which see Wakelnig, *Feder*). The earliest known reference to the *Ldc* in Arabic may be found in Qusṭā ibn Lūqā (see Daiber, 'Qosṭa', pp. 127–8).

143 See also the *R. fī l-Radd*, which has a parallel to *PR* but also quotes two sections of the *Ldc*.

composed around 985–1030, which has not survived in its entirety.[144] There are two abridgements of the *Ṣiwān*, the *Muntakhab* (*Selection*) abridged by an anonymous editor, and the *Mukhtaṣar* (*Abridgement*) of ʿUmar ibn Sahlān al-Sāwi (d. *c.* 1145). As al-Qāḍī points out, both works not only summarise the original *Ṣiwān*, but also add material of their own.[145] There further exists the *Mukhtār min Kalām al-ḥukamāʾ al-arbaʿa al-akābir* (*Selection from the Sayings of the Four Major Philosophers*), which draws on the *Ṣiwān* material for its presentation of Pythagoras, Socrates, Plato and Aristotle, and seems to be more closely related to the *Mukhtaṣar* than the *Muntakhab*. Finally there are two later additions to the *Ṣiwān*. The first is entitled *Tatimmat Ṣiwān al-ḥikma* (*Supplementation of the Cabinet of Wisdom*) and was composed in the twelfth century by Ẓahīr al-Dīn al-Bayhaqī. It is Bayhaqī who claimed that the *Ṣiwān* was by Abū Sulaymān al-Sijistānī, but internal evidence of the extant abridgements as well as external evidence invalidates his claim.[146] The second is the *Itmām Tatimmat Ṣiwān al-ḥikma* (*Completion of the Supplementation of the Cabinet of Wisdom*), which was composed between the end of the twelfth and the middle of the thirteenth century by the same person who abridged the *Ṣiwān* in the *Muntakhab*.

A number of sources of the *Ṣiwān* as established by al-Qāḍī, i.e. the works of al-Tawḥīdī, Miskawayh and al-ʿĀmirī, make it likely that the *Ṣiwān* was composed at the same place and time as the *PR*. As its possible author, al-Qāḍī suggests Abū l-Qāsim al-Kātib, who was related to al-ʿĀmirī and described as his *ghulām*.[147] Although the *Ṣiwān* and the *PR* obviously stand in the same tradition and discuss common topics – see, for example, passages (53), (89) and (214) – they do not share a single word-to-word parallel.

The *Mukhtār min Kalām al-ḥukamāʾ* is, as Gutas (p. 431) shows, 'a recension of the Pythagoras – Aristotle section' of the *Ṣiwān* into which the compiler interpolated borrowings from other gnomological sources, like the Ps.-Platonic *Taqwīm al-siyāsa al-mulūkīya*. The date of compilation must be assumed to fall between the second half of the eleventh and the beginning of the fourteenth century.[148] The *Mukhtār min Kalām al-ḥukamāʾ*, in its material additional to the *Ṣiwān*, may quote from the same unidentified Aristotle Source as the *PR* in passage (9).

[144] For this and the following see Gutas, '*Ṣiwān*', pp. 645–7; and al-Qāḍī, '*Ṣiwān*'.

[145] al-Qāḍī, '*Ṣiwān*', pp. 89–93.

[146] *Ibid.*, pp. 99–105.

[147] *Ibid.*, pp. 106–19.

[148] See Gutas, *Literature*, pp. 431–5.

VII Marsh 539: its physical appearance

The manuscript belongs to the Bodleian Library in Oxford which received it through a bequest by Narcissus Marsh (1638–1713), Archbishop of Armagh and fellow of Exeter College, in 1714. It thus belongs to the Marsh collection within which it is numbered 539, whereas it appears as no. 484 in Uri's catalogue.

The given title and author

Two different titles are indicated in the manuscript, but both are probably inaccurate. First there is a label apparently added to the flyleaf of the manuscript at the time when it was in Marsh's possession, reading: '539, *De Demonstratione Sapientiae*, Auth: Abuʾl Kasem, Arab., mutilus ab initio et [in] fine'. A recent statement in pencil on the facing page claims that this label actually belongs to Marsh 251, but as Cottrell has already stated, Marsh 251 is incomplete neither at the beginning nor at the end. Therefore she reasonably concludes that one should not doubt the originality of this eighteenth-century label.[149] Its information is, in any case, also taken up on the binding where the gilded inscription 'ABULCASEM' appears on the spine. However, the questions of where this information about title and author given on the label comes from and whether it is reliable still remain. It may derive from the statement on the verso page of the flyleaf which reads: '*Hādhā l-Kitāb Ṣiwān al-ḥikma li-l-Bayhaqī Abī l-Qāsim Ẓahīr al-Dīn*'. As Cottrell points out, Bayhaqī is not called Abū l-Qāsim which is his father's name, but rather Ibn Abī l-Qāsim. Further, he is not the author of the *Ṣiwān*, but its later addition *Tatimmat Ṣiwān*.[150] It can also be easily ascertained that the *PR* contained in the Marsh 539 is not the *Tatimmat Ṣiwān*. The statement on the flyleaf is thus completely incorrect and if the information on the label derives from it, picking not the name Bayhaqī, but the erroneous Abū l-Qāsim and translating *Ṣiwān* rather freely as *Demonstratio*, the label must be disregarded as well.

Apart from these rather dubious indications of title and author, the manuscript contains, due to textual loss, no further characteristics such as *incipit*, *explicit*, colophon, date and place of composition or dedication. There are also no chapter headings.

[149] See 'L'Anonyme', p. 417.
[150] See *ibid.*, p. 418.

The assumed date of composition

Rosenthal dated the manuscript to the thirteenth century and Savage-Smith narrowed it down to the first half of this century, as the existence of catchwords points towards 1200, whereas the absence of chain lines places it before 1250.[151]

The folios

The paper of the main text block is non-European with eight to nine laid lines per centimetre, but no visible chain lines. The dimensions of the page are 200 x 150 millimetres, and of the written area 145 x 110 mm. There are usually thirteen lines per page, but in some cases a line is deliberately left blank as a break marker (fols. 17a, 89a, 101a, 102b) or some lines are added to finish a certain passage (fols. 20b, 33a with fifteen lines; fols. 21a, 24a, 109a with fourteen lines). In a few cases there are fewer lines for no apparent reason (eight lines on fol. 74b and eleven lines on fols. 74a and 107b). In the Marsh manuscript no ruling is visible.

There are two sets of foliation in the upper left-hand corner, one in Hindu-Arabic numerals written in ink and the other in European numerals in pencil. There are six additional pages of European paper at the beginning (numbered i–iii and 1–3) and at the end (numbered 134–9), probably added during rebinding. The actual body of the manuscript starts on fol. 4a, ends on fol. 133b and misses fols. 90, 91, 104–6 and 121–2. The flyleaf numbered fol. 4, which may or may not belong to the rest of the manuscript, has an Arabic note in the upper right-hand corner, Marsh's motto Πανταχῇ τὴν ἀλήθειαν in Greek in the upper left-hand corner, an ownership statement below, a Persian poem in the centre of the page, and two further ownership statements which are hard to decipher at the bottom.[152]

The type of quire originally used is probably quaternion, but as the rebinding is tight and as there are, due to the missing folios, a number of anomalous quires it is hard to tell. The individual quires are not visibly

[151] See Rosenthal, 'Aš-Šayḫ', I, p. 465, and personal communication of Emilie Savage-Smith.

[152] The decipherable part of the note reads as follows:

لا يتجزا جوز يعني بابه باله اصا ومالز ...

On some of the other statements, see Cottrell, 'L'Anonyme', pp. 416–18. Many manuscripts in Marsh's collection are inscribed with his motto (see Wakefield, 'Marsh's oriental collections', p. 77).

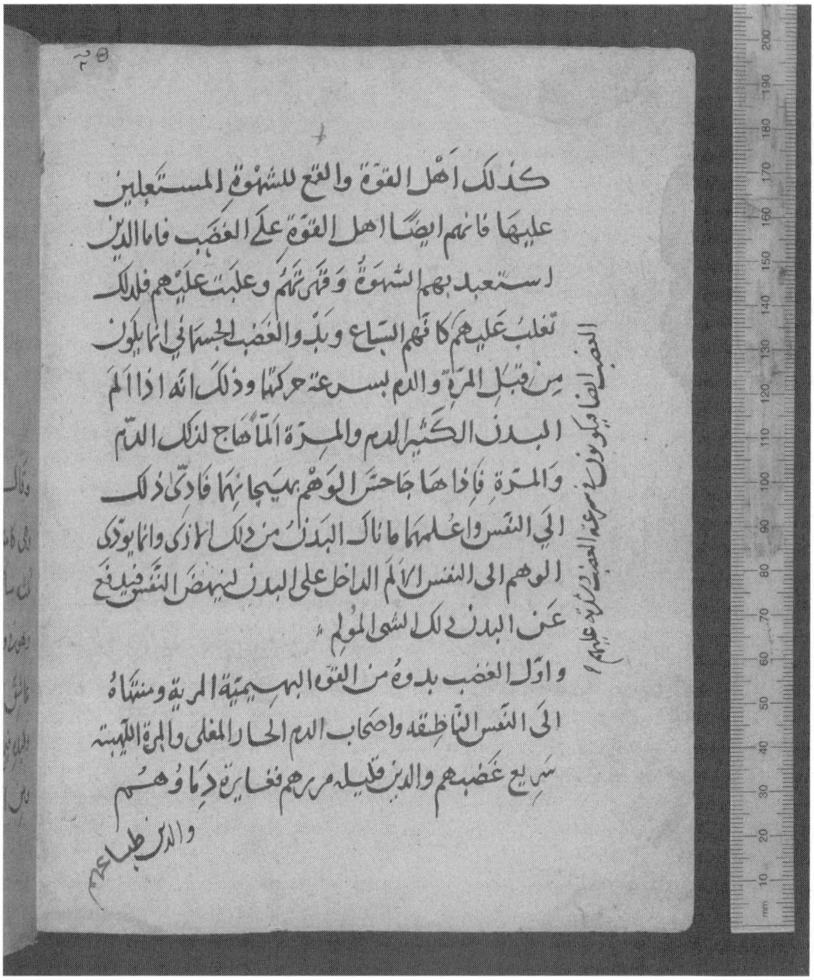

MS Marsh 539, fol. 103b (page before the second lacuna).

numbered. There are catchwords on the bottom left-hand corner of each verso-page.

The lacunae

The body of text contains three lacunae deliberately left blank, the first of which occurs on fol. 89b, where after seven written lines the bottom half of the page is empty and the catchword reads '*bayāḍ*' (blank space). The folios 90–1 are completely missing, as can be concluded from the folio following fol. 89 which is numbered 92. Its recto page is empty, but the writing resumes on fol. 92b, line 8, calculated on the basis of the usual thirteen lines per page. In the same way fols. 104–6 are completely missing,

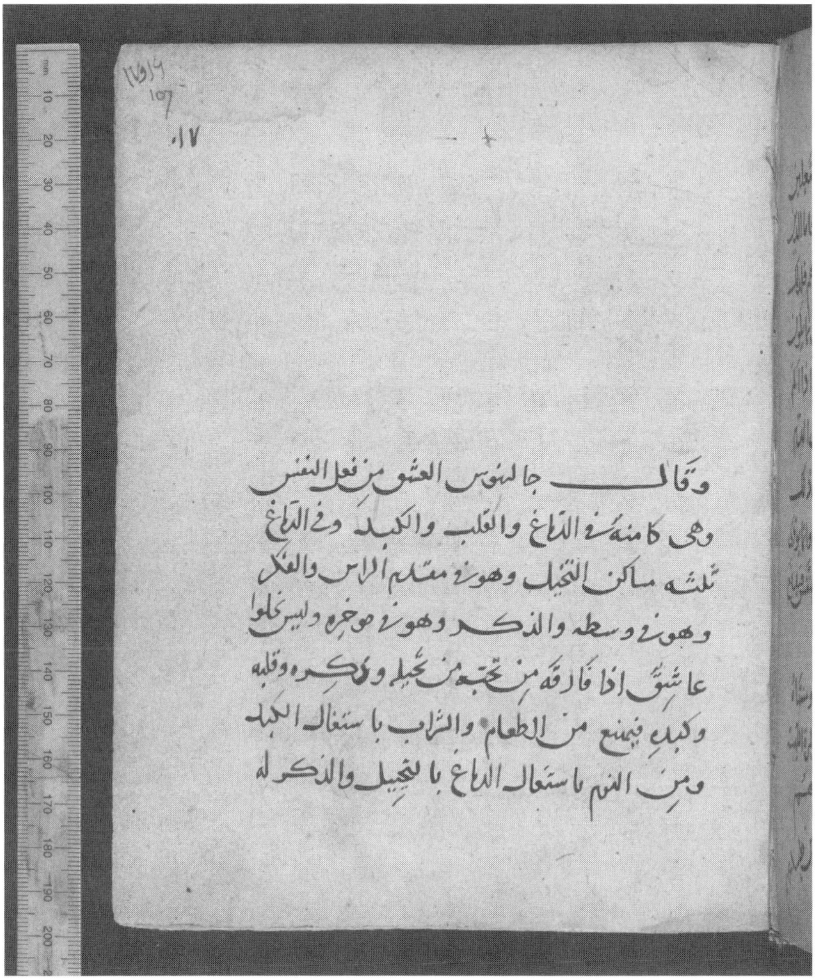

وقال جالينوس العشق من فعل النفس
وهي كامنة في الدماغ والقلب والكبد وفي الدماغ
ثلثة مساكن التخيل وهو في مقدم الراس والفكر
وهو في وسطه والذكر وهو في موخره وليس يخلو
عاشق اذا فارقه من يحبه من تخيله وفكره وذكره وقلبه
وكبده فيمتنع من الطعام والشراب باشتغال الكبد
ومن النوم باشتغال الدماغ بالتخيل والذكر له

MS Marsh 539, fol. 107a (end of the second lacuna).

although fol. 103b ends with a catchword which seems to indicate that some lines before the deliberate lacuna are lost due to the removal of fol. 104. The upper half of fol. 107a is empty and the writing resumes on 107a, line 7. This seems to indicate that the empty folios still existed at the time from which the foliation originates, but were removed at a later stage, probably during a process of rebinding.[153] The third lacuna is not apparent in the text, yet is detectable from the catchword '*bayāḍ*' on fol. 120b and

[153] Alternatively, the empty folios may have already been removed in the Islamic world and their absence would have only been apparent from the foliation in Hindu-Arabic numerals which originate from a time before the removal, and were simply followed by the European foliation.

the absence of numbers 121–2 in the foliation. In this case, some lines which may have immediately followed the lacuna on fol. 122b could be lost.

The script

The entire text is written in *naskh* by one hand, except for fol. 74 which may be a replacement of the original folio and thus written in a different hand. However, the slightly different appearance of the script may also be due to the usual scribe writing in haste or being unusually negligent. On a few occasions the original reading is corrected by a later hand (always indicated in the critical apparatus), and especially where the original scribe wrote one word over two lines, the second half of the word is added in the previous line, thus doubling this second half (for example 51a6–7, 54b13–55a1). The diacritical points which exist for most words must have been supplied by at least two different hands, one of which sometimes corrects the other. Vocalisation occurs inconsistently, but throughout. The text is written in monochrome black ink which is partly faded to a shade of sepia. The manuscript has no rubications or illustrations, but occasionally three dots arranged triangularly appear as break markers (indicated in the critical apparatus).

The binding

The manuscript has been rebound in a European, light brown leather binding without flap. It has a gold-stamped double frame on the front and the back and seven horizontal double lines on the spine, which also bears the gilded inscriptions 'ABULCASEM' (50–55 mm from the top) and 'MARSH. 539.' (190–200 mm from the top). The dimensions of the covers are 280 x 165 x 30 mm.

VIII The manuscript: an autograph?

The question of whether the Marsh manuscript is an autograph, i.e. the original copy of the compiler, still needs to be discussed. The palaeographic evidence does not allow for a definite answer due to the particular character of a compilation. Lacunae in the text and sentences which were first omitted accidentally, and later, probably during a revision process, added in the margin, do not necessarily indicate that the manuscript is a copy of an already existing text. In the case of a compilation these features may well be explained by incomplete source texts and by a final collation of the compilation with its source texts by the compiler. However, the lacunae are also no clear indication of an autograph as the copyist may well have reproduced them exactly as he had found them in his model, which could have been the autograph or a copy of it. The scenario in which the lacunae are due to some physical damage which befell the original compilation is less likely as all of them seem to occur within one source text, i.e. Miskawayh's alternative *Fawz*.[154]

The fact that the text occurring on fol. 23b9–13 is crossed out and cited later on fol. 26b9–13 is difficult to interpret. If the Marsh manuscript were an autograph it could be explained by the compiler starting with an excerpt at a position where he had not planned to insert it, so that he had to cross it out and add it anew in a different location. If the Marsh manuscript were a copy one might assume that, in turning a page, the copyist skipped a folio or two, then realised his mistake and went back to the correct folio.

However, based on the assumption that the *PR* was probably compiled during Miskawayh's lifetime or shortly after, i.e. in the eleventh century if not earlier, the Marsh manuscript, which is dated to the first half of the thirteenth century, cannot be the autograph of the compiler. The manuscript may, however, originate from the geographical part of the Islamic world in which the *PR* was compiled, i.e. in the Persian-speaking East, as Plato's *Phaedo* is spelt with a *fāʾ* with three dots, and on one occasion the *tāʾ marbūṭa* is written as plain *tāʾ*, as in Persian (fol. 43a12).

IX The edition: some remarks

The Arabic edition aims to provide a reproduction of the single manuscript of the *PR*, Marsh 539 (indicated as 'M' in the critical apparatus), which is as faithful as possible without being a facsimile or a diplomatic edition.

The manuscript is dotted throughout, yet with frequent omission of the dots on the *tāʾ marbūṭa* and inconsistency with regard to the final *yāʾ*, which appears dotted as well as undotted whether or not the letter represents a final *yāʾ* or *alif*. In these cases the dotting has been corrected according to the modern standard without indication of the original reading.

[154] See above, p. 5.

However, words in the critical apparatus are represented as they appear in the manuscript, i.e. with, for example, undotted *tāʾ marbūṭa.*

The personal prefixes of verbs in the third person singular in the imperfect are often dotted twice, i.e. as masculine and feminine at the same time, most probably by the original scribe and by a later hand.[155] In such cases and cases where the dotting is masculine, but should, according to modern grammar, be feminine or vice versa, the dotting has been corrected in the edition without recording the manuscript reading. In cases where the first person plural is used, the edition either follows the manuscript reading or reports it, if it has been changed, in the critical apparatus.

Missing or incorrect dotting of other words is only indicated where it provides a meaningful, but obviously unlikely, alternative, or where it occurs with remarkable consistency (e.g. النـــخاع occurs three times as الثجاع).

The orthography of the *hamza* has been adapted to the modern standard. However, all changes which have required a correction of the *hamza* carrier or an addition or omission of a tilde are reported in the critical apparatus. Exceptions have been made for isolated final *hamza* which is never supplied in the manuscript and therefore always an editorial addition, and in all occurrences of *shay'an.* This word appears in the manuscript as شيا, in which the *yāʾ* is often dotted as well as carrying a *hamza,* in which cases the *hamza* seems to be supplied by a later hand as in many other cases of occurrence of *hamza.*

All changes of the original *rasm* are reported in the critical apparatus, and even cases in which the scribe crossed out one or two incorrect letters are indicated. The only exceptions are incorrect ligatures of *alif* with a following letter (eg. حرلكـة instead of حراكـة) which are only mentioned where reading either *alif* or *lām* is possible. In general the *rasm* has only been changed in the main text of the edition where the original reading is without sense. Suggestions for alternative readings are given in the critical apparatus and introduced by the words 'or read'.

Inconsistencies in the spelling of names (for example, Aristotle) and technical terms (for example, *usṭuquss*) have not been removed as they clearly stem from the use of different source texts, and it is a remarkable

155 As mentioned above (p. 56), the manuscript is almost entirely written by one scribe, whereas the diacritics must be supplied by the original and at least one later hand. Since it is often impossible to distinguish between them, only in exceptional cases are diacritics reported to be by a later hand in the critical apparatus. The few later corrections of the *rasm* by a different hand are, however, always indicated in the apparatus.

feature of the *PR* that the compiler made no effort to harmonise his excerpts. This observation sits well with the general impression that the compiler did not tamper with his sources, but reproduced them faithfully. Consequently, the English translation does not aim at consistency in such a way that the same Arabic term would always be translated by the same English equivalent.

Entries in the critical apparatus report changes of the manuscript reading by previous editors – Rosenthal, Badawī, Zimmermann, D'Ancona and Cottrell – who edited selected passages, mainly those attributed to the *GS*. Where their reading has been adopted in the present edition, the first three letters of their name are put beside the emended reading. In cases where no name is given, emendations are mine.[156]

In two instances the lacunae in the Marsh manuscript (89b8–92b7 and 121a-122b) can be filled by text of the partly parallel *K. al-Ḥikma* and the corresponding critical apparatus thus refers to MS Esad Efendi 1933 indicated as 'E'.

My sectioning of the text of the *PR* often deviates from the sections introduced by the scribe by putting three dots and/or leaving blank space. However, all original breaks are reported in the critical apparatus.

The apparatus of parallel passages indicates the source quoted by means of an Arabic short title transcribed in Latin characters. In cases where more than one edition or manuscript of a text is used, the source is indicated by its short title followed by the abbreviated name of the editor or manuscript.

At times the editor of a source text adopts a reading which deviates from the reading of the *PR*, but indicates in his critical apparatus that the reading of the *PR* is also found in at least one of his manuscripts. In such cases no deviation is indicated between the source and the *PR*.

The footnotes to the translation indicate sources of – as 'cf. source' – and parallels to – as 'cf. parallel'– the *PR*. Where the relation between a text and the *PR* is not entirely clear, a simple 'cf.' is used. All texts which are introduced by 'cf.' in the notes are taken into account in the apparatus of parallel passages.

[156] Since the present edition is the first edition of the complete text of the *PR*, the previous passages edited are not numerous and so the emended reading always comes to stand next to the manuscript reading in the critical apparatus, I have not deemed it necessary to specify whether an emendation corrects a miswritten *rasm*, an incorrect dotting or vocalisation (on which see Gutas in his edition of Theophrastus, *On First Principles*, p. 92). The reader may easily judge for himself.

X Abbreviations used in the apparatus to the Arabic text

Sigla

Bad	partial edn Badawī in *Aflūṭīn*, pp. 184–94
Cot	partial edn Cottrell in 'L'Anonyme', pp. 435–9
D'Anc	partial edn D'Ancona in *Plotino*, pp. 239–46
E	Istanbul, Süleymaniye Kütüphanesi, MS Esad Efendi 1933
M	Oxford, Bodleian, MS Marsh 539
Ros	partial edn Rosenthal in 'Aš-Šayḫ'
Ros2	partial edn Rosenthal in 'Knowledge', p. 396
Zim	partial edn Zimmermann in 'Origins', p. 221
〈 〉	editorial additions to the text of the *PR*
[ ]	in the Arabic text, indicate editorial deletions from the text of the *PR*; in the English translation, indicate editorial addition to improve literal translation

Abbreviations

add.	added
ind.	indicated
marg.	margin
om.	omitted
secl.	secluded

Abbreviations of the sources quoted in the apparatus of parallel passages

Ādāb	al-Anṣārī, *Ādāb al-falāsifa* (ed. Badawī)
Fawz	Miskawayh, *al-Fawz al-aṣghār* (ed. ʿUḍayma)
Dhamm	Ibn al-Jawzī, *Dhamm al-hawā* (ed. ʿAṭā)
Firdaws	ʿAlī ibn Rabban al-Ṭabarī, *Firdaws al-ḥikma* (ed. al-Ṣiddīqī)
Fī Ṭabīʿa	Nemesius, *Kitāb al-Abwāb fī ṭabīʿat al-insān* (ed. Haji-Athanasiou)
GS	Greek Sage
Ikhtiṣār	*Ikhtiṣār al-Iskandarānīyīn – Summa Alexandrinorum* (ed. Badawī)
Istithmār	Ibn al-Ṭayyib, *Istithmār li-Maqālat Fīthāghūrus al-maʿrūfa bi-l-dhahabīya, tafsīr Buruqlus* (ed. Linley)

K. al-Akhlāq	*Arabic Version of the Nicomachean Ethics* (ed. Akasoy and Fidora)
K. al-Ḥaraka	Anonymous, *Kitāb al-Ḥaraka* (partial edn appendix 2)
K. al-Ḥikma	Anonymous, *Kitāb al-Ḥikma* (MS Esad Efendi 1933, fols. 56b-109b)
K. al-ḥukamāʾ	Anonymous, *Mukhtār min Kalām al-ḥukamāʾ* (ed. Gutas)
K. al-Mujtanā	Ibn Durayd, *Kitāb al-Mujtanā*
	ed. Ros – partial edn Rosenthal in 'Sayings'
	ed. ʿAw – ed. al-ʿAwwād
	ed. Dāl – ed. al-Dālī
K. al-Nafs	(Aristotle), *De anima*-paraphrase
	ed. Arn – *Aristoteles'* De anima, ed. Arnzen
	MS A – MS Aya Sofya 4156, fols 54a-106a
K. al-Samāʿ	Isḥāq ibn Ḥunayn's translation of Aristotle's *al-Samāʿ al-ṭabīʿī* (*al-Ṭabīʿa*, ed. Badawī)
K. al-Zahra	Muḥammad Ibn Dāwūd, *Kitāb al-Zahrah* (ed. Nykl)
Kāmil	al-Majūsī, *Kitāb Kāmil al-Ṣināʿa al-ṭibbīya* (Cairo edn)
K. fī l-Firāsa	Ps.-Aristotle, *Physiognomonica – Kitāb fī l-Firāsa* (ed. Ghersetti)
K. fī Manāfiʿ	Galen, *Kitāb fī Manāfiʿ al-aʿḍāʾ* (partial edn Savage-Smith or appendix 2)
K. Ṣaḥāʾif	ʿAlī al-Riyāḍī, *Kitāb Ṣaḥāʾif al-laṭāʾif* (MS Aya Sofya 4130)
Malady of Love	'Malady' (ed. Biesterfeldt and Gutas)
M. al-Ḥikam	Mubashshir, *Mukhtār al-Ḥikam* (ed. Badawī)
M. fī l-ʿAql	*Maqālat al-Iskandar al-Afrūdīsī fī l-ʿAql*
	ed. Bad – ed. Badawī in *Commentaires*
	ed. Fin – ed. Finnegan in 'Texte arabe du ΠΕΡΙ ΝΟΥ'
Masāʾil	Ḥunayn ibn Isḥāq, *al-Masāʾil fī l-Ṭibb* (ed. Abū Rayyān, ʿArab and Mūsā)
PR	*Philosophy Reader*
R. al-Siyāsa	Ps.-Aristotle, *Risāla al-Siyāsa al-ʿāmmīya* (ed. Maróth)
Rasāʾil	*Kitāb/Rasāʾil Ikhwān al-Ṣafāʾ* (Beirut edn)
R. fī Ārāʾ	Anonymous, *Risāla fī Ārāʾ al-ḥukamāʾ al-yūnānīyīn* (partial edn Badawī)
R. fī l-Farq	Qusṭā ibn Lūqā, *Risāla fī l-Farq bayn al-rūḥ wa-l-nafs* (ed. Troupeau)
R. fī l-Radd	Ps.-Plato, *Risāla fī l-Radd ʿalā man qāla inna al-insān talāshā wa-fanā*
	MS Ü – Istanbul, Üniversite Kütüphanesi, MS 1458, fols. 105b-106b

	ed. Tür – ed. Türker in 'Traité'
	ed. Cot – partial edn Cottrell in 'L'Anonyme'
R. fī l-Ṭibb	Abū Saʿīd ibn Bakhtīshūʾ, *Risāla fī l-Ṭibb wa-l-aḥdāth al-nafsānīya* (ed. Klein-Franke)
Sh. li-ḥarf al-lām	abridgement (chs. 6–9) of *Sharḥ li-ḥarf al-lām*, i.e. Themistius' paraphrase on Aristotle's *Metaphysica*, Book *Lambda* (ed. Badawī)
Tahdhīb	Miskawayh, *Tahdhīb al-Akhlāq* (ed. Zurayk)
Tartīb al-saʿādāt	Miskawayh, *Tartīb al-saʿādāt* (ed. Awjabī)
ThA	*Theology of Aristotle*
Uthūlūjiyā	(Plotinus), *Uthūlūjiyā Arisṭāṭālīs*, i.e. *Theology of Aristotle* editions are only cited when the two differ
	ed. Bad – ed. Badawī in *Aflūṭīn*
	ed. Die – ed. Dieterici in *Uthūlūjiyā*
	Lew – MSS readings indicated by Lewis in *Re-examination*
ʿUyūn	Ibn Abī Uṣaybiʿa, *ʿUyūn al-anbāʾ fī ṭabaqāt al-aṭibbāʾ* (ed. Müller)
W. Fūthāghūrus	*Waṣāyā Fūthāghūrus* as lemmata in Iamblichus' commentary (ed. Daiber)

Arabic Text and English Translation

5a (1) ... أن يوقعوا على الأوّل شيئاً ما لا شكّ أنّه غير لائق به .

(2) وقال في رسالته إلى ديانسيوس لا توقع على الأوّل صفات لا تشاكله لكَيلا يخرج من معنى الواحد إلى معنى الأنفس أو الصور أو القوى أو يضيف إليه كثرة البتّة أو شيئاً موصوفاً أو معلوماً ، بل ينبغي أن تفسّره لا موصوفاً ولا معلوماً . وتكون النفس أبداً معلّقة بمحبّته ، تقرب منه بالنقص من الكثرة التي فيها وتسارع إليه بهذا النحو من الترقّي وتخلّف الإحساس والأوهام والخيالات والأهواء والآراء كلّها أسفل وتثبته على الإعظام رئاسة وشرفاً . فتدنو بالواحد إلى الواحد ليس لتعرفه ولكن ليتّصل به ، وأمّا كثرة الفحص والتنقير فلا تستعملها البتّة ولا تتناوله بالأوهام أو بشيء من الأشياء التي تشبهنا وتليق بنا . فإنّك إذا فعلت ذلك خرجت من مشاركة الواحد وانحططت إلى الحياة الفانية التي هي حياتنا .

5b (3) وقال بان لي بالقياس والتجربة أنّي إذا كنت مفكّراً في الأمر الإلهيّ فإنّ عناية لله مظلّة عليّ وإذا انقطعت عن ذلك بالأمور البدنيّة كانت عنايته متبرّئةً منّي ، فمن أراد أن تكون عناية لله غير مفارقة له فلا يَفْتُر عن الفكر فيه .

(4) وقال ارتقيت إلى السموات الثلاثة وطلبت الترقّي إلى السماء الرابعة فقالت

2 وقال في : وقال الحكيم في *R. fī l-Radd* | رسالته : رسالته *R. fī l-Radd*, ed. Cot, MS Ü : رسالة ed.
Tür | ديانسيوس : وساسوس *R. fī l-Radd* | توقع : يوقع *R. fī l-Radd* | الأوّل : الواحد *R. fī*
l-Radd 3 الأنفس : النفس *R. fī l-Radd* | أو[1] ... القوى : والصور والقوى *R. fī l-Radd* 4 البتّة
om. *R. fī l-Radd* 4–5 بل ... معلوماً[2] om. *R. fī l-Radd* 5 معلّقة : متعلّقة *R. fī l-Radd* | بمحبّته :
بمحبة *R. fī l-Radd*, ed. Tür, MS Ü 5–8 تقرب ... والتنقير om. *R. fī l-Radd* 8–9 فلا ... البتّة :
ولا يستعمل كثرة البحث والسنة ed. Cot : ولا يستعمل كثرة البحث والصفة *R. fī l-Radd*, ed. Tür :
ولا يستعمل كثرة البحث والسفة MS Ü 10 فإنّك : لأنك *R. fī l-Radd* | ذلك om. *R. fī l-Radd*
مشاركة الواحد : الفاعل *R. fī l-Radd* | وانحططت : وانحططت *R. fī l-Radd*, MS Ü : وانخططت ed.
Tür, Cot | الحياة : الحيات *R. fī l-Radd* 16 after وقال add. أفلاطون قد *R. fī Ārāʾ* | الثلاثة :
الثلاث أمّا الأولى فهي علم الفلسفة الصناعيّة وأمّا الثانية فهي المعرفة الطبيعيّة و أمّا الثالثة فالصورة
العقليّة *R. fī Ārāʾ*

1 after به blank space of three letters B 2 ديانسيوس : Cot ديانسيوش M 4 تفسّره : M
نفسّره Cot 6 الإحساس : M الاحَساس : Cot الأحساس : or read الأحاسيس ? 7 والأهواء :
Cot والاهْوآ M | الإعظام : M الأعظم Cot | فتدنو : فتدفوا M فيدفّ Cot 8 لتعرفه M :
ليعرفه Cot | والتنقير : والنقير M واليقين Cot 9 تستعملها : يستعمله M, Cot | تشبهنا :
illegible M : يُستعملها Cot 10 وانحططت : وانخططت Cot | الحياة : الحيوه M 15 after فيه
blank space of three letters M 16 الثلاثة : الثلثه M

(1) ... that they[1] apply to the First something that certainly does not do 5a
Him justice.

(2) He says in his letter to Dionysius[2]: do not apply to the First attributes which are unlike Him, in order not to move away from the notion of the One towards the meaning of souls, forms or powers, or to relate any kind of multiplicity or something describable or knowable to Him. Rather, you must interpret Him as something indescribable and unknowable. The soul is always adhering to His love. It gets close to Him by diminishing the multiplicity in itself and in this way of ascension it hastens to Him. It leaves behind all the sensations, concepts, imaginings, desires and opinions and confirms Him by magnifying His leadership and eminence. Thus it comes close to the One through Oneness, not in order to come to know Him, but in order to be joined to Him. As to the multiplicity of investigation and interpretation, it does not employ them at all. It does not reach Him by concepts or any of the things which resemble and are suited for us [i.e. as man]. If you do that nevertheless, you move away from participation in the One and descend to the perishable life which is our life.

(3) He says: reason and experience have shown me that whenever I am 5b
contemplating divine matters, God's providence is sheltering me, and whenever I stop doing that owing to bodily matters, His providence keeps itself from me. So he who does not want to lose God's providence, should not cease contemplating Him.

(4) He says:[3] I had ascended to the three heavens and sought to ascend

[1] As the beginning of the MS is missing and the following passage does not seem to be directly connected with this sentence, it is hard to guess who 'they' might be.

[2] See partial parallel in *R. fī l-Radd*, MS Ü, fol.106a15–18; Türker, 'Traité', 61.6–12; Cottrell, 'L'Anonyme', p.435.
Cottrell suggests reading 'Dionysius', i.e. *Diyānusiyūs* based on Proclus' *Theologia Platonica* II, 51.21–2, where Proclus refers to Plato's second letter as 'Letter to Dionysius' and she offers a highly probable explanation for the corruption of the name ('L'Anonyme', p. 423, n. 29).

[3] Cf. parallel in *R. fī Ārāʾ*, 318.16–19, attributed to Plato and arranged slightly differently.

ليّ النفس والطبيعة طلبت ما قد حجب العقل عنه . أراد بالسماء الأولى الفلسفة الصناعيّة وبالثانية المعرفة الطبيعيّة وبالثالثة الصور العقليّة .

(5) وقال هرمس صعب أن يوقف على حقيقة أمر البارئ جلّ ذكره وغير ممكن أن يوصف ، وذلك أنّه ليس بمدرك ولا يدرك التامّ ما ليس بالتامّ ويصعب أن يعرف الأزليّ ما ليس بأزليّ فإنّ الأزليّ باق أبداً وغير الأزليّ فان والفاني خيال 6a وظلّ . فعلى قدر ما بين الضعيف والقويّ وما بين الجيّد والرديء كذلك بين الفاني والأزليّ .

(6) وقال فوثاغورس الأعداد الإلهيّة أربعة وكلّهنّ خير لا شرّ فيهنّ وأعداد الطبيعة أربعة لا خير فيهنّ . وواحد قائم بذاته وهو الخير الأوّل الذي ظهر منه الأعداد الأربعة وكوّن من حركاتها الأربعة الأخرى . والواحد الأوّل هو رأس الأعداد الثمانية ومحرّكها ومدبّرها ومقوّمها وحافظها وبارئها . فأمّا الأربعة الأولى فالعقل والنفس والصورة 〈الأولى〉 والقوّة الخيريّة المتمّمة لجميع الموجودات و〈أمّا〉 الأربعة الواقعة في الثواني فالطبيعة والهيولى والصورة الثانية والزمان وحركاته .

(7) وقال أرسطالس في كتابه المسمّى ما بعد الطبيعة إنّه قد وضح أنّ لله واحد أزليّ ليس بجسم ولا متحرّك وهو من العلم والحكمة والجود والقدرة في الغاية . أراد بقوله ليس بجسم نفي التشبيهات عنه لأنّ الصفات والنعوت حالّة 6b بالأجسام وبأنّه غير متحرّك نفي التغييرات والتشبّهات كلّها عنه لأنّ الحركة عند أرسطالس تجمع الحركة المكانيّة والانتقالات والتغيّرات .

(8) وقال ليس للزمان زمان أوّل قبله ، والزمان إنّما هو عدد مقادير الحركة المتّصلة المستوية الدائمة ، والحركة التي هي للجرم المستدير الدائم وإنّما تشكل

1 قد om *R. fī Ārā’* 1–2 أراد ... العقليّة om *R. fī Ārā’* 3 صعب : إنّه لصعب *K. al-Mujtanā* | البارئ : الخالق *K. al-Mujtanā* | جلّ ذكره om *K. al-Mujtanā* 4 after أنّه add. غير ممكن أن يصف جسم مدرك ما *K. al-Mujtanā* | بالتامّ : بتامّ *K. al-Mujtanā* 5 يعرف : يقرن *K. al-Mujtanā* | ما : بما *K. al-Mujtanā* 6 الجيّد والرديء : الدون والأشرف *K. al-Mujtanā* 7 والأزليّ : والإله الذي لا يموت *K. al-Mujtanā* كذلك : فكذلك *K. al-Mujtanā*

2 after العقليّة : blank space of three letters M 3 هرمس : هُرمس M 5 والفاني : والفان M 6 والرديء : الردي M 11 الثمانية : الثمَنية M 14 after وحركاته blank space of three letters M 19 والانتقالات M : or read والانفعالات ? | after والتغيّرات blank space of three letters M 20 للزمان : الزمان M

towards the fourth heaven, when it told me that soul and nature had sought what intellect prevented [them] from entering. He means by the first heaven skilful philosophy, by the second one natural science and by the third one the intellectual forms.

(5) Hermes says:[4] it is difficult to learn the truth about the Creator – exalted be His name – and impossible to describe Him. For He is imperceptible.[5] The imperfect does not perceive the perfect. It is difficult for the non-eternal to know the eternal, for the eternal is everlasting, whereas the non-eternal is perishable and the perishable is an imagining and a shadow. As the weak is to the powerful, and the faultless to the vicious, so is the 6a
perishable to the eternal.

(6) Pythagoras says: the divine numbers are four. All of them are good, there is no evil in them. The numbers of nature are also four, but there is no good in them. One is self-subsistent and It is the first good from which the four numbers arise. From their motions It generates the other four. The first One is the head of the eight numbers and their mover, their leader, their constitutor, their keeper and their creator. The first set of four are intellect, soul, ⟨first⟩ form and the beneficent power which brings all beings to perfection. The set of four which comes second are nature, matter, second form and time with its motions.

(7) Aristotle[6] says in his book called *Metaphysics* that it has already become clear that God is one, eternal, not a body and unmoved. His knowledge, wisdom, generosity and power are absolute. By saying 'He is not a body', [Aristotle] means to exclude anthropomorphisms from Him, because attributes and characteristics are inherent in bodies. By saying 'He is 6b
unmoved', he means to exclude all changes and imitations from Him, because motion, according to Aristotle, comprises locomotion, translocations[7] and alterations.

(8) He says: there is no first time prior to time [itself]. Time is the measure and number of motion which is continuous, regular and perpetual. The motion which belongs to the perpetual and rotating body is shaped in the

[4] Cf. Ibn Durayd, *K. al-Mujtanā*, ed. Rosenthal, p.183; ed. al-ʿAwwād, 194.1–6; ed. al-Dālī, 144.5–10.

[5] Here the text of the *K. al-Mujtanā* is corrupted, which may be the reason for the shortening of the text by the compiler of the *PR*.

[6] 'Arisṭālīs', though with a long i, is the form of Aristotle's name attested, e.g., in the *Fihrist*.

[7] The comparison with Aristotle, *Met.* 1073a11 (ἀλλὰ μὴν καὶ ὅτι ἀπαθὲς καὶ ἀναλλοίωτον·) and its Arabic translation *lā yanfaʿilu wa-lā yataghayyaru* suggests the emendation of 'translocations' (*al-intiqālāt*) to 'affections' (*al-infiʿālāt*).

بشكل الاستدارة من البارئ وهو العلّة الأولى . ولا يعرف له بدء في المواضع من أيّ أجزائه مبدأ جرمه لانعطاف الأجزاء بعضها على بعض . ومن أجل هذه الأشياء التي ذكرناها نضطرّ إلى أن نفعل العلّة الأولى المحرّكة دائماً التي هي علّة دوام حركة الأجرام المستديرة وأنّ جوهرها قبل جوهر الأجرام المستديرة والأجرام المتحرّكة بحركة مستديرة كثيرة ، وافتراقها وفصولها تضطرّنا إلى أن نجعلها أوائل كثيرة لأنّ مقادير حركاتها مختلفة . فأمّا العقل فيأبى أن يكون الأوائل كثيرة يخالف بعضها لبعض ، فإنّه لو كان ذلك كذلك لكان بالحرى
7a أن تكون الأشياء المتحرّكة التي تحتها مختلفة لا شرح لها فكان يكون هذا العالم
غير ذي نظم ولا شرح ، ولو كانت الأوائل مختلفة غير ذات شرح لاحتاجت إلى أوّل آخر معلول ينظمها ويؤلّفها ولم تكن الأوائل حينئذ أوائل . فأمّا إذ صار العالم كلّه ذا نظم وشرح والنظم والشرح لا يكونان باتّفاق وبلا علّة ، فلا بدّ إذن من أن تكون علّة النظم موجودة اضطراراً . فإن كان هذا هكذا فلا محالة إذاً أنّه يعرف من هذا القول أن تكون الأوائل موجودة وأن تكون علّة إبداع الأول والأشياء كلّها ونظمها واحدة .

(9) وقال في كتاب ما بعد الطبيعة فوق جوهر السماء جوهر لا عظم له ولا قدر من الأقدار غير مستحيل بنوع من الاستحالات لا نهاية لقوّته ومن أجل ذلك يفعل فعله بلا زمان وهو فعّال بذاته . فكذلك هو دائم الفعل وليس فعله بحركة ، ولا فيه شيء بالقوّة لكنّ الأشياء فيه بالفعل وقوّته منبثّة في العالم دائماً .

7b (10) وقال إنّ الأوائل المحرّكة بحركة طبيعيّة إثنان وإن أحدهما ليس هو بذاتيّ

15 في ... ما : فيما K. al-ḥukamāʾ | ²جوهر : شيء K. al-ḥukamāʾ | عظم له : أعظم منه
K. al-ḥukamāʾ 16 after الأقدار add. بقدره K. al-ḥukamāʾ | الاستحالات : الأنواع
K. al-ḥukamāʾ 17 فعله : أفعاله K. al-ḥukamāʾ | فكذلك : فلذلك K. al-ḥukamāʾ 18 بحركة :
يحرّكه K. al-ḥukamāʾ | ¹فيه : منه K. al-ḥukamāʾ | ²فيه : منه K. al-ḥukamāʾ 20 وقال : فقال
K. al-Ḥaraka 20–70.1 وقال ... يتحرّك : والمبادئ المحرّكة على المجرى الطبيعيّ ضربان ، وأحدهما
ليس هو طبيعياً ، وذلك أنّه ليس فيه مبدأُ حركة له . والذي هو بهذه الصفة هو شيءٌ إن كان يُحرّك
من غير أن يتحرّك ، مثل الذي هو غير متحرّك على وجه من الوجوه أصلاً وهو أول الأشياء كلها

1–3 ولا ... الأولى add. in marg. M 1 بدء : بدو M 3 ¹التي : الذي M | ذكرناها : ذكرنا M
7 بالحرى : بالجرْي M 8 يكون : يلون M 15 وقال : M ارسطانوس add. in different hand
before *wa-qāla* M | جوهر السماء add. in marg. M | ²جوهر : جوهرا M 20 بذاتيّ M : or
read بطبيعيّ ?

shape of rotation by the Creator, who is the First Cause. Among the
positions of any of its parts no beginning for it is known which could be
the starting point of its body as its parts are inclined towards each other.
On account of these things we have mentioned, we are obliged to make the
First Cause perpetually the mover who causes the perpetuity of the
rotating bodies' motion; and [we are oblidged to make] Its substance prior
to the substance of the rotating bodies and to the bodies which are moved
in multiple rotary motions. Their inconsistency and differences oblige us
to make [their] principles many as the measures of their motions are
different. Intellect, however, denies that the principles are many and vary
from each other. If it were like that, it would be likely that the moved
things which were below them should differ without having any order,[8] as 7a
if this world possessed neither organisation nor order. If the principles
were different and did not possess any order, they would need another
principle which would be caused to organise and combine them, and the
principles would then not be principles. However, since the whole world
possesses organisation and order, and [since] this organisation and order
are not by chance or without cause, the consequence is inevitable that the
cause of orgnisation exists necessarily. If it is like that, then it is positively
known on the basis of this statement that the principles exist, that the
cause for the creation of the first [principles], all things and their organ-
isation is one [and the same].

(9) [Aristotle] says in the *Metaphysica*:[9] above the substance of heaven there is a substance which has no magnitude nor any extension, which is not subject to any kind of change and the power of which is infinite. Therefore it performs its action without time, as it is active by itself. Thus it is perpetually acting and its action involves no motion. In it there is nothing in potentiality, but the things in it are in actuality. Its power is perpetually spread in the world.

(10) He says[10] that the principles which cause natural motion are two. One 7b

[8] *sharḥ* as equivalent to the Greek τάξις – see Endreß, *Proclus Arabus*, Arabic p.58.

[9] Cf. parallel in *Mukhtār min Kalām al-ḥukamāʾ*, 164.7–12, where the passage is simply introduced by 'He says'. An introduction similar to that in the *PR*'s source may be the reason why the name 'Aristotle' (or rather *Arisṭānūs*) has been supplied in the Marsh MS at a later stage, squeezed in at the end of the previous quotation on fol.7a8.

[10] Cf. Aristotle, *Physica* 198a35–198b2 in Isḥāq's translation in *al-Ṭabīʿa*, 140.1–5; parallel to this and the following three passages in *K. al-Ḥaraka*, appendix 2.

وذلك أنّه ليس فيه مبدأ لحركة وهو الشيء المحرّك دائماً من غير أن يتحرّك .

(11) وقال إنّ الحركة تسلك في جميع الأشياء دائماً وإنّ الحركة التي لا تعب فيها إنّما تكون في الأجرام المتحرّكة حركة مستديرة . ثم بيّن الأجرام المستوية والمستديرة وقال إنّه لا يمكن أن يكون عظم ما ولا شيء من الأجرام الطبيعيّة [لا] المستوية ولا الأجرام المستديرة لا نهاية له في العظم .

(12) وقال إنّ الجرم المتناهي في العظم لا يمكن أن يكون له قوّة لا نهاية لها لمدّة حركة لا نهاية لها وإنّ القوّة المتناهية لا تقوى على أن تحرّك حركة لا نهاية لها .

فلمّا قدّم هذه المقدّمات استنبط منها نتائج شريفة .

(13) فقال إنّ هاهنا جوهراً غير هذه الجواهر اضطراراً هو فوقها وأعلى منها ، لا عظم له ولا قدر من الأقدار ولا يستحيل بشيء من الاستحالات البتّة ولا غاية لقوّته ولا نهاية . فلذلك يفعل أفاعيله بلا زمان أعني أنّه لا يفعل في زمان 8a دون زمان لكنّه فعّال دائماً لأنّه فعّال بذاته لا بحركة ولأنّه ليس فيه شيء بالقوّة أصلاً لكن الأشياء فيه بالفعل وقوّته منبثّة في العالم كلّه دائماً . والحركة الدائمة أعني حركة السماء والعالم كلّه ، بها يقوى على قبول بتلك القوّة الشريفة لتدوم حركتها وتبقى . وكلّ واحد من تلك الأشياء إنّما يقبل من تلك القوّة على قدر تهيّؤه لقبولها ولا يقوى شيء من الأشياء أن يكون هذه الهيئة فيه باتّفاق وبلا علّة فاعلة للكلّ ، ومحال أيضاً أن تكون أنّيّة شيء من

al-Ṭabīʿa 20 after إثنان add. أعني بالحركة الطبيعيّة الشيء المتحرّك بذاته لا بحيلة ولا بإرادة
K. al-Ḥaraka | ثم قال إنّ أحد الأوّلين المحرّكين : وإن أحدهما K. al-Ḥaraka | بطبيعيّ : بذاتيّ
K. al-Ḥaraka

1 لحركة : حركة K. al-Ḥaraka 2 وقال : فقال K. al-Ḥaraka | دائماً : دائمة with additional text after it K. al-Ḥaraka | التي om. K. al-Ḥaraka 3 after إنّما add. يمكن أن K. al-Ḥaraka | ثم بيّن : وجعل قياسه في إثبات ذلك ... ثم وصف بعد ذلك K. al-Ḥaraka 4 وقال : فقال K. al-Ḥaraka 6 وقال : فقال K. al-Ḥaraka 6–7 لمدّة[2] ... لها om. K. al-Ḥaraka 9 جوهراً om. K. al-Ḥaraka | وأعلى منها : واعلاها K. al-Ḥaraka | after فوقها add. كلّها K. al-Ḥaraka 10 البتّة om. K. al-Ḥaraka 12 دون زمان : في زمان ويبطل K. al-Ḥaraka | ولأنّه : وأنّه K. al-Ḥaraka 13 أصلاً om. K. al-Ḥaraka | وقوّته منبثّة : وقواه مبثوثة K. al-Ḥaraka 14 بتلك القوّة : تلك الحركة K. al-Ḥaraka 15 تلك[1] om. K. al-Ḥaraka 16 هذه : بهذه K. al-Ḥaraka

1 after يتحرّك blank space of two letters M 2 الأشياء : الاشيآ M 3–4 ثم ... الطبيعيّة add. in marg. with صح notation M 5 لا المستوية : ولا المستوية after a blank space of three letters and in faint ink which may hint at an attempted erasure M 6 لا[2] : ولا M 7 تحرّك : تتحرك M 14 الدائمة : دائماً M

of them is not essential,[11] that is to say there is no principle of motion in it. This is what moves perpetually without being moved.

(11) He says that motion enters all things perpetually and that the motion which is effortless can only be in the moving bodies as circular motion. Then he distinguishes the linearly [moving] and the rotating bodies and says that it is not possible that there is any magnitude or anything among the natural linearly [moving] bodies or the rotating bodies which is infinite with regard to magnitude.

(12) He says that the body of finite magnitude cannot have infinite power for the duration of an infinite movement and that the finite power is not able to move infinitely. Thus after he has set forth these premises, he deduces noble results from them.

(13) He says[12] that there is necessarily a substance which is unlike these substances and which is above them and more elevated than they are. It has neither magnitude nor any extension, is by no means subject to any kind of change and its power has neither limit nor end. Therefore it performs its actions without time, that is it does not at times act and 8a
others not. Rather it is perpetually active, because it is active by its essence, not by motion, and because in it there is absolutely nothing in potentiality, but the things in it are in actuality. Its power is perpetually spread in the whole world. Due to the perpetual motion, i.e. the motion of heaven and the whole world, the reception of that noble power is possible, so that its motion persists and remains. However, every one of those things receives of that power according to its receptivity. Not a single thing would be able to have this disposition in it by chance and without an efficient cause for the whole universe. It is also unthinkable that the being of

[11] The Marsh MS reads ‘essential’ (*dhātī*) whereas the Greek, Isḥāq’s Arabic translation and the *K. al-Ḥaraka* read ‘natural’ (*ṭabīʿī*).

[12] The first part of this passage repeats, with minor variations, passage (9), which is explicitly ascribed to Aristotle, *Metaphysics*. The parallel in the *K. al-Ḥaraka* is closer to the present version.

الأشياء بالاتّفاق ومن تلقاء نفسه ، لكن إنّما علّة أنّيّة كلّ شيء في الجوهر الشريف العالي من جميع هذه الجواهر المستحيلة ذي القوّة التي لا غاية لها ولا نهاية . وهي علّة أنّيّة الأشياء وحركتها وثباتها ، والدليل على أنّ ذلك الجوهر
8b الكريم علّة أنّيّة الأشياء أنّه لم يكن ليمكن الشيء الذي لم يحتج في كونه إلى العلّة الإلهيّة أن يقبل منها شرحاً ولا تدبيراً ولا ترتيباً .

(14) قال ابرَقلسْ في كتابه المعروف بستكس الصغرى إنّ الخيرات إنّما صارت إلى جميع الموجودات من العلّة الأولى . فإذا اعتقدنا في هذه العلّة أنّها معدن الخيرات ومعطيها وأنّها الخير بالحقيقة ووصفناها بذلك فقد أصبنا الحقّ بحسب ما يمكن في أنفسنا تصوّره وذكرناها بما قدرنا عليه من العبارة . فأمّا بحسب ما يستحقّه فإنّ الأولى بنا أن نقول إنّ هذا الوصف في غاية النقصان عن استحقاقه . وذلك مثل ما أورده أفلاطن فإنّه قال إنّ المبدأ الأوّل وهو الإله الأوّل أعلى وأشرف وأشدّ تعرّفاً من المعقولات ومن كلّ جوهر وإنّه حقّ قائم بذاته فوق جميع الأشياء الموجودة ، وشبّه مثاله الذي لا يدركه الأوهام ولا
9a يحيط به الأوصاف بأن تكون الشمس في الأشياء المحسوسة كذلك البارئ في الأشياء المعقولة .

(15) وقال أولى الأشياء التي نظنّ أنّها لائقة بأوصافه هو أن يقال إنّه لا يدرك ولا يعلم ، لا الحواس تدركه ولا العقول تحيط به ، ليس في الأسماء ما يليق به ولا في الأوصاف ما ينبئ عن حقيقته ، متفرّد بذاته متعال بجوهره عن خلائقه حقّ ثابت فوق كلّ علم وعالم .

وقولنا فيه إنّه الخير والجود وبصيرتنا ذلك من أسمائه فهو أنّه لما كان أشرف

17 فيه om. *K. al-Ḥaraka* | وبلا ... للكلّ : ولا علة للكلّ *K. al-Ḥaraka*
1 في : من *K. al-Ḥaraka* 2 من : على *K. al-Ḥaraka* | هذه om. *K. al-Ḥaraka* | ذي القوّة : وحركتها وثباتها وأنّه *K. al-Ḥaraka* 3 وهي : فهي *K. al-Ḥaraka* 4 أنّه : المتحرّكة والقوة *K. al-Ḥaraka* | لم يحتج : لا يحتاج *K. al-Ḥaraka*

1 تلقاء : تلقآ M 2 ذي : ذو M 5 after ترتيباً blank space of half a line M 6 بستكس : نستكس *tāʾ* undotted in M : بسطوخوسيس Bad 7 اعتقدنا : أعقدنا M 8 ومعطيها : M ومعطيتها Bad 10 يستحقّه : M تستحقّه Bad 13 يدركه : يُد ك لُهَ M : تدرك Bad 14 يحيط : M تحيط Bad | بأن تكون : illegible M: بأنه كما تكو⟨ن⟩ Bad 15 after المعقولة blank space of three letters M 16 نظنّ : M يُظَنَّ Bad 19 after وعالم blank space of three letters M 20 وبصيرتنا : وبصيرتا M : وتصييرُنا Bad : or read وبصرنا ?

anything would be by chance and spontaneous. Rather the cause of everything's being is only in the noble substance which is elevated above all these changeable substances and which possesses a power without limit or end. This [power] is the cause of things' being, their movements and their stability. The proof that that excellent substance causes things' being
is that it could not possibly happen that something the generation of 8b
which has not required the divine cause could receive order, direction or arrangement from [this divine cause].

(14) Proclus says in his book known as the *Minor Elements*[13] that goods only come to all existents through the First Cause. Thus when we believe this cause to be the source and giver of goods and to be the true good and describe It like that, we have already reached the Truth to the extent our souls can conceive of and have spoken of It by expressing what we are capable of. However, compared to what He is entitled to, we would do better to say that this description falls extremely short of His entitlement. That accords with Plato's exposition that the First Principle which is the first God is more elevated, nobler and more knowledgeable than the intelligibles and every substance, and that He is true, self-sufficient and above all existing things. He likened His mode [of being], which concepts
do not grasp and descriptions do not encompass, to the sun['s position] in 9a
relation to the sensibles, and likewise is the Creator in relation to the intelligibles.

(15) He says: the most appropriate among the things we find suitable to describe Him is to say that He is neither perceivable nor knowable. Neither do the senses perceive Him nor the intellects comprehend Him. Among all possible names none is suited for Him and among the descriptions none bespeaks His reality. He is unique in His essence and, by His substance, sublime above His creation, true, lasting, above every knowledge and knower.

We say about Him that He is the good and generosity and conceive this as pertaining to His names, because since He is the noblest thing, we call Him

[13] See Badawī's edition in *al-Aflāṭūnīya*, 257.3–258.12 of passages (14–16). The attribution may refer to Proclus Procleius' commentary on Pythagoras' *Dicta aurea*.
The Arabic has *s-t(?)-k-s* which may be a transcription of either στοιχείωσις or στοιχεῖον (reading *s-t-k-ī-n* instead of *s-t-k-s*). The emendation *usṭūkhūsīs*, already suggested by Badawī (*Aflāṭūnīya*, p.257, n.1) and Endreß (*Proclus Arabus*, p.23), follows Ibn al-Nadīm and Ibn al-Qifṭī.

الأشياء سمّيناه بأشرف ما وجدناه عندنا من الأسماء لأنّ الذي هو لائق بجوهره يقصر عنه كلّ وصف .

(16) وقال إنّ لله عزّ وجلّ قبل كلّ شيء وسبب الخيرات في كلّ شيء ، فإنّ الخيرات التي عندنا هي منه لا غير كالضوء الذي هو في الشمس ويضيء منه كلّ شيء . فلذلك يجب قبل كلّ شيء أن يبجّل الذي هو سبب سائر الخيرات وسبب كلّ شيء فاضل إذ كانت الخيرات إنّما حصولها لنا منه ، 9b ويجب أن نعلم أنّ تبجيل لله يكون بشيئين أحدهما أن يكون اعتقادنا فيه الاعتقادات الصحيحة النفيسة المصيبة والآخر أن تكون عباداتنا له موافقةً لِما أمرت به الشريعة وأن نواصل ذلك مواصلةً دائمةً ونأخذ أنفسنا به .

والآراء التي يجب اعتقادها في لله ثلاثة الأوّل أنّه جواد معطي الخيرات ، ولا يظنّ أحد أنّه سبب السيئات إذ الإله لا يكون سبباً للشرور ، والثاني أنّه غير متغيّر إذ كان كلّ متغيّر إلى الأمر الأدون إنّما يتغيّر لضعفه والإله ليس بضعيف بل هو في الغاية من القوّة ، والثالث أنّه عالم بالأشياء على حقائقها علماً لا خطأ فيه .

(17) وقال ثامسطيوس إنّ لله تعالى هو علّة نظام العالم وترتيبه وحقّ أوّل في الغاية القصوى وهو المبدأ الأوّل وإنّه يعلم ذاته وجميع الأشياء التي هو لها مبدأ 10a معاً وإنّه مالك لذاته ومالك لجميع الأشياء التي قوامها به ، والقول أيضاً بأنّه

3–5 وقال[1] ... شيء : أوّل الوصايا الذهبيّة الإخبار بأنّ المعظّم أوّلاً على فرائض السنّة من غير المائتين هو الله فالله تعالى هو سبب جميع الموجودات وأصل كلّ الخيرات الموجودة في الكلّ وتعظيم *Istithmār* 5–6 فلذلك ... منه : ومن هذه صفته فواجب تعظيمه *Istithmār* 7–9 ويجب[2] ... به : بفعل الخير قدّامه الإله يكون بالاعتقاد الصحيح فيه ‹و› *Istithmār* 10–11 والآراء ... للشرور : والاعتقاد الصحيح في الله أنّه خير وواهب الخيرات وأنّه ليس بعلّة للشرور *Istithmār* 11–13 والثاني ... القوّة : وأنّه لا يناله التغيّر لأنّه في الغاية والتغيّر انتقال من شيء إلى شيء *Istithmār* 13–14 والثالث ... فيه : وأنّه العالم بالشيء على حقيقته من غير أن يشوب علمه *Istithmār* 15–16 وقال ... القصوى : وذلك أنّ الله ناموس وسبب نظام الأشياء الموجودة كذب *Sh. li-ḥarf al-lām* 16 وهو : هو أن الله ذلك جميع من تبيّن وقد *Sh. li-ḥarf al-lām* وترتيبها وهو ناموس حيّ *Sh. li-ḥarf al-lām* | ومالك : مالك أيضاً فهو *Sh. li-ḥarf al-lām* 17 مالك : إذا كان مالكاً *Sh. li-ḥarf al-lām*

2 after وصف blank space of five letters M 5 يبجّل Bad : ينحل M 8 والآخر : والاخر M 10 والآراء : والارا M | اعتقادها Bad: اعتقادنا M | ثلاثة : ثلثه M 11 الإله : الآله M | سبباً : شيا M 13 after القوّة blank space of three letters M 14 after فيه blank space of five letters M 15 ثامسطيوس : ثامسطُوس M 17 بأنّه : فانه M

by the noblest names we find among us, on the grounds that every description must fail to reach what is suited for His substance.

(16) He says[14] that God, to Whom belong might and majesty, is prior to everything and the cause of goods in everything. For goods which are among us are from no other but Him as [is the case of] the light which is in the sun and from which everything is lit. Therefore, before everything else, it is necessary to venerate Him Who is the cause of all goods and the cause of everything virtuous, since our [disposition to] attain goods is only because of Him. It is also necessary for us to know that God is venerated 9b
through two things. One of them is for us to hold beliefs about Him which are correct, excellent and appropriate. The other is to worship Him according to the instructions of the law, to persevere in this continuously and to admonish ourselves to do it.

The beliefs we must hold about God are three. The first is that He is generous and the giver of goods. Let nobody assume that He is [also] the cause of bad things, since the God is not the cause of evil. The second is that He is unchanging, since everything changing towards a lower condition does so only because of its weakness and the God is not weak, but at the peak of power. The third is that He knows, without fail, things as they really are.

(17) Themistius says[15] that God the Sublime is the cause of the order and arrangement of the world; He is First Truth to the ultimate and utmost degree. He is the First Principle. He knows together His Self and all things for which He is the principle. He is the master of both, His Self and all things the subsistence of which is due to Him. The statement that He sees 10a

[14] Cf. the strikingly similar beginning of Ibn al-Ṭayyib's *Istithmār* of Pythagoras' *Dicta aurea* in his *Proclus' commentary*, ed. Linley, 8.2–10; and for this (as well as for the two previous passages) see Badawī's edition in *al-Aflāṭūnīya*, 258.3–12.

[15] Cf. abridgement of Themistius' *Sharḥ* on Book *Lambda*, ed. Badawī in *Arisṭū*, 18.4–5, 21.10–12.

يرى ويعقل ذاته قول صواب .

(18) وقال أرسطالس في رسالة 〈السياسة〉 العامّيّة للاسكندر كن متمسّكاً بعروة الحقّ مستبصراً بالله الواحد الذي لا أوّل له ولا شبيه ولا ضدّ ولا زوال لملكه القاهر القادر الجواد الذي يصيّر الأمور كلّها إلى إرادته .

(19) ثم قال والشكر لله واجب والمنّ له على البريّة ، إنّه واحد لا أوّل له ولا انتهاء لملكه ، أنشأ الخليقة لا من موجود وأحدثها لا من متقدّمات ، خلق الرؤوس على ما شاء . فالرؤوس أول الخليقة وابتداء ما أنشأ البارئ والطبائع وما كان من اختلاف الطبائع تفرّع من تلك الرؤوس . فكلّ ما ينطق به لا يحيط بعلوّه وعلوّه بلا مكان وجوده بلا زمان وعظمته لا بمقدار ، لا السكون من صفاته ولا الحركات من سماته . لم تبلغه حاسّة ولم ينله وهم ولم يعرف منه الفكر
10b ولم يقصد إليه إلا بالعقل . لا يتجزّأ ولا يتكيّف ولا يلزمه المائية ولا الأينيّة ، سبق الأزليّة وعلا التغيّر ، يجري ملكه دائماً وبه يدوم الدوام وبه يبقى

2 وقال ... للاسكندر : وسار الإسكندر يلتمس بلاد فارس فكتب إليه أرسطوطاليس رسالة سمّاها السياسة العامّيّة ... *R. al-Siyāsa* | كن متمسّكاً : خذ يا دارا إن كنت غير متهيئ للقاء واعلم أنّي غير منتظر بالشخوص إليك واعمل على أنّك تلقى قوماً عدّتهم قليلة لكن قوّتهم بقدر ما وهب الله عزّ وجلّ لهم متمسّكين مستنصرين *R. al-Siyāsa* 3 بعروة الحقّ : بالحقّ *R. al-Siyāsa* | مستبصراً : قادر قاهر جواد بصير ترجع *R. al-Siyāsa* 4 القاهر ... يصيّر : شبه *R. al-Siyāsa* | شبيه : *R. al-Siyāsa* | كلّها om *R. al-Siyāsa* 5 ثم ... لله : الشكر واجب لله *R. al-Siyāsa* | after البريّة add. والطول من عنده إياه أحمد وهو ملجئي وبه أستعين على المهمّ من كبير أمري وصغيره . *R. al-Siyāsa* 6 انتهاء : وشكري له شكر مَن يعرف مِنَنه عليه ولا يحصى نعمه لديه ومَن يقول زوال *R. al-Siyāsa* | موجود : موجودات *R. al-Siyāsa* 7 after الرؤوس add. كيف شاء الأوائل *R. al-Siyāsa* | الخليقة : والرؤوس *R. al-Siyāsa* | فالرؤوس : وبري الطبائع الكلّيّة من تلك الرؤوس الخلقة *R. al-Siyāsa* | after البارئ add. عزّ وجلّ *R. al-Siyāsa* 8 after اختلاف add. خلق *R. al-Siyāsa* | after الرؤوس add. فالرؤوس ثلاثة لا محالة أوّلها وأكرمها الصورة ، والثاني الهيولى والثالث العدم لا بزمان ولا بمكان . والهيولى بالقوّة والصورة بالفعل والعدم هو انقلاب الهيولى إلى ما هو بالفعل .... *R. al-Siyāsa* | after ما Maróth's edition of the *R. al-Siyāsa* has an erroneous negation 9 بلا زمان : لا بزمان *R. al-Siyāsa* | after زمان add. وصفاته لا بإثبات *R. al-Siyāsa* | after بمقدار add. علا فقدر وانفرد فقهر وعلى ما شاء تجرى ، وقوّته لا بتشبيه *R. al-Siyāsa* 10 after سماته add. دبّر الأشياء بلا ضعف وأنشأ بلا تحرّك الأمور *R. al-Siyāsa* | يعرف : يقرب *R. al-Siyāsa* 11 after بالعقل add. ولم يلحقه ذهن *R. al-Siyāsa* | after وهم add. هو القويّ الذي لا يعجز والحكيم الذي لا يجهل والجوّاد الذي لا يبخل . تقصر الألفاظ عن عظمته ويضيق المنطق عن بلوغ كنهه واحد *R. al-Siyāsa* | يلزمه : تلزمه *R. al-Siyāsa* | المائية : الماهية *R. al-Siyāsa* 12 after الأينيّة add. ولا يحصى بكمّ ولا يحاط بأين *R. al-Siyāsa* | after وعلا add.

5 after قال blank space of three letters M 7 الرؤوس : النوس M | فالرؤوس : فاكرؤس M 8 الرؤوس : الروس M

and knows His Self is also correct.

(18) Aristotle says in the *Letter ⟨On Gouverning⟩ in General* to Alexander:[16] hold on to the ties of truth and seek to perceive God intellectually, the One, Who has no first [principle], no like, no opposite and Whose sovereignty does not come to an end, He is the subduer, the omnipotent and the generous who causes all matters to pursue His will.

(19) Then he says:[17] thanks be to God, as He is benevolent towards creation. He is one, has no first [principle] and His sovereignty has no end. He has caused the origin of created beings *ex nihilo*, brought them forth from what had not existed before and created the principles[18] according to whatever He wills.[19] The principles are the first created beings and the beginning of what the creator has originated. The natures and what stems from the differences between the natures branch out from those principles. Whatever is uttered does not comprise His exaltedness, as His exaltedness is without place, His generosity without time and His majesty without measure. Rest does not belong to His attributes nor motion to His characteristics. Sense perception does not reach Him, nor does imagination attain to Him, nor does reflection know of[20] Him. One only strives for Him through intellect. He is neither divisible nor qualifiable and neither 'what- 10b
ness' nor 'whereness' are attached to Him. He precedes eternity and is elevated above alteration. His sovereignty goes on perpetually, through

[16] Cf. source in *R. al-Siyāsa al-ʿāmmīya*, ed. Maróth in *Correspondence*, (25.1–3), 48.12–16. The shortening reading of the *PR*, *al-risāla al-ʿāmmīya*, is also found in one of the three MSS Grignaschi uses for his edition, namely Paris 2422 (see his 'Siyāsatu-l-ʿāmiyya', p.97, n.1).

[17] Cf. source in *R. al-Siyāsa*, ed. Maróth in *Correspondence*, 49.19–50.7, 50.15–51.2; see parallel in Mubashshir, 185.2–7, whose reading corresponds almost exactly with the one in the *R. al-Siyāsa*.

[18] The text, which the *R. al-Siyāsa* and Mubashshir have in addition to the *PR*, may have dropped out *per homoeoteleuton*. However, the reading of the *PR* may also result from an intentional shortening by the compiler as the text is fully understandable as it stands.

[19] End of the parallel text in Mubashshir.

[20] The reading of the *R. al-Siyāsa*, which is *wa-lam yaqrub minhu l-fikr* (and thus the *lectio difficilior*), saves one from reading *yaʿrif min* as in the Marsh MS.

البقاء .

(20) والفرقة الثانية أصحاب الرواق وهم الرواقيون . فإنّهم وافقوهم في التوحيد وخالفوهم في أنّه ليس بجسم ، وقالوا إنّه جسم إلا أنّه ليس بشيء من الأجسام الباقية وإنّه من اللطافة والرقّة في أقصى الغاية ، فهو ينفذ في كلّ جسم وكلّ جزء من جسم بلا دافع ولا مانع وهو في كلّ مكان على هذه الجهة .

(21) قال برمانيدس إنّ العلّة القصوى واحدة بسيطة ، هي الأشياء كلّها بأنّها علّة لها وليست بشيء من الأشياء لأنّها أعلى من معلولاتها وأرفع .

(22) قال إنّ الأوهام لا تقع على كنه معرفتها ولا العقول على صفتها لأنّ العقول معلولة منها والمعلولة لا تقدر على أن تحيط بعلّتها علماً . فأمّا الأوهام 11a فإنّها معلولة من معلول فمن أجل ذلك قصرت وبعدت عن أن يتوهّم العلّة القصوى .

(23) قال فيثاغورس إنّ في هذا العالم عقلاً نقياً صافياً مبايناً لكلّ شيء وإنّه هو الفاعل لكلّ شيء . ثم قال إنّ الأوّل مبسوط مباين للأشياء وقال إنّ الأوّل واحد محض لا مثنوية فيه البتّة .

(24) وقال ارقليطوس إنّ الأوّل واحد محض روحانيّ دائم لم يزل ، لا بدء له ولا نهاية . وقال إنّ الأوّل ليس ببدن لأنّ لا بدّ أن يسيل سيلاناً دائماً ولا يثبت على حال واحدة ولا ينبغي للأوّل أن يكون على هذه الصفة بل ينبغي أن يكون واحداً مبسوطاً روحانيّاً ثابتاً قائماً على حال واحدة .

(25) وقال افلاطن في خطبة له : لله المجد والعظمة والكبرياء والعزّة والجبروت والقدرة وهو الواحد الفرد الذي تفرّد بأزليّته وتوحّد بوحدانيّته ، مبدع المبدعات 11b بالخير الذي بثّ في الكلّ وسرى في العوالم ، وأحيا به ما خلق وذرأ وأنشأ وبرأ

R. al-Siyāsa البقاء باقٍ : الدوام | *R. al-Siyāsa* فجرى : يجري | *R. al-Siyāsa* بالتعزّز عن

1 after البقاء blank space of half a line M 2 الرواق : رَبيَس undotted M 3 بشيء : لَشيءٍ M 5 جزء : جزٍ M 6 after الجهة blank space of three letters M 8 بشيء : لشي M | after وأرفع blank space of six letters M 12 after القصوى blank space of five letters M 15 after البتّة blank space of six letters M 16 ارقليطوس : ارفلنطوس M | بدء : بدو M 19 after واحدة blank space of three letters M 20 after له blank space of three letters M 22 وأحيا به : واحيايه M

Him perpetuity persists and through Him permanence remains.

(20) The second sect[21] are the people of the porch,[22] who are the Stoics. They agree with them on the *tawḥīd*, but disagree with them on His non-corporeality. They say that He is a body, yet He is not [like] any of the rest of the bodies and that He is of utmost fineness and delicacy. So He permeates every body and every part of the body without resistance or hindrance and in this way He is everywhere.

(21) Parmenides says that the Ultimate Cause is one and simple. It is all things by being their cause and no thing, because it is more elevated and more exalted than its effects.

(22) He says that neither the imaginative [faculties or their acts] meet with the core knowledge of [the Ultimate Cause] nor the intellects with Its description, for the intellects are caused by It and effects are not capable of comprising their cause knowingly. As to the imaginative [faculties or their acts], they are caused by the effect and therefore fail to reach and are 11a
far away from imagining the Ultimate Cause.

(23) Pythagoras says that in this world, there is intellect, pure, unmixed and separate from everything and that it is the agent for everything. Then he says that the First is simple and separate from all things. He says that the First is one, pure and without any duality in It at all.

(24) Heraclitus says that the First is one, pure, spiritual, eternal and never vanishing. He has neither beginning nor end. He says as well that the First is not a body, for [a body] inevitably flows in a perpetual flow and is never stable in a single condition. The First does not meet this description, rather He must be one, simple, spiritual, stable and remaining in a single condition.

(25) Plato says in one of his discourses: God has glory, majesty, magnificence, might, power and omnipotence. He is the unique One, who is unique in His eternity and one in His unity. He has created the created beings by the good which is spread in the universe and pervades the 11b
worlds. By it, He has enlivened what He has created, made, originated and brought into existence and He has put it into motion by the motion of the

[21] The passage may be taken from some doxographical account on God's Oneness (*tawḥīd*) and corporeality. The reference to the first group and thus their identity is not preserved.

[22] The reading of the Marsh MS is defective here. Goldziher ('Säulenmänner', p.503) lists among different Arabic terms for the Stoics the following ones beginning with *aṣḥāb*: *aṣḥāb al-uṣṭuwān* and *al-usṭuwān*, *a. al-riwāq* and *a. al-miẓalla*. In al-Anṣārī's *Ādāb al-falāsifa*, p.40, we find the latter two and *aṣḥāb al-arwaqa*.

وجعله متحرّكاً بحركة الخير والنور الساطع الفائض على الكلّ . فبتلك الحركة تحرّكت المتحرّكات وبذلك الخير اشتاق الكلّ إلى التأحّد بالخير الأزليّ .

فقويت نفوسهم بإشراق النور على ذواتهم فصارت الذوات عقولاً فعّالةً ونفوساً متحرّكةً علّامةً . ثم إنّ النفس لما قويت بالعقل وتهوّيت بالخير وتحرّكت بالنور المشرق عليها صارت فعّالة خيّرة عاقلة متحرّكة حيّة ، وانطبعت فيها الطبيعة الأحديّة العقليّة وصارت تلو العقل في الهيئة والصورة والشكل والفعل والخير ، واحتوت على السماء العليّ والعوالم السفليّة وتحرّكت بذوات حركاتها الأجرام .

واعطي الكلّ النظام وقبل الكلّ من الخير الفعل الذاتيّ وقبلت الطبيعة الصور
وصارت هيولى لها ذات طقس ونظم وترتيب إلاهيّ أحديّ سرمديّ . 12a

ثم إنّ النفس بخيرها البسيط وفرط شوقها الإلاهيّ ترقّت إلى عالم الخير فكساها الخير الحلية الخيريّة وملّكها ما دونها من الذوات ، وتأحّدت بالعقل الفعّال فصارت هي والعقل في قرن واحد لا يختلفان ولا يتباينان . فالعقل مطلع على النفس ناظر إلى الربّ والنفس متأحّدة بالخير ومتشكّلة بالعقل الفعّال ، والكلّ شيء واحد في العالم الأزليّ والحيّز السرمديّ . ثم إنّ هذه النفس اطّلعت إلى هذا العالم الذي هو عالم الطبيعة فرأتها مظلمة غير ذات نور ولا حياة ولا خير فنزلت إليها رحمة من خالقها وعطفاً عليها .

فأحيت العالم وتحرّكت بحركتها الطبيعة وصارت ذات صور وأشكال وهيئات وأفعال متشابهة بعضها ببعض ، ولبثت النفس في عالم الطبيعة وتحرّكت

5 حيّة : حيئة M 6 تلو : M or read نتلو ? | الهيئة : الهية M | والخير : M or read الحيّز ?
7 السفليّة : السفليّ M 8 الكلّ[1] : للكل M 10 الإلاهيّ : اللاهي M 12 يتباينان : ينبايتان M
14 والحيّز : الخير M 15 فرأتها : فرأهَا M 17 وهيئات : وهياتٍ M

good and the light radiating and emanating into the universe. Due to that motion, the moving beings have been moved and due to that good the universe has come to desire to become one with the eternal good.

Then their souls became strong by the radiation of the light towards their essences, so the essences became active intellects and souls capable of movement and knowledge. When the soul became strong through the intellect, was given the essence of goodness[23] and moved by the light radiating towards it, it then became active, good, intellective, moving and alive. The singular, intellectual nature was imprinted on it and it came to follow the intellect in pattern, form, shape, action and goodness. It encompassed the elevated heaven and the lower worlds and the bodies were moved by the essences of its motions.

Organisation was given to the universe and it received [its] essential activity from the good. Nature received the forms and became matter for them possessing a divine, singular and endless order, organisation and 12a
arrangement.[24]

Thereafter, through its simple good and excessive desire for the divine, the soul ascended to the world of good, whereupon the good clothed it in the adornment of goodness and made it have dominion over the essences below it. The soul became one with the active intellect. Thus it and the intellect occur in one union, they neither differ nor vary. As it gazed at the Lord, the intellect informed the soul and the soul became one with the good and assumed the shape of the active intellect. The universe was one single thing in the eternal world and the endless realm.[25] Then this soul looked from above at this world, which is the world of nature, and considered [nature] to be dark and without light, life and good. So [the soul] descended towards [nature and the world] by way of mercy [caused] by its creator and sympathy with [nature].

Therefore it enlivened the world and by its motion nature was moved and started to have forms, shapes, patterns and activities corresponding to each other. The soul abided in the world of nature and the moving beings

[23] The verb may either be understood as belonging to the root *h-w* of *huwa, huwīya* (used, e.g., by al-Kindī, see Afnan, *Terminology*, p. 123) or the root *h-w-ā*, which would have to be translated as 'became infatuated by, desiring the good' or even 'became blown around'.

[24] The sentence is ambiguous. One may also read: 'A matter occurred, the essence of which has divine order ...' or 'Nature became a matter the essence of which has divine order'.

[25] The Arabic text has 'good' (*khayr*), but as 'realm' (*ḥayyiz*) is parallel to 'world' (*ʿālam*) and *khayr* and *ḥayyiz* are often mixed up in manuscripts, I opt for the emendation.

12b المتحرّكات بقواها وذواتها وقسمت على الكلّ حياتها وخيراتها واختارت لقوّتها وذاتها صورةً عقليّةً وجسّدت تلك الصورة وجسّمتها ، وظهرت فيها لتبدى من ذاتها وحركتها الفعل الخيريّ والصنع الإلاهيّ والكلمة العقليّة وتتكلّم بالنطق الروحانيّ ، واتخذت لإظهار حركاتها وخيراتها هذه الأجساد آلةً لصور كلماتها وأدّلة لصنائعها وجعلت صور الأجساد صورةً عقليّةً بالذات والشكل والحلية والزينة والحسن والجمال .

ولو لا أنّ هذه النفس الفعّالة أبدعت من الهيولى هذه الأجساد على أمثلة الصور العقليّة لما أمكنها أن تسكن فيها ⟨...⟩ وما اتّخذتها آلةً عقليّةً وهيئةً نفسانيّةً . وتباينت هذه الصور الإنسانيّة الصور البهيميّة التي لا نطق لها ولا كلمة عقليّة إلاهيّة . ولما صارت صورة الإنسان على هذا الشكل والهيئة

13a والصورة قبلت النفس المثل الإلهيّ والنطق الروحانيّ وصارت في قوّة الذات علّامة عاقلة فعّالة درّاكة . وحباها العقل الهويّة النفسيّة التي لا تدثر ولا تندرس ولا تعفى ولا تنطمس وتبقى بهويّتها النفسانيّة وصورتها العقليّة ، وصارت في الشبه والصورة أمثلة لتلك الروحانيّات الأزليّة الأحديّة وكملت بها العالم الطبيعيّ وتمّ الكلّ ، وأزال الخير عنها النقص والجهل وأعطاها البقاء والدوام تشبّهاً بالجواهر الأزليّة الخالدة غير المائتة ولا البائدة .

فصار الإنسان بهذه النفس مختاراً فاضلاً ناطقاً عاقلاً وانتهى إليه الفعل العقليّ

4 واتّخذت : واتحدت M | لإظهار : لاطهار M 4–5 آلةً ... لصنائعها : M or read آية لصور كلماتها ... لصانعها ?: or read وأداة لصنائعها ... ? 8 أمكنها : امكنت M | اتّخذتها ... فيها ? فيها ولا أن تتخذها : or read M | اتّخذتها : اتحدها M | وهيئةً : وهَيةً M 11 المثل : للمثل M | الإلهيّ : اللاهي M 12 وحباها : وحياها M 16 غير المائتة : الغير الماتية M 17 بهذه : بهذا M

moved by their powers and essences.[26] [The soul] distributed its life and its goodness over the universe. 12b
It chose an intellectual form for its power and essence, and corporealised and embodied that form. [The soul] manifested itself in [that form] so that by its essence and movement, the good activity, the divine making and the intellectual word (*logos*) are revealed and expressed by spiritual speech. For the manifestation of its motions and goods, [the soul] made use of these bodies as instrument for the forms of its words (*logoi*) and as proofs for its skilfully made products,[27] and made the forms of the bodies intellectual form[s] in essence, shape, adornment, grace, goodness and beauty.

If this active soul had not created these bodies out of matter as images of the intellectual forms, then it would neither have been possible that it resided in them ⟨...⟩[28] nor would it have made use of them as intellectual instrument[s] and psychic pattern[s]. These human forms differed from the beastly forms which have neither reason nor an intellectual, divine word (*logos*). When the form of man had attained to this shape, pattern and form, the soul received resemblance to the Divine and spiritual reason. 13a
It became fully knowledgeable, intellective, active and strongly perceptive in [its] essential power. The intellect gave it the psychic essence which neither becomes effaced or erased, nor perishes or gets destroyed, and through its psychic essence and intellectual form [the soul] remains. In likeness and form, it resembled those eternal, singular spiritual beings and through them it completed the natural world,[29] and the universe reached perfection. The good removed deficiency and ignorance from [the soul] and gave it permanence and perpetuity in imitation of the eternal, perennial, immortal and imperishable substances.

So by the means of this soul, man became free to choose, virtuous, rational, intellective, and the intellectual act and the divine good finally

[26] An alternative reading could be 'and the moving beings moved by [the soul's] powers and its essence' – changing *dhawātihā* to *dhātihā* simply by deleting the *waw.*

[27] The parallel expressions 'as instrument for the forms of its words and as proofs for its skilfully made products' (*ālatan li-ṣuwar kalimātihā wa-adillatan li-ṣināʾiʿihā*) sound somehow mismatched. Maybe the original reading was 'as instrument ... and as tool (*adātan*)...' or 'as sign (*āyatan*) ... and as proofs ...', in which case it may have been 'as proofs for its Creator (*li-ṣāniʿihā*)'.

[28] Here some problem with the Arabic as it stands (negation continues with *wa-mā* instead of the expected *wa-lā an*) seems to indicate that the compiler has omitted some material of his source text.

[29] It is possible that the subject was originally *al-jawāhir* – parallel to the 'eternal substances' in what follows – but has dropped out. In that case the sentence would have read: 'and through them [or it, i.e. the soul] <the substances> of the natural world were completed'.

والخير الألاهيّ ، وجعل لله عزّ وجلّ الإنسان في عالم الطبيعة صنو الأرواح العقليّة الملكيّة في عالم الخير ونظير الروحانيّات وأنزل على نفس الإنسان السكينة الروحانيّة والقوّة الإلاهيّة والروح الخيريّة فالإنسان أمثلة الملائكة الساكنين في العليّين وأشباه السكينات الخيريّات الساريّات في العالم. 13b

فما أعظم وأكبر ما أنعم البارئ سبحانه على هذا الإنسان من إعطائه اياه الفهم والفكر والنطق والعلم وتقويته له بهذا العقل المنوّر ، وما أشرف وأكرم ما أسداه وخوّله من التصرّف في أمور العالم ، وميّزه من البهائم وسخّر له ما أنشأه من الحيوانات وأنبته من النبات وملّكه ما أخرجه من معادن الأرض ، ودعاه إلى التأحّد بجوهره والتبرّؤ من عنصره والتعدي من هيولى جسده وتنقية ذاته من وضر طبيعته.

يأَيُّها الإنسان إنّك صوّرت وخلقت وجعل في هيكلك هذه النفس الناطقة لأمر جسم . قد غاب عنك أغراض الخير فيك وحضرك ما قرب من هذه الطبيعيات الداثرات .

فيا لها سعادة إن أدرك الإنسان ما خُلق له ودعي إليه وسعى إلى الفعل الذي يراد منه وأتعب بدنه وحرّك نفسه في طلب ذلك الخير المعدّ له المدعو إليه 14a واشتاق إلى نيل تداركه والوقوف على مناهجه والثبات على ما ندب له وشاق نفسه إلى موقف النجاة والفلاح ومقام الفوز والنجاح .

ويا لها شقوة إن مكث في حركات الطبيعة ولبث في ديجور الحيوانيّة ، وأشقى الأشقياء من ينوّم نفسه الناطقة الفاضلة عن الأمر الذي يكمل به دلّها ويتمّم

3 الإلاهيّة : اللاهية M 9 والتبرّؤ : والتبرى M | وتنقية : وتيقه M 11 الإنسان : الناس M وخلقت : وخلفت M | وجعل : وفجعل M 12 لأمر : M or read لا من ? 15 المعدّ له : المعدّلة M 19 الأشقياء : الاشقيآ M

got to him. God, to Whom belong might and majesty, made man in the world of nature the twin brother of the intellectual, angelic spirits in the world of good and the counterpart of the spiritual beings. He bestowed upon the soul of man spiritual tranquillity, divine power and good spirit, so human beings are images of the angels abiding in the loftiness above 13b loftiness and the likes of the good, tranquil [beings] (*daemons*) pervading the world.[30]

How great and magnificent are the gifts that the Creator – praise be to Him – bestowed upon this human being by His giving him understanding, thinking, reason, knowledge and the power He imparted to him through this enlightening intellect. How noble and excellent are the favours He conferred on and granted to him for his free conduct in the affairs of the world. He set him apart from the beasts, made the animals He created and the plants He caused to grow subservient to him, gave him dominion over the minerals in the ground He produced,[31] urged him to be one with his substance, rid himself of his material element, transcend the matter of his body and cleanse his essence of the filth of his nature.

O Man, you were formed and created and this rational soul was put in your bodily frame for [the] body's affair[s].[32] Yet the aims of the good in you have already been hidden from you and the perishable natural beings surrounding you have become present to you.

O what happiness, if man grasps what he was created for and to what he was urged, strives for the action which is wanted from him, tires his body, sets himself in motion in pursuit of that good which is intended for him 14a and to which he is urged, desires to attain his rectification, to learn about his [correct] manners and to adhere to what he is called for and renders his soul desirous for the place of salvation and prosperity and the position of triumph and success.

O what misery, if [man] remains in the agitations of nature and abides in the darkness of animality. The most miserable of the miserable is he who makes his rational and virtuous soul neglect the matter by which its proper conduct is completed and its refinement perfected and which

[30] This reading is possible as *insān* can be used as singular and plural. If, however, *insān* is to be read as singular here, the most likely emendation would be: 'man is most similar (*amthal* instead of *amthila*) to the angels abiding in the loftiness above loftiness and most like (*ashbah* instead of *ashbāh*) the good tranquil [beings] pervading this world'.

[31] It is also possible to read: '[The Creator] ... made the animals [man] reared and the plants he cultivated subservient to him, gave him dominion over the minerals in the ground he extracted ....'.

[32] The readings 'for [the] body's affair' (*li-amr jismin*) and '[the rational soul being] not from [the] body' (*lā min jismin*) are both difficult. Some text may have dropped out.

〈به〉 فضلها ويذهب جهلها ويضيء عقلها ، ويختار الضارّ الخسير على النافع الربيح ويتبع شهواته الدنية ولذّاته البهيميّة ويتشبّه بالحيوان الغاوي ويتشكّل بشكل البهائم العادية الضاريّة .

فيأيّها الإنسان إنّك مسؤول عمّا تؤثّر في ذات نفسك ومعاقب على ما تفسد به جوهرك عقوبة شديدة . فإنّك إذا أفسدت ذاتك صرت مفسداً شريراً
14b خبيثاً ، فابدأ بإصلاح نفسك وذاتك ، فإن قدرت على ذلك هان عليك إصلاح غيرك وإن عجزت عنه كنت بالحرى أن تعجز عن تقويم أمورك .

يأيّها الإنسان قس نفسك بما تعشقه من شأنك وتشتاق إليه من طبيعة نفسك . فبالحقّ أقول إنّ هذه النفس إذا تذوّتت بذات طبيعيّ أو إلاهيّ بعثت بعد مماتها على تلك الذات التي كانت تختارها في حياتها ، ألا أنّما طبيعة النفس دالّة على الشيء الذي تبعث عليه يوم البعث والنشر وتحيى به في عالم الحياة ومجمع الخير والشرّ .

إنّ الإنسان لا يكون أنساناً أذا خالف الشريعة وفارق السنّة وأفسد العقيدة وباوش الصلحاء من أهل المدينة . فقوّموا بشراً يعلم ، فإنّها الأمر الثابت المحفوظ بالسكينات والروحانيّات .

ومن ألحد في شريعته وحاد عن سنّته فليسمع قول اميروس حيث يقول العادل
15a عن شريعته تعاديه سكينته فيورّده بمورد الهلكى . استعينوا في كلّ ما يعزّ عليكم من الأمور بالخير الأوّل وابتهلوا إليه وقدموا نذوركم وصلواتكم تنالوا الفلح من شأنكم .

(26) وقال ارسطاطالس في خطبة له : سبحان الله الأزليّ القديم الذي كلّ الأزل مبدع من أزليّته علّة العلل والمعلولات ومذوّت الذوات العقليّات وفاعل الأفعال الخيريّات الذي توحّد بربوبيّته واستكان الكلّ بوحدانيّته . فالكلّ مفتقر

1 ويضيء : ويُضي M   2 البهيميّة : البرهيمية M   10 كانت : كان M   12 after والشرّ blank space of three letters M   13 العقيدة : العقدة M   14 الصلحاء : الصُّلحَا M   16 after يقول blank space of three letters M   17 بمورد : مورد M | after الهلكى blank space of three letters M   18 عليكم : لكم M   19 after شأنكم blank space of half a line M   20 الأزليّ : الازل M   21 الأزل M : or read الأزليّ ?   22 بوحدانيّته : لواحدانيته M

removes its ignorance and illuminates its intellect. [The most miserable is further he] who freely chooses the unprofitable and harmful over the profitable and beneficial, follows his base passions and his beastly pleasures, imitates the stray animal and takes the shape of aggressive, harmful beasts.

O Man, you are thus accountable for acting upon the essence of your soul and severely punishable for what you do to corrupt your substance. Whenever you corrupt yourself, you become corruptive, evil and wicked. So begin improving your soul and yourself. If you are capable of that, the 14b
improvement of others becomes easy for you and if you are incapable of that, you are all the more incapable of rectifying your own affairs.

O Man, judge yourself[33] by the things you love due to your character and you strive for due to the nature of your soul.[34] So I truly say that if this soul assumes either a natural or a divine essence,[35] after its death it will be resurrected in accord with the essence it has chosen in life. Indeed, the nature of the soul is indicative of how it will be resurrected on the Day of Resurrection, and how it lives in the world of life and in the meeting place of good and bad [i.e. in this life].

Man is not man if he conflicts with the [Islamic] law, parts with [prophetic] tradition, undermines his creed and attacks the righteous among his fellow citizens. So raise men who know! For these [i.e. the law, tradition and creed] are the matters which are established and preserved by the tranquil (*daemons*) and the spiritual [beings].

He who deviates from his law and swerves from his tradition, should listen to Homer when he says: his tranquillity (*daemon*) turns against him who turns away from his law, thus bringing him to the road of the damned. 15a
Seek help from the First Good in everything which distresses you, beseech Him and offer your pledges and your prayers, so you may be successful in your affairs.

(26) Aristotle says in one of his discourses: Praise be to God, Who is eternal without beginning or end, from Whose eternity the whole eternity is created, the Cause of the causes and effects, the Giver of essence to the intellectual essences, the Agent of the good actions with Whose divinity the universe has become one and to Whose unity it has submitted itself.

[33] Reading *qis*; an alternative reading is 'pursue [the wellbeing of] your soul by ...' (*qussa* ...).

[34] This reading disregards the possible *mā-min*-construction, in which case we would have to read: '[the things] you love about your character and [the parts of] your soul's nature you strive for'.

[35] Literally 'an essence of [someone] natural or divine'.

إلى خيره وهو مفيض الخير على العوالم بلا صفة ولا نعت .

هو الذي لا يدركه العقل إلا ما يطّلع إليه بخيره المستفيض فيه ولا يعقله إلا بما يكون في جوهره من أحديّته الأزليّة ، وهو الدائم الوجود الذي لا تتصوّره النفس ولا يقدر النطق أن يصفه إلا بآثار أفعاله . فالعقل يعقل خيره فقط ويتّحد بذلك الخير .

15b والنفس تتشكّل بالعقل بالكلمة الإلهيّة فصارت للنفس في هذه الطلعة هويّة وجوديّة وبهذه الهويّة الوجوديّة فعلت الأفعال النفسانيّة ونطقت بها وعلمت معلولاتها وتذوّتت بذات العلم من جهة جوهرها ، وصارت عاقلة بهويّتها العقليّة وتشبّهت بالعقل الفعّال لأنّها من سنحها وجوهرها ، وكانت متأحّدة بالخير كتأحّد العقل بعلّته لأنّ جوهر العقل والنفس جوهر واحد ويختلف بالنطق والكلمة الإلاهيّة ، وأفعالهما شيء واحد لا تفاوت في طبيعة أفعالهما والاختلاف في جوهريّة عقلهما وسنح ذاتهما .

ولما اطّلعا إلى بارئهما تهوّيت فيهما الهويّة الحقّة المحضة ، فصارت العقلُ متأحّداً بالخير والنفس موافقة له في جوهره . فتحرّكت النفس بحركتها الذاتيّة المجعولة فيها من أوّل إبداعها وقبلت من العالم الربوبيّ الصور المشاكلة لذاتها وتصرّفت
16a في عالمها . وصارت من حركتها ضرورة الهيولى غير الجرميّة ولا ذات أقطار .

ثم لما أبدعت هذه الهيولى تحرّكت النفس بصورها فتجسّمت من حركاتها الهيولى الطبيعيّة وقبلت الصورة . وصارت النفس محرّكةً للطبيعة وكثّرت الحركات والصور ووقّعت الحدود على الأشياء الهيولانيّة ، وكانت الهيولى مظلمة غير ذات صور ولا ذات كلمة . فصوّرت فيها النفس هذه الصور المختلفة

4 بآثار : باثار M   8 معلولاتها : M or read معلوماتها ?   10 بعلّته : لعلته M   12 وسنح : وسَنحَة M   13 فيهما : فيها M   16 غير : الغير M   20 صور : M or read نور ? | فصوّرت : فتصوّرت M

The universe is in need of His good, whereas He pours forth the good over the worlds without distinction or qualification.

It is He Whom intellect does not grasp, except to the extent to which it obtains knowledge of Him by means of His good spread in it, and Whom [the intellect] does not know, except through the extent to which His eternal singularity is also present in its substance. He is the always Existent, Whom the soul does not imagine, nor can reason describe Him except by the impressions of His actions. So the intellect knows only His good and is united with that good.

The soul is shaped like the intellect by the divine word (*logos*).[36] In this aspect, the soul has come to have an existential being and by this existen- 15b
tial being it has performed the psychic actions, brought reason to them, known their effects and taken on the essence of knowledge in its substance. [The soul] has come to know its intellectual being and imitated the active intellect, because [the soul] belongs to [the intellectual being's] origin and substance. It has become one with the good as the intellect has become one with its cause, for the substance of intellect and of soul are one single substance, but differ in reason and the divine word (*logos*). The activities of both of them are one and the same and there is no dissimilarity in the nature of their activities. The difference, however, is in the substantiality of their intellection and in the origin of their essence[s].

After both of them had obtained knowledge of their Creator, the true pure essence arose in both of them. So the intellect became one with the good and the soul came to accord with the intellect regarding [intellect's] substance. The soul moved in its essential motion which had been implanted in it at the very moment it had been created. It received the forms which resemble its essence from the divine world and acted freely in its world. From its motion there necessarily occurred non-corporeal,
dimensionless matter. 16a

Thereupon, after this matter had been created, the soul moved by its forms. Then, by its motions, the natural matter was embodied and received form. The soul became the mover for nature, multiplied movements and forms, and inflicted limits upon material things. Matter was dark and without either forms or word (*logos*). Therefore the soul formed in it these different forms, acted freely in it, took matter as its instrument and gave it power which resembles privation. [Matter]

[36] The inverse reading 'shaped like the divine word (*logos*) by the intellect' is possible as well.

وتصرّفت فيها واتّخذت الهيولى آلة لها وأعطتها القوّة الشبيهة بالعدم ، وتلوّنت
بألوان الصور والحركات المختلفة . ثم لما رأت النفس الهيولى ضعيفة ذليلة
نطقت بمنطقها وعملت أعمالها واختارت لها طينية هي صورة الجسد الإنسانيّ
ليتمّ بوجودها فيها الناقصات من أمور العالم . واعطاها البارئ سبحانه وتعالى
الخير البسيط والعلم الوافر والعقل الكامل . فأدركت بالعقل الهويّات وعلمت
بالعلم الذوات وفعلت بالخير الأفعال .

16b فهذه النفس جعلها لله عزّ وجلّ موطن العلم والعقل . وصارت فعّالة بالنطق
والفكر ومتحرّكة بذاتها المتحرّكة وحياتها النفسانيّة وقوّتها الإلاهيّة . وتشبّهت
بنوع الفعل بالعقل الفعّال لأنّها منحدرة من عالم العقل إلى هذا العالم بنوع
الطبيعة الفعّالة وامتزجت بهذه الهيولى وجاورتها بنوع المجاورة والمشابكة لا بنوع
المخالطة والذوات الطبيعيّة . وإنّها ظهرت في هذا العالم لتصنع الصنع اللائق
بها الشبيه بجوهرها وتؤثّر في عالمها الأثر الإلاهيّ الذي يبقى ويدوم .

(27) ومما دلّ على إيمانهم بالبعث والمعاد ما قاله سقراط في الكتاب المنسوب
إلى ڤاذن ، وقد قيل له عند موته ما لنا لا نرى عندك كراهية للموت ووحشة
لمفارقة الإخوان . قال إنّي لأرجو أن أصير إلى آلهة حكماء خيار وإلى أناس
17a هم أفضل ممن هاهنا ، وأثق غاية الثقة بأنّ مصيري إلى آلهة موال لي في غاية
الجود والخيرورة .

(28) وقوله في هذا الكتاب إنّا سمعنا من المتقدّمين أنّ من مضى إلى الآخرة بعد
النقاء والاستبصار فإنّه يساكن الآلهة ومن مضى وليس عنده شيء من
الفضائل كان مأواه الحمأة .

1 وأعطتها : واعطاها M   1–2 وتلوّنت بألوان : M or read وتكوّنت بأكوان ?   2 after المختلفة blank space of six letters M   3 بمنطقها : منطقها M   5 وعلمت : وعلم M   8 الإلاهيّة : اللاهية M   10 وجاورتها : وجاورها M   11 والذوات : M or read بينها والذوات ?   12 الإلاهيّ : اللاهي M   16 آلهة : الهة M   17 after والخيرورة blank space of half a line M   18 الكتاب : لكتاب M | الآخرة : الاخرة M   19 النقاء : النقآ M   20 after الحمأة an entire line is left blank M

assumed different sorts of forms and motions.[37] Whereupon, after the soul had deemed matter to be weak and low, it brought reason to it by its reason, performed its activities and chose for them [i.e. its activities] something material, which was the form of the human body, so that by [this form's] existence in [matter], the deficient ones among the affairs of the world became perfect. The Creator, praise to Him, the Sublime, gave it the simple good, the comprehensive knowledge and the complete intellect. By the intellect it grasped the beings, by means of knowledge it knew the essences and by the good it performed its actions.

Thus God, to Whom belong might and majesty, has made this soul the 16b
home of knowledge and intellect. It has become active by means of reason and thinking and moved by means of its moving self, its psychic life and its divine power. By means of activity it imitated the active intellect as it descended from the world of intellect to this world in the way of active nature, became mixed with this matter and close to it by vicinity and insertion, not by commingling and the way of the natural essences [becoming mixed].[38] It manifested itself in this world to make the making [of the world] suitable for it and resembling its substance, and to produce the divine impression, which remains and persists, in its world.

(27) Among the indications for their belief in resurrection and the hereafter is what Socrates says in the book attributed to Phaedo:[39] on the occasion of his death, it was said to him: why do we not see you loathing death and being filled with anxiety facing separation from [your] fellow men? He said:[40] I truly anticipate coming to wise and superior gods and to people who are better than [those who are] here, and I am absolutely 17a
confident that my destination [is to get] to the gods who will be my masters in utmost goodness.[41]

(28) He also says in this book[42] that we have learnt from the forefathers that he who passes away to the Hereafter after having attained purity and insight will live in the neighbourhood of the deities, whereas he who passes away without possessing any virtue will have his resting place in the mud.

[37] A possible emendation is to read *wa-takawwanat bi-akwān* instead of *wa-talawwanat bi-alwān* and translate 'Matter was generated in beings of different forms and motions.'

[38] The addition of *baynahā* before *wa-dhawāt* would allow for the reading 'by commingling with the essences of the natural things'.

[39] The Arabic seems to be written with a ڤ, which might indicate a Persian milieu of the scribe or his model.

[40] Cf. *Phaedo*, 63b6–8 and 63c2–4.

[41] The hendiadys *al-jūd wa-khayrūra* renders the Greek ἀγαθός.

[42] Cf. *Phaedo*, 69c3–7.

(29) وقال ارسطالس العالم العلويّ منوط بالعالم العقليّ ومماس به ومنوّر بنوره ومستفيض من فيضه ومتحرّك بحركته ، والعالم العقليّ مناط بالعالم الإلاهيّ ومتطلّع إليه ومتحرّك بقوّته الجاعلة فيه ومستمدّ من علوّه الخير ، والخير منزّل عليه غير ممنوع منه ولا مقطوع عنه .

والنفس واقعة في وسائط مراكز الأجرام العلويّة والعالم العقليّ لا بعيدة من 17b مركزه ولا قريبة من عالمه ، غير أنّها متطلّعة إلى عالم العقل وناظرة إليه ومقتبسة من نوره النور النفسيّ ، وقربها من العالم الطبيعيّ أقرب من قربها إلى العالم العقليّ والعالم الإلاهيّ . فلما صارت مواقع النفس في الوضع الإلاهيّ في الوسائط مالت قواها إلى الأقرب منها فنوّر ما لاصقها وتمّم ما قرب منها .

وكان الناقص من مصاقب أطرافها العالم الطبيعيّ فانجذبت إليه الجواهر النفسانيّة وأشرقت على العالم الجسميّ وأبدت الخير الفيّاض والعقل الفعّال . فمالت إلى إتمام الناقص وإكماله وإلباسه الجمال تشبّهاً بما مثل في سنح جوهرها من الجود الخيريّ والفيض العقليّ .

ولهذا قال افلاطن حياة العالم روحة نفسانيّة .

(30) قال الشيخ اليونانيّ الفاعل الأوّل ساكن غير متحرّك بشيء من أنواع الحركة ، وسكونه يحدث مثاله أعني العقل لا في شيء لأنّ مثاله أنّيّة شريفة 18a قويّة تعلو في الشرف والقوّة على جميع الأنّيّات التي تحتها . وفي ذلك المثال كلّ علم ومعرفة لأنّه لما انبجس من الفاعل الأوّل التفت إلى علّته ونظر إليها لنحو قوّته ، فصار عقلاً وجوهراً وهذا الجوهر هو العقل . فإذا أضيف إلى ما تحته من الجواهر كان أوّلاً وإذا أضيف إلى ما فوقه كان ثانياً .

(31) قال ومن أراد أن يعلم كيف أبدع الواحد الحقّ الأشياء الكثيرة فليلق بصره على الواحد الحقّ فقط وليخلّف الأشياء كلّها خارجاً منه وليرجع إلى

21 قال ومن : ونحن مبتدئون وقائلون من *Uthūlūjiyā* 22 وليخلّف : ولا يختلف إلي *Uthūlūjiyā*, ed. Die | وليرجع : ويرجع *Uthūlūjiyā*, ed. Die

2 مناط : لمناط M | الإلاهيّ : اللاهي M 3 الجاعلة M : or read الحاصلة ? 6 مركزه M : or read مركزها ? 8 الإلاهيّ[1] : اللاهي M | الإلاهيّ[2] : اللاهي M 11 الخير : بالخير M 12 مثل : مثلث M | سنح : سنحة M 14 after نفسانيّة blank space of six letters and three dots (∴) in the marg. M 17 تعلو Ros, Bad : تَعْلوا M 19 لنحو M, Ros : بنحو Bad 20 after ثانياً blank space of six letters M

(29) Aristotle says: the upper [i.e. celestial] world is dependent on the world of intellect, touched by it, enlightened by its light, spreading superabundantly due to its emanation and moved by its motion. The world of intellect is dependent on the divine world, striving for it, moved by the effective power in it[43] and taking the good from its exaltedness. The good is sent down to it and neither held back nor cut off from it.

The soul lies in an intermediary position between the location of the upper bodies and the world of intellect, and is neither remote from its location[44] nor close to its world. However, it strives for the world of intellect, gazing at it and seeking to take the psychic light from its light. As 17b
for closeness, it is closer to the natural world than to the world of intellect and the divine world. So after the positions of the soul had settled into the intermediary positions in the divine set-up, its powers inclined to what was closest to it and thus what was next to it was enlightened and what was close to it was brought to perfection.

The deficient among the neighbour[s] at its borders is the natural world, so the psychic substances were drawn to and radiated towards the physical world and revealed the emanating good and active intellect. They inclined to the perfection of the deficient, its completion and its garbing with beauty in imitation of the generous good and the emanating intellect by the [part] in their substance[s'] origin which is similar [to these two].[45]

Therefore Plato says: the life of the world is a psychic spirit.

(30) The Greek Sage says:[46] the First Agent is at rest and unmoved by any kind of motion. His rest brings forth His image, i.e. the intellect, but not in anything, because His image is a noble, powerful being which is elevated in rank and power above all beings below. All knowledge and cognition is in 18a
that image, for after it had emanated from the First Agent, it turned to its Cause and gazed at It according to its power. So it became intellect and substance, and this substance is the intellect. In relation to the substances below it, it is first, whereas in relation to What is above it, it is second.

(31) He says:[47] he who wants to know how the true One created the multiple things shall cast his glance only on the true One, leave behind all

[43] A possible emendation of *bi-quwwatihī al-jāʿila* to *al-ḥāṣila* would translate as 'its power arising in it'.

[44] Probably read *markazhā* instead of *markazhū*, i.e. 'not remote from their [i.e. the upper bodies'] location'.

[45] Literally 'in making themselves like that what resembles the good generosity and the intellectual emanation in the origin of their substance[s]'.

[46] Belonging to *GS*; see ed. Rosenthal, p.474; and ed. Badawī, 184.3–9.

[47] Belonging to *GS*; see ed. Rosenthal, p.474; and cf. parallel in *ThA*, ed. Badawī, 114.7–14; and ed. Dieterici, 111.15–112.4.

ذاته وليقف هناك ، فإنّه سيرى بعقله الواحد الحقّ ساكناً واقفاً عالياً على الأشياء كلّها العقليّة منها والحسّيّة ويرى سائر الأشياء كأنّها أصنام منبثّة منه ومائلة إليه . فبهذا النوع صارت الأشياء تتحرّك أعني أنّه واجب أن يكون لكلّ متحرّك شيء ما يتحرّك إليه وإلا لم يكن متحرّكاً البتّة ، وإنّما يتحرّك المتحرّك شوقاً إلى الشيء الذي كان منه لأنّه إنّما يريد نيله والتشبّه به . فمن
18b أجل ذلك يلقي بصره عليه فيكون ذلك علّة حركته اضطراراً .

(32) وينبغي للفاعل الأوّل أن يكون ساكناً غير متحرّك ، إذ كان واجباً أن يكون شيء ما ثانياً بعده وأن يكون فعله من غير رويّة ولا حركة ولا إرادة مائلة إلى المفعول . والمفعول الأوّل وهو العقل انبجس من شدّة سكون الفاعل وقوّته ، ثمّ انبجست منه سائر الأشياء العقليّة والحسّيّة بتوسّط العقل ، وإنّما انبجست منه الأشياء وصارت ذات أنّيّات من قبل إفراط قوّة أنّيّته وشدّتها . وذلك أنّ القوّة إذا كانت شديدة انبجس منها إمّا شيء جوهريّ وإمّا شيء عرضيّ ، أمّا الشيء الجوهريّ فكالحرارة التي تنبجس من النار وكالبرودة التي تنبجس من الثلج وأمّا الأشياء العرضيّة فكذوات الروائح الطيّبة . فإنّ الشيء ذا الرائحة الطيّبة ما دام ثابتاً قائماً فتلك الرائحة تنبثّ منه إلى خارج فيلتذّ بها الحيوان القريب منه .

19a (33) وقال إنّ العقل إنّما صار هو الأشياء كلّها لأنّ مبدعه ليس كشيء من الأشياء ، وإنّما صار المبدع الأوّل لا يشبه شيئاً من الأشياء لأنّ الأشياء كلّها منه ولأنّه لا حلية له ولا صورة له خاصّة لازمة . وذلك أنّ المبدع الأوّل واحد

1 وليقف : فليقف *Uthūlūjiyā*, ed. Die | سيرى : يرى *Uthūlūjiyā* 2 منه om. *Uthūlūjiyā*
3 after تتحرّك add. إليه *Uthūlūjiyā* | واجب أن om. *Uthūlūjiyā* 6 حركته : حركة *Uthūlūjiyā*

2 منبثّة منه M : or read مثبتة ? 7 إذ M : إذا Ros, Bad 8 يكون[1] Ros, Bad : يلون M
11 ذات M : or read ذوات ? Ros 15 تنبثّ Ros, Bad : تُنَشبُ M | فيلتذّ Ros, Bad : فلتذ M
16 after منه blank space of ten letters M

things outside him, return to himself[48] and stop there. For he will see with his intellect the true One at rest, stationary and elevated above all things, be they intelligible or sensible, and he sees the other things as if they were images spread from Him[49] and inclining towards Him. In this way things come to move, that is, it is necessary that everything moving has something towards which it moves, or else it would not be moving at all. The moving thing only moves in the desire for its origin because it wants to attain and imitate it. Therefore it casts its glance on it and that is then necessarily the cause of its motion. 18b

(32) The First Agent[50] must be at rest and unmoved, since[51] it is necessary that something is second and posterior to Him, and that His action is without deliberation, motion and volition inclining towards the [resulting] effect. The first effect, which is the intellect, emanated from the force of the Agent's rest and His power. Then, by the mediation of the intellect, all other intelligibles and sensibles emanated from Him. The things emanated from Him and the essence of beings came about by the excessive power and force of His being. For if a power is forceful, either something substantial or accidental emanates from it. The substantial is like the heat which emanates from fire and like the cold which emanates from ice. The accidental is like the essences of pleasant fragrances. As long as something possessing a pleasant fragrance is lasting and subsisting, that fragrance spreads out from it, so that the living being close to it may delight in it.

(33) He says[52] that the intellect only became all things, because its creator 19a
is not like any of them. The First Creator never comes to resemble any of them, because all things are from Him and because He has neither shape nor form specific and inherent to Him. For the First Creator is one alone, that is, He is only being [and this being] has no fitting attribute, because

[48] The Arabic is ambiguous here. The possessive pronouns in *minhū* and *dhātihī* may refer either to God or to him who casts his glance on the true One. The latter seems in line with the Neoplatonic reversion to one's own self. Dieterici (*Theologie*, p.113) translates: 'er kümmere sich nicht um alle Dinge, die ausserhalb seiner sind, und wende sich zur Betrachtung seines (Gottes) Wesens', whereas Lewis ('Plotiniana', p.273) has: 'Let him who wishes to know how the true One originated the many things *cast his gaze on the true One alone*, and leave behind all things outside it, and return to himself.'

[49] Lewis, reading *muthbata* instead of *munbaththa*, translates ('Plotiniana', p.273) '*He will see all other things standing as if they were images* and ...'.

[50] Belonging to *GS*; see ed. Rosenthal, p.476; and ed. Badawī, 184.10–185.2. Rosenthal ('Aš-Šayḫ', p.477, n.2) also draws attention to the inner textual parallel in the *PR*, fols. 28b10–29a1, i.e. passage (44) below.

[51] Rosenthal and Badawī emend *idh* to *idha* based on the Greek (εἰ) and thus read 'if it is necessary that'.

[52] Belonging to *GS*; see Rosenthal, pp.478–80; and ed. Badawī, 185.3–19.

وحده أعني أنّه أنّيّة فقط ليس لها صفة تليق بها لأنّ الصفات كلّها منبثّة
منها . فمن أجل ذلك صارت الأشياء كلّها فيها وليست هي في شيء من
الأشياء إلا بنوع علّة . فأمّا العقل فإنّ الأشياء فيه وهو في الأشياء ، وإنما
صارت الأشياء في العقل لأنّ صورها فيه ومنه انبثّت في الأشياء لأنّه علّة
الأشياء التي تحته ، غير أنّه وإن كان العقل علّة للأشياء التي تحته فإنّه ليس
بعلّة تامّة للشيء لأنّه إنّما هو علّة صورة الشيء فقط لا علّة هويّة . فأمّا
الفاعل الأوّل فإنّه علّة تامّة وذلك أنّه علّة هويّة الشيء وصورته بلا توسّط
19b وعلّة هويّة النفس وصورة الأشياء بتوسّط العقل والنفس والأشياء كلّها مصوّرة
في العقل من غير أن تكون مصوّرة في العلّة الأولى بل إنّما هي منبجسة منها .
وكلّ شيء من الأشياء العقليّة محدود وإنّما حدّ الشيء حليته وصورته ، وذلك
أنّ العلّة الأولى لما أبدعت هويّة ألاشياء لم تدعها محمولة على غير حدّ بل هي
التي تحدّها وتحيط بها وإنّما تحدّها بصورها . فحدّ الشيء المفعول هو صورته
وسكونه ، والسكون إنّما هو حلية للعقل وحدّ له وبالحلية والسكون يكون قوام
العقل وثباته وسائر الأشياء العقليّة .

(34) ويقال إنّ الواحد المحض يشبه الضوء والواحد الثاني المنسوب إلى شيء
آخر يشبه الشمس والشيء الثالث يشبه القمر الذي ينال ضوءه من
الشمس . فالنفس فيها عقل مكتسب ينيرها بنوره ويصيّرها عقليّة ، والعقل
ففيه نور ذاتيّ وليس هو نور فقط لكنّه جوهر قابل للنور . فأمّا الشيء الذي
20a ينير العقل ويفيض عليه النور فهو نور فقط وليس هو شيء آخر غير النور
لكنّه نور مبسوط صرف محض يفيض قوّته على العقل فيصيّره عقلاً منيراً

1 تليق : يليق Ros   5 للأشياء M, Bad : الأشياء Ros   6 هويّة M, Bad : هويته Ros   7 وصورته
M : وصورته secl. Ros   8 هويّة Ros, Bad : هويته M | وصورة : or read وصور ? Ros
11 هويّة Ros, Bad : هويته M   12 تحدّها[1] . . . بها Bad, suggested in footnotes by Ros : يحده
ويحيط به M, Ros   15 ويقال M : وقال Ros, Bad | الضوء Ros, Bad : الضور M | الثاني Ros,
Bad : الناقي add. above *al-mansūb* M | after المنسوب add. النانى undotted and crossed out
M   16 آخر Ros, Bad : أخر M | ضوءه Ros, Bad : ضوه M   18 نور[2] Ros, Bad : انورٌ M
19 فهو : وهو M, Ros, Bad | آخر Ros, Bad : أخُر M

all attributes spread from it. Therefore all things came to be in [His being], whereas it is in none of the things except in the manner of a cause.[53] As for the intellect, things are in it and it is in things. Things only came to be in the intellect, because their forms are in it and spread from it into the things. For the intellect is the cause of the things below. Yet, even if the intellect is a cause for the things below, it is no final cause for anything,[54] as it is only the cause of a thing's form, not the cause of its being. The First Agent is a final cause, that is He is the cause of the [first created] thing's being and its form without mediation and the cause of the soul's being and of the form of things by mediation of the intellect. The soul[55] and all things 19b
are formed in the intellect, without being formed in the First Cause. Rather they only emanate from It. All intelligibles are defined and the thing's definition is its shape and form. That is to say after the First Cause has created the things' being, It does not leave [them] with something undefined predicated of them,[56] but It is what defines and comprises them. It defines them by their forms. Thus the definition of the thing, which is [Its] effect, is its form and rest. Rest is shape and definition for the intellect and through [this] shape and rest the intellect's subsistence and stability and all other intelligibles exist.

(34) It is said[57] that the pure One resembles light, the second one which is related to something else resembles the sun, and the third thing resembles the moon which attains its light from the sun. Thus in the soul, there is an acquired intellect which illuminates it with its light and makes it intellectual. In the intellect, there is essential light, and it is not only light, but also a substance receiving light. As to what illuminates the intellect and pours light over it, it is only light and nothing other than light, but simple, 20a
absolute, pure light which pours its power over the intellect and makes it

[53] For the expression 'in the manner of a cause (*per modum causae*)' see D'Ancona, 'Profilo', pp.128–9; and Adamson, *Arabic Plotinus*, p.86.

[54] The Arabic term, which I translate with 'final cause', is *ʿilla tāmma*. It may also be translated as 'perfect, perfecting cause'. However, in the present context, I think it permissible to use this rather Aristotelian terminology.

[55] It is not obvious whether the text reads as indicated (also in Rosenthal, 'Aš-Šayḫ', pp.479–81; and Lewis, 'Plotiniana', p.281) or 'by mediation of the intellect and the soul. All things are formed in the intellect ...' which would result in the problem that the soul is somehow said to help to cause its own being, unless we assume that the mediation of intellect *and* soul only refers to the forms of the things. In the *Ldc*, where the theory of mediation is also expounded, only the mediation of intellect is ever mentioned.

[56] Rosenthal ('Aš-Šayḫ', p.481), followed by Lewis ('Plotiniana', p.281), translates: 'It did not leave them, in their dependence on a substratum without definition.'

[57] Belonging to *GS*; see ed. Rosenthal, pp.480–82; and ed. Badawī, 186.1–9.

مضيئاً ، غير أنّ النور الذي في العقل هو شيء في شيء ، فأمّا النور الذي ينير العقل فليس هو في شيء آخر بل هو نور وحده قائم ثابت بذاته ينير جميع الأشياء غير أنّ من الأشياء ما يقبل نوره أكثر ومنها ما يقبل نوره أقلّ .

(35) وقال إنّ الواحد الحقّ هو مبدع الأشياء وليس هو ببعيد منها ولا بمفارق لها ، بل هو مع الأشياء كلّها إلا أنّه معها كأنّه ليس معها ، وإنّما تستبين معيّته مع الأشياء التي تقوى على قبوله .

(36) ويقول إنّه واحد عظيم أعظم الأشياء لا بالجثّة لكن بالقوّة . وكذلك إذا قلنا إنّه لا نهاية له لا يعني أنّه لا نهاية له بأنّه جثّة أو عدد لكنّا نعني أنّه لا يحيط بقوّته شيء ، وذلك أنّه فوق وهم المتوهّم ثابت قائم بذاته ليس فيه 20b شيء من الصفات ، وهو خير ليس لذاته لأنّ ذاته هي الخير المحض الحقّ لكنّه خير لسائر الأشياء التي تقوى على قبول الخير الذي يفيضه عليها . وليس له حركة لأنّه قبل الحركة وقبل الفكرة وقبل العلم ، وليس فيه شيء يريد أن يعلمه كما يعلم العالم ، بل هو العلم الذي لا يحتاج إلى أن يعلم بعلم آخر لأنّه هو العلم المحض الأقصى المحيط بكلّ علم وعلّة العلوم .

(37) فإن قال قائل إن كان العقل إنّما يستفيد الصورة العقليّة والحسّيّة من الفاعل الأوّل إذا ألقى بصره عليه ثم يفيضها على ما تحته وإنّما يفيض الصور على ما تحته صورة بعد صورة أفتراه كذلك يستفيدها من الفاعل الأوّل واحدة بعد واحدة ؟ فإن كان يستفيد الصور واحدة بعد واحدة كان الفاعل [على] الأوّل لا محالة إنّما يفعل أفاعيله فعلاً متجزئاً . فإن كانت تلك الأنّيّة الحقّة تفعل أفاعيلها واحداً بعد واحد فقد بقي فيها إذن أفاعيل لم تظهر بعد . فإن كان كذلك فإنّها إذن لم تفعل الأشياء بأنّها فقط لكنّها فعلها برويّة وحركة ما وهذا محال قبيح .

21a (38) قلنا إنّ الأنّيّة الأولى فعلت فعلها كلّه دفعة واحدة بغاية الفعل وليس من

1 مضيئاً Bad : مضياً M, Ros 3 الأشياء[2] Ros, Bad : الاشيآ M | after أقلّ blank space of six letters M 4 الأشياء Ros, Bad : الاشيآ M 7 ويقول M, Bad : ونقول Ros 8 يعني M : نعني Ros, Bad | جثّة M, Ros : جهة Bad 14 after العلوم blank space of half a line M 15 الصورة M, Bad : الصور(ة) Ros 16 يفيض M, Ros : تفيض Bad 17 كذلك M : لذلك Ros, Bad 18 على del?. Ros; Bad secl. 19 الحقّة : الحق M, Ros, Bad 22 after قبيح blank space of six letters M 23 الأولى Bad : الاول M

an illuminating, enlightening intellect. Yet, the light in the intellect is something in something else, whereas the light which illuminates the intellect is not in anything else, but is light alone, subsisting and lasting in its essence. It illuminates all things, but there are things which receive its light more abundantly and others which receive it less.

(35) He says[58] that the True One is the creator of things. He is neither remote nor separated from them. Rather He is with all things, yet He is with them as if He were not with them. However, His togetherness with things which are able to receive Him is evident.

(36) He says[59] that He is one, great, the greatest of things, not in mass, but in power. Thus whenever we say that He is infinite, it does not mean that He is infinite in mass or number, but we mean that nothing encompasses His power. That is He is above the imagination of anyone who imagines, lasting, self-subsisting and not having any attribute in Him. He is good not 20b
for Himself, as His essence is the pure, true good, but He is good for all other things which are able to receive the good which He pours over them. He has no motion, as He is prior to motion, prior to thinking and prior to knowledge. In Him there is nothing He wants to know in the way in which a knower knows. Rather He is the knowledge which does not need to be known by another knowledge. For He is the pure, ultimate knowledge comprising every knowledge and the cause of all [branches of] knowledge.

(37) Someone may say:[60] if the intellect acquired the intelligible and sensible form[s] from the First Agent while casting its glance on Him, and then poured them over what is below, but only poured the forms over what is below one by one, would you then deem it to be in the same manner that it acquired them from the First Agent, namely one by one? So if it acquired the forms one by one, the First Agent would, most certainly, perform His actions only in parts. And if that true Being performed His actions one by one, there would consequently remain in His actions which have still not become manifest. If it were like that, He would consequently not make things by only being [Himself], but would make them with some sort of deliberation and motion, and that is unthinkable and unseemly.

(38) We say[61] that the First Being performs all His action at one stroke and 21a

[58] Belonging to *GS*; see ed. Rosenthal, p.482; and ed. Badawī, 186.10–13.

[59] Belonging to *GS*; see ed. Rosenthal, pp.482–4; and ed. Badawī, 186.14–187.3.

[60] This passage may or may not belong to *GS*. It has, however, been taken up in the editions of Rosenthal and of Badawī, who explains (*Aflūṭīn*, p.187, n.1) that the passage is not Plotinian, but a commentary on the preceding text. See ed. Rosenthal, p.484; and ed. Badawī, 187.4-10.

[61] As with the previous passage, this one may or may not have been part of *GS*; see ed. Badawī, 187.11–18.

وراء فعلها فعل آخر ، لا أنّها لم تقو على أن تأتي بفعل آخر لكنّها أنّما فعلت فعلها بغاية الحكمة التي لا يمكن أن يكون المفعول بنوع آخر على غير ما هو عليه الآن . فنهاية الفعل إنّما هو من تلقاء المفعول لا من تلقاء الفاعل . فلما فعل الفاعل فعله دفعة واحدة بغاية الإحكام أحسن العقل بفعله كلّه معاً وفي دفعة واحدة من غير أن يكون عالماً بفضائله كلّها . فإن كان هذا هكذا وكان العقل أوّل مفعول فإنّما يعلم فضائل الفاعل الأوّل شيئاً بعد شيء وكذلك يفيضها على ما تحته شيئاً بعد شيء . فالعقل يحسن بفضائل الفاعل الأوّل دفعة واحدة من غير أن يعلمها في دفعة واحدة ويفيضها على النفس دفعة بعد دفعة .

(39) قال الشيخ اليونانيّ العقل الأوّل المبتدع ليس بذي صورة ⟨و⟩إذا أضيف إلى المبدع الأوّل كان ذا صورة لأنّه يتناهي فيشكّل فيكون ذا حلية وصورة . 21b فأمّا المبدع الأوّل [فاما المبدع الاول] فإنّه لا صورة له لأنّه ليس من فوقه شيء آخر يريد أن يتناهي إليه ولا شيء مما تحته يقدر أن يتناهي إليه ، فهو من كلّ الجهات غير متناه ، فلذلك صار لا حلية له ولا صورة . ولو كان المبدع الأوّل صورة لكان العقل كلمة ما وليس العقل بكلمة ولا فيه كلمة ، لأنّه ابتدع ابتداعاً من غير أن يكون مبدعه ذا صفة أو صورة فيجعل تلك الصفة كلمة فيه . فالعقل ليس بكلمة ولا فيه كلمة بل هو فاعل الكلم في الأشياء لأنّه ذو صفة وحلية . فإذا فعل الشيء أثّر فيه من بعض صفاته شيئاً ، فيكون ذلك الأثر هو الكلمة الفاعلة في الشيء .

(40) وينبغي للأوّل ألا يكون كثيراً من جهة من الجهات وإلا كانت الكثرة التي فيه معلّقة في واحد آخر قبله ، بل ينبغي أن يكون واحداً خيراً محضاً وأن

1 آخر[1] Bad : اخر M 3 تلقاء[2] Bad : تلقا M 4 أحسّ Bad : أحسن 7 يحسُّ Bad : يحسن 9 after دفعة[2] blank space of half a line M 10 ⟨و⟩إذا Ros, Bad : وإذا 11 إلى . . . حلية add. in marg. M 12 [أما المبدع الأول] Ros : فاما . . . الاول 13 آخر Ros, Bad : آخر M 14 حلية Ros, Bad : حيلة M 20 ألا Bad : ان M : ⟨لا⟩ أن Ros 21 آخر Ros, Bad : اخر M

perfectly. Beyond this action of His there is no other action. It is not the case that He was incapable of bringing forward another action, rather that He performed His action with the utmost wisdom which makes it impossible that the effect was in any way different from what it is now. The limit of the action is only on the part of the effect, not on the part of the agent. When the Agent performed His action at one stroke with the utmost accuracy, simultaneously the intellect became good[62] by His entire action at one stroke, but without knowing all His virtues. If it is like that and the intellect is the first effect, then it only knows the virtues of the First Agent one after the other and pours them likewise over what is below, namely one after the other. So the intellect becomes good by the virtues of the First Agent at one stroke, but without knowing them at one stroke, and pours them over the soul one after the other.

(39) The Greek Sage says:[63] the first created intellect does not possess form. When regarded in relation to the First Creator, it possesses form, because it is finite and thus shaped, hence possessing form.[64] As to the First Creator, He does not have form, because there is nothing else above Him 21b
which He wants to reach as His limit and nothing below Him is capable of reaching Him as its limit. So He is infinite in all ways. Therefore He has not come to have form. If the First Creator were a form, the intellect would be a word (*logos*), but the intellect is not a word (*logos*) nor is there a word (*logos*) in it. For it was created without its Creator possessing attribute or form to make into a word (*logos*) in [the intellect]. So the intellect is not a word (*logos*) nor is there a word (*logos*) in it, rather it is the agent of words (*logoi*) in the things, as it possesses attribute and form. Whenever it makes something, it produces an impression of some of its attributes on it. That impression is the word (*logos*) active in the thing.

(40) The First[65] must not be multiple in any way, otherwise the multiplicity in Him would depend on another One prior to Him. So He must rather be one, pure good and the Creator for one thing, which is good and possesses

62 The reading of the MS is not sufficiently clear here and in what follows. However, it seems to me to read *ḥ-s-n* IV. rather than *ḥ-s-s* VI., which Badawī has edited and which Lewis must have read, since he translates ('Plotiniana', p.475) 'mind perceives all its action at once, simultaneously, without knowing all its virtues'. It would be odd if the Arabic Plotinus with its dicotomy between intellect and sense perception used the expression that intellect perceived. In the *ThA* the soul is, on rare occasions, said to perceive, but never the intellect. The root *ḥ-s-n* is, however, explicitly linked to intellect, e.g., in the heading of the fourth chapter of the *ThA* which is called *Fī sharaf ʿālam al-ʿaql wa-ḥusnihī*.

63 Belonging to *GS*; see ed. Rosenthal, p.486; and ed. Badawī, 188.1–10.

64 I translate the Arabic hendiadys *ḥilya wa-ṣūra* simply by 'form'.

65 Belonging to *GS*; see ed. Rosenthal, pp.488–91; and ed. Badawī, 188.11–189.12.

يكون مبدعاً لشيء واحد خير ذي صورة خيريّة إمّا يكون أثراً من المبدع الأوّل وإمّا أن يكون أثر أثره . فأثر المبدع هو العقل وأثر العقل هو الحياة ، والحياة 22a هي عقل أيضاً كما أنّ أثر النار هو نار أيضاً ، غير أنّه ينبغي أن يكون في الأثر ما في المؤثّر وما ليس فيه ، وذلك أنّ المؤثّر الأوّل واحد وأثره العقل والعقل إثنان لأنّه مبتدع والمبتدع بعد المبدع . وأثر العقل النفس ففي العقل ما في النفس وفيها ما ليس في العقل وذلك أنّها أكثر من إثنين ، فلذلك صارت النفس تميل وتتصرّف وتصرّفها هو فكرتها . فلا يزال كلّ واحد من الصور يؤثّر أثراً فيكون أثره أكثر منه إلى أن يأتي إلى أثر لا يؤثّر البتّة . فأوّل الأوائل مؤثّر لا يتأثّر وآخر الأشياء أثر لا يؤثّر ، وما بينهما مؤثِّر ومؤثَّر ، مؤثِّر لما تحته ومؤثَّر مما فوقه ، غير أنّ الأثر الأوّل وهو العقل يتحرّك حركة عقليّة مستوية لا عوج فيها البتّة لأنّه أثر من مؤثّر ساكن لا يتحرّك بشيء من الحركات البتّة . والأثر الثاني وهو النفس يتحرّك حركة نفسانيّة مائلة قليلاً لأنّها من مؤثّر متحرّك ، فلذلك صارت النفس متفكّرة . والأثر الثالث هو الجرم الشريف 22b الأوّل السماويّ وحركته حركة موضعية مستوية في الموضع وإنّما صار يتحرّك حركة موضعية لأنّه مؤثّر من متحرّك مائل إلا أنّ حركته ليست بموضعية . والأثر الرابع هو الأجرام الأرضيّة وحركتها حركة موضعية غير مستوية في الموضع والكيف ، فلذلك صارت تتضادّ في الموضع والقوى . وكلّ الآثار معلّق في الفعل الأوّل وهو العقل والعقل معلّق في الفاعل الأوّل والفاعل الأوّل هو مبدعها كلّها وحافظها إلا أنّه مبدع بعضها بلا توسّط وبعضها بتوسّط .

(41) قال افلاطن إنّ كلّ عقل ساكن في المكان العقليّ وهو العالم الذي سمّيناه المعقول الروحانيّ وهذا العالم محيط بالقوى الكلّيّة والقوى الجزئيّة فإنّ العقل

1 إمّا M, Bad : ⟨أن⟩ إما Ros 2 الحياة Bad : الحيوة M, Ros | والحياة Bad : والحيوة M, Ros
5 إثنان Ros, Bad : اثنين M 8 يأتي M, Ros : نأتي Bad 9 وآخر Ros, Bad : واخر M
15 after بموضعية blank space of five letters M 16 مستوية Ros, Bad : المستوية M 19 after بتوسّط blank space of half a line M 21 الجزئيّة D'Anc : الجزية M

the form of goodness, either being an impression of the First Creator or an impression of His impression. The impression of the Creator is the intellect 22a
and the impression of the intellect is life. Life is also intellect just as the impression of fire is also fire. However, in the impression there must be what is in the producer of the impression as well as what is not in it. That is the First Producer of an impression is one and His impression is the intellect, which is two, as it is created and the created is after the Creator. The impression of the intellect is the soul, thus in the intellect there is what is in the soul, but in [the soul] there is [also] what is not in the intellect, as the soul is more than two. Therefore the soul starts to incline and to act freely. Its free action is its thinking. Each single form continues to produce an impression, its impression always being more multiple than it until an impression is reached which does not produce any impression at all. The first of the first [principles] is the producer of an impression without receiving one, whereas the last thing is an impression without producing one. [The things] between these two are producing and receiving impressions, producing impressions on what is below and receiving impressions from what is above. Yet, the first impression which is the intellect moves in an intellectual, regular motion which does not deviate at all, because it is an impression produced by a producer at rest Who does not move in any way at all. The second impression which is the soul moves in a psychic, slightly inclining motion, because it is produced by a moving producer. Therefore the soul becomes thinking. The third impression is 22b
the first noble heavenly body and its motion is a regular, local motion in a defined place. It starts moving in a local motion, because it is an impression produced by something inclining and moving, though not in a local motion. The fourth impression is the earthly bodies and their motion is a local motion which is irregular in regard to place and quality. Therefore they start to oppose each other in position and powers. All impressions depend on the first [product of] action, which is the intellect. The intellect depends on the First Agent and the First Agent is the Creator and Keeper of them all. However, He is the Creator of some of them without mediation and of some by means of mediation.

(41) Plato says[66] that every intellect rests in the place of intellect, which is the world we call the spiritual intelligible. This world encompasses the universal and particular powers, for intellect is not only one, as it is one

[66] Belonging to *GS*; see ed. D'Ancona, pp.239–41.

23a ليس بواحد فقط لأنّه واحد وكثير . فإن كان العقل واحداً وكثيراً فالنفس إذن
واحدة وكثيرة ، والأنفس الكثيرة المختلفة إنّما كانت من النفس الواحدة
كالصور الكثيرة من الجنس الواحد وكما أنّ الصور الكثيرة المختلفة وإن كانت
من جنس واحد فإنّ فيها ما هو أفضل وأكرم وفيها ما هو أخسّ وأدنى ،
كذلك الأنفس الكثيرة المختلفة وإن كانت من نفس واحدة ، فإنّ بعضها
أفضل من بعض وبعضها أخسّ من بعض وذلك أنّ بعض الأنفس ما هو
عقليّ بالفعل فقط وبعضها ما هو عقليّ لا بالفعل لأنّها عقل ضعيف ،
وكذلك العقل في العالم العقليّ منه ما هو أشرف من غيره ، وذلك أنّ العقل
منه ما هو عقل عظيم القوّة يحيط بقوّته سائر العقول التي في ذلك العالم كأنّه
حيّ عظيم يحيط بجميع الحيوان وكلّ واحد من سائر العقول إنّما هو ما هو
بالفعل من غير أن يحيط بشيء آخر غيره كالحيوان الجزئيّة ، وإنّما يشبّه العقول
23b الكثيرة والعقل الواحد المحيط بها بمدينة حيّة ذات نفس محيطة بأشياء حيّة
ذوات أنفس . فالمدينة أتمّ وأكمل وأقوى من الأشياء الحيّة التي فيها وكلّ واحد
من تلك الأشياء الحيّة التي فيها إنّما هو لنفسه لا لغيره أعني أنّه لا يحيط إلا
بذاته فقط ، والمدينة وإن كانت تحيط بالأشياء التي فيها ، فإنّ تلك الأشياء
لا تمتنع من أن تكون من طبيعة المدينة ، كما لا تمتنع الأشياء الناريّة وإن
كانت كثيرة مختلفة بعضها كبير وبعضها صغير أن تكون هي والنار من طبيعة
24a واحدة أكثر . [وانما يغلب على النفس ... وبين الحس فرق] كذلك العقول
التي في العالم الأعلى وإن كانت كثيرة مختلفة فإنّها لا تمتنع أن تكون هي
والعقل الواحد المحيط بها من طبيعة واحدة لأنّ جوهر ذلك العقل الواحد
وسائر العقول إنّما هو جوهر واحد لا اختلاف فيه ، وإنّما اختلفت بالقوّة

1 after واحد[1] add. in marg. as an explanatory gloss probably by a different hand جنس العقل M | after وكثير add. in marg. (see above) أشخاص العقول M 2 after واحدة add. in marg. (see above) جنس نفس M | after وكثيرة add. in marg. (see above) أشخاص نفوس M 11 آخر D'Anc : اخر M | الجزئيّة D'Anc : الجزوية M 12 محيطة M : محيط D'Anc | بأشياء D'Anc : باشيآ M 13 الأشياء D'Anc : الاشْيآ M 14 تلك M : ذلك D'Anc 15 وإن M : وإذ D'Anc 18 أكثر M : أكبر D'Anc : after أكثر blank space of half a line and three dots (∴) in the marg. M | وانما . . . فرق crossed out in M, where it appears correctly on fol. 26b 19 الأعلى M : العالي D'Anc 20 المحيط M : المخط D'Anc 21 هو M : هي D'Anc | لا اختلاف D'Anc : لاختلاف M

and many.[67] If intellect is one and many, then the soul is also one and 23a
many.[68] However, the many different souls are from the one soul, just as the many forms are from one genus, and although the many different forms are from one genus, there are more virtuous and excellent ones and baser and viler ones among them. In the same way there are the many different souls, although they are all from one soul. For some of them are more virtuous and some viler than others, because some souls are solely intellectual in actuality, while others are intellectual, but not in actuality, as they are [of] weak intellect. In the same way some intellect in the world of intellect is nobler than the rest. For there is an intellect of great power which encompasses all other intellects in that world with its power, as if it were a great living being encompassing all living beings, whereas each of the other intellects is only what it is [itself] in actuality, without encompassing anything other than itself like the particular living beings. The
many intellects and the one intellect encompassing them may be 23b
compared to an ensouled living city encompassing ensouled living things. The city is more perfect, complete and powerful than the living things in it, and each of those living things in it is only for itself, not for anything else, i.e. it only encompasses itself. Although the city encompasses [all] the things in it, those things are not prevented from belonging to the city's nature. In the same manner the fiery things, even if they are many and different, some big and some small, are not prevented from belonging,
with fire, to the same more abundant nature.[69] In the same way the 24a
intellects in the upper world, although they are many and different, are not prevented from belonging, along with the one intellect encompassing them, to the same nature. For the substance of that one intellect and the other intellects is only one substance without difference in it. However, they differ in power and [what they] encompass, because the one intellect

67 Below the word 'one' the words 'genus of intellect' are added in the margin, and below 'many', 'individual intellects'.

68 Above the word 'one' the words 'genus of soul' are added in the margin, and above 'many', 'individual souls'.

69 Here (fol. 23b9–13) the scribe of the Marsh MS started to write a passage which should appear later on fol.26b9–13 and was thus crossed out. This, however, supplies us with a second textual witness of fol.26b9–13.

والإحاطة لأنّ العقل الواحد يحيط بقوّته سائر العقول والعقول لا تحيط بشيء آخر غيرها لأنّها كالأشخاص عند صورها أو عند أجناسها . فالعقل إذن واحد وكثير واحد بأنّه محيط بأشياء كثيرة عقليّة وكثير بأنّه كشخص لا يتميّز بل هو جوهر كثير الثمرة ، وأوّل ثمرته النفس الشريفة العقليّة الناطقة ولا يمتمع العقل الواحد من أن يصوّر الصور العقليّة وإنّما يصوّرها في الأنّيّة المبتدعة من الأنّيّة الأولى مبدعة الأنّيّات كلّها .

فأمّا عمل النفس الناطقة فهو أن يعقل ويعلم ويفعل فعلاً آخر غير العلم وإلا لم يكن بينها وبين العقل فرق ، وذلك أنّ النفس تعقل وتعلم بأنّها عقليّة أي تحت العقل والعقل فيها مستفاد ، ولها من ذاتها عمل آخر لأنّ لكلّ شيء من
24b الأشياء عملاً خاصاً لا يشركه في عمله غيره . فالنفس تعقل من تلقاء العقل وتعمل عملاً آخر من ذاتها ، وذلك أنّها إذا ألقت بصرها على ما فوقها أي على العقل كانت عاقلةً عالمةً وإذا ألقت بصرها على ذاتها رتّبت ما تحتها ودبّرته واستولت عليه وكانت هي القيّمة عليه ، وهي التي تفيده الصورة والحلية الحسنة ، وإنّما صارت النفس تفعل الأفاعيل الشريفة وتصوّر الصور الحسنة في الشيء الذي يليها سفلاً لأنّه لم يكن من الواجب أن يقف الخليقة في العالم العالي العقليّ وهي مبتدعة من البارئ الأوّل ممتلئة قوّة ونوراً وحياة تغلى وتفور . فلم يكن بدّ من أن تسلك تلك الحياة وتلك القوّة وذلك النور سفلاً فيكون منه عالم آخر أخسّ منه وأدنى وهو العالم السفليّ الحسّيّ ، غير أنّه وإن كان أخسّ منه وأدنى فإنّه مضطرّ أن يكون إذ كانت علّته مضطرّة . وإنّما
25a صار حدوث العلّة العقليّة مضطرّاً لأنّ فاعلها علّة محضة ليس فيها شيء معلول البتّة وإذا كانت العلّة علّة فقط ليس فيها علّة ولا فيها شيء معلول كان الحدوث منها اضطراراً . ولست أعني بالاضطرار الاضطرار الطبيعيّ بل أعني الاضطرار الإراديّ فإنّه يعلو كلّ اضطرار طبيعيّ . والعلّة المحضة تفعل فعلها فوق الاضطرار الطبيعيّ والإراديّ لأنّها علتهما جميعاً ، فإن كانت العلّة

4 الثمرة : للثمرة M, D’Anc 4–5 الناطقة . . . العقليّة add. in marg. M 5 يصوّر M : يصار D’Anc 7 آخر D’Anc : اخر M 8 أي M : أعني D’Anc 10 عملاً M : عاملاً D’Anc 11 آخر D’Anc : اخر M | أي M : أعني D’Anc 13 تفيده M : يفيض D’Anc 15 الخليقة M : الخليق D’Anc 18 آخر D’Anc : آخرُ M 22 ولست D’Anc : وليست M 23 يعلو : يعلوا M, D’Anc 24 علتهما M : علّتها D’Anc

encompasses the other intellects with its power, whereas [those] intellects encompass nothing but themselves, as they are like individuals among their forms or genera. Therefore the intellect is one and many, one by encompassing many intellectual things and many by being like an individual[70] which is not distinct [in substance], but a substance of many fruits. The first of its fruits is the noble, intellectual rational soul. The one intellect is not prevented from forming the intellectual forms. However, it forms them within the being which is created by the First Being which creates all beings.

The action of the rational soul is to think, to know and to perform activities other than knowing [as well], otherwise there would be no distinction between it and the intellect. For the soul thinks and knows by being intellectual, i.e. below the intellect, and the intellect in it is acquired. Due to its essence, the soul does something else, for every single thing does something specific in which no other participates with it. Thus the soul 24b
thinks by virtue of the intellect and does something else due to its own self. For whenever it casts its glance on what is above, namely on the intellect, it is intellective and knowing, and whenever it casts its glance on its own self, it orders, directs and masters what is below and is its custodian. It is the soul which gives it good form. The soul has started doing its noble activities and forming good forms in what follows it below, because the creation which is created by the First Creator, filled with power, light and life excessive and welling forth,[71] must not stop in the elevated world of intellect. It is inevitable that that life, that power and that light proceed downwards, so that from it another world comes to be, viler and baser than it, which is the lower, sensible world. Although it is baser and viler than it, it is still necessarily existent, since its cause is necessary. The emergence of the intellectual cause has only become neces- 25a
sary, because its Agent is a pure cause in which there is nothing caused at all. If the cause is solely a cause, in which there is neither a cause nor anything caused, the emergence from it is a necessity. By 'necessity' I do not mean natural, but wilful necessity, for it exceeds every natural necessity. The Pure Cause performs its action above the natural and wilful necessity, for it is the cause of both of them together. If the First Cause is

[70] Lewis ('Plotiniana', p.235) corrects the text to invert the meaning and translates: 'one because it is like an undifferentiated individual, and multiple because it encompasses many intellectual things'.

[71] Lewis ('Plotiniana', p.237) translates 'seething and boiling'. I read *tuġlā* (instead of the vocalised *taġlī*) and *tafūr* in the sense of '(water) welled, came forth', according to Lane.

الأولى علّة الاضطرارات الطبيعيّة والإراديّة فإنّما تحدث الأشياء منها بنوع أعلى وأرفع من الطبيعة والإرادة .

(42) وقال إنّ الأنفس الجزئيّة الوحدانيّة تشتاق إلى العقل شوقاً عقليّاً وتحرص على لزومه والتمسّك به والرجوع إليه ، ولها قوّة تسلك إلى هذا العالم وتنبثّ فيه وتنبسط عليه . وإنّما هي بمنزلة ضياء الشمس فإنّه متعلّق بقرص الشمس علواً وهو منبثّ في جمع العالم الأرضيّ يفيض عليه ضوءه ونوره بلا حسد 25b كذلك الأنفس متعلّقة بالعقل في العالم الأعلى وقوّتها منبثّة في العالم الحسّيّ السفليّ تضيء على الأجسام كلّها إلا أنّ نورها في الجسم على نحو قبول الأجسام لقوّتها ونورها . وهي وإن كانت تفيض نورها على الأجسام الأرضيّة فإنّها لا تفارق عالمها العقليّ بل تكون فيه مع النفس الكلّيّة وتدبّر معها الأجرام السماويّة كأنّها ملك عظيم مستول على الأشياء يدبّرها ويسوسها وهي معها من غير أن ينحدر من ذلك الموضع الشريف الملكيّ العالي ولا يفارقه . وذلك أنّ الأنفس الكثيرة لا تفارق النفس الكلّيّة ولا ذلك الموضع الشريف العالي لكنّها فيه مع النفس الكلّيّة دائماً ما دامت في ذاتها أعني ما دامت تشتاق إلى الأشياء الكلّيّة وتدبّرها ويسوسها مع النفس الكلّيّة . فإن خرجت من ذاتها أي من الكلّ إلى الجزء واشتاقت إلى أن تكون فيه وتتفرّد بذاتها ولم ترد أن تكون مع النفس الكلّيّة فإنّها حينئذ تفارق عالمها وتنحدر إلى هذا العالم الحسّيّ . 26a فيصير كلّ واحد منها في جزء من أجزاء هذا العالم وإذا صارت الأنفس في هذه الجزئيّات وطال مكثها فيها ومفارقتها النفس الكلّيّة وعدمت العالم العقليّ ولم تلق بصرها إليه وصارت كأنّها شيء جزئيّ ضعفت ولم تجد مَن يقوّيها لأنها تبقى منفردة وحدها وليست النفس الكلّيّة معها . فيكثر حينئذ فكرها ورويّتها ولا تلقي بصرها إلا على الأشياء الجزئيّة . وذلك

2 والإرادة : M 〈و〉الإرادة D’Anc   3 الجزئيّة D’Anc : الجزية M   5 هي : هو M, D’Anc   6 ضوءه D’Anc : ضوه M | حسد D’Anc : جسد M   7 متعلّقة : متعلّقت D’Anc   8 تضيء : يضي M : يضيء D’Anc   11 الأشياء D’Anc : الاشيآ M   12 معها : معه M, D’Anc   15 ويسوسها : ويوسها D’Anc   16 الجزء D’Anc : الجزو M   18 جزء D’Anc : جزوٍ M | من M : ومن D’Anc | أجزاء : اجزا M : أخزاء D’Anc   19 الجزئيّات : الجزيات M : الجزءيات D’Anc   20 تلق : يلو M : يلقي D’Anc | جزئيّ ضعفت : جزوي ضعفَت M : جزء وكضعفت D’Anc   21 تبقى M : بقي D’Anc   22 الجزئيّة : الجزوية M : الجزءية D’Anc

the cause of the natural and the wilful necessities, things emerge from It in a more elevated and exalted way than [from] nature and will.

(42) He says[72] that the particular, singular souls desire the intellect intellectually and strive for attachment, adherence and reversion to it. They have a power which proceeds to this world, spreads and expands over it. It is similar to sunlight. For as [sunlight] is attached to the sun's disc above and also spread over all the earthly world pouring its brightness and light over it ungrudgingly, so the souls are attached to the intellect in 25b
the upper world, whereas their power is spread over the lower, sensible world and casts light upon all the bodies. Yet [unlike sunlight] their light is in [each] body in accord with the bodies' receptivity of their power and light. Although they pour their light over the earthly bodies, they do not depart from their world of intellect, but remain there together with the universal soul. With it they direct the heavenly bodies, as if it were a great sovereign mastering [all] things, directing and ruling them while being with them, but without descending and separating from that noble, elevated, sovereign position. For the many souls do not depart from the universal soul or that noble, elevated position, but always remain there, together with the universal soul, as long as they remain in themselves, i.e. as long as they desire the universal things, and direct and rule them together with the universal soul. If they forsake themselves, i.e. the universal for the particular,[73] desire to be in [the particular] and to be alone by themselves and do not wish to be with the universal soul, then they depart from their world and descend to this sensible world. Each of 26a
them thus comes to be in one of the parts of this world. When the souls come to be in these particulars, their stay in them and their separation from the universal soul endure, they lack the world of intellect, do not cast their glance on it and become like a particular, [then] they become weak and find no one to strengthen them, as they remain isolated and alone and the universal soul is not with them. So their thinking and deliberation then multiply and they only cast their glance on the particular things. This happens because of their separation from the whole and [because] they

[72] Belonging to *GS*; see ed. D'Ancona, pp. 242–4.

[73] Lewis' translation ('Plotiniana', p. 237) omits a portion of the text, namely 'as long as they desire the universal things, and direct and rule them together with the universal soul. If they forsake themselves, i.e. the universal for the particular'.

من أجل مفارقتها الكلّ وأنّها لزمت شيئاً واحداً جزئيّاً واستولت عليه وتركت الواحد الكلّيّ وفارقته وأقبلت على الواحد الجزئيّ 〈الذي〉 تدبّره بتعب ونصب . فإذا تعبت التعب كلّه ذهب حينئذ عنها النشاط وعرض لها الكسل والهموم فتسقط ريشها وتكون في البدن كأنّها في حبس وثيق ليس لها مخرج .

فإذا طال فيه مكثها وقويت على أن ترفع بصرها إلى فوق أي إلى العقل ورجعت إلى ذاتها أطلقت من وثاقها وحلّ عنها رباطها وصعدت إلى عالمها .

فلم يقدر الجسم الذي هي فيه أن يمنعها من ذلك وإنّما تبدأ في الصعود إلى 26b عالمها إذا ألقت عنها أوساخ الجسم الجزئيّ وتذكّرت العالم العقليّ . فإنّها حينئذ تتوهّم ذلك العالم والأنّيّات التي فيه وتبصرها ببصرها الشريف العالي فإنّ فيها شيئاً شريفاً لازماً لها دائماً كانت علواً أو سفلاً وهو العقل . فإذا رفعت النفس بصرها إلى العالم الأعلى وهي في هذا البدن كانت بمنزلة الحيّ البرّيّ البحريّ اضطراراً ، وذلك أنّها تشتاق إلى ذلك العالم وإلى هذا العالم السفليّ الدنيّ إلا أنّ منها ما يسير بسيرة العالم الأعلى أكثر ويسير بسيرة العالم السفليّ أقلّ ومنها ما يسير بسيرة العالم السفليّ أكثر ويسير بسيرة العالم الأعلى أقلّ . وإنّما تغلب على النفس السيرة العالية إذا ألقت بصرها على العقل كثيراً وإذا قلّ نظرها إلى العقل غلب عليها العالم الجزئيّ السفليّ .

وكلّما طال نظر النفس إلى العقل كانت الفضائل فيها أكثر وأظهر وكلّما طال نظرها إلى الجسم كانت الفضائل أقلّ وأخفى حتّى أنّه ربّما لم يكن بينها وبين الجسم فرق لشدّة اتّصالها به وبُعدها عن العقل . 27a

وقد ذكر ذلك افلاطن غير أنّه إنّما ذكره بلغز ، فإنّه قال إنّ البارئ لما قسّم الأنفس من الكائن الأخير وصيّرها كالأشياء الجزئيّة أهبطها إلى الكون

1 الكلّ M, D'Anc : or read الكلّيّ ? | جزئيّاً : جزويا M : جزءياً D'Anc | وتركت : ونزلت M, D'Anc 2 الجزئيّ : الجزوي M : الجزءي D'Anc 4 والهموم : والمهوم M : D'Anc | وثيق : ضيق M, D'Anc 8 الجزئيّ : الجزوي M : الجزءي D'Anc 8–9 فإنّها . . . العالم add. in marg. M 13 الدنيّ : الذي M : الأدنى D'Anc 15 العالية : الغالبة M crossed-out section on fol. 23b9 M | ألقت : الفت M crossed-out section on fol. 23b9 M 16 وإذا crossed-out section on fol. 23b10 M : او اذا M, D'Anc | قلّ : أقّل M : أقّل D'Anc | الجزئيّ : الجزي M : الجزءي D'Anc | السفليّ : الأدنى M : الأدنى D'Anc 19 الجسم : الحس M : D'Anc | after العقل blank space of two letters M 21 الكائن : الكاين M : الكأس D'Anc | الجزئيّة : الجزيه M : الجزءية D'Anc

are attached to only one particular thing and master it, give up the universal one and depart from it, and turn to the particular one which they direct with toil and fatigue. When they are completely tired, then activity leaves them and inactivity and anxiety befall them, so they let drop their feathers and are in the body as if they were in a solid prison inescapable for them. When they have remained there for long, have gathered the strength to raise their glance upwards, namely towards the intellect, and have returned to themselves, they are set free from their fetters, their bonds are untied from them and they ascend to their world. Then the body 26b
in which [the souls] are [imprisoned] is no [longer] capable of preventing them from [doing] that. They begin the ascent to their world when they cast away from themselves the filth of the particular body and remember the world of intellect. For they then imagine that world and the beings in it and see them with their noble and elevated glance, because in [the souls] there is something noble, always attached to them, be they above or below, and that is intellect. Thus when the soul raises its glance to the upper world while being in this body, it is necessarily similar to an amphibian. That is it desires that world as well as this base lower world, yet some of it lives the life of the upper world to greater and the life of the lower world to a lesser degree, while some of it lives the life of the lower world to a greater and the life of the upper world to a lesser degree. The elevated life[74] prevails in the soul whenever it casts its glance much on the intellect, whereas whenever its gaze at the intellect becomes less, the particular lower world prevails in it. The longer the soul's gaze at the intellect lasts, the more abundant and manifest are the virtues in it, and the longer its gaze at the body lasts, the less and the more imperceptible are the virtues. This may go so far that there is no longer a distinction between [the soul] and the body due to the force of its connection with it and its remoteness 27a
from the intellect.

Already Plato has mentioned that, yet he only mentioned it in enigmatic language. For he said that after the Creator had split the souls from the last being[75] and made them become like particular things, by necessity, He

[74] The scribe of the Marsh MS started to write this section of the text (until the end of fol.26b) on fol.23b9–13 by mistake, then realised his mistake and crossed it out. Thus we have a second textual witness for the passage indicated in the apparatus.

[75] Lewis ('Plotiniana', p.239) and D'Ancona (p. 244) translate 'goblet' apparently reading *al-kaʾs* after the Greek of Plotinus and inspired by Plato: 'when the creator divided the souls out of the last goblet' (*Enn.*, tr. Armstrong) – which may well be correct here, as on fol.43b where the MS reads in succession first *kaʾs* and then what might be either *bi-l-kāʾin* or *bi-l-kaʾs*. However, in the present case the MS reads unmistakably *al-kāʾin* which makes sense in its own terms.

اضطراراً إذ صارت جزئيّة ملائمة للكون . وقال إنّ لله تبارك وتعالى أهبط الأنفس إلى هذا العالم وألقاها فيه كما يلقي الأكّار البذر في بطن الأرض ، وإنما عنى الفيلسوف بقوله إنّ لله تعالى ألقى الأنفس في هذا العالم السفليّ كالبذور أي خلقها وصيّرها في الأجرام السفليّة والعالية ، وإنّما لفظ بهذا اللفظ كعادته فإنّ أكثر عادته كانت في وصف الأشياء إنّما كانت بحكاية وأمثال .

(43) فإذا سمعتَه يقول إنّ لله تعالى ألقى الأنفس إلى هذا العالم الحسّيّ كالبذور الذي يلقى في الأرض ثم سمعته يقول إنّ البارئ أرسل الأنفس إلى العالم
27b الحسّيّ ليضمنها فيكون الكلّ الجرميّ تامّاً كاملاً ، ثم سمعته يقول إنّ النفس في هذا البدن إنّما هي كأنّها في مغارٍ أو قبر ثم سمعته يقول إنّ النفس انهبطت إلى هذا العالم اضطراراً وإنّ النفس انهبطت إلى هذا العالم طوعاً ، فانسب هذه الأصناف كلّها إلى معنى واحد وأضفها إليه وهو أنّ الأنفس إنّما صارت في هذا العالم من تلقاء أنفسها وإرادة البارئ ليدبّر بها هذا العالم ويحسّنه ويزيّنه لأنّها علّة الصورة الحسنة الكريمة ، وليس من الواجب أن تكون العلّة تفارق معلولها وترفضه بل أن تلزمه وتمسكه ، وكينونة العلّة أي النفس مع معلولها البدن اضطراريّة استطاعيّة لأنّها علّة عقليّة . فإذا كانت العلّة عقليّة فعلت فعلها اضطراراياً استطاعيّاً أو استطاعيّاً اضطراريّاً ولا يمكن أن تفعل فعلاً مضطرّاً محضاً لأنّ ذلك من حيز الطبيعة ولا تفعل فعلاً استطاعياً لأنّ ذلك من شأن الفاعل الذي مرّة يفعل ومرّة لا يفعل . وليس هذا باستطاعة تامّة
28a وذلك أنّ الفاعل المستطيع التامّ لا يكون موجوداً وفعله غير موجود بل يكون وجوده مع وجود فعله سواء ، فهذه هي الاستطاعة التامّة التي يشبه الاضطرار في فعلها من غير أن يكون في حيز الاضطرار . فلهذه العلّة ربّما قال الفيلسوف إنّ الأنفس انهبطت إلى هذا العالم طوعاً وربّما قال إنّ الأنفس إنّما انهبطت إلى هذا العالم اضطراراً وليست تلك الاستطاعة معرّاة من الاضطرار

1 جزئيّة : جَزية M جزءية D’Anc   7 سمعته M : سمعت D’Anc   8 ليضمنها : ليضمرها M ليضمّرها D’Anc | الجرميّ M : الجزءي D’Anc   9 النفس ... في¹ add. in marg. M   10 انهبطت M : أهبطت D’Anc   12 أنفسها M : نفسها D’Anc   13 الصورة : لصورة M للصورة D’Anc   14 وكينونة : وكينونه M, D’Anc   15 اضطراريّة استطاعيّة : اضطراري استطاعي M, D’Anc   15–16 فعلها ... اضطراريّة add. in marg. M   16 اضطراراياً M : اضطراراً D’Anc | اضطراريّاً M : اضطراراً D’Anc   17 حيز : خير D’Anc   18 شأن : شان M ثبان D’Anc   20 وجوده M : وجده D’Anc | التامّة M : التماة D’Anc | التي : الذي D’Anc   21 حيز M : خير D’Anc

consigned them to [the realm of] generation, since they had become particular and consistent with generation. He said that God the Blessed and Sublime had consigned the souls down to this world and cast them into it just as the ploughman casts the seed into the depths of the soil. By saying that God the Sublime had cast the souls into this lower world like seeds, the philosopher meant He had created them and let them come to be in the lower and upper bodies. This type of expression was characteristic of him, for his chief characteristic was to describe things using stories and parables.

(43) So if you hear him say[76] that God the Sublime cast the souls into this sensible world like seeds which are cast into the soil, then you hear him say that the Creator sent the souls to the sensible world so that it includes 27b
them and the bodily universe is then perfect and complete, next you hear him say that the soul in this body is only as if it were in a cave or grave, and again you hear him say that the soul descended to this world by necessity and that the soul descended to this world voluntarily, then ascribe and assign all these kinds [of description] to one and the same meaning which is that the souls come to be in this world of their own accord and by the Creator's will so that through them He arranges, beautifies and adorns this world. For [the souls] are the cause of the beautiful and noble form. The cause must not depart from its effect and abandon it, rather it must be attached to it and hold on to it. The togetherness of the cause, namely the soul, with its effect, i.e. the body, is by necessity as well as willingly, because it is an intellectual cause. When the cause is intellectual, it performs its act either by necessity, yet willingly, or willingly, but restrained by necessity. It is not possible that it performs a purely necessitated act, because that belongs to the realm of nature; nor does it perform a free act, because that belongs to the matter of the agent who acts sometimes and sometimes does not act. This is not by perfect freedom of action, for the perfectly free agent is not existent when his act is inexist- 28a
ent, rather his existence is inseparably together with the existence of his act. This is the perfect freedom of action which resembles necessity in its action, yet does not belong to the realm of necessity. Therefore the philosopher sometimes said that the souls descended to this world voluntarily and sometimes said that the souls descended to this world by necessity. Neither is that freedom of action devoid of necessity nor this neces-

[76] Belonging to *GS*; see D'Ancona, pp. 244–6.

ولا هذا الاضطرار معرًّى من تلك الاستطاعة . وكذلك إذا سمعت اختلاف أقاويل الأوّلين في هبوط النفس إلى هذا العالم فلا تتوهّمنّ علته الاختلاف في علّة هبوط الأنفس إلى هذا العالم ولا أنّهم إنّما اختلفوا لأنّهم جهلوها ولكن ينبغي أن ينسب أقاويلهم كلّها إلى معنى واحد .

28b وقال امبذوقليس إنّي إنّما صرت في هذا العالم فرّاراً من سخط الله تعالى لأنّي أخطأت فخفت العقوبة ، وقد صرت إلى الشيء الذي كنت أخافه أي العقوبة وذلك أنّ هذا العالم إنّما هو حبس النفس وعقوبة لها .

وقال ارقليطوس إنّي لما أخطأت هويت إلى هذا العالم لأستريح فصرت منه إلى التعب الأكثر والنصب .

فهذه الأقاويل كلّها إنّما تجرّ إلى معنى واحد وهو الهبوط إلى هذا العالم طوعاً وكرهاً، طوعاً من أجل ذاتها لأنّها اشتاقت إلى أن تكون مع أفاعيلها وكرهاً لأنّها أرسلت من العلّة الأولى وهو بارئها ومبدعها .

(44) وقال الشيخ اليونانيّ انبجاس العقل من الأوّل لشدّة قوّته وسكونه لا لحركته ، لأنّ القوّة الشديدة ينبجس منها إمّا شيء جوهريّ كالحرارة من النار والبرودة من الثلج أو عرضيّ كالرائحة من ذي الرائحة ، لأنّه ما دام ثابتاً فتلك
29a الرائحة تنبثّ منه .

(45) وقال الاسكندر العقل الهيولانيّ بمنزلة قوّة على اقتناء الصناعة ، والمستفاد بمنزلة من اقتنى الصناعة ، والفعّال هو الذي به تصير الهيولانيّ له ملكة .

17–18 وقال ... الصناعة : قال العقل عند أرسطو على ثلاثة أضرب أحدها العقل الهيولاني .... وللعقل ضرب آخر ، هو الذي قد صار يعقل وله ملكة أن يعقل ، وقادر أن يأخذ (يأخذ ed. Bad : يدرك ed. Fin) صور المعقولات بقوّته في نفسه ، وقياسه قياس الذين يكون فيهم ملكة الصناعات القادرين بأنفسهم على أن يعملوا أعمالها . فإن الأوّل ما كان شبيهاً بهؤلاء ، بل بالذين فيهم قوّة يعلمون (يعلمون ed. Bad : يقبلون ed. Fin) بها الصناعة ، حتى يصيروا صُنّاعاً . وهذا العقل هو العقل الهيولانيّ بعد أن صارت له ملكةٌ واستعدادٌ (واستعداد ed. Bad : واستفاد ed. Fin) أن يعقل وأن يتقبّل (يتقبّل ed. Bad : يفعل ed. Fin) . وإنما يكون في الذين قد استكملوا وصاروا يعقلون . وأما الثالث – وهو غير الإثنين الموصوفين — *M. fī l-ʿAql*   18 والفعّال هو : وهذا هو العقل الثاني فهو الفعّال *M. fī l-ʿAql*

1 هذا : هذه M, D'Anc   2 علته : عليه M : على D'Anc   3 الأنفس M : النفس D'Anc
4 after واحد blank space of six letters M   5 امبذوقليس : اميذوقليس M, D'Anc | فرّاراً : فرار M, D'Anc   7 after لها blank space of six letters M   8 لأستريح M : لاستراحة D'Anc
12 after ومبدعها blank space of half a line M   16 after منه blank space of half a line M
17 اقتناء : اقنا M   18 به added above the line M

sity devoid of that freedom of action. Likewise when you hear the discrepant statements of the ancients about the soul's descent to this world, then do not imagine that they are caused by a disagreement with regard to the cause of the souls' descent to this world, nor that they only disagree because they do not know [better], but all their statements must rather be ascribed to one and the same meaning.

Empedocles said: I came to be in this world as a fugitive from the wrath of 28b
God the Sublime, because I erred and thus feared punishment, but [in doing so] had already got to what I feared, namely punishment. For this world is a prison and punishment for the soul.

Heraclitus said: after I had erred, I fell into this world to find rest, but through that I met with greater toil and fatigue.

All these statements are reduced to a single meaning which is that descent to this world is voluntary and unwilling, voluntary on account of the soul's essence for it has desired to be with its actions, and unwilling, because it has been sent by the First Cause, which is its creator and originator.

(44) The Greek Sage says:[77] the intellect's emanation from the First is due to the force of His power and rest, not due to His motion. For from the forceful power either something substantial emanates, like heat from fire and cold from ice, or something accidental like the fragrance from something possessing fragrance. For as long as it lasts, that fragrance spreads out from it. 29a

(45) Alexander says:[78] The material intellect is similar to the potential to appropriate some skill. The acquired [intellect] is similar to someone who has appropriated the skill. The active [intellect] is the one due to which the material intellect comes into possession of a certain habitus. Like light

[77] Belonging to *GS*; see Badawī, 189.13–17; and cf. partial parallel in passage (32) above.

[78] Cf. Alexander of Aphrodisias, *Fī l-ʿAql*, ed. Badawī, 31.15–34.2; and ed. Finnegan, 181.3–184.10, whose deviating readings are only given in the apparatus when crucial.

كالضوء الذي هو علّة للألوان المبصرة بالقوّة في أن تبصر بالفعل ، كذلك هذا الفعّال يجعل الذي بالقوّة عقلاً بالفعل بأن يثبت فيه ملكة التصوّر العقليّ .

(46) وقال الفارابي العقل الإنسانيّ هيولى القبول ، والمستفاد حصول المعقولات فيه ومصيرها قنية له ، والفعّال هو الذي يخرجه إلى الفعل ويجعل المستفاد في النفس ، وبه يحكم على جميع المعقولات وإليه يرجع في القضايا والأحكام . وهو مبدأ لهذين وبه يقول الإنسان إن في فطرة العقل كذا وأوّل في العقل . فالأوّل مادّة والثاني صورة والثالث فاعل .

(47) وقال بعضهم الوجود الأوّل الذي ظهر من قدرة الله تعالى العقل الفعّال ، 29b
ولذلك هو تامّ الوجود باق أبداً ثابت على حالة واحدة لا يتغيّر لأنّ الفيض متّصل به أبداً . ثم النفس بوساطة العقل ، فلذلك صارت ناقصة الوجود بإضافتها إلى العقل واحتاجت إلى الحركة شوقاً إلى تمامها . ثم الفلك بوساطة النفس ، فصار ناقص الوجود بالإضافة إلى النفس واحتاج إلى الحركة التي يستطيعها الجسم وهي حركة المكان ، فصارت حركته هي التي تتمّم وجوده له الدائم الذي قدّره الله تعالى له . ولما انتهى الوجود إلى أجسامنا كان بتوسّط الفلك وأجزائه فضعف جدّاً ، وحصلنا من الوجود الجسميّ على التكوّن الذي

1 كالضوء الذي : وقياس هذا الفاعل (الفاعل ed. Bad : الفعل ed. Fin) كما يقول أرسطو أيضاً
M. fī l-ʿAql, ed. Bad الألوان : للألوان | M. fī l-ʿAql — قياس الضوء لأنه كما أن الضوء
للتصوّر : التصوّر | M. fī l-ʿAql العقل الهيولانيّ add after يجعل | M. fī l-ʿAql العقل : الفعّال 2
منه : من ... تعالى | Fawz ونحن الآن قائلون إنّ : om. K. al-Ḥikma وقال بعضهم 9 M. fī l-ʿAql
لأزليّة مفيضه ، وسعة جوده ، فالعقل .add أبداً 11 after Fawz إنما حصل للعقل الأوّل المسمّى
إذن أبديّ الوجود وهو تمام الوجود ، بالإضافة إلى سائر الموجودات التي دونه ، وبالإضافة إلى المفيض
فلذلك ... | Fawz ولما كان وجود : ثم | Fawz عليه الوجود فإنّه ناقص عنه بالضرورة كما قلنا
Fawz فاحتاج : واحتاجت | Fawz بإضافته : بإضافتها 12 Fawz حصل ناقص : ناقصة
ولما : ثم الفلك | Fawz تمامه وتشبّها بالعقل وهو تمام بالإضافة إلى الأجسام الطبيعيّة : تمامها
Fawz فاحتاج : واحتاج | Fawz كان : فصار 13 Fawz حصل الفلك موجوداً
K. al-Ḥikma, Fawz له وجوده : وجوده له | Fawz الدوريّة وهي التي تتمّم : هي ... تتمّم 14
إلى .add الوجود after | Fawz وقلّ .add جدّاً after | Fawz وكواكبه .add وأجزائه 16 after
K. al-Ḥikma

3 after العقليّ blank space of six letters M 8 after فاعل vertical black line to indicate break M 11 after أبداً blank space of five letters M 15 after له blank space of three letters M

which causes the potentially visible colours to be actually seen, the active [intellect] renders [the material intellect], which is in potentiality, an actual intellect by establishing within it the habitus for intellectual conception.

(46) Al-Fārābī says: The human intellect is matter for the reception [of the intelligibles]. The acquired [intellect] is [the human intellect] attaining the intelligibles and [the intelligibles] becoming a habitus for it. The active [intellect] is the one which leads [the human intellect] to actuality and brings about the acquired [intellect] in the soul. Through it all intelligibles are judged, and it is consulted for assertions and judgements. It is the principle for these two [other intellects]. With regard to it man says that it is like that in the disposition of the intellect and it is principal among intellect. Thus the first [mentioned] is matter, the second form and the third the agent.

(47) One of them says:[79] the first existence which arose from the omnipotence of God the Sublime is the active intellect. Therefore it is perfect in 29b
existence, everlasting, remaining in the same condition and unchanging, because [its] emanation is always joined to Him. Next is the soul by intermediation of the intellect. In relation to the intellect, it therefore came to be deficient in existence and in desiring its perfection it needed motion. Next is the sphere by intermediation of the soul. So in relation to the soul, it became deficient in existence and needed the [kind of] motion bodies can perform, i.e. locomotion. So its motion came to be what perfects for it its perpetual existence which God the Sublime has ordained for it. When existence finally got to our bodies, it was by the mediation of the sphere and its parts, so it was very weak. Due to [our] corporeal existence we have received [a mode of] being that does not remain in the

[79] Excerpted from Miskawayh, alternative *Fawz*; cf. parallels in *K. al-Ḥikma*, MS E, fol. 56b2–13; and *Fawz*, 55.2–56.3.

لا يثبت على حال واحدة طرفة عين .

(48) وأجرام الموجودات متّصلة بعضها ببعض والكلّ واحد إذا أخذ من مركز
الأرض إلى أن ينتهي إلى السطح الأقصى من الفلك التاسع ، وهو حيوان 30a
واحد ذو أجزاء كثيرة مختلفة . وتقسيم أجزاء هذا الكلّ بالقسمة الأولى
ينقسم إلى قسمين إلى عالم الكون والفساد وهو عالمنا وإلى العالم الذي لا
كون فيه ولا فساد وهو السماء والأفلاك بما فيها من الكواكب .

(49) والموجودات كلّها تنقسم إلى قسمين جسمانيّ وروحانيّ ، فأمّا الجسمانيّة
فالأجرام العلويّة والأسطقسات وهي مخلوقة كرات إذ كان شكل الكرة أفضل
الأشكال وأشرفها وأبعدها من قبول الآفات . ولم يمكن أن تكون متفرّقة لأنّ
الكرات إذا تميّزت وتباعد بعضها من بعض وجب أن يكون بينها جسم آخر
أو خلاء ، والجسم الذي يحصل بين الكرات لا يكون كريّاً والخلاء ممتنع
وجوده أعني أبعاداً في غير مادّة ، فوجب الضرورة أن تحيط الكرات بعضها
ببعض على حسب ما هو موجود . وذلك أنّ كرة الأرض تحيط بها كرة الماء 30b
إلا ما انحسر عنها من شقّ الشمال ، وذلك لحكمة عظيمة جعل لها مركز
الشمس خارجاً من مركز الكلّ ، فقربت من ناحية الجنوب فجذبت الرطوبات
إلى هناك فحصل البحر في الجنوب وانجرّ الماء من الشمال بقدر ما تمّت به
العمارة ونشأ فيها الحيوان . وكرة الماء تحيط بها كرة الهواء وكرة الهواء تحيط بها
كرة النار وكرة النار تحيط بها كرة فلك القمر . ثم تحيط الأفلاك بعضها ببعض

1 لا يثبت : حصل هو كلّه وجود إذ كان غير باق ولا ثابت *Fawz* | واحدة om. *K. al-Ḥikma* : after واحدة add. ولا *Fawz* 2 وأجرام . . . متّصلة : فنقول أمّا اتّصال أجرام الموجودات … *Fawz* | والكلّ : وأنّ الكلّ *Fawz* 3 الأرض : الأصل *K. al-Ḥikma* | وهو : وأنّه *Fawz* 4 كثيرة om. *Fawz* | وتقسيم : فهو أمر قد فرغ منه الحكيم واستقصاه فأمّا تقسيم *Fawz* | after الكلّ add. فإنّه *Fawz* 6 بما : لما *Fawz* 7 والموجودات : فنقول إن الموجودات *Fawz* | فأمّا : أمّا *K. al-Ḥikma* 7–8 الجسمانيّة . . . وهي : الجسمانيّة فإنّها *K. al-Ḥikma* : الجسمانيّ فإنّها *Fawz* 8 إذ : إذا *Fawz* 10 بينها : بنمهما *K. al-Ḥikma* 11 أو خلاء : وخلاء *K. al-Ḥikma, Fawz* | بين : من *K. al-Ḥikma* 12 في : من *K. al-Ḥikma* | الضرورة : بالضرورة *K. al-Ḥikma, Fawz* 14 لحكمة : بحكمة *K. al-Ḥikma* 15 من[1] : عن *Fawz* 15–16 فجذبت . . . وانجرّ : وانحسر *K. al-Ḥikma* 16 وانجرّ : وانحسر *Fawz* | من : عن *Fawz* 17 after العمارة add. في الأرض *Fawz* 18 after الأفلاك add. المكوكبة *Fawz*

2 إذا M : or read إذ ? 4 أجزاء[2] : اجزآ M | بالقسمة : بالقسم M 6 السماء : السمآ M after الكواكب blank space of five letters M 9 الآفات : الافات M 10 آخر : اخر M 11 خلاء : خلآ M 18 فلك : الفلك M

same condition for a single moment.
(48) The bodies of the existents[80] are joined to each other and [all of them, i.e.] the universe is one, stretching[81] from the centre of the earth until it comes, at last, to the ultimate plane of the ninth sphere. It is one single 30a
living being possessing many different parts. In a first division, the parts of this universe are divided into two parts, namely the world of generation and corruption which is our world, and the world in which there is neither generation nor corruption and which is the heaven and the spheres with the stars in them.
(49) All existents are divided[82] into two parts: corporeal and spiritual. The corporeal are the upper bodies and the elements which are created spherical, since the shape of a sphere is the best, the noblest and the least damageable of all shapes. It is impossible that they are separate, because if spheres were set apart and became remote from each other, there would have to be either another body or a void between them. The body which might occur between spheres would not be spherical and the existence of a void, i.e. dimensions existing in non-matter, is impossible. So it is absolutely necessary that the spheres encompass each other according to 30b
what [spheres] exist.[83] For the sphere of earth is encompassed by the sphere of water with the exception of the northern part which became uncovered. That happened due to great wisdom because of which the centre of the [sphere of the] sun was put outside the centre of the universe. So [the sun] was close to the region of the south and drew moistures there. Thus the sea came to be in the south and water was drawn from the north to an extent which allowed habitation and cultivation to take place, and in there living being[s] were brought forth. The sphere of water is encompassed by the sphere of air, the sphere of air by the sphere of fire, the sphere of fire by the celestial sphere of the moon.

[80] Excerpted from Miskawayh, alternative *Fawz*; cf. parallels in *K. al-Ḥikma,* MS E, fols.56b13–57a1; and *Fawz*, 111.10–112.5.

[81] It would be tempting to read *uḥḥida* – ‘the universe is united from ... to’ as Sweetman seems to do, translating (p. 158): ‘all existing things from the centre of the earth to the topmost surface of the ninth heaven are one’. However, the Marsh MS, *K. al-Ḥikma* and the edition of the *Fawz* provide the dots for the indicated reading.

[82] Excerpted from Miskawayh, alternative *Fawz*; cf. parallels in *K. al-Ḥikma,* MS E, fols.57a1–58a8; and *Fawz*, 100.1–101.13, 102.4–15.

[83] I.e. the innermost sphere does not encompass anything else.

إلى أن ينتهي إلى فلك تاسع غير مكوكب ، يقال له فلك الأفلاك وهو يحرّك
الأفلاك الثمانية بحركة نفسه إلى خلاف جهات حركاتها ويديرها في كلّ يوم
وليلة دورة واحدة . ثم إنّ كلّ واحد من هذه الكرات بالإضافة إلى ما فوقها
31a هو كالثفل له وكالكدر ، وذلك أنّ الأرض بالإضافة إلى الماء كدورة وكالثفل
له ، وكذلك حال الماء عند الهواء وحال النار عند فلك القمر ، وعلى هذا
القياس فلك القمر عند ما فوقه إلى أن يبلغ إلى فلك الأفلاك غير المكوكب .
فهذه صورة الموجودات الجسمانيّة .

فأمّا القسم الآخر من الموجودات أعني الروحانيّة فإنّها وإن لم تكن مجسّمة فهي
أيضاً بنوع من الاعتبار اللائق بها محيط بعضها ببعض ولكن إحاطة روحانيّة
لأنّها غير محتاجة إلى مكان . فلذلك ينبغي أن يعتقد فيها أنّ إحاطتها إحاطة
اشتمال وتدبير ، وذلك مثل الطبيعة التي نقول فيها إنّها محيطة بالأجسام
الكريّات ، ولكن نريد إحاطة تحريك وتقدير واشتمال تصوير وتدبير لأنّها قوّة
31b إلهيّة سارية في الأجسام كلّها تدبّرها حتّى لا يفوتها شيء منها لا ظاهراً ولا
باطناً . ومن عرف كيف إحاطة الطبيعة بالأجسام كلّها عرف كيف إحاطة
النفس بالطبيعة وكيف إحاطة العقل الفعّال بالجميع وكيف يشتمل على الكلّ
مدبّر واحد .

وهذا القسم الروحانيّ من الموجودات من أجل أنّها ليست أجساماً وغير محتاجة إلى مكان فإنّ اتّصالها إذا اتّصلت لا يضيّق بعضها على بعض ولا يزيد فيها ولا ينقص أعني زيادة جسميّة . وإنّ ما عرض للأجسام أن يضيّق بعضها على بعض أمكنتها لأنّها تتّصل إمّا بالاختلاط ومجاورة الأجزاء وإمّا بالنهايات

2 إلى : وإلى *Fawz* 3 واحد : واحدة *Fawz* 4 كدورة : كدرة *Fawz* 5 after الهواء add. وحال الهواء عند النار *Fawz* 6 فلك[1] : نظن بفلك *Fawz* | عند : إلى *Fawz* | فوقه : فوق *Fawz* 8 فأمّا : وأما *K. al-Ḥikma* | الروحانيّة : الروحانيّات *K. al-Ḥikma* 9 من om. *K. al-Ḥikma* | بها om. *Fawz* 10 يعتقد : تعتقد *K. al-Ḥikma* : نعتقد *Fawz* 11 مثل : أن *Fawz* | التي om. *Fawz* 12 ولكن : ولكنّا فانا نعود : ولسنا نريد الإحاطة التي بيّنّاها في الأجسام الكريات *Fawz* 17 وهذا : فنقول إنّ هذا *Fawz* | الروحانيّ : أعني الروحانيّات *Fawz* 18 فيها : فيهم *Fawz* 19 جسميّة : جسمانيّة *K. al-Ḥikma* | وإنّ ما : وانّما *K. al-Ḥikma, Fawz*

4 الماء : المآ M 7 after الجسمانيّة blank space of three letters M 8 الآخر : الاخر M 11 نقول : يقول M | إنّها : النها M 12 نريد : يريد M 13 سارية : شارية M 16 after واحد blank space of half a line M 19 وإنّ ما : وانما M 20 الأجزاء : الاجزآ M

Then the celestial spheres encompass each other until the ninth, starless sphere is reached at last. It is called sphere of the spheres, moves the eight celestial spheres by its own motion towards their differently directed motions and rotates them once every day and night. Every single one of these spheres is in relation to what is above like sediment and turbidity. For earth is turbid in relation to water and like sediment. The situation of 31a
water with regard to air and of fire with regard to the celestial sphere of the moon is the same. According to this analogy the celestial sphere of the moon relates to what is above [and so on] until the starless sphere of the spheres is reached. This is then the form of the corporeal existents.

As to the other division of existents, i.e. the spiritual ones, they too, though not embodied, may be considered as encompassing each other in a manner appropriate for them, yet in a spiritual encompassment, as they are in no need of space. Therefore one must believe about them that their encompassment is by way of comprise and direction. This is like nature about which we say that it encompasses the spherical bodies. Yet we mean an encompassment by way of setting in motion and ordainment and comprising of formation and direction, for [nature] is a divine power pervading all bodies, which it directs so that none of them escapes it neither openly nor hidden. He who knows how nature encompasses all 31b
bodies knows how the soul encompasses nature, how the active intellect encompasses everything and how one director comprises the universe.

As the existents of this spiritual division are neither bodies nor in need of space, when they join, they join without constraining each other and without being augmented or diminished, i.e. by corporeal augmentation or [diminution]. It is the positions of the bodies which make them liable to mutual constraint, because they join either by the mingling and proximity of their parts or by the edges and the mutual contact of surfaces. In both

ومماسة السطوح ، وفي كلتا الحالتين تزداد مساحة بما تتّصل بها وذهاباً في
الجهات الثلاث . وإذا لم يكن هذا القسم الذي نحن في ذكره جسماً ولا
32a حامل طول وعرض وعمق لم يعرض له ذلك . ومثاله بالقياس الحسّيّ أنّ أنوار
الكواكب وشعاعاتها لا محالة كثيرة وهي واصلة إلى الهواء واختلافها بحسب
اختلاف ما تصدر عنه وليس يظنّ أحد أنّها تتضايق في الهواء ، ولو كانت
أضعافاً مضاعفة على ما هو عليه وإنّما يزداد الهواء بها استنارة فأمّا في أنفسها
فلا تزداد مساحة ولا يضيّق بعضها على بعض .

(50) قال ونحن واصفون النفس وكيف تبثّ الحياة في الكلّ وفي الأجرام الجزئية
وقائلون إنّ النفس البدنيّة ينبغي لها أن تكون هي الواصفة النفس الكلّيّة
العظيمة القوّة ، وأن تكون عظيمة الهمّة قد رفضت عنها جميع الأشياء الحسّيّة
ونفت عنها الأغاليط العارضة فيها وأن تكون ساكنة لا يهيّجها شيء من
32b الشهوات البدنيّة فيخرجها من حرّيتها ويستعبدها بل ينبغي لها أن تسلب
الحركات البدنيّة وتنقلها آلتها وأن يكون بدنها ساكناً لا يقبل شيئاً من الآثار
الحسّيّة الدنيّة المذمومة التي تردّ عليه ، وأن تكون الأشياء المحيطة به ساكنة لا
تهيّجه ولا تحرّكه بحركاتها الكثيرة المختلفة أعني أن تكون الأرض ساكنة والبحر
والهواء والسماء . فإذا توهّم المرء الأشياء الحسّيّة كلّها ساكنة ولم تتحرّك
بحركاتها الكثيرة الضوضاء فليتوهّم حينئذ النفس منبثّة عليه من خارج من جميع
النواحي مطلّة عليه بنورها كالشمس التي تلقي شعاعها على سحابة مظلمة ،
فتنيّرها فتخرّج نورها من خلال السحاب كالخيوط الذهبانيّة فتضيئ البصر
وتنيّره . كذلك النفس إذا ألقت شعاعها على جرم ما أضاءته وأنارته وصيّرته
33a حيّاً مضيئاً نيراً مملوءاً فضائل على نحو قوّة قبوله . وذلك أنّ النفس لما أشرفت

3 ومثاله . . . أنّ : وأنا أمثل هذا المعنى بمثال حسّيّ ليقرب إلى الفهم فأقول إنّ Fawz 5 اختلاف om. K. al-Ḥikma | عنه : عنها K. al-Ḥikma 6 هو : هي K. al-Ḥikma, Fawz | فأمّا : وأمّا K. al-Ḥikma 7 مساحة : ساحة Fawz | يضيّق . . . بعض : يضايق بعضها بعضاً Fawz

1 كلتا : كلتى M 2 الثلاث : الثلث M | ولا : مَالا M 5 يظنّ : نَطُن M 7 after بعض blank space of half a line M 8 after النفس add. لبدَنيَه and crossed out M | الجزئية : الجزوية M 9 النفس الكلّيّة : للنفس للكلية M 10 عظيمة : عظيم M 13 وتنقلها آلتها : ونيقلها اليها M : or read وتفعلها آلتها ? 19 فتضيئ : فيضي M 20 أضاءته : اضاتَه M 21 مضيئاً : مُضياً M | مملوءاً : مملواً M | أشرفت M : or read أشرقت ?

situations they increase in surface due to what is joined to them and by stretching in the three directions. But if this division which we mention is not a body and not bearing length, breadth and depth, that does not happen to it. To compare it to something sensible, it is like the lights and 32a
rays of the stars which are most certainly many when they enter air and differ according to their different origins, yet no one assumes that they become constrained in the air. Even if they were multiplied many times more than they are and the illumination of the air increased by them, they themselves would neither increase in surface nor constrain each other.

(50) He says:[84] we are describing the soul and how it spreads life in the universe and in the particular bodies, saying that the embodied soul must describe the universal soul which is great in power. It must be great in endeavour, having already rejected all sensibles and expelled from itself any faults which have befallen it. It must be at rest and none of the bodily passions must stir it, make it abandon its autonomy and enslave it. It must 32b
rather deny bodily motions and turn them into its instrument.[85] Its body must be at rest and not receive any of the base and blameworthy sensible impressions which are brought to it. The things which encompass [the body] must [also] be at rest and not stir or move it by their many different motions, that is earth must be at rest and [also] sea, air and heaven. If man imagines all sensibles at rest and not moving in their very noisy motions, he shall then imagine the soul spreading over him[86] from outside, from all directions, besprinkling him with its light like the sun,[87] which casts its rays on a dark cloud. So it illuminates it, then lets its light emerge through the clouds like golden threads and so lights and illuminates sight.[88] When the soul casts its rays upon some body, it likewise lights and illuminates it and lets it become living, luminous, illuminated and filled with virtues according to its receptiveness. For when the soul looms over[89] heaven, it 33a

[84] Belonging to *GS*.

[85] The reading is dubious here, and Lewis ('Plotiniana', p.263) indicates 'duo uocabula dubia'. The emendation of *wa-n-y-q-l-hā ilayhā* to *wa-tanqula* (or *wa-tafʿala*) *ālatahā*, however, seems to make some sense.

[86] Lewis ('Plotiniana', p.263) interprets the pronoun to refer to 'man' which is the most probable option in the Arabic. However in the Greek version, Plotinus is here already referring to heaven as he does later, which the Arabic then renders faithfully.

[87] Lewis ('Plotiniana', p.263) omits some text ('like the sun ... So it illuminates it,') here.

[88] Here *baṣar* seems to be a mistranslation for ὄψις (*Enn.* V 1, 2, 21) which is meant to express the golden 'look' which a dark cloud acquires when the sun's rays light it up.

[89] The Marsh MS clearly has *ashrafat*, but the reading expected in this context and translated by Lewis is *ashraqat* – 'shines, radiates upon'.

على السماء أفادتها نوراً ساطعاً وحياة بلا موت وحرّكتها بحركة دائمة . فلما أثّرت النفس في السماء هذه الآثار الشريفة وحرّكتها بحركة متّصلة دائمة صارت حيواناً سعيداً وإنّما صارت السماء سعيدة شريفة من أجل النفس الممتلئة عقلاً وحِلماً التي صارت فيها .

(51) والنفس وإن كانت في الشرف والفضل على الحال التي وصفنا بأنّها تؤثّر في الأجرام الآثار الحسنة وتفيض عليها الفضائل الكريمة فإنّها مثال للعقل ، وكما أنّ المنطق الظاهر إنّما هو مثال للمنطق الباطن الذي في النفس وكذلك النفس إنّما هي منطق العقل وفعلها كلّه إنّما هو بمعونة العقل والحياة التي تفيضها على الأشياء إنّما هي أثر من الآثار للعقل . وفعل العقل بأسره هو النفس ، والعقل والنفس إنّما هو بمنزلة النار والحرارة أمّا العقل فكالنار وأمّا النفس فكالحرارة المنبثّة من النار على شيء آخر ، غير أنّه وإن كان العقل
33b والنفس إنّما هو بمنزلة النار والحرارة فإنّ الحرارة إنّما تسيل من النار سيلاناً وتسلك سلوكاً إلى أن تأتي الشيء القابل لها فتكون فيه فأمّا العقل فإنّه ينبثّ في النفس من غير أن يسيل منه قوّة من قواه .

(52) والنفس إذا صارت في الأبدان لم تر الأشياء كما كانت تراها قبل وهي مجرّدة ، فتتحيّر وتتيه وتمتلئ هموماً وغموماً فتضعف وتلتجي إلى الفكرة والقياس المنطقيّ ، فإنّ الفكرة إنّما هي نقصان العقل وذلك أنّ العقل إذا احتاج إلى القياس والفكرة كان ناقصاً غير تامّ . وكذلك الصناعة التامّة لا يحتاج صاحبها إلى الفكرة لكنّه يفعل فعله بلا رويّة ولا فكرة ، فأمّا الصناعة الناقصة فإنّ صاحبها يحتاج إلى الفكرة والرويّة وذلك أنّه إذا أراد أن يفعل فعلاً ما وكان صانعاً ضعيفاً روّى وفكر كيف يعمل . وإذا كانت في عالمها استغنت عن

6 للعقل : من العقل *Uthūlūjiyā* 7 وكما : كما *Uthūlūjiyā* 7–8 إنّما ... هي : هو *Uthūlūjiyā*, ed. Bad : إنما هو ed. Die 8 بمعونة : بمعرفة *Uthūlūjiyā*, ed. Die 9 وفعل ... أثر : من *Uthūlūjiyā* | بأسره : بأسرها *Uthūlūjiyā* 9–10 هو النفس om. *Uthūlūjiyā* 10 إنّما هو : هما *Uthūlūjiyā* | after العقل add. الكليّ *Uthūlūjiyā* 10–11 وأمّا ... فكالحرارة : والنفس كالحرارة *Uthūlūjiyā* 11 وإن : إن *Uthūlūjiyā* 12 إنّما هو : هما *Uthūlūjiyā* 13 after تأتي add. إلى *Uthūlūjiyā* | فأمّا : وأما *Uthūlūjiyā* 14 قواه : قواها *Uthūlūjiyā*

2 الآثار : الاثار M 6 الآثار : الاثار M 9 الآثار : الاثار M 11 آخر : اخر M 13 فإنّه ينبثّ add. in marg. M : or read فإنّه يثبت ? 15 الأشياء : الاشيآ M 16 وتمتلئ : تمتلي M 21 after يعمل blank space of ten letters M

gives it radiant light and immortal life and moves it in perpetual motion. When the soul produces these noble impressions in heaven and moves it in a continuous, perpetual motion, [heaven] becomes a fortunate living being.[90] Heaven only becomes fortunate and noble on account of the soul full of intellect and insight which has come to be in it.

(51) For although the soul is in the condition of nobleness and refinement,[91] which we have described as producing good impressions in bodies and pouring the excellent virtues over them, it[92] is an image of the intellect. Just as the outer speech is only an image of the inner speech in the soul, in the same way the soul is the speech of the intellect, its entire activity is by the help of intellect and the life which it pours over the things is only one of the intellect's impressions. The intellect's activity in its entirety is the soul. The intellect and the soul are similar to fire and heat. The intellect is like fire and the soul like heat spreading from fire towards something else. However, even though the intellect and the soul are similar to fire and heat, heat flows and proceeds from fire, flowing and 33b
proceeding until it comes to its recipient and is then in it, whereas the intellect spreads[93] in the soul without any of its powers flowing from it.

(52) When the soul comes to be in bodies,[94] it does not see things as it saw them before when it was bodiless. Then it becomes confounded, perplexed and full of anxieties and worries, so it grows weak and takes refuge in thought and logical conclusion.[95] For thought is the deficiency of intellect, because when intellect needs logical conclusion and thought, it is deficient and imperfect. It is the same in the perfect craft: its master does not need thought, but performs his action without deliberation and thought, whereas the master of a deficient craft needs thought and deliberation. For if the weak craftsman wants to perform an action, he deliberates and thinks how to do it. But when [the soul] is in its own world, it is able to

[90] ζῷον εὐδαίμον in *Enn.* V 1, 2, 24.

[91] Belonging to *GS*; partial parallel to the second part in *ThA*.

[92] Cf. parallel to *ThA*, which continues until the end of passage (51), ed. Badawī, 108.17–109.6; and ed. Dieterici, 105.5–13. Minor differences between the two editions are not indicated in the apparatus.

[93] It is difficult to decide whether the Marsh MS reads *fa-innahu yanbaththu* or *yathbutu*, especially as it is added in the margin. Lewis ('Plotiniana', p.267) opts for the latter and translates 'whereas mind stays within the soul'.

[94] Belonging to *GS*.

[95] 'Thought and logical conclusion' (*al-fikra wa-l-qiyās al-mantiqī*) seems to render the Greek λογισμός.

34a الفكرة واستعمال القياس . فإن قال قائل إن كانت الأنفس في عالمها المجرّد
ليست بمفكّرة فكيف تكون ناطقة ، قلنا إنّما نعني بالنطق هاهنا النطق الكائن
بالقوّة وبالفكرة والرويّة فأمّا النطق العقليّ الكائن بالفعل 〈فهو〉 الذي لا
يفارق النفس بل هو معها دائماً ولا يفكر ، فالنفس ناطقة في العالم الأعلى
أيضاً . ولا ينبغي أن يظنّ ظنّ أنّ النفس وإن كانت في العالم الأعلى العقليّ
ناطقة تستعمل الأصوات لأنّه ليست هناك حاجة تدعوها إلى أن يكلّم
بعضها بعضاً بمثل هذا الكلام الذي يتكلّم به هاهنا ، ولا بينها خلاف في
درك الأشياء فيحتاج أن يعلّم بعضها بعضاً وأن يؤدّي أحدها إلى آخر ما
عنده من العلم بالنطق والصوت لكنّها كلّها هناك درّاكة بالسواء ، فإن احتاج
أحدهما إلى أن تخبر بما عنده علمه صاحبه من ساعته من غير أن احتاج إلى
34b أن يسمع منه مثل ما يحتاج هاهنا . وذلك أنّه إذا همّ أحدها بشيء علم
سائر الأنفس ما همّ به . فإن قال قائل فما يقول في الأنفس الكائنة في
الأجرام السماويّة فإنّها في أجرام لا تنطق كنطقها هاهنا فكيف يعلم بعضها
وَهم بعض وهي لا تستعمل النطق كاستعمالها هاهنا وليست بمجرّدة فيعرفه
بعضها ما يهتمّ به الآخر ، قلنا إنّ الأنفس السماويّة لا تحتاج إلى أن تنطق
ولا يعبر بعضها عن بعض بالكلام لأنّها إنّما تفعل أفاعيلها هناك فعلاً طبيعيّاً لا
اختلاف فيه بشرح وترتيب لا ينتقل من ذلك الشرح إلى غيره لأنّه في غاية
الصواب ، وأجرامها لا تنتقل ولا تستحيل وكذلك صارت أفاعيلها مستوية لا
اختلاف فيها ، فلذلك لا يحتاج بعضها إلى بعض فإذا لم يحتج بعضها إلى
بعض لم تحتج إلى النطق ولا إلى الكلام .

35a (53) والنفس إذا كانت في البدن لم تكن محصورة فيه ولا هي في عضو من
أعضائه لكنّها كلّها في كلّ عضو من غير أن تتجزّأ بتجزّؤ البدن ، وليست
في البدن بل البدن فيها لأنّها أبسط وأوسع من البدن وهي محيطة بالبدن
والبدن لا يحيط بها .

1 المجرّد add. above line M   5 ظنّ أنّ : ظنَان M   8 آخر : اخر M   9 بالسواء : بالسوآ M   14 فيعرفه M : or read فيعرف ?   15 الآخر : الاخر with blank space of three letters after it M   20 after الكلام blank space of half a line M   22 تتجزّأ : يتجزّى M | بتجزّؤ : يتجزّى M

dispense with thought and employing logic. If someone says: if the souls in their bodiless world are not thinking, how are they rational? We say: here by rationality we mean a rationality in potentiality,[96] one by thought and deliberation. As to the intellectual rationality in actuality, it is the one which does not depart from the soul, but is always with it and does not think. Thus soul is also rational[97] in the upper world. One must not assume that even though the soul is rational in the upper intellectual world, it employs voices, because no need there urges [the souls] to speak to each other with speech similar to what they use here. There is no difference between them in the perception of things, so that they would have to teach each other and convey their knowledge to each other by way of [rational] articulation and voice. Rather, there they are all equally perceptive. Thus if one of them needs to inform the other about what it knows, the other knows it at the same time without the need to hear it from [the first] as is required here. For if one of them is concerned about something, the other souls know what it is concerned about. Someone may say: what does one say about the souls which are in the heavenly bodies? For they are in bodies, but do not employ articulation like the one [employed] here. So how does one know what is in the mind of the other, when they do not employ articulation as they do here, yet are not bodiless, so that one of them may know what the other cares for? We say that the heavenly souls do not need to articulate, nor does one of them address the other by speech, for they perform their actions there naturally, neither differing in order and arrangement nor shifting from that order to another one, as it is most correct. Their bodies neither shift nor change and likewise their actions are equal and without difference. So none of them therefore needs the other and when they do not need each other, they do not need [rational] articulation or speech.

34a
34b

(53) When the soul is in the body,[98] it is neither confined in it nor is it in any of the body parts, but its entirety is in every body part without being divided as a body would be. It is not in the body, rather the body is in it, for it is simpler and more extensive than the body. It encompasses the body whereas the body does not encompass it. 35a

[96] Lewis' ('Plotiniana', p.37) translation, 'We reply: They can do without reason there', must be based on reading the text as *qulnā innahā taghnā ʿan ...* or similar, instead of *qulnā innamā naʿnī bi-...*

[97] The term *nāṭiq* is ambiguous in Arabic as it may express being either endowed with reason or with speech.

[98] Belonging to *GS*.

(54) ولكلّ قوّة من قوى النفس موضع معلوم من مواضع البدن تكون فيه ، لا لأنّها تحتاج إلى الموضع لثباتها وقوامها لكنّها إنّما تحتاج إليه لظهور فعلها من ذلك المكان المتهيّئ لقبول ذلك الفعل . والنفس هي التي صيّرت ذلك العضو متهيئاً لقبول فعلها لأنّها إنّما تهيّئ العضو بالهيئة التي تريد‹ها› أن يظهر فعلها منه ، فإذا هيّأت النفس العضو على الهيئة الملائمة لقبول قوّتها لظهرت قوّتها من ذلك العضو ، وإنّما تختلف قوى النفس على نحو اختلاف هيئة الأعضاء وليس للنفس قوى مختلفة ولا هيئة مركّبة منها بل هي مبسوطة ذات قوّة شريفة تعطي الأبدان القوى إعطاءً دائماً وذلك أنّها فيها بنوع بسط لا بنوع تركيب . 35b فلما صارت النفس تعطي الأبدان القوى نُسبت تلك القوى إليها لأنّها علّة لها ، وصفات المعلول أحرى أن تنسب إلى العلّة من إلى المعلول ولا سيّما إذا كانت شريفة تليق بالعلّة أكثر مما تليق بالمعلول .

ونقول إنّه ليس جزء من أجزاء النفس في مكان البتّة كانت النفس داخلة في البدن أو خارجة منه ، وذلك أنّ المكان يحيط بالشيء الذي فيه ويحصره وإنّما يحيط المكان بالشيء الجسمانيّ وكلّ شيء يحصره المكان ويحيط به فهو جسم . والنفس ليست بجسم ولا قواها بأجسام ، فليست إذن في مكان لأنّ المكان لا يحيط بالشيء الذي لا جسم له ولا يحصره ، وإنّما قلنا إنّ قوى النفس في أماكن معلومة من البدن نريد بذلك أنّ كلّ قوّة من قوى النفس تظهر فعلها من بعض أعضاء البدن .

ونقول إنّه ليست النفس في البدن كما يكون الشيء في الظرف فإنّها لو 36a

1 ولكلّ : فنقول إن لكلّ *Uthūlūjiyā* | قوى om. *Uthūlūjiyā* | موضع ... البدن : موضعاً معلوماً
Uthūlūjiyā 2 لأنّها : أنها *Uthūlūjiyā* | الموضع : الموضع *Uthūlūjiyā*, Lew : المواضع ed. Bad,
Die | إنّما om. *Uthūlūjiyā* 5 لظهرت : أظهرت *Uthūlūjiyā*, ed. Bad : وتظهر ed. Die 6 هيئة :
هيآت *Uthūlūjiyā* 7 هيئة : هي *Uthūlūjiyā* | شريفة : شريفة *Uthūlūjiyā*, Lew : om. ed. Bad,
Die 8 بسط : بسيط *Uthūlūjiyā*, ed. Die 9 نُسبت ... إليها : تنسب إليها تلك القوى
Uthūlūjiyā 10 من : منها *Uthūlūjiyā* | ولا : لا *Uthūlūjiyā* 12 after ونقول add. بقول
مستقصىً *Uthūlūjiyā* 13 يحيط : محيط *Uthūlūjiyā*, ed. Die 14 بالشيء الجسمانيّ : بشيء
جسمانيّ *Uthūlūjiyā*

3 المتهيّئ : المتهي M 4 متهيئاً : متهياً M | تهيّئ : تهي M | بالهيئة : بالهيَة M 5 الهيئة : الهية
M 7 هيئة : هيه M 8 بسط : بسيطٍ with *yāʾ* undotted M | تركيب : تركبيب M 10 من :
منها M 11 after بالمعلول blank space of five letters M 12 ونقول M : or read ويقول ?
جزء : جُزءٌ M 19 ونقول M : or read ويقول ?

(54) Each of the soul's powers[99] occupies a certain, definite position in the body where it is located, not because it needs the position to last and subsist, but only [because] it needs it to manifest its activity from that place which is prepared to receive such activity. It is the soul which has prepared that body part to receive its activity, for it prepares the body part in the shape in which [the soul] desires to manifest its activity. When the soul has prepared the body part according to a shape well receptive to its power, it manifests its power from that body part. The powers of the soul only differ with regard to the different shapes of the body parts [they employ], whereas the soul does not have different powers nor a shape which is composite of them. Rather it is simple, possesses a noble power and perpetually gives the bodies its powers. For [the powers] are [in the 35b soul][100] in a simple, not composite manner. So when the soul has started giving powers to the bodies, those powers are still linked to it, because it is their cause. The attributes of the effect are linked to the cause to a greater extent than to the effect, and especially when they are noble they are more befitting for the cause than for the effect.

We say that none of the parts of the soul is in any place at all, whether the soul be inside or outside the body. For place encompasses and confines the thing in it, but place only encompasses something bodily. Every thing confined and encompassed by place is a body. But the soul is not a body, nor are its powers bodies. Consequently they are not in a place, because place does not encompass and confine anything which has no body. However, we said that the powers of the soul were in certain places of the body, meaning by that that each of the soul's powers manifests its activity from some part of the body.

We say that the soul is not in the body as a thing is in a receptacle, 36a because, if it were like that, the body would not be ensouled. For if the

[99] Belonging to GS; cf. parallel in *ThA*, ed. Badawī, 41.7–16, 42.4–10, 42.14–18; and ed. Dieterici, 27.18–28.10, 28.19–29.7, 29.12–17. Lewis (*Re-examination*, pp. 56–7) lists two instances in which the reading of the Istanbul MS of *ThA* which according to him (*ibid.*, p. xx) is 'the oldest, embodies a far truer tradition than the rest, which all go back to one archetype' corresponds with the Marsh MS against all other MSS (given as 'Lew' in the apparatus).

[100] The sentence is ambiguous. Lewis ('Plotiniana', p. 43) translates 'For she is in them', and similarly Dieterici (*Theologie*, p. 28) 'weil sie in ihnen ... ist', which allows for two interpretations, i.e. the soul is in its faculties or in the bodies.

كانت كذلك لكان البدن غير ذي نفس ، وذلك أنّه لو كان البدن محيطاً
بالنفس كإحاطة الظرف بما فيه لزم من ذلك أن تكون النفس تسلك إلى
البدن قليلاً قليلاً كسلوك الماء إلى الظرف ولكان بعض النفس يضمحلّ كما
يضمحلّ بعض الماء الذي ينشّفه الظرف ، وهذا قبيح جداً .

(55) ويقول إنّ من الأوّلين من قال إنّ محلّ النفس من البدن محلّ الملّاح من
السفينة وما أحسن لما شبّه النفس بالملّاح ، وذلك أنّه كما أنّ الملّاح يدبّر
السفينة وهو مباين لها كذلك النفس هي مباينة للبدن وهي مدبّرة . فأمّا النوع
الذي نحن نطالبه فقائل هذا القول مخطئ فيه ، وذلك أنّ الملّاح في السفينة
بنوع عرضيّ لا بنوع طبيعيّ ، والدليل على ذلك أنّه لا يكون في السفينة كلّها
36b بدفعة واحدة مثل النفس في البدن لكنّه يكون في السفينة كلّها بالانتقال من
موضع إلى موضع . فأمّا نحن فنقول إنّ النفس في البدن مثل الصناعة في آلتها
أو مثل الملاحة في السُكّان ، لو كان السكّان متنفّساً والملاحة في داخله
تحرّكته حركة محكمة ، فأمّا إذا صارت الملاحة خارجة من السكان لا تقدر أن
تكون في داخله فليس الملاحة في السكّان بمنزلة النفس في البدن ، وذلك أنّه
لا تقدر قوّة الملاحة أن تنفذ في جميع السكّان لأنّ الملاح في السفينة بنوع
عرضيّ لا طبيعيّ . والقوّة التي في الشيء إذا كانت طبيعيّة نفذت في جميع
الشيء الذي هي فيه ، فالنفس إذا في البدن مثل ما الصناعة الطبيعيّة في آلتها
والنفس إنما هي بمنزلة الصناعة الطبيعيّة والبدن بمنزلة الآلة لها غير أنّ بين
37a النفس والصناعة الطبيعيّة فصلاً ، وذلك أنّ الصناعة الطبيعيّة ليست تباين
آلتها فأمّا النفس فبائنة من البدن ، فالنفس وإن كانت مباينة للبدن لكنّها لا
تقوى أن تفعل أفاعيلها إلا بآلة ما .

(56) وإنّما يكون ابتدأ فعلها حيث ابتدأ الآلة ، فإن كان هذا هكذا قلنا إنّ
الأوّلين إنّما جعلوا مكان النفس في العلو من البدن وذلك أنّهم رأوا الحسّ
والحركة والوهم إنّما هنّ قوى من قوى النفس الحسّيّة ، ثم رأوا الفكرة والمنطق

2 after النفس add. مما *Uthūlūjiyā* 4 ينشّفه : يشّفه *Uthūlūjiyā*, ed. Die

1 أنّه : انها M 3 كسلوك : لسلوك M 4 after جداً blank space of three letters M 7 مدبّرة M : or read تدبّره ? 13 تحرّكته : تحركه M 19 فصلاً : فضلا M 21 after ما blank space of three letters M

body encompassed the soul as the receptacle encompasses its contents, it would follow from this that the soul entered the body little by little as water enters a receptacle and some of the soul would disappear, just like some of the water that the receptacle absorbs disappears, and this would be very unseemly.

(55) He says[101] that one of the Ancients said that the soul is to the body as the navigator to the ship. How well [he did] when he compared the soul to the navigator. For just as the navigator directs the ship while being separate from it, the soul is equally separate from the body while directing [it]. Yet, as to the manner [of the soul's presence in the body], which we are after, the author of this comparison was mistaken. For the navigator is in the ship accidentally, not naturally. The proof for this is that he is not in the entire ship at one time, as the soul is in the body, but he is in the entire 36b
ship only by shifting from place to place. We, however, say that the soul is in the body as a skill could be in its instrument or as navigation in the rudder, if the rudder were ensouled, with navigation inside it so it would move it accurately. But as navigation occurs outside the rudder without the ability of being inside, navigation is not in the rudder like the soul is in the body. For the power of navigation cannot permeate all of the rudder, because the navigator is in the ship accidentally, not naturally. When the power in a thing is natural, it permeates the whole thing in which it is. Therefore the soul is in the body as a natural skill is in its instrument. The soul is similar to a natural skill, whereas the body is similar to an instrument of it, yet between the soul and a natural skill there is a difference. For a natural skill does not separate from its instrument, whereas the soul is 37a
separable from the body. So although the soul is separate from the body, it cannot perform its activities without some instrument.

(56) The beginning of [the soul's] activity[102] is where the instrument begins. If it is like that, we say that the Ancients placed the soul in the uppermost part of the body: for they deemed sense perception, motion and imagination to be powers of the sensible soul and then they deemed thought and rationality to be neighbours of these powers, but nobler than

[101] Belonging to *GS*.
[102] Belonging to *GS*.

مجاورين لهذه القوّة وهما أشرف منها فصيّروها أعلى منها وأرفع ، فلذلك قالوا إنّ النفس في أعلى طرف من أطراف الحيوان وهو الرأس . وليس أنّ النفس محصورة في الدماغ لكنّه لما كانت قوّة النفس الحسّيّة والوهميّة والفكريّة راتبة في الدماغ وكانت هذه القوى أفضل من سائر قوى النفس وأشرف قالوا إنّ مسكن النفس الدماغ من أجل أنّها تفعل أفاعيلها الشريفة من ذلك الموضع .

37b (57) فإن قال قائل إن كانت النفس تعلم الشيء الواحد المبسوط والمركّب الكثير القشور دفعةً واحدةً فكيف صارت ذات قوى كثيرة وصار بعض قواها أوّلاً وبعضها آخراً ، قلنا إنّ قوّة النفس واحدة مبسوطة وإنّما تتكثّر قواها في غيرها لا في ذاتها ، والدليل على أنّ قوّتها مبسوطة واحدة فعلها فإنّه واحد أيضاً والنفس وإن كانت تفعل أفاعيل كثيرة لكنّها إنّما تفعلها كلّها معاً . وإنّما تتكثّر أفاعيلها وتتفرّق في الأشياء التي تفعل فعلها ، فإنّها لما كانت جسمانيّة متجزّئية لم تقو على أن تقبل أفاعيل النفس كلّها معاً لكنّها قبلتها قبولاً متجزّئياً ، فكثرة الأفاعيل إذن في الأشياء لا في النفس .

(58) قال ولما اشتاقت النفس إلى أن تتعلّق بهذا العالم الحسّيّ لكثرة جولانها مع 38a جرم السماء وكنّا نحن هناك كنّا مُلائمين لها فتعلّقت بنا . وكذلك صارت في هذا العالم الحسّيّ وقد كنّا نحن في هذا العالم أيضاً وكان لكلّ واحد منّا صنم فيه إلا أنّه لا يظهر ذلك الصنم إلا في الزمان الذي يليق به أن يظهر فيه . فلمّا تعلّقت النفس بواحد من الناس العقليّين فعلق به جميع الناس العقليّين حتّى صاروا في حيّز النفس ومثل ذلك الصوت ، فإنّ الصوت واحد والكلمة واحدة فيأتي كلّ واحد منّا ، فيضع سمعه إرادة أن يسمع ذلك الصوت أو تلك الكلمة فإذا سمعت الصوت الواحد المسامع الكثيرة وصار الصوت في السمع كان بالفعل سمعاً وإنّما كان السمع بالفعل لأنّ الفاعل صار فيه ،

7 بعض قواها : بعضها *Uthūlūjiyā* 8 إنّ : لأن *Uthūlūjiyā* 9 قوّتها . . . واحدة : قواها واحدة مبسوطة *Uthūlūjiyā* 10 والنفس : فالنفس *Uthūlūjiyā* 11 تفعل : تقبل *Uthūlūjiyā* 12 متجزّئية : متحرّكة *Uthūlūjiyā* | على om. *Uthūlūjiyā* 13 متجزّئياً : متحرّكاً *Uthūlūjiyā*

3 راتبة : رايته M 7 الكثير : الكبير *bā'* also dotted as *thā'* M | كثيرة : كبيرة *bā'* also dotted as *thā'* M 8 آخراً : اخراً M 12 متجزّئية : مُتجَزيةً M 13 متجزّئياً : متجزياً M | after النفس blank space of half a line M 18 به M : or read بها ? 19 واحد[1] : واحدة M

[these others]. So they made them more elevated and exalted than them and therefore said that the soul was in the most elevated part of the animal, which is the head. It is not that the soul is confined in the brain, but since the sensible, imaginative and cogitative faculty of the soul are established in the brain, and these faculties are more virtuous and nobler than all other faculties of the soul, they said that the dwelling place of the soul was the brain because it performed its noble activities from that place.

(57) If someone says:[103] if the soul knows at one stroke the thing which is 37b
one and simple and the composite which has many pieces, how has it come to possess many powers, some of them occurring first and some last? We say that the power of the soul is one and simple and its powers are only multiplied in others, not in itself. The proof that its power is simple and one is its activity, because it is one as well. Although the soul performs many activities, it performs all of them together. Its activities become multiplied and separated only in the things which perform[104] [the soul's] activity. For as they are corporeal and divided, they are not able to receive all activities of the soul together, but receive them only in part. Therefore the multiplicity of activities is in the things, not in the soul.

(58) He says:[105] since the soul desired to be attached to this sensible world because of its many wanderings in the company of heaven's body, and
[since] we were there being in accord with it, it became attached to us. In 38a
this manner it came to be in this sensible world, where we already were as well, and for each of us there already existed an image in it, although that image only manifests itself at the appropriate time. When the soul has been attached to one of the intellectual people, then all other intellectual people cling to him[106] until they get to be in the realm of the soul. An example of this is voice. For voice is one and speech is one as well, yet it [voice as the composite of sound and speech] comes to each of us. So one sets up one's hearing willing to hear that voice or that speech; and when many ears have heard one voice and the voice has occurred in the hearing, there has been hearing in actuality. There has only been hearing in actuality, because the agent has occurred in [the hearing] and by the agent I

[103] Belonging to *GS*; cf. parallel in *ThA*, ed. Badawī, 32.3–9; and ed. Dieterici, 18.4–11.

[104] The following sentence and *ThA* suggest the reading *taqbalu* – 'the things which receive [the soul's] activity', instead of the actual reading of the Marsh MS, i.e. *taf'alu* – 'the things which perform'.

[105] According to Lewis belonging to *GS*, but the content does not sound Plotinian.

[106] A possible emendation of *bihī* to *bihā*, based on the very similarly structured sentence below, would allow for the reading 'then all other intellectual people cling to it [i.e. the soul]'.

أعني بالفاعل الصوت . وكذلك لما تعلّقت النفس بإنسان واحد من الناس العقليّين فعلقوا بها كلّهم أيضاً . فصرنا نحن منهما جميعاً غير أنّ أحد الجزئين الذين فينا لم يزل يحسّ بدئاً وبه تمّمنا أخيراً وصرنا على هذه الجهة الأخرى . 38b

(59) وقال النفس واحدة وكثيرة لأنّ في النفس الواحدة الأنفس كلّها ، لا أن الأنفس الكثيرة اختلطت بالنفس الواحدة لكنّها لما كانت من نفس واحدة فصارت كلّها واحدة ، يعني بذلك أنّ مخرجها واحد وليس أنّها خرجت من شيء بل كذلك أبدعت واحدةً وكثيرةً والكثرة في الواحد . وليس في عالم الأنفس ولا في جميع العالم الأعلى شيء واقع تحت الكون ولا تحت القسم والتجزّؤ . وإنّما يظنّ أنّه يقبل التجزّؤ إذا صار في الشيء القابل للتجزّؤ ، فأمّا إذا رجع الشيء المبسوط إلى ذاته فهو على ما هو من أوّل الدهر إلى الآن على حال واحدة.

(60) قال والنفوس الكثيرة تنبعث من النفس الواحدة الشريفة من غير أن تفنى النفس ولا تتجزّأ أو تباين النفوس بعضها بعضاً كالخاتم الذي ينطبع منه خواتيم كثيرة من غير أن يفنى نقش الخاتم . 39a

(61) قال والدليل على أنّ الأنفس كلّها واحدة أنّا قد يرحم بعضنا بعضاً ، فإذا رأينا أحداً قد نالته ضرورة أو آفة رثينا له وتوجّعنا ، وربّما أحببنا المرء الذي لم نره قطّ إلا لحظةً واحدةً . فلو لا أنّ الأنفس متّحدة لما رحم أحد أحداً ولا توجّع له عند ضرورته ولا أحبّه قبل معرفته .

(62) فإن قال قائل إن كانت أنفس الناس كلّها واحدة لنفس واحدة فمن الواجب إذا حسست شيئاً أن يحسّ بذلك الشيء إنسان آخر وإذا كنت خيراً فاضلاً أن تكون أنت خيراً فاضلاً وإذا وجدتَ من الآثار أثراً لذيذاً كان أو

2 الجزئين : الجزين M 3 يزل يحسّ : يزل نحن M | after الأخرى blank space of half a line M 5 الأنفس الكثيرة : النفس الكبيرة M 9 والتجزّؤ : التجزّى M | التجزّؤ : التجزي M | للتجزّؤ : للتجزِي M 13 تتجزّأ : يتجزّى M 14 يفنى : يعني M 16 آفة : افةً M | رثينا : رتينا M 17 أحد : احداً M 20 حسست M : or read حسنت ? | المرء : المرأ M | أحببنا : اجتينا M | آخر : اخر M | يحسّ : يحسن M

mean the voice. Likewise, when the soul has been attached to one man among the intellectual people, then all of them have become attached to it as well. So out of the two [parts?] we have become a whole, yet one of the two parts in us did not cease to perceive at first[107] and by it we are brought to perfection at last and become in accord with this other side. 38b

(59) He says:[108] the soul is one and many, because in the one soul there are all souls. It is not that the many souls mingle with the one soul, but since they are from one soul, all of them have become one. He means by that that where they come from is one, not that they come from a thing, but that they are created in that manner, i.e. as one and many and the multiplicity is in the one. There is nothing in the world of the souls nor in all the upper world which falls under generation or under particularisation and divisibility. He assumes that [something] becomes divisible, when it comes to be in a divisible thing, yet when the simple thing returns to itself, it is in the same condition in which it has been from the first eternity until now.

(60) He says:[109] many souls are sent out from the one noble soul, without the soul perishing and becoming divided or the souls separating from each other, just like the seal-ring from which many seals are imprinted, without the seal-ring's engraving perishing. 39a

(61) He says:[110] the proof for all the souls being one is that we have compassion for each other. When we see someone who is troubled by distress or harm, we pity him and feel sorry, and sometimes we love a man whom we have only seen for a moment. Were it not that the souls are united, one would not have compassion with the other and feel sorry for him at the time of his distress, and one would not love him before knowing him.

(62) Someone may say:[111] if all the souls of people are one because of one soul, it is necessary that whenever I feel something, another human being will feel that thing [as well], and when I am good and virtuous, you will be

[107] The reading is not obvious here, and the reference to the two parts is especially dubious. Does it still refer to the soul and the human being in which it manifests itself or is there a sudden change of subject and the soul and body in each of us are referred to? Lewis' translation ('Plotiniana', p.255) reads: 'Similarly, when the soul attaches herself to one of the men of mind, they all become attached to him, and we all become of their number. But one of the two parts which are in us is always ... (tria uocabula legi nequeunt) ... and we come to be on this other side'.

[108] Belonging to *GS*.

[109] Belonging to *GS*.

[110] Belonging to *GS*.

[111] Belonging to *GS*.

محزناً وجد جميع الناس مثله ، قلنا إن كانت النفس واحدة فالآثار الواقعة علينا
ليست بواحدة ، وذلك أنّي وإن كنتُ وأنت انساناً واحداً بالإنسانيّة فإنّه قد
يمكن أن تكون الإنسانيّة المتحرّكة فيّ واقعة فيك ساكنةً ، وليس ذلك بمحال
39b ولا يعجب أن أكون وأنت شيئاً واحداً بنوع من الأنواع وإن أكون قابلاً لبعض
الأشياء وأنت غير قابل له وإن أحسّ ببعض الأشياء وأنت لا تحسّ ، لأنّ كلّ
واحد منّا يقبل من ذلك النور على نحو قوّته وتهيئة لقبوله . ومثال ذلك ضوء
الشمس فإنّ الشمس تلقي ضوءها على جميع الأشياء ، فما كان منها أشدّ
صفاءً كان أكثر قبولاً لذلك الضوء ، وما كان منها مظلماً قليل الصفاء كان
أقلّ قبولاً لذلك الضوء ، فعلى نحو قبول كلّ واحد من الأشياء ضوء الشمس
يكون الضوء فيه .

(63) قال والنفس المدبّرة لهذا الكلّ الأرضيّ أشرف من الطبيعة لأنّها أشدّ بسطاً
منها ، ولذلك تلزم الطبائع كلّها وتدبّرها . فكما أنّ الطبيعة الواحدة تلزم
أجزاء الحيوان الكثيرة وتدبّرها كذلك النفس الواحدة الكلّيّة تلزم الطبائع
40a الكثيرة وتدبّرها . والطبائع تبع لهذه الأنفس متعلّقة بها وخارجة منها كأنّها
أغصان نبتت من شجرة واحدة .

(64) فإن سأل سائل عن الفصل بين النفس والطبيعة ، قلنا إنّ النفس هي

1 فالآثار : فَالاثَارُ M 5 أحسّ : احسن M | تحسّ : تحسن M 7 ضوءها : ضوها M
10 after فيه break (∴) ind. and blank space of more than half a line M 11 الطبيعة : الطبيعية M 13 وتدبّرها : مدبرهَا M 14 وتدبّرها M : may also originally read ومدبّرها in which a later hand dotted the mīm as a *yā*ʾ

good and virtuous [too], and whenever you[112] experience any impression, be it pleasurable or sad, all people will experience the same. We say: the soul is one, yet the impressions which befall us are not one. For even though I and you are one man by human nature, it is still possible that in me there is the nature of a moved human being, whereas in you there happens to be the nature of a human being at rest.[113] That is not unthinkable nor is it astonishing that you and I are the same thing in some respect,
even if I receive some things which you do not receive and although I feel 39b
some things which you do not feel. For each of us receives of that light according to his power and receptivity. An example of that is sunlight, for the sun casts its light on all things. Yet purer things receive more of that light, whereas darker and less pure ones receive less of that light. So the light in every single thing is in accord with its reception of sunlight.

(63) He says:[114] the soul directing this earthly universe is nobler than nature, for it is simpler than it. Therefore it is attached to all natures and directs them. Just as the one nature is attached to the many parts of the living being[115] and directs them, so is the one universal soul attached to the many natures and directs them. The natures follow these souls[116] and
are attached to them, yet are outside them as if they were branches 40a
growing from one tree.

(64) If someone asks[117] about the difference between soul and nature, we

[112] The Marsh MS is vocalised as *wajadta*, whereas the Greek has the corresponding verb in the first person singular, which could, according to the *rasm* alone, also be the intended meaning of the Arabic.

[113] I.e. the nature of a human being in motion versus the nature of a human being at rest. Cf. *Enn.* IV 9, 2, 4–6: ὡς ἄνθρωπος ὁ ἐν ἐμοὶ κινουμένῳ· ἐν ἐμοὶ γὰρ κινουμένῳ καὶ ἐν σοὶ μὴ κινουμένῳ ἐν ἐμοὶ μὲν κινούμενος, ἐν σοὶ δὲ ἑστὼς ἔσται·

[114] Belonging to *GS*.

[115] This statement probably results from an incomplete rendering of the Plotinian text (*Enn.* IV 4, 11, 8–9), in which all the parts of the universe are compared to the parts of a single living being. Replacing 'living being' by 'the realm of the living beings' would easily restore the intended meaning of the Greek and might even be the original Arabic.

[116] Cf. *Enn.* IV 4, 11, 9–11: Πάσας γὰρ τὰς φύσεις κρατεῖ μία, αἱ δὲ ἕπονται ἀνηρτημέναι καὶ ἐξηρτημέναι καὶ οἷον ἐκφῦσαι, ὡς αἱ ἐν κλάδοις τῇ τοῦ ὅλου φυτοῦ. – 'For one nature rules all the natures, and they come after it, depending on and from it, growing out of it so to speak, as the natures in branches grow out of that of the whole plant' (tr. Armstrong). As the Arabic is shorter, the question arises whether the passage suffered textual loss or was deliberately altered and shortened. If one assumes that what we have in the *PR* of IV, 4 is most of what there was, i.e. that, e.g., ch. 11 was followed by ch. 13, the textual change may be explicable as an attempt to connect the different parts of the text.

[117] Belonging to *GS*.

الأولى والطبيعة هي الأخرى ، والنفس هي الشيء الثابت القائم والطبيعة هي الشيء المستحيل . والطبيعة إنّما هي صنم النفس والنفس هي التي تبيّن الطبيعة . والطبيعة تشبّه بالشمع الذي يطبع به الخاتم فيكون بعض الطبع ناتئاً وبعضه منخفّضاً ، فالناتئ منه ظاهر بيّن والمنخفّض ضعيف لا يكاد يتأثر ، فالطبيعة تشبه رسم الطبع المنخفّض . فلذلك إنّما يفعل فقط من غير أن تعلم ما تفعل لأنّها تفعل الشيء بلا إرادة منها ، كما يفعل الحارّ في الشيء الملامس له ، فإنّه يفيض عليه من حرارته فيسخنه غير أنّه لا تكون الحرارة في المسخن مثلها في المسخن بل هي في المسخن أضعف وأقلّ . ولهذه العلّة صارت الطبيعة لا وهم لها ، فإن قال قائل ولا للعقل وهم أيضاً ، قلنا إنّ 40b العقل أعلى وأرفع من الوهم وأكرم منه وأفضل والوهم إنّما هو بين الطبيعة والعقل ، وذلك أنّ الطبيعة لا تعلم شيئاً والعقل يعلم علماً جوهريّاً . وأمّا الوهم فإنّه إنّما يعلم باكتساب لأنّه يفيد المتوهّم معرفة ما أثّر فيه . فإن كان هذا هكذا رجعنا فقلنا إن لنفس الكلّ عقلاً استفادته من فوقها ، والعقل هو حياتها لأنّها به تعلم ما تعلم وهو الذي أفادها الصورة النفسانيّة . فلمّا أفاض العقل على هويّة النفس قوّته تصوّرت النفس عند ذلك ، ثم أفاضت النفس قوّتها على الهيولى وأشرقت بشعاعها عليه تصوّرت الطبيعة حينئذ ، والطبيعة إنّما هي توهّم النفس وشعاع من شعاعها . والطبيعة آخر الأشياء العقليّة وذلك أنّ العقليّة تأتي من العقل الأوّل إلى أن تنتهي إلى الطبيعة فتقف هناك . 41a فأمّا الأشياء التي بعد الطبيعة فإنّما هي أصنام ورسوم الأشياء العقليّة غير أنّه ينبغي لنا أن نعلم أنّ الطبيعة تفعل وتنفعل من ذاتها ، فأمّا الشيء الذي قبلها هي النفس فإنّها تفعل ولا تنفعل لا من ذاتها ولا من الطبيعة .

1–2 الشيء . . . الثابت[2] add. in marg. M 2 هي[1] : هو M 4–5 ضعيف . . . المنخفّض add. in marg. M 4 يتأثر : يتئثر M 5 فالطبيعة : فاالطبيعة M 8 ولهذه : ولهذاه M 13 لنفس : النفس M 17 آخر : اخر M 21 after الطبيعة blank space of five letters M

say that soul is first and nature second.[118] Soul is the lasting, subsisting thing and nature the changeable thing. Nature is only the image of the soul and it is the soul which makes nature evident. Nature is compared to wax in which a seal-ring is imprinted: part of the imprint is proud and part sunken. The proud part is clear and evident, whereas the sunk part is shallow with hardly any impression, and nature resembles the pattern of the sunken imprint. Therefore it only acts without knowing what it does, because it acts without any will of its own, just as something hot acts on the thing in contact with it. For it pours some of its heat over it and thus heats it, yet the heat in the heated [object] is not like the heat in the thing heating [it], but it is weaker and there is less of it in the heated [object]. For that reason nature does not happen to have imagination. If somebody says: intellect also has no imagination, we say that intellect is more elevated and refined than imagination, nobler and better than it. Imagination lies between nature and intellect. For nature does not know anything, 40b
whereas intellect knows with its substantial knowledge. As for imagination, it only knows by acquisition, for he who imagines acquires knowledge of the impression which has been produced on him. If it is like that, we return [to where we were] and say that the soul of the universe has an intellect which it acquired from above. The intellect is its life, for, due to it, [the soul] knows what it knows, and it is [the intellect] which gives it the form of the soul. When the intellect has poured its power over the being of the soul, the soul has been formed at that [time]. Thereupon the soul has poured its power over matter and shone its rays over it, and then nature has been formed. Nature is the soul's imagining and one of its rays. Nature is the last of the intellectual things, for the intellectual things come from the first intellect until they finally reach nature where they stop. As for things which are posterior to nature, they are images and traces of the 41a
intellectual things, yet we must know that nature acts and is acted upon by itself.[119] As for the thing prior to it, namely the soul, it acts and is not acted

[118] The Greek has πρῶτον – ἔσχατον, for which the correct translation would be *ūlā* – *ākhira*. The Arabic version, however, seems here to stress the metaphysical hierarchy of being and thus the relation between soul and nature which may well be described as 'first and second'. The Plotinian text focuses more on the intellectual power of the higher principles in contrast to the somehow unintellectual power of nature, between which imagination is placed, which explains why the Greek has 'first – last'.

[119] Cf. *Enn.* IV 4, 13, 19–25: τὸ δὲ ἐξ αὐτῆς ἐμφαντασθὲν εἰς ὕλην φύσις, ... Ἀλλ' ἡ φύσις εἰς αὐτὴν ποιοῦσα καὶ πάσχουσα, ἐκείνη δὲ ἡ πρὸ αὐτῆς καὶ πλησίον αὐτῆς ποιοῦσα οὐ πάσχει, ἡ δ' ἔτι ἄνωθεν εἰς σώματα ἢ εἰς ὕλην οὐ ποιεῖ. According to the Greek 'nature acts on and is acted upon by matter', which the Arabic translation may have originally rendered, but which may have been lost in the transmission process.

وأمّا الشيء الذي فوق النفس وهو العقل فإنّه لا يفعل في الهيولى ولا في الأجرام لأنّ الهيولى إنّما هو أثر لأثره أي للنفس والأجرام أثر لأثر أثره أي للطبيعة ، لأنّ الطبيعة هي التي كوّنت الأجرام الأولى أعني الأصطقسات . فأمّا الحيوان والنبات فكيف كوّنت وهل للطبيعة فيها حظّ أم لا ، وإن كان لها حظّ فكيف ذلك ؟ أتراه كحظّ ضياء الشمس في الهواء ؟ فإنّه ما دامت الشمس مشرقة على الهواء فإنّ الهواء يستضئ بضوءها وإذا غابت الشمس لم يكن في الهواء ضياء البتّة ، لأنّ الشمس تفارق الهواء وتكون قائمة بحياله ، 41b وكذلك الهواء إذا فارقته الشمس قائم بحيالها وحده ، أم حظّ الطبيعة في الحيوان والنبات كالنار المسخنة للشيء ؟ فإنّ النار وإن فارقت الشيء الذي أسخنته لكن السخونة باقية في الشيء المسخن غير أنّ تلك السخونة هي غير السخونة في النار .

(65) ونريد أن نعرف كيف ينبغي أن يكون قبول البدن للنفس والطبيعة ، أمثل الهواء الذي استثار بضوء الشمس أم كالهواء المسخن من حرارة النار ؟ فيقول إنّه لا يحيا البدن من النفس إلا بتوسّط الطبيعة ، لأنّ الطبيعة معلول النفس والبدن معلول الطبيعة ، فالعلّة القصوى لا تكون في المعلول الأقصى إلا بتوسّط العلّة الدنيا . ويقول إنّه ينبغي للبدن أن يكون قابلاً للنفس مثل الهواء القابل حرارة النار ، فإنّه إذا أثرت النار في الهواء وسخنته ثم غابت عنه بقيت 42a حرارة النار في الهواء كالرسم ، وإذا كان الهواء على هذه الحال كان سريع القبول للحرارة أيضاً . كذلك ينبغي أن يكون البدن القابل للنفس متهيّئاً لقبول النفس والتهيّئة أن يكون في البدن رسم وظلّ من ظلّ النفس وهو الطبيعة كأبدان النبات والحيوان البهيميّة فإنّ [فان] فيها ظلّ النفس العاقلة . والدليل على ذلك أنّ أبدان النبات والحيوان تقبل الآثار والآلام ولا سيّما أبدان الحيوان البهيميّة فإنّها تقبل الآلام واللذات من غير أن يعرف الألم واللذّة . فأمّا نحن فإنّا نعرفها معرفة لا تؤثّر فينا ، أعني بقولى نحن النفس

4 فكيف : وكيف M 5 أتراه : اثراه M 6 بضوءها : بضوهَا M 7 بحياله : بخيالهَا M 8 بحيالها : بخياله M 10 أسخنته : اسخنه M 11 after النار blank space of more than half a line M 12 والطبيعة M : or read وللطبيعة ? 13 استثار M : or read استنار ? 16 after الدنيا blank space of three letters M 17 حرارة M : or read لحرارة ? | وسخنته : وسخته M 22 الآثار : الاثار M

upon either by itself or by nature.

As for the thing above soul, namely intellect, it acts neither on matter nor on bodies, for matter is only a trace of its trace, namely of soul, and the bodies are a trace of the trace of its trace, namely of nature. For it is nature which generated the first bodies, i.e. the elements. As for animals and plants, how were they generated? Has nature a share in them or not? And if it has a share, how so? Do you deem it to be like the share of sunlight in the air? For as long as the sun radiates towards the air, the air is lit by its light, and when the sun vanishes there is no light in the air at all. For the sun departs from air and stands opposite to it. In the same way, air stands 41b
opposite to it and alone whenever the sun separates from it. Or is the share of nature in animals and plants like fire which heats something? For even if fire departs from the thing which it has heated, heat remains in the heated thing, although that heat is different from the heat in the fire.

(65) We want to know[120] how the body must receive soul and nature. Is it like the air which is excited[121] by sunlight or as the air heated by the heat of fire? So he says that the body is enlivened by the soul only by the mediation of nature. For nature is the effect of soul and body the effect of nature, and the remotest cause is in the remotest effect only by the mediation of the more proximate cause. And he says that the body must be receptive to the soul as air is receptive to the heat of fire. For when fire has produced an impression on air and heated it, and has then vanished from it, fire's heat remains in the air as a trace. When air is in this condition, it is 42a
also quick to receive heat. In the same way the body which is receptive of the soul must be prepared for receiving the soul, and this preparation [consists of] the existence of a trace and a shadow of the soul's shadow, i.e. nature, in the body, just as in the bodies of plants and beasts there is a shadow of the intellective soul. The proof for this is that the bodies of plants and animals receive impressions and pains, especially the bodies of beasts, for they receive pains and pleasures without knowing pain and pleasure. As to us, we know them in a way that does not produce an impression on us. By saying 'we', I mean the rational soul which employs

[120] Belonging to *GS*.

[121] The Marsh MS reads *istathār* 'to be excited' which can easily be emended to *istanār* 'to be lit, illuminated' which would correspond to the Greek ὁ πεφωτισμένος (*Enn.* IV 4, 18, 5–6).

الناطقة التي تستعمل البدن والآثار الواقعة عليه واللذّات التي يلتذّ بها . وإنّما نألم بألم البدن لأنّه لنا وليس بغريب منّا ولسنا نحن البدن فقط ولا هو نحن بل البدن متعلّق بنا فأمّا نحن فإنّا بالشيء الأفضل الأشرف وهو النفس . 42b

(66) وقال إنّ أوّل أثر تؤثّره النفس إنما تؤثّره في الهيولى لأنها أول الأشياء الحسّيّة ، فلمّا كانت أول الأشياء الحسّيّة استوجبت أن تنال الخير من النفس أوّلاً ، وإنما أعني بالخير الصورة . ثم ينال بعد ذلك كلّ واحد من الأشياء الحسّيّة من ذلك الخير على نحو قوّته لقبول ذلك الخير . قال ولما قبلت الهيولى الصورة من النفس حدثت الطبيعة ، ثمّ صوّرت الطبيعة وصيّرتها قابلة للكون اضطراراً وإنما صارت الطبيعة قابلة للكون بما جعل فيها من القوّة النفسانيّة والعلل العالية ، ثم وقف الفعل عند الطبيعة ومبدأ الكون . فالطبيعة آخر العلل العقليّة المصوّرة وأول العلل المكوّنة ، ولم يكن من الواجب أن تقف العلل الفواعل المصوّرة الجواهر قبل أن تاتي الطبيعة . وإنما كان ذلك كذلك من 43a أجل العلّة الأولى التي صيّرت الأنّيّات العقليّة عللاً فواعل مصوّرة الصور العرضيّة الواقعة تحت الكون والفساد .

(67) وقال الهيولى آخر الأشياء الإلهيّة العقليّة ومبدأ الأشياء الجرميّة الحسّيّة غير أنّها بدء هيولانيّ وليست ببدء فاعل .

(68) وكذلك قال افلاطن إنّ الأشياء التي انبجست وابتدعت من ملك الأشياء كلّها وهو الملك الحقّ أولى وثانية وثالثة . أمّا الأشياء الأولى فهي التي انبجست من الأوّل الحقّ ، وأمّا الأشياء الثانية فهي التي أثّرتها [من] الأشياء الأولى ، وأمّا الأشياء الثالثة فهي التي أثّرتها الأشياء الثانيّة . أمّا الأشياء الأولى

4 وقال : فنقول *Uthūlūjiyā* 7 قوّته : قوّتها *Uthūlūjiyā*, ed. Die | قال ولما : ونقول لما *Uthūlūjiyā* 9 بما : لما *Uthūlūjiyā* 10 الفعل : فعل العقل *Uthūlūjiyā* | فالطبيعة : فالكون *Uthūlūjiyā* 11 من الواجب : يجب *Uthūlūjiyā* 12 الجواهر : للجواهر *Uthūlūjiyā* | قبل : من قبل *Uthūlūjiyā* 13 فواعل : وفواعل *Uthūlūjiyā*, ed. Die | الصور : للصور *Uthūlūjiyā*

1 والآثار : والاثار M : or read والآلام ? 2 نألم : يالم M | بغريب : بقريب M 3 after النفس blank space of half a line M 6 أوّلاً : الاول M 10 ومبدأ : وتبد M | آخر : اخر M 14 after والفساد blank space of half a line M 16 بدء : بدوٌ M | ببدء : ببدوٍ M 17 after وكذلك blank space of three letters M 18–19 أولى . . . الحقّ add. in marg. M 20 أثّرتها : اثر بها M | after الثانيّة blank space of five letters M

the body, the impressions befalling it, and the pleasures in which it delights. We only suffer the pain of the body because [the body] belongs to us and is not alien to us. However, we are not solely body and the body is not us but rather attached to us, whereas we are the more virtuous and 42b
nobler thing, namely the soul.

(66) He says[122] that the soul produces the first effect it produces in matter, because it is the first of the sensibles. Since it is the first of the sensibles, it deserves to attain the good from the soul first, and I mean by 'the good' [its] form. After that each one of the sensibles then attains from that good according to its potential to receive that good. He says: when matter received form from the soul, nature occurred. Then [the soul] formed nature and made it receptive to generation by necessity, and nature only became receptive to generation through the psychic power and the elevated causes put in it. Thereupon the activity ⟨of the intellect⟩[123] stopped at nature and the beginning of generation. So nature[124] is the last of the intellectual, formative causes and the first of the generative causes. The efficient causes, which form the substances, must not stop before they reach nature. It is only like that on account of the First Cause Who made 43a
the intellectual beings become efficient causes forming the accidental forms which fall under generation and corruption.

(67) He says:[125] matter is the last of the divine intellectual things and the starting-point of corporeal sensible things, yet it is a material, not an efficient beginning.

(68) Similarly Plato said[126] that the things which have been poured forth and originated from the King of all things, Who is the true King, are first, second and third. The first things are those which have been poured forth from the true First. The second things are those which the first things have produced as effects. The third things are those which the second things have produced as effects. The first things are the intellect and the

[122] Belonging to *GS*; cf. parallel in *ThA*, ed. Badawī, 86.15–87.6; and ed. Dieterici, 78.17–79.10.

[123] The *PR*'s reading *al-fiʿl* may result from a merging of an original reading *fiʿl al-ʿaql* preserved in *ThA*.

[124] *ThA* has 'generation' (*kawn*) instead of 'nature'. The *PR*'s reading seems to be more probable, but as there is no parallel Greek text it is impossible to be certain.

[125] According to Lewis, belonging to *GS*.

[126] Belonging to *GS*. Whether the introductory 'Similarly' is meant to belong to this passage is not clear, as it appears at the end of the preceding passage and before the gap which separates the two passages. To read it with the previous passage, however, does not make any sense.

فهي العقل والعالم العقليّ وإنّما انبجست من المبدِع الأوّل . وأمّا الأشياء الثانية فهي الهيولى والأجرام السماويّة وإنّما أثّرت من النفس بمعونة العقل . وإمّا الأشياء الثالثة فهي الأجرام المستحيلة والعالم المستحيل وإنّما أثّرت من الأجرام 43b السماويّة بمعونة النفس . ثم قال إنّ للعلّة علّة وإنّما عنى بالعلّة العقل وأنّ له علّة هي علّة العلل التي هي العلّة الحقّة لأنّه ليس من فوقها علّة أخرى . وإنّما سمّى افلاطن العقل علّة لأنّه علّة النفس ، وذلك أنّه قال إنّ العقل أثّر النفس في الكأس العظيمة الشريفة وإنّما عنى بالكأس الأنّيّة المبتدعة من العلّة الأولى وفي تلك الأنّيّة صوّر العقل النفس . فمن أجل ذلك قال إنّ العقل وإن كان علّة فإنّه معلول أيضاً وعلّته الخير المحض الذي فوق العقل وفوق الجوهر . فقد بان ووضح أنّ افلاطن علم علماً صحيحاً أنّ الخير المحض هو علّة العقل والعقل علّة للنفس والنفس علة للهيولى والأشياء الهيولانيّة .

(69) وقال إنّ افلاطن أراد بقوله إنّ لله تعالى خلق الأنفس أوّلاً ثم اضطرّها إلى 44a أن دخلت في هذه الأبدان أن الأنفس كانت عنده بأجمعها دائماً ، ثم أنّه ترك الأنفس تفعل أفاعيلها . فأوّل ما فعلت فإنّما فعلت الصورة في الهيولى ، ثم فعلت الصوّر الطبيعيّة ، ثم تركها لتكون هي التي تؤثّر أفاعيلها وآثارها وتمسكها وتلزمها وتدبّرها . فلهذه العلّة اضطرّت الأنفس إلى أن تصير في الأجرام وتتمسك بها .

(70) وقال ارسطالس في السماع الطبيعيّ الطبيعة مبدأ حركة وسكون فيما فيه ذلك على الأمر الأوّل بالذات لا بالعرض ، يعني به جميع انحاء الحركات والسكنات ووقوف الأسطقسات والمراكز والكلّيّات . والطبيعة تحرّك فتقصد غاية تقف عندها وتسكن . قوله على الأمر الأوّل فصل بينه وبين النفس ،

18–19 الطبيعة ... بالعرض : فتكون الطبيعة مبدأً ما وسبباً لأن يتحرّك ويسكن الشيء الذي هي فيه أولاً بالذات لا بطريق العرض *al-Ṭabīʿa*

1 المبدِع : المبدَع M | after الأوّل blank space of three letters M 2 بمعونة : بمعونت M after العقل blank space of three letters M 5 الحقّة : الحق M 7 الأنّيّة : الآنية M 11 والأشياء : والاشيأ M : or read ? الهيولى | للهيولى M : or read ? النفس | للنفس M : or read ? وللأشياء | after الهيولانيّة blank space of five letters M 12 اضطرّها : اضطَرّاها M 15 أفاعيلها : افاعيلُها M | وآثارها : واثارها M 17 after بها blank space of more than half a line M 18 after الطبيعيّ blank space of three letters M | مبدأ : مبداء M 19 انحاء : انحا M 21 after وتسكن blank space of three letters M

world of intellect, which have been poured forth from the First Creator. The second things are matter and the heavenly bodies, which have been produced as effects by the soul with the help of intellect. The third things are the changeable bodies and the changing world, which have been produced as impressions by the heavenly bodies with the help of soul. Then he said that the cause has a Cause, meaning by 'the cause' intellect, 43b
and that it has a Cause which is the Cause of the causes, i.e. the True Cause, because above it there is no other cause. Plato called intellect a cause, because it is the cause of soul. For he said that intellect produced soul as an effect in the great, noble goblet, meaning by 'the goblet' the being originating from the First Cause. In that being intellect formed soul. On account of that he said that although intellect is a cause, it is also caused, its Cause being the pure Good which is above intellect and above substance. Thus it has already become evident and clear that Plato correctly knew that the pure Good is the cause of intellect, intellect a cause of soul, and soul a cause of matter and the material things.

(69) He says[127] that by saying that God the Sublime created the souls first and then compelled them to enter these bodies, Plato meant that the souls 44a
were always with Him altogether, and that He then left the souls to perform their actions. So their first action was to effect form in matter, then they effected the natural forms. Thereupon He left them to be the ones to impress their actions and effects, to retain them, to be attached to them and to direct them. For this reason the souls were obliged to come to be in bodies and to adhere to them.

(70) Aristotle says in the *Physics*:[128] Nature is the principle of motion and rest in something in which that [principle] is primarily essential, not accidental. By that he means all sorts of motion and rest, the static quality of elements, centres and universals. Nature moves, pursues an end at which it stops and rests. He says 'primarily' to distinguish between

[127] According to Lewis, belonging to *GS*.

[128] Cf. Aristotle, *Phys.* 192b20–3 in Isḥāq's translation *al-Ṭabīʿa*, 79.7–9.

لأنّا إذا رأينا شيئاً تحرّكنا إليه فإنّ البصر تحرّك الاشتياق والاشتياق تحرّك القوى
الطبيعيّة والقوى الطبيعيّة تحرّك الحيّ ، فليس البصر حرّك أوّلاً . فإذا كانت 44b
النفس غير الناطقة لا تحرّك بالأمر الأوّل فكم بالحريّ النفس الناطقة ليست
محرّكة على الأمر الأوّل أي ليس بين العلّة والمعلول واسطة .

(71) وقال الطبيعة حياة تنفذ في الأجسام فيفعل فيها التخلّق وتدبّرها وهي
مبدأ حركة وسكون .

(72) وقال ابقراط الطبيعة محرّكة الأشياء وهي التي توصّل كلّ طبيعة إلى
جوهرها الملائم لها .

(73) وقال جالينوس الطبيعة هي القوّة الفعّالة الموجودة في كلّ شيء وهي
العقل الخاصّ به .

(74) وقال بعضهم الطبيعة اسم مشترك ، يقال بمعنى الذات كما يقال طبيعة
الحرارة أي ذاتها ، وعلى صورة كلّ شيء بها يفعل فعله الخاصّ به وهي مبدأ
حركته وسكونه كالنطق الذي يفعل الإنسان أفعاله الخاصّيّة به ، وعلى المادّة
القريبة لشيء شيء من المركّبات ، وهي أيضاً مبدأ الحركة والسكون . 45a

(75) وقال إنّ النفس تصير في العالم الحسّيّ الجسمانيّ من غير أن تفارق العالم
العقليّ . فكيف يمكن أن تفارق ذلك العالم وهي عقليّة ؟ وقد قلنا إنّ الشيء
لا يفارق عالمه إلا بفساد منه ، والنفس لا تفسد فهي إذاً غير مفارقة لعالمها
العقليّ وإن كانت في العالم الحسّيّ . وإذا قيل إنّ النفس فارقت البدن فليس
يعني أنّها فارقته فراقاً موضعياً لأنّها مباينة للبدن مباينةً طبيعيّةً والمفارقة إنّما هي
من حيّز الأجسام لا من حيّز الأشياء الروحانيّة . وذلك أنّ البدن إنّما يفارق
البدن وأمّا الشيء الروحانيّ فلا يقع عليه اسم المفارقة إلا بنوع عارية . فإن قال
قائل إن كانت النفس لا تفارق عالمها العقليّ فكيف يقول إنّها في هذا
البدن ؟ قلنا إنّ النفس مباينة لهذا البدن وإنّما تكون معه أو فيه ببعض صورتها 45b
التي بالفكرة والوهم والميل إليه . وأمّا سائر صورة النفس فإنّها راكدة قد

4 after واسطة blank space of five letters M 5 حياة : حيوه M | التخلّق : or read الخلق ? 6 after وسكون blank space of three letters M 8 after لها blank space of five letters M 14 after والسكون blank space of more than half a line M 21 after عارية blank space of three letters M 22 فكيف : وكيف M 24 التي : لتي M

[nature] and soul. For when we see something, we move towards it because vision has moved desire, desire moved the natural powers and the natural powers moved the living being. Thus vision has not moved 'primarily'. So 44b
if the irrational soul does not move 'primarily', then how much less does the rational soul move 'primarily', that is without an intermediary between the cause and the effect.

(71) He says: nature is life which permeates the bodies so that a natural constitution is formed in them, and directs them. It is the principle of motion and rest.

(72) Hippocrates says: nature is the mover of things and it is [nature] which joins each nature to its appropriate substance.

(73) Galen says: nature is the active power existing in everything and the intellect specific to it.

(74) One of them says: 'Nature' is a homonym. It is used to mean 'essence', like one speaks of the 'nature of heat', i.e. its essence. [It also refers] to the form of each thing by which it performs its specific activity and which is the principle of its motion and rest, just like rationality by which man performs his specific actions. [It also refers] to the proximate matter of something belonging to the composites. It is also the principle of motion and rest. 45a

(75) He says[129] that the soul comes to be in the sensible, corporeal world without departing from the intellectual world. Since it is intellectual, how could it depart from that world? We have already said that something only departs from its world by being corrupted. The soul does not corrupt, therefore it does not depart from its intellectual world, even if it is in the sensible world. When it is said that the soul departs from the body, it does not mean that it departs from it spatially, because it is naturally separate from the body, and departure only belongs to the realm of bodies, not to the realm of spirituals. For a body only separates from a body, whereas to the spiritual thing the term 'departure' applies only metaphorically. If someone says: unless the soul departs from its intellectual world, how can one say that it is in this body? We say that the soul is separate from this body and only with or in it by virtue of some of its form which is in 45b
thought, imagination and the inclination to it. As to what remains of the soul's form, it is motionless, having already been poured forth[130] in the

[129] Belonging to *GS*.

[130] Lewis ('Plotiniana', p.285) translates 'firmly fixed in the upper intellectual world'.

انبجست في العالم العالي العقليّ . وإنّما يكون منها في البدن الشيء الذي يقدر على تصوير وحيلة وليس عمله ولا فعله إلا في البدن فقط ، أعني أنّ النفس تصوّر البدن ببعض قواها وتحرّكه وتحيله وتقيمه وتلزمه وتحسّنه وتدبّره بغاية التدبير الملائم . فبهذه القوّة تكون النفس في هذه الأبدان وإنّما سائر قواها فإنّها لا تفارق العالم العقليّ بل تكون فيه دائماً .

(76) وقال إن جعلنا النفس ساكنة في مكان ما فإنّا لا نجد مكاناً أشدّ ملائمةً بها من الرأس لأنّ الرأس يحيط بها ، لكنّه إنّما عنى بذلك أنّ أفاعيلها إنّما تكون
46a هناك لأنّه مكان يتهيّئ لقبول أفاعيلها الفكريّة والمنطقيّة ، وأمّا أفاعيلها الحيوانيّة فإنّها تكون في القلب ، وأمّا أفاعيلها النباتيّة فإنّها تكون في الكبد . فلمّا صارت أفاعيل النفس الشريفة تكون في الرأس قالوا إنّ مسكن النفس الرأس ، ولم يعنوا بذلك أنّ النفس تحتاج إلى مكان تكون فيه بل عنوا أنّ النفس لما صارت في هذا البدن احتاجت إلى آلات تظهر بها أفاعيلها . فأمّا هي بعينها فليست بمحتاجة إلى آلة ولا أن تكون في مكان لأنّ المكان معلولها ، فالعلّة لا تكون في معلولها بل المعلول يحتاج إلى أن يكون في علّته .

(77) وقد قال افلاطن إنّ النفس داخلة في هذا العالم وخارجة منه لأنّها روحانيّة وليست بجسم . فهي داخلة في الأجسام وخارجة منها لأنّ الأجسام لا تحيط
46b بها بل هي التي تحيط بالأجسام لأنّها علّتها ، فالعلّة تحيط بمعلولها من غير أن يفوتها منه شيء .

(78) وقال ارسطالس النفس تمام لجسم طبيعيّ ذي حياة بالقوّة ، أراد بقوله تمام أنّ النفس صورة جوهريّة لا عرضيّة ، وبقوله جسم طبيعيّ أن يفصل

15 وقد . . . لأنّها : وهي بعده والنفس ... العقل أخرى نكتة *K. al-Ḥikma* 19 وقال ارسطالس . . . وذي : ذي ed. Arn | فحدّ النفس فقال : فانصرف إلى تحديد النفس فقال إن *K. al-Nafs*, MS A أمّا قوله تمام فأراد به : فقوله تمام ارادتُه *K. al-Nafs*, MS A 19–20 تمام . . . أراد[2] : وأمّا قوله لجسم *K. al-Nafs*, MS A 20 وبقوله . . . طبيعيّ : وقوله اسم طبيعي اراد به *K. al-Nafs*, MS A : طبيعيّ فأراد ed. Arn

2 وحيلة : or read وحلية ? 5 after دائماً blank space of more than half a line M 7 بها[1] : or read لها ? 8 after والمنطقيّة blank space of three letters M 9 after القلب blank space of three letters M 10 قالوا إنّ : قالوان M 11 عنوا أنّ : عنوان M 12 آلات : الاتٍ M 18 after شيء blank space of more than half a line M 19 حياة : حيوة M

elevated, intellectual world. In the body there is only that part of it which is capable of giving form and transforming, and its action and activity are solely in the body. That is, the soul forms the body by one of its powers, moves it, changes it,[131] sustains it, is attached to it, renders it good and directs it with the utmost appropriateness. So by this power the soul is in these bodies, whereas all its other powers do not depart from the intellectual world, but are always in it.

(76) He says:[132] if we believe the soul to be resident in some place, then we do not find a place more appropriate for it than the head, because the head encompasses it. Yet, he means by that that only [the soul's] activities are there, for it is a place receptive to its cogitative and logical activities. 46a
Its animal activities are in the heart and its vegetative activities in the liver. So as the activities of the noble soul have come to be in the head, they say that the soul's dwelling place is the head. By that they do not mean that the soul needs a place in which to be. Rather they mean that when the soul came to be in this body, it needed instruments to manifest its activities. As to the soul itself, it needs neither instrument nor being in a place, for place is caused by it, and the cause is not in its effect, but the effect needs to be in its cause.

(77) Plato has already said[133] that the soul is inside and outside this world, for it is spiritual and not a body. It is inside and outside bodies, because bodies do not encompass it, but it is [the soul] which encompasses the bodies. For it is their cause and the cause encompasses its effect[s] without any of them escaping it. 46b

(78) Aristotle says:[134] the soul is a perfection of the natural body possessing life potentially. He means by saying 'perfection' that the soul is a substantial, not an accidental form; by saying 'natural body' that he distinguishes

131 Lewis ('Plotiniana', p.285) translates 'adorns it'.

132 Belonging to *GS*.

133 According to Lewis, belonging to *GS*; cf. parallel in *K. al-Ḥikma*, MS E, fols.79b17–80a3, which, however, puts it in a different context.

134 Cf. source in *DA*-paraphrase, ed. Arnzen, 215.4–217.8; 219.1–4, 7–16; and a second recension in MS A, fols.67a4–b15; 68a11–b1, 68b4–69a3. As the Aya Sofya recension has not been edited yet, I give the reading variants of both, Arnzen's edition and the Aya Sofya (as MS A), and, in cases of only minor differences between the two, the latter.

الجسم الطبيعيّ من الجسم الصناعيّ مثل الباب والسرير ، وبقوله ذي حياة بالقوّة أن يفصله من الحديد والحجارة والأشياء التي ليست متهيّئة لقبول الحياة . وحدّ النفس بحدّ آخر فقال النفس كمال لجسم طبيعيّ آليّ ذي حياة بالقوّة ، أراد بآليّ أنّ له آلة جسمانيّة تقوى على قبول الحياة مثل القلب والكبد وما شاكلها من الأعضاء .

وقسم التمام فقال التمام على نوعين ، أحدهما مثل المرء العالم بالكتابة الفارغ فإنّه إذا شاء كتب ، والنوع الثاني مثل المرء لا يحسن أن يكتب وقد يمكنه أن
يتعلّم فيكون كاتباً . فالنفس تمام للجسم الطبيعيّ الآليّ بالنوع الأوّل مثل 47a
الكاتب الحاذق الذي إذا شاء كتب .

وقسم التمام بقسمة ثانية فقال التمام على نوعين ، مفارق وغير مفارق . أمّا التمام المفارق فمثل الملاح فإنّه تمام السفينة وهو مفارق لها من غير أن يدخل عليه فساد . وأمّا التمام الذي ليس بمفارق فمثل حرارة النار فإنّها إذا فارقت النار فسدت وبادت . فالنفس تمام للجسم الطبيعيّ تمام مفارق ، أعني أنّها مفارقة للجسم من غير أن تفسد أو تبيد .

1 مثل الباب : كالباب *K. al-Nafs*, ed. Arn | الباب والسرير : الاب وما اشبهه *K. al-Nafs*, MS A 1–2 وبقوله ... بالقوّة : وأمّا قوله ذو حياة بالقوّة أراد *K. al-Nafs* 2 والحجارة : الحجر *K. al-Nafs* | والأشياء ... متهيّئة : فإنهما وأن كانا من الأجسام الطبيعيّة فليسا متهيّين *K. al-Nafs*, MS A : وما أشبههما فإنهما وإن كانا جسمين طبيعيّين فليسا متهيأين ed. Arn 3 وحدّ النفس : فان النفس | فقال ... كمال : تمام لجسم طبيعيّ ذي حياة بالقوّة ثم قال إن نحن حدّدنا النفس إذا كان ذلك جائزاً لانّا إنّما عنينا بالآلة جسماً قلنا إنّها تمام *K. al-Nafs* 3–4 آليّ ... جسمانيّة : له آلات *K. al-Nafs* 4 قبول الحياة : أن يكون منها حياة *K. al-Nafs*, MS A : أن يكون بها حياً ed. Arn 5 شاكلها : أشبههما *K. al-Nafs* 6 after فقال add. إن *K. al-Nafs*, MS A | بالكتابة : بالكتاب *K. al-Nafs*, MS A 7 after المرء add. الذي *K. al-Nafs* | يحسن أن om. *K. al-Nafs*, MS A | وقد : فقد *K. al-Nafs*, ed. Arn | يمكنه : يمكن *K. al-Nafs*, MS A 8 للجسم ... الآليّ : لجسم طبيعي *K. al-Nafs*, MS A 9 الحاذق om. *K. al-Nafs*, MS A 10 وقسم : ثم قسم *K. al-Nafs*, MS A | بقسمة : قسمة *K. al-Nafs*, MS A | على نوعين : نوعان *K. al-Nafs* | أمّا : فأما *K. al-Nafs*, MS A 10–11 أمّا ... الملاح : فالمفارق كالملاح *K. al-Nafs*, ed. Arn 12 فساد : الفساد *K. al-Nafs* | after فإنّها add. تمام للنار *K. al-Nafs* | إذا : وانها إذا *K. al-Nafs*, MS A : فإذا ed. Arn 13 تمام[1] om. *K. al-Nafs* 14 مفارقة للجسم : تفارق الجسم *K. al-Nafs* | أو تبيد : وتبيد *K. al-Nafs*, ed. Arn

1 حياة : حيوة M 3 الحياة : الحيوة with blank space of five letters after it M | آخر : اخر M | لجسم : الجسم M | حياة : حيوة M 5 after الأعضاء blank space of six letters M 7 شاء : شآ M 9 after كتب blank space of ten letters M

the natural body from the artificial body, like a door and a bed; and by saying 'possessing life potentially' that he distinguishes it from iron, stones and things which are not receptive to life. And he defines the soul by another definition saying: the soul is the completion of a natural, instrumental body possessing life potentially. He means by 'instrumental' that [this body] has bodily instrument[s] which are able to receive life, like heart, liver and similar organs.

He divides perfection saying: perfection is of two kinds. One of them is like a man who knows how to write, [but] is idle [at the moment]. Yet, whenever he wants to, he writes. The second kind is like a man who is not capable of writing, but might possibly learn it and [could] then write. The 47a
soul is a perfection for the natural, instrumental body of the first kind, like a proficient writer who writes whenever he wants to.

In a second division he divides perfection saying: perfection is of two kinds, separable and inseparable. Separable perfection is like the navigator, for he is the perfection of the ship and separable from it without corruption coming upon him. Inseparable perfection is like the heat of fire, for whenever it separates from fire it dies down and perishes. The soul is a separable perfection of the natural body, i.e. [the soul] is separable from the body without corrupting or perishing.

فإن قال قائل إن كانت النفس هي التي تتمّ الجسم فتصيّره حيّاً ، فما بالها لا تسمى وحدها حيواناً إلا مع الجسم ؟ قلنا ذاك كالقوّة التي في العين ، فإن بها يتمّ العين فتصيّرها مبصرة ، ولا يقال للقوّة وحدها بصر إلا مع العين .

وقال التمام على أنواع وذلك أنّه إذا كان التمام لعضو واحد لا يعدوه صار
47b صورة لذلك العضو فقط مثل البصر ، فإنّه تمام للعين لا يعدوها إلى غيرها فهو صورة لها . وإذا كان التمام لعضو واحد ونفذ في جميع الأعضاء كان نفساً إلا أنّها نفس لا تفارق جسمها إلا عن فسادها مثل القوّة النامية والقوّة الحسّيّة ، فإنّ مسكنهما في أعضاء معلومة وينفذان في جميع البدن ، فإذا فارقا أبدانهما فسدا وبادا ، ولا يفعلان أفاعيلهما إلا مع حواملهما . وإذا كان التمام للجسم كلّه لا لعضو كان نفساً أيضاً إلا أنّها مفارقة لحاملها بلا فساد يدخل عليها للمفارقة مثل النفس المنطقيّة .

(79) وقال النفس جوهر مبسوط ، أتراها جوهراً كالهيولى أو كالصورة ؟ ثم قال الحياة لكلّ متنفّس والحياة تكون بالنفس ، والنفس هي التي تميّز بين الأشياء المتنفّسة وغيرها ، فالنفس إذاً صورة الأجسام المتنفّسة والجسم كالهيولى ،
48a فالنفس في الجسم المتنفّس جوهر كالصورة وليست فيه كالهيولى . وحدّ النفس

1 فإن : إن K. al-Nafs, ed. Arn | هي om. K. al-Nafs, MS A | تتمّ : تتمم K. al-Nafs, ed. Arn | فتصيّره : فيصير K. al-Nafs, MS A 2 وحدها : منفردة K. al-Nafs, MS A | ذاك : ذلك K. al-Nafs 2–3 فإن . . . يتمّ : فإنها تتم K. al-Nafs 3 العين¹ om. K. al-Nafs, MS A | فتصيّرها مبصرة : وبها تكون بصيرة K. al-Nafs | بصر : إنّها تبصر K. al-Nafs, ed. Arn 4 وقال : ثمّ قسم التمام قسمة رابعة فقال K. al-Nafs | أنواع : الأنواع K. al-Nafs, MS A | صار : كان K. al-Nafs, MS A 5 مثل : كفعل K. al-Nafs, ed. Arn | كان : صار K. al-Nafs | لعضو : بعضو K. al-Nafs, ed. Arn 6 لها : للعين K. al-Nafs, ed. Arn | يعدوها . . . غيرها : يعدوا العين K. al-Nafs, MS A 7 فسادها : فساد K. al-Nafs, MS A | ونفذ : فغلب K. al-Nafs, MS A | وإذا : فإذا K. al-Nafs, ed. Arn 8 مسكنهما : مسكنها K. al-Nafs, MS A 9 يفعلان أفاعيلهما : تفعلنّ افاعيلها K. al-Nafs, MS A 10 after لعضو add. واحد K. al-Nafs 10–11 بلا . . . للمفارقة : لأنها إذا فارقت جسمها لم تفسد ولم تبد K. al-Nafs 11 المنطقيّة : الناطقة K. al-Nafs 12 وقال النفس : ولا محالة أنها K. al-Nafs, MS A : فلا محالة أنّ النفس إذاً ed. Arn | أتراها : أفتراها K. al-Nafs | أو : أم K. al-Nafs, MS A | ثم قال : فقال إن K. al-Nafs, MS A : وبيّن ذلك بأن قال ed. Arn 14 فالنفس إذاً : فان النفس K. al-Nafs, MS A | الأجسام : للأجسام K. al-Nafs | after كالهيولى add. لها K. al-Nafs 15 الجسم : النفس K. al-Nafs, MS A | كالهيولى وحدّ : فيما أخبر أنّ النفس جوهر كالصورة حدّ K. al-Nafs

1 فما add. above line M 3 after العين² blank space of three letters M 6 لعضو : العُضوِ M 8 وينفذان : ويفذان M 11 after المنطقيّة blank space of six letters M 12 أتراها : اثرأهَا M | جوهراً : جوهرٌ M 13 الحياة : الحيوة M | والحياة : والحيوة M 14 المتنفّسة² : المنفسة M

If someone says: if it is the soul which brings the body to perfection and lets it come alive, why is it that the soul alone is not called a living being, but only together with the body? We say: that is like the power in the eye. For the eye is brought to perfection by it and it lets the eye see, but this power alone is not said to be vision except together with the eye.

He says: perfection is of [several] kinds, for when the perfection belongs to one body part which it does not leave, it becomes the form solely of that 47b
body part, e.g. vision. For it is the perfection of the eye which does not leave it to turn to something else and is thus its form. When the perfection belongs to one body part, but permeates all body parts, it is the soul, yet a soul which only departs from its body upon corruption like the growing and the sensible power. For they both have a dwelling place in certain body parts and permeate the entire body, but when they depart from their bodies, they corrupt and perish. They only perform their actions together with their bearers. When the perfection belongs to the whole body and not to [just one] body part, it is also the soul, yet it is separate from its bearer, without corruption coming upon it because it is separate, like the rational soul.

(79) He says:[135] the soul is a simple substance. Do you deem it to be a substance like matter or like form? Then he says: life belongs to every ensouled being and exists due to the soul. It is the soul which sets the ensouled things apart from the others. Therefore the soul is the form of the ensouled bodies, and the body is like matter. So the soul in the ensouled body is a substance as form, and it is not in [the body] as matter. 48a

[135] Cf. source in *DA*-paraphrase, ed. Arnzen, 213.10–215.4; and MS A, fols. 66b6–67a3.

بحدّ عامّ إلا أنّه نقل اسم الصورة إلى اسم التمام لأنّ التمام يَسِمُ الأشياء الجوهريّة فقط والصورة والحلى ربّما وَسَمَت الأشياء العرضيّة مثل الأشكال والحلى . فلذلك وصفت النفس بالتمام ولم يصفها بالصورة لئلا يظنّ ظانّ أنّ النفس عرض .

(80) وقال كلّ جسم من الأجسام فإنّ فعل جزئه غير فعل الجزء الآخر وفعل كلّه غير فعل جزء من أجزائه ، وذلك أنّ فعل البصر غير فعل الأذن وفعل الأنف غير فعل الفم . وليست النفس كذلك لأنّ الجزء منها يفعل فعل الكلّ والكلّ يفعل فعل الجزء لأنّ الكلّ والجزء منها يفعل فعّالاً واحداً لأنّه شيء واحد . وذلك أنّها تفعل بجوهرها كلّه على حالة واحدة وليس فيها أجزاء مختلفة فتكون أفاعيلها مختلفة ، ولا أجزاءها متشابهة فيكون فعلها أكثر وأقلّ
48b على نحو فعل الأجسام المبسوطة المتشابهة الأجزاء ، بل هي جوهر مبسوط وليست بجسدانيّة . فلذلك يفعل الجزء منها ما يفعل الكلّ ويفعل الكلّ ما يفعل الجزء ، فليست النفس بجسم لامتناع هذا في الجسم .

وقال لكلّ جسم بدء في نشوئه وغاية ينتهي إليها ، وأجود ما يكون الجسم إذا

1 يَسِمُ : متمّم K. al-Nafs, MS A : يسمّى ed. Arn 2 وَسَمَت ... والحلى[1] : ربّما سمت K. al-Nafs, MS A : إنّما تسمى ed. Arn 3 فلذلك : ولذلك K. al-Nafs, MS A | وصفت : وصف K. al-Nafs | ولم : وترك أن K. al-Nafs, ed. Arn 3–4 ولم[2] ... النفس om. K. al-Nafs, MS A 5 وقال : فقال إن وردّ عليهم الفيلسوف أيضاً K. al-Nafs | كلّ[1] : لكلّ K. al-Nafs, MS A | فإنّ om. K. al-Nafs 6 كلّه : على كليّته K. al-Nafs, MS A | جزء : واحد K. al-Nafs, MS A : جزء واحد ed. Arn | فعل[2] om. K. al-Nafs, MS A | الأذن : السمع K. al-Nafs, MS A 7 الأنف : اليد K. al-Nafs, ed. Arn | after الفم add. وفعل اليد غير فعل الرجل وفعل الرجل غير فعل البدن كلّه K. al-Nafs, MS A : البدن كلّه ed. Arn | after الجزء add. الواحد K. al-Nafs, MS A : add. فيها ed. Arn | الكلّ : الجزء الآخر K. al-Nafs, MS A 8 after الكلّ add. منها K. al-Nafs | لأنّه ... منها om. K. al-Nafs 9 وذلك ... واحدة om. al-Nafs, MS A | حالة : حال K. al-Nafs, ed. Arn | وليس : وليست K. al-Nafs 10 أجزاءها : أجزاء K. al-Nafs, ed. Arn 12 وليست : ليست K. al-Nafs | بجسدانيّة : بجسمانيّة K. al-Nafs, MS A : بذات أجزاء جسمانيّة ed. Arn | ويفعل : وفعل K. al-Nafs, MS A | ما يفعل[1] : فعل K. al-Nafs | يفعل[1] : فعل K. al-Nafs, MS A 12–13 ما يفعل[2] : منها ما فعل K. al-Nafs, MS A : فعل ed. Arn 13 فليست : وليست K. al-Nafs, MS A | بجسم ... الجسم : إذاً بجسم K. al-Nafs, ed. Arn 14 after وقال add. أيضاً إن K. al-Nafs, MS A : فقال وردَّ أيضاً عليهم ed. Arn | بدء : بدءاً K. al-Nafs, MS A | نشوئه :

15 after كالهيولى blank space of six letters M
4 النفس : للنفس M 5 جزئه : جزه M 9 أنّها تفعل : انه يفعل M 10 أجزاءها : اجزاوهَا M
13 after الجسم blank space of ten letters M 14 بدء : يَدْ with a *hamza* above the *dāl* M
نشوئه : نَشوِهِ M

He defines soul generally, yet he changes the term 'form' to the term 'perfection', as perfection marks only substantial things, whereas form and shapes may [also] mark accidental things like outward appearances and shapes. Therefore the soul is described as perfection, and he does not describe it as form, lest somebody may think that the soul is an accident.

(80) He says:[136] in each of the bodies the activity of [one of] its parts is unlike the activity of the other part, and the activity of its whole is unlike the activity of one of its parts. For the activity of vision is unlike the activity of the ear, and the activity of the nose unlike the activity of the mouth. The soul is not like that, because a part of it performs the activity of the whole, and the whole performs the activity of the part. For the whole [soul] and a part of it perform as one agent, because they are the same. For it acts by its entire substance in one condition, and in it there are neither different parts, so its activities would be different, nor are its parts homogeneous, so its activity would be more or less in the manner of the activity of the simple homoeomerous bodies. Rather, it is a simple 48b
substance and not corporeal. Therefore a part of it does what the whole does, and the whole does what the part does, and so the soul is not a body, because that is impossible in a body.

He says: In its development, every body has a beginning and an end which it ultimately reaches. The best a body [can] be, [is reached] when it has

[136] Cf. source in *DA*-paraphrase, ed. Arnzen, 203.9–205.5; and MS A, fols.62b10–63b3.

انتهى إلى غايته ، فإذا أخذ في النقصان ضعف . وليست النفس كذلك لأنّا نرى أنفس المكتهلين أكثر ضياءً وأجود فعلاً وأبدانهم أضعف من أبدان الأحداث . ولو كانت النفس جسماً لنقص فعلها بنقصان البدن ، وإذا كان الأمر كذلك فليست النفس جسماً .

(81) وقال إن كانت النفس مركّبة من الأسطقسات وإنّما تعلم الأجسام بالأجسام التي فيها ، قلنا إنّ كلّ جسم من الأجسام إنّما هو مركّب من الأسطقسات ، فما باله لا يكون ذا نفس علامة ؟ فإن لم تكن للأجسام أنفس علامة فلا محالة أنّ النفس ليست علامة لأنّها مركّبة من الأسطقسات 49a
بل إنّما تعلم الأجسام بعلّة أخرى . وقال إن كانت النفس مركّبةً من الأسطقسات والأسطقسات علّة هيولانيّة ، فما العلّة الفاعلة التي ركّبت الأسطقسات فحدثت النفس منها ؟ فإنّه لا يمكن أن يكون الشيء هو علّة انفعاله وتركيبه وإلا كان الشيء علّةً ومعلولاً وفاعلاً ومنفعلاً معاً وهذا غير ممكن . فإن لم تكن الأسطقسات علّة تركيب أنفسها فلا محالة أنّ لها علّة فاعلة هي التي تركّبها ، والعلّة الفاعلة هي أشرف من سائر العلل وليس في هذا العالم أشرف ولا أكرم من النفس . فالنفس هي التي تركّب الأسطقسات

نفس K. al-Nafs, MS A
1 فإذا : وإذا K. al-Nafs, ed. Arn | ضعف : ضعفت قوته K. al-Nafs, MS A 2 المكتهلين :
المتكهلين K. al-Nafs, MS A | وأجود : وافقه K. al-Nafs, MS A : وأنور ed. Arn 3 ولو : فلو
K. al-Nafs | بنقصان : بقدر نقصان K. al-Nafs, MS A 3–4 وإذا ... كذلك : فإذا كان هذا
كما ذكرنا K. al-Nafs, MS A 4 جسماً : بجسم K. al-Nafs 5 وقال ... كانت : فليست
K. al-Nafs, MS A : وردّ عليهم أيضاً فقال إن كانت ed. Arn | after النفس add. الآن K. al-Nafs,
MS A 6 قلنا إنّ : لنا K. al-Nafs, MS A : قلنا ed. Arn | إنّما هو : فهو K. al-Nafs, MS A : هو ed.
Arn 7 الأسطقسات : هيئة الاستقسات K. al-Nafs, MS A : هذه الأصول ed. Arn | فإن : وإن
K. al-Nafs | للأجسام : الأجسام لها K. al-Nafs, MS A 8 علامة[1] om. K. al-Nafs, MS A
علامة[2] : بعلامة K. al-Nafs, MS A | after الأسطقسات add. والاستقسات هيولانية K. al-Nafs,
MS A : أستقصات ed. Arn 9 بعلّة : لعلة K. al-Nafs, MS A | وقال : وردَّ عليهم أيضاً فقال
K. al-Nafs 12 after ومعلولاً add. معاً K. al-Nafs | ومنفعلاً : ومنفعلة K. al-Nafs, MS A
13 after ممكن add. قد بيّن الفيلسوف امتناع ذلك في كتاب سمع الكيان K. al-Nafs, MS A: بيّنا
كيف لا يمكن ذلك في كتاب سمع الكيان ed. Arn 14 فاعلة om. K. al-Nafs | هي[2] om.
K. al-Nafs, ed. Arn 15 أكرم ... أشرف : أكرم K. al-Nafs, MS A : أكرم ولا أشرف ed. Arn

2 أنفس : النفس M | المكتهلين : المتكهلين M | أبدان : الابدان M 4 after جسماً blank
space of three letters M 9 after أخرى blank space of three letters M 10 after
الأسطقسات add. بل انما and crossed out M 13 لها علّة : الفاعله M

ultimately got to its end. When it starts to fall short [of its end], it becomes weak. The soul is not like that, for we see the souls of men of mature age having more light and acting better, while their bodies are weaker than the bodies of the youth. If the soul were a body, its activity would fall short when the body falls short. Since the matter is like that, the soul is not a body.

(81) He says:[137] if [somebody claimed that] the soul was a composite of the elements and only knew the bodies through the bodies within itself, we [would] say that each of the bodies is, however, a composite of the elements. So why would it then not possess a fully knowledgeable soul? If the bodies do not have fully knowledgeable souls, then the soul is most certainly fully knowledgeable not because it is a composite of the 49a
elements, but it [must] know the bodies through another cause. He says: if the soul were a composite of the elements and the elements were a material cause, what would be the efficient cause which has composed the elements to give rise to the soul? For it is not possible that something is the cause of its [own] being acted upon and its being composed, otherwise it would be cause and effect, acting and being acted upon altogether, and this is impossible.[138] So if the elements are not the cause of their own composition, then they have most certainly an efficient cause which has composed them. The efficient cause is nobler than all other causes and in this world there is nothing nobler and more excellent than the soul. Thus it is the soul which composes the elements, and if the soul effects [their

[137] Cf. source in *DA*-paraphrase, ed. Arnzen, 207.18–209.13; and MS A, fols.64b8–65a14.

[138] Here the *DA*-paraprhase refers the reader to Aristotle's *Physica*, but, as with all other references in his source text, the compiler of the *PR* does not reproduce it.

فإن كانت النفس فاعلة فليست إذن بمركّبة من الأسطقسات .

(82) وقال النفس ليست بمشوبة بشيء من الأجسام أي ليست مختلطة بالجسم
49b كالأجسام والمركّبة ، ولا هي ممزوجة بالجسم ، ولا هي محمولة في الجسم
كالصور الهيولانيّة أعني الأشكال وأشباهها ، لكنّها في الجسم كالجوهر في
الجوهر أي كالملاح في السفينة . ولو كانت النفس في البدن كالأجسام
المركّبة والأشياء الممتزجة لما علمت شيئاً البتّة ، وذلك أنّ الشبه لا ينفعل من
شبهه وإذا لم ينفعل منه لم يحسّ به ولم يعلمه لأنّ الحسّ والعلم إنّما يكونان
بحركة المحسوس للحاسّ وبحركة المعلوم للعالم . والشيء الطبيعيّ لا يتحرّك من
شيء آخر مثله لأن ينفعل ، فإذا لم يتحرّك لم ينفعل وإذا لم ينفعل لم يحسّ
ولم يعلم .

(83) وردّ على من قال إنّ النفس تحدث من ائتلاف البدن ، فقال لو كان
كذلك لوجب أن يكون إذا نقص عضو من أعضاء البدن أن ينقص ائتلاف
البدن ، فإذا نقص ائتلاف البدن نقصت النفس بنقصان ائتلاف البدن ، فإذا
50a نقصت نقص فعلها وكان علمها بالأشياء خطأ ، وهذا باطل لأنّه ربّما نقص

1 بمركّبة : مركّبة *K. al-Nafs* 2 وقال : إنّ ولئلا يظنّ ظانّ أنّ الفيلسوف إنّما عنى بقوله *K. al-Nafs* بمشوبة بشيء : شيء *K. al-Nafs, MS A* | أي... مختلطة : لكنه عنى أيضاً وليست في الجسم أنها ليست في جسم لاكنه عنى بذلك أنّها ليست مختلطة : بذلك أنّها ليست *K. al-Nafs, MS A* بالجسم : بمختلطة *ed. Arn* | ممزوجة : ممتزجة *K. al-Nafs, MS A* | والمركّبة : المركّبة *K. al-Nafs* 3 هي[2] om. *K. al-Nafs, MS A* | بجسم *K. al-Nafs, ed. Arn* 4 كالصور : تجد في الصورة *K. al-Nafs, MS A* : كالصورة *ed. Arn* | وأشباهها : في أشباهما *K. al-Nafs, MS A* | لكنّها : التي كانها *K. al-Nafs, MS A* 4–5 في[2]... كالملاح : والجوهر وكالملاح *K. al-Nafs, MS A* 5 ولو... البدن : وكالأجسام المركّبة *K. al-Nafs, MS A* 5–6 كالأجسام... الممتزجة : يدرك الأشياء المتزّجة *K. al-Nafs, MS A* : كالأشياء الممزوجة وكالجسم المركّب *ed. Arn* 6 البتّة om. *K. al-Nafs, MS A* : بتة *ed. Arn* | الشبه : العقل *K. al-Nafs, MS A* 7 وإذا : فإذا *K. al-Nafs, MS A* | منه om. *K. al-Nafs* 8 للحاسّ : المحسوس للمحسوس الحاسّ *K. al-Nafs, ed. Arn* | المعلوم للعالم : للمعلوم العالم *K. al-Nafs, ed. Arn* | والشيء... لا : فلا *K. al-Nafs, MS A* 9 after آخر add. هو *K. al-Nafs* | لأن ينفعل : بالفعل *K. al-Nafs* | فإذا : وإذا *K. al-Nafs, ed. Arn* | وإذا لم : ولم *K. al-Nafs, MS A* 11 after وردّ add. أيضاً *K. al-Nafs* 11–12 فقال... لوجب : قلنا يجب من هذا *K. al-Nafs* 12 after عضو add. واحد *K. al-Nafs* | ينقص : ينتقص *K. al-Nafs* 13 فإذا[1] : وإذا *K. al-Nafs, ed. Arn* | فإذا[2] : وان *K. al-Nafs, MS A* : وإذا *ed. Arn* 14 after نقصت add. النفس *K. al-Nafs* | فعلها وكان : علمها وكان *K. al-Nafs, MS A* : فصار ⟨أيضاً⟩ عقلها *ed. Arn*

1 after الأسطقسات blank space of three letters M 2 ليست[1] : ليس M 6 الشبه : الشبئة M 9 آخر : اخر M | لأن : فان M 10 after يعلم blank space of six letters M 13 نقصت : نقص M 14 بالأشياء : بالاشيآ M

composition], then it is not a composite of the elements.
(82) He says:[139] the soul is neither mixed with any of the bodies, i.e. not mingled with the body like the bodies and composites [are], nor blended 49b
with the body nor borne in the body like material forms, i.e. shapes and the likes of it. Rather it is in the body like a substance in a substance, i.e. like the navigator in the ship. If the soul were in the body like composite bodies and blended things, it would not know anything at all. For like is not acted upon by like, and when it is not acted upon by it, it neither perceives nor knows it, because sense perception and knowledge are only through the motion of the perceptible for the perceiver and through the motion of the knowable for the knower.[140] The natural thing is not moved by another thing like it, so that it is acted upon. When it is not moved, it is not acted upon, and when it is not acted upon, it neither perceives nor knows.
(83) He refutes[141] someone who says that the soul is brought forth by the harmony of the body and says: if it were like that, it would necessarily follow that when one of the body parts is deficient, the harmony of the body is deficient; and when the harmony of the body is deficient, the soul is deficient because of the deficiency of the harmony of the body; and when [the soul] is deficient, its activity is deficient and its knowledge of things faulty. This is absurd, because body parts may be deficient, while 50a

139 Cf. source in *DA*-paraphrase, ed. Arnzen, 315.8–18; and MS A, fols.95b15–96a14.

140 The *PR* agrees with the Aya Sofya version of the *DA*-paraphrase here, whereas the version edited by Arnzen has 'the motion of the perceiver' and 'the motion of the knower'. Thanks to the ambivalence of *li-* the meaning can still be interpreted similarly, cf. Arnzen's translation, p.314: 'mittels der Bewegung des Wahrnehmungsorgans durch den Wahrnehmungsgegenstand bzw. mittels der Bewegung des Erkennenden durch den Erkenntnisgegenstand'.
A similar passage in the Arabic translation of Aristotle's *DA* 424a30 (60.12–13) reads:
الحركة التي تصل من المحسوس إلى الحاسّ .

141 Cf. source in *DA*-paraphrase, ed. Arnzen, 209.14, 211.5–10, (211.12–15); and MS A, fols.65a14–15, 65b11–66a3, 66a7–10.

أعضاء البدن والعقل صحيح على حاله ، فليست النفس إذن من قبل ائتلاف البدن .

وقال إن كانت النفس من قبل مزاج البدن فزاغ البدن عن حال اعتداله ، وجب أن تزيغ النفس وتنتقل عن حالها والأمر بخلاف ذلك ، فليست النفس من قبل مزاج البدن .

(84) وقال أبو علي مسكويه النفس جوهر بسيط وليست بجسم ولا عرض ، وهي محرّكة ذاتها باقية أبداً لا تبطل ولا تموت . فأمّا البرهان على أنّها ليست بجسم فإنّ الجسم إذا قبل صورة لم يمكنه أن يقبل صورة غيرها من جنسها إلا بعد أن يخلع الصورة الأولى ويفارقها على التمام ، ومثال ذلك أنّ الفضّة إذا قبلت صورة الجام لم يمكنها أن تقبل صورة الكوز إلا بعد أن تزوّل عنها صورة الجام وتخلعها خلعاً تماماً ، وكذلك الشمع إذا قبل صورة نقش ما لم يمكنه أن [50b] يقبل صورة نقش آخر إلا بعد أن تنمحي عنه صورة النقش الأوّل وتفارقه مفارقةً تامّةً . وإذا وجدنا شيئاً يخالف حاله هذه الصفة ، أعني أنّه يقبل صوراً كثيرةً من غير أن يبطل شيء منها ، تبيّن لنا أنّه ليس بجسم .

ومن الدليل على أنّ هذا المعنى ليس بجسم أنّ جميع أعضاء الحيوان من الإنسان وغيره ، صغر منه أم كبر ، ظهر منه أو بطن ، إنّما هو آلة مستعملة لغرض لم يكن ليتمّ إلا به . وإذا كان البدن كلّه آلات ولكلّ آلة منها فعل

بعض .add after نقص[2] | K. al-Nafs, MS A وذلك أنه : لأنّه | K. al-Nafs, MS A خطأ : باطل K. al-Nafs

1 قبل om. K. al-Nafs, MS A 3 وقال om. K. al-Nafs, MS A | البدن[2] om. K. al-Nafs, MS A عن حال : من K. al-Nafs, MS A 4 وجب ... تزيغ : كان واجباً أن تولع K. al-Nafs, MS A وتنتقل : وترجع K. al-Nafs, MS A | والأمر ... ذلك : وهذا خطأ K. al-Nafs, MS A 6 وقال ... مسكويه om. K. al-Ḥikma 6–7 وقال ... تموت : في اثبات النفس وأنها ليست بجسم ولا عرض Fawz 7 محرّكة : متحرّكة K. al-Ḥikma 7–8 فأمّا ... فإنّ : فأقول إن من الأشياء البيّنة الواضحة أن Fawz 8 فإنّ : أن K. al-Ḥikma 10 الجام : الخاتم K. al-Ḥikma 11 الجام : الخاتم K. al-Ḥikma | تماماً : تامّاً K. al-Ḥikma, Fawz 13 وإذا : وعلى هذا جميع الأجسام وهذه قضية صادقة مشهورة لا يحتاج فيها إلى دليل فإنّ نحن Fawz | يخالف ... الصفة : حاله بالضدّ من حال فيه Fawz 15 ومن الدليل : ومما يدلّ أيضاً Fawz 16 منه[1] : الأجسام في المعنى الذي ذكرناه K. al-Ḥikma, Fawz | أم : أو Fawz

1 حاله : حالة M 5 after البدن blank space of three letters M 7 after تموت blank space of six letters M 8 فإنّ : ان M | غيرها : غيره M 11 وتخلعها : وتخعلها M 12 آخر : اخر M | النقش : النفس M 14 after بجسم blank space of three letters M 17 إلا به : الايه M

the intellect is correct and in its [proper] condition. Therefore the soul does not exist through the harmony of the body.
He says:[142] if the soul existed through the mixture of the body and the body deviated from its balanced condition, the soul would necessarily deviate and shift from its condition; but the matter is different. So the soul does not exist through the mixture of the body.
(84) Abū ʿAlī Miskawayh says:[143] the soul is a simple substance and is neither a body nor an accident. It is self-moving, always remaining and neither perishes nor dies. The proof that it is not a body is that when a body has received a form, it is only able to receive another form of the same genus after it has cast off the first form and departed from it completely. An example of this is that when silver has received the form of a vessel, it is only able to receive the form of a drinking cup after it has removed the form of the vessel from it[self] and cast it off completely. Likewise when wax has received the form of some imprint, it is only able to receive the form of another imprint after the form of the first imprint 50b
has vanished and departed from it completely. If we find something the condition of which is at variance with this description, i.e. which receives many forms without any of them perishing, it is clear to us that it is not a body.
Among the proofs for this notion, [i.e. the soul] not being a body, is that all body parts of a living being, no matter whether man or other, whether they are small or big, external or internal, are only instrument[s] employed for an aim which is only accomplished through them. Since the whole body consists of instruments and each of these instruments has a

[142] This passage is missing in the Arabic text, on which Arnzen bases his edition, so he supplies his retranslation from the Persian version instead. Therefore I only give the reading variants of the Aya Sofya in the apparatus.
[143] Excerpted from Miskawayh, alternative *Fawz*; cf. parallels in *K. al-Ḥikma*, MS E, fols. 62a11–b14; and *Fawz*, 61.3, 62.12–63.8, 64.4–14.

خاصّ لا يتمّ ⟨إلا⟩ بها اقتضى مستعملاً ، كما نجد آلات الصايغ والنجّار وغيرهما . وليس يجوز أن يقال إنّ بعض البدن يستعمل بعضه هذا الاستعمال لأنّ ذلك البعض الذي يشار إليه ويظنّ أنّه يستعمل الآلات الباقية هو أيضاً آلة أو جزء من آلة وجميعها مستعملة ، فمستعملها غيرها . وإذا كان 51a مستعملها غيرها ولم يكن جزءاً منها وجب أن يكون غير جسم ليتمّ له ألا يشغل مكان الجسم ولا يزاحم الآلات الجسميّة في مواضعها ، لأنّه لا يحتاج إلى مكان وتستعملها كلّها على اختلاف الأغراض المستعملة فيها في حال واحدة من غير غلط ولا عجز ليتمّ من الجميع أمر واحد . فإنّ هذه الأحوال ليست أحوال الأجسام ولا موضوعة في أحكامها .

(85) وعلى أنّها ليست بعرض أنّ العرض لا يحمل عرضاً لأنّه في نفسه محمول أبداً ، أي موجود في غيره لا قوام له بذاته . وهذا الجوهر الذي وصفنا حاله هو أبداً حامل بأتمّ وأكمل من حمل الأجسام للأعراض . فإذاً النفس ليست جسماً ولا جزءاً من جسم ولا عرضاً .

(86) والدليل على أنّها بسيطة أنّ ما عدم كلّ صورة فهو بسيط ، إذ المركّب هو ما يركّب من موضوع وصورة . ولذلك تصوّرت كلّ معقول وقبلته قبولاً 51b واحداً بالسويّة كما أنّ العقل لما لم يكن له صورة يختصّ بها قبلت الأشياء بالحقيقة والسويّة وكما أنّ الهواء لما كان موضوعاً لقبول الألوان وجب أن يكون في ذاته عادماً لكلّ لون ، ولو كان يختصّ بلون لكان قبوله لما يجانسه

2–3 يستعمل ... البعض om. *K. al-Ḥikma* 7 الأغراض : الأعراض *K. al-Ḥikma* 8 غلط : غلط *K. al-Ḥikma* 9 موضوعة : مشروطة *Fawz* 10 وعلى : فأمّا *Tahdhīb* | after بعرض add. فهو بيّن من قبل *Tahdhib* | لأنّه : لأن العرض *Tahdhīb* 12 حامل : قابل وحامل *Tahdhīb* 14 والدليل على : ولأجل ذلك قلنا أنها *Fawz* | أنّ : لأنّ *Fawz* 15 يركّب : تركب *K. al-Ḥikma, Fawz* | ولذلك : ولما كانت النفس العاقلة عادمة لكلّ صورة *Fawz* | كلّ : بكل *Fawz* 16 كما ... له : ولو كان لهذا العقل *Fawz* 16–17 قبلت ... والسويّة : لم يكن قابلاً لكل حقيقة ولا كان قبوله إياها بالسوية على التمام *Fawz* 17 وكما أنّ : فإن *Fawz* 18–164.1 يجانسه ... ولما om. *Fawz*

4 آلة : الة M with blank space of three letters after it M | جزء : جزوٌ M 5 جزءاً : جزا with a faint *hamza* added above the *zāʾ* M 7 الأغراض : الاعراض M 9 موضوعة : مو موضوعة M | after أحكامها blank space of half a line M 13 جزءاً : جزاً with a faint *hamza* added above the *zayn* by a later hand M | after عرضاً blank space of ten letters M

specific activity which is only accomplished through it, [the body] requires someone to employ [the bodily instruments] just as we find with the instruments of the jeweller, the carpenter and others. It is not permissible to say that a portion of the body employs another in this way, for that portion which is indicated and believed to employ the remaining instruments is also an instrument or part of an instrument and they are all employed. So the one employing them is something else. When the one 51a
employing [the bodily instruments] is something else and not part of them, it is necessarily not a body so that it does not have to occupy the space of a body nor constrict the bodily instruments in their positions. For it does not need any space. It employs them all for the different aims for which they are employed in one single condition, without error or weakness, so that one single matter is accomplished through the[ir] entirety. These are not the conditions of bodies nor are they implied in [the bodies'] principles.

(85) [The proof] that [the soul] is not an accident[144] is that an accident does not bear [another] accident for it is in itself always borne, that is it exists in something else and has no subsistence by itself. This substance, the condition of which we have described, is always a more perfect and complete bearer than the bodies' bearing of accidents. Therefore the soul is neither body, nor part of a body, nor accident.

(86) The proof that [the soul] is simple[145] is that what lacks every form is simple, because the composite is what is composed of substratum and 51b
form. Therefore [the soul] conceives every intelligible and receives it in the same and equal manner, just as things are truly and equally received [by the intellect], since intellect does not have any form by which it is specially distinguished, and just as the air necessarily lacks in itself any colour since it is the substratum for the reception of colours. If it were distinguished by one [particular] colour, its reception of what is of the same genus as [this colour] would be easier and more comprehensive

[144] Cf. Miskawayh, *Tahdhīb al-Akhlāq*, 5.10–14.

[145] Excerpted from Miskawayh, alternative *Fawz*; cf. parallels in *K. al-Ḥikma*, MS E, fols.62b14–63a1; and *Fawz*, 74.5–6, 74.4–5, 74.1–3, 73.8–10, where the order of the sentences is reversed.

أيسر وأوفر ولما يخالفه أصعب وأعسر .

(87) وعلى أنّها محرّكة ذاتها قول افلاطن جوهر النفس هي الحركة ، وهذه الحركة هي حياة النفس ولما كانت ذاتيّة كانت الحياة ذاتيّة ، والحركة المنسوبة إلى النفس غير الحركات الستّ التي يختصّ بها الجسم الطبيعيّ وهي الكون والفساد والنموّ والنقصان والاستحالة والنقلة ، فليس يليق شيء منها بهذا الجوهر ، بل هذه حركة الرويّة وهي جولان النفس الموجود لها دائماً . فإنّك لا تجد النفس خالية من هذه الحركة في حال من الأحوال . وهذه الحركة لما لم تكن جسمانيّة 52a
لم تكن مكانيّة ولما لم تكن مكانيّة لم تكن خارجة عن ذات النفس . وحركة النفس التي أشرنا إليها على نحوين ، أحدهما نحو العقل والآخر نحو الهيولى . فإذا تحرّكت نحو العقل استنارت واستفادت منه وإذا تحرّكت نحو الهيولى أفادتها وأنارتها ، ولما كانت الحركة ذاتيّة للنفس قلنا إنّها هي تحرّكت نحو الهيولى ، فأمّا الهيولى فإنّها لا تتحرّك ولا الحركة من شأنها .

(88) وافلاطن يقول الحياة حياتان والموت موتان ، ويسمّي الحياة التي بحسب حركة النفس الناطقة نحو العقل حياةً طبيعيّةً ، والحياة التي نحو الهيولى حياةً إراديّةً ، وكذلك الموت المقابل لها . ومن هاهنا قال مت بالإرادة تحي بالطبيعة .

(89) وقالوا الموت موتان ، طبيعيّ وهو ترك النفس استعمال الجسد ، وإراديّ وهو إماتة شهوات النفس النامية وترك متابعة الأهواء الطبيعيّة .

1 أصعب وأعسر : أعسر *Fawz* 2 وعلى ... قول : ولذلك قال *Fawz* 3 after الحياة add. لها *K. al-Ḥikma, Fawz* 3–5 والحركة ... فليس : فإنّا قد وينبغي أن ننظر إلى هذه الحركة التي للنفس قلنا إنّ النفس جوهر وليست بجسم والحركات التي كنّا أحصيناها أعني الستّ هي حركات الجسم وليس شرحنا من أمرها ما شرحنا *Fawz* 5–6 بل هذه : فنقول إن هذه الحركة هي *Fawz* 7–8 جسمانيّة ... تكن[3] om. *K. al-Ḥikma* 8 وحركة : ونعود فنقول إن حركة *Fawz* 9 أشرنا إليها : أشرنا إليها *Fawz* | نحو[2] om. *K. al-Ḥikma* 13 وافلاطن ... ويسمّي : وكذلك قال افلاطن إن الفلسفة هي التدرّب بالموت الإرادي لأن عنده أن الموت موتان والحياة حياتان ، وإذا كانت إحدى الحياتين بحسب هذه الحركة من النفس والأخرى بحسب تلك فوجب لذلك أن يكون أيضا الموت موتين ليكون كلّ واحد يقابل صاحبه ، وهو يسمّي *Fawz* 14 والحياة : ويسمّي الحياة *Fawz* 15 ومن هاهنا : فلذلك *Fawz* 17 وقالوا : وقال *K. al-Ḥikma*

1 after وأعسر blank space of five letters M 7 خالية : حَالتَه M 9 والآخر : واخر M 10 فإذا ... الهيولى add. in marg. M 11 after وأنارتها blank space of five letters M | للنفس : النفس M 18 after الطبيعيّة blank space of five letters M

whereas its reception of what is different from it would be more difficult and harder.

(87) [The proof] that [the soul] is self-moving[146] is Plato's saying: the substance of the soul is motion. This motion is the soul's life and since it is essential, life is essential. The motion which is linked to the soul is unlike the six motions by which the natural body is specifically distinguished, namely generation, corruption, growth, decrease, change and transposition. None of them suits this substance [i.e. soul]. This is rather the motion of deliberation which is the ever existent revolving of the soul. Thus you do not find the soul devoid of this motion under any condition. Since this 52a
motion is not bodily it is not locomotion, and since it is not locomotion it is not outside the soul's self. The motion of the soul indicated by us is in two directions: one of them is towards intellect and the other towards matter. Whenever it moves towards intellect, it is illuminated and acquires benefits from it, and whenever it moves towards matter, it benefits and illuminates it. Since motion is essential to the soul, we say that it moves towards matter, whereas matter does not move and motion is none of its concern.

(88) Plato says:[147] life is of two kinds as well as death. He calls the life which is by the motion of the rational soul towards the intellect 'natural life', and the life which is towards matter 'voluntary life', and similarly death which is the opposite. Hence he says: die voluntarily, so you will live by nature!

(89) And they [i.e. the philosophers] say:[148] death is of two kinds, a natural one, which is the soul's abandonment of [its] use of the body, and a voluntary one, which is killing the appetites of the growing soul and the abandonment of the pursuit of natural passions.

[146] Excerpted from Miskawayh, alternative *Fawz*; cf. parallels in *K. al-Ḥikma,* MS E, fol.63a1–13; and *Fawz*, 84.16–85.1, 84.11–16, 85.15–86.4, arranged differently.

[147] Excerpted from Miskawayh, alternative *Fawz*; cf. parallels in *K. al-Ḥikma,* MS E, fol.63a13–17; and *Fawz,* 89.11–90.1.

[148] Excerpted from Miskawayh, alternative *Fawz*; cf. parallel in *K. al-Ḥikma,* MS E, fol.63a17–19. See also Miskawayh, *Tahdhīb*, 211.23–212.4, which ascribes it to the philosophers and may thus explain why our passage starts with 'And they say'.

52b (90) وقال ارسطاطاليس الدليل على أنّ النفس لا تفسد بعد مفارقتها للبدن
فعلها ، فإنّها تفعل في الشيء النائي عن جسمها وتروّي فيه وتفكّر وتعرف
الشيء من غير أن تكون بذاتها حاضرة لذلك الشيء . فإذا كان فعلها يبعد
من بدنها فلا محالة أنّ جوهرها يبقى بعد فراقها للبدن ، وإلا كان فعلها أكرم
من جوهرها وهذا خطأ ، لأنّه لا يمكن أن يكون فعل الجوهر أكرم من الجوهر
لأنّ الفعل إنما يظهر عن الجوهر لا الجوهر عن الفعل .

(91) وقال أبو علي الكلام في بقاء النفس ما قاله افلاطن فإنّه أورد فيه ثلاث
حجج ، إحداها أنّ النفس تعطي كلّ ما توجد فيه حياة ، والثانية أنّ كلّ
فاسد فإنما يفسد من قبل رداءة فيه ، والثالثة أنّ النفس متحرّكة من ذاتها .

فأمّا الحجّة الأولى فهي النفس تعطي الحياة أبداً [على] ما توجد فيه وكلّ ما
يعطي الحياة أبداً ما يوجد فيه فالحياة جوهريّة له ، وما كانت الحياة جوهريّة له
53a فلن يمكن أن يقبل ضدّها وضدّ الحياة هي الموت . فالنفس إذاً لا تقبل
الموت .

وأمّا الحجّة الثانية فإنّها مبنيّة على أنّه لا رداءة في النفس . فينبغي أن نشرح
حقيقة الرداءة وما يراد بها ليتمّ لنا سياقة البرهان بعد ذلك ، فنقول إنّ الرداءة

1 وقال ... الدليل : والدليل *K. al-Nafs* | أنّ النفس : أنّها *K. al-Nafs* | للبدن : البدن *K. al-Nafs*, MS A 2 جسمها : بدنها *K. al-Nafs*, ed. Arn | وتروّي : تروّي *K. al-Nafs*, MS A 3 بذاتها : حاضرة بدنها حاضراً *K. al-Nafs*, ed. Arn 3–4 يبعد من : ينفذ من *K. al-Nafs*, MS A : يتعدي ed. Arn 4 فلا : ولا *K. al-Nafs*, MS A | وإلا ... فراقها : ولا فارقة البدن ما *K. al-Nafs*, MS A 5 وهذا خطأ : وذلك باطل *K. al-Nafs*, ed. Arn | يمكن أن om. *K. al-Nafs*, MS A | الجوهر[1] : الجواهر *K. al-Nafs*, ed. Arn | الجوهر[2] : الجواهر *K. al-Nafs*, ed. Arn 7 وقال ... الكلام : والكلام *K. al-Ḥikma* 7–8 وقال ... حجج : اعتمد افلاطن في بقاء النفس على ثلاث حجج *K. al-Ḥikma* 10 فهي : فسياقتها على هذا *Fawz* | على : كلّ *Fawz* 10–11 وكلّ ... فيه om. *Fawz* 11 له[1] : لها *Fawz* | وما : ولما *Fawz* | له[2] : لها *Fawz* 12 فلن : فلم *K. al-Ḥikma*, *Fawz* | ضدّها وضدّ : ضرها وضر *K. al-Ḥikma* | هي : هو *K. al-Ḥikma* : om. *Fawz* 13 after الموت add. قد أطنب أصحاب أفلاطن في تفسير هذا الفصل وأكثروا شرحه وبينوا صحّة مقدّماته وتركيبها وصحّة النتيجة منها وسنذكر بعض ذلك إذا فرغنا من إيراد الحجج الثلاث *Fawz* | لا تقبل : لا يمكن أن تقبل *Fawz* 15 ليتمّ : لتتمّ *Fawz*

6 after الفعل[2] blank space of three letters M 7 ثلاث : ثلث M 8 إحداها : احديها M | الحياة[2] : الحيوة M | فالحياة : فالحيوة M 10 الحياة : الحيوة M 11 الحياة[1] : الحيوة M 9 رداءة : رَدأةٍ M 13 after الموت blank space of three letters M 14 رداءة : ردأه M 15 الرداءة[1] : الردأه M | الرداءة[2] : الرداةَ M | فنقول : فيقول M | بعد ذلك : بعدلك M | ليتمّ : اليتم M

(90) Aristotle says:[149] the proof that the soul does not corrupt after its separation from the body is its activity. For it acts on something distant 52b
from its body, looks into it, contemplates and knows the thing without being itself in the presence of that thing. So when its activity is remote from its body,[150] its substance, most certainly, remains after its separation from the body; otherwise its activity would be more excellent than its substance, which is a faulty [assumption]. For the activity of a substance cannot be more excellent than the substance, because the activity arises from the substance, not the substance from the activity.

(91) Abū ʿAlī says:[151] what can be said about the soul's immortality is what Plato says, for he lists three arguments for it. One of them is that the soul gives life to everything in which it is found. The second is that everything corruptible only corrupts through the badness in itself. The third is that the soul is self-moved.

As to the first argument, it is [as follows]: the soul gives life perpetually to that in which it is. Everything which gives life perpetually to that in which it is has life substantially. That which has life substantially cannot receive 53a
the opposite [of life], and the opposite of life is death. Therefore the soul does not receive death.

As to the second argument, it is based on [the premise] that there is no badness in the soul. We must then explain the true nature of badness and what is meant by it so that the course of the argument is afterwards completely [clear] to us. We say that badness is linked to corruption,

[149] Cf. source in *DA*-paraphrase, ed. Arnzen, 219.17–221.5; and MS A, fol.69a3–12.

[150] Here part of the argument – along the lines '[when its activity is remote from its body,] its substance must be remote and separable from the body as well' – must have dropped out, most probably *per homoeoteleuton*, as Arnzen convincingly argues in the case of the *DA*-paraphrase (*Aristoteles' DA*, pp.124–5 and 383).

[151] Excerpted from Miskawayh, alternative *Fawz*; cf. parallels in *K. al-Ḥikma,* MS E, fols.63a19–64a14; and *Fawz,* 81.10–82.13, 83.2–84.3.

مقترنة بالفساد والفساد مقترن بالعدم والعدم مقترن بالهيولى ، فالرداءة مقترنة بالهيولى . وبيان ذلك أنّه حيث لا هيولى فلا عدم وحيث لا عدم فلا فساد وحيث لا فساد فلا رداءة . فالهيولى معدن الرداءة وينبوع الشرّ وأصله الذي منه يتفرّع ، ويقابل هذه الرداءة الجودة والجودة مقترنة بالبقاء والبقاء مقترن بالوجود والوجود هو أوّل صورة أبدعها البارئ عزّ وجلّ . فلذلك هو خير محض لا يشوبه شرّ ولا عدم واختصّ به العقل الفعّال ، وذلك أنّ الوجود الحقّ الذي ليس فيه هيولى بتّةً ولا معنى الانفعال هو العقل الأوّل . والنفس صورةٌ تكمل البدن بوجودها ، فليست إذن هيولى وقد بيّنّا أيضاً أنّها 53b ⟨ليست⟩ صورة هيولانيّة ، أي محتاجة إلى الهيولى في وجودها . فالنفس ليس فيها شيء من رداءة ، فالنفس ليس لها فساد والنفس ليس لها عدم ، فالنفس إذا باقية . وسياقة البرهان النفس ليس فيها شيء من الرداءة وكلّ ما ليس فيه شيء من الرداءة فليس بفاسد ، فالنفس ليست بفاسدة .

فأمّا الحجّة الثالثة فهي هذه النفس متحرّكة من ذاتها وكلّ ما كانت حركته من ذاته فهو غير فاسد ، فالنفس غير فاسدة .

وقال برقلس في بيان الحجّة الأولى كلّ أمر ضادّ أمراً صادراً عن قوّة ما فهو مضادّ للقوّة التي عنها صدر ذلك الأمر ، مثال ذلك البرودة فإنّها مضادّة للحرارة الصادرة عن النار وهي أيضاً مضادّة لما صدرت عنه الحرارة أعني النار .

فإذا كان هكذا قلنا إنّ الموت إذا كان مضادّاً للحياة التي في البدن فهو أيضاً مضادّ لحياة النفس التي عنها صدرت حياة البدن . فإذا كانت النفس العاقلة 54a

2 ذلك : هذا الكلام *K. al-Ḥikma, Fawz*   3 فالهيولى : والهيولى *K. al-Ḥikma*   4 ويقابل : ومقابل *K. al-Ḥikma, Fawz*   5 عزّ وجلّ : تعالى *K. al-Ḥikma, Fawz*   7 والنفس : ونعود الآن فنقول إن ... النفس *Fawz*   8 بوجودها فليست : بوجودها فيه وليست *Fawz*   10 فالنفس¹ ... لها² om. *K. al-Ḥikma* | فالنفس : والنفس *Fawz*   11 وسياقة البرهان : فأمّا سياقة البرهان فهكذا *Fawz*   13 فأمّا : وأمّا *K. al-Ḥikma, Fawz*   15 وقال : وأمّا ما أورده *Fawz* | after الأولى add. الذي وعدنا بذكره فهو هذا *Fawz*   18 after كان¹ add. هذا *Fawz*   18–19 أيضاً مضادّ : مضادّ أيضاً *K. al-Ḥikma, Fawz*

1 فالرداءة : فالرّدأةُ M   3 رداءة : رَدأة M | الرداءة : الردة with an *alif* add. above the *dāl* M   4 ويقابل : ويتقابل M | الرداءة : الردأةُ M   7 after الأوّل blank space of ten letters M   10 رداءة : ردأه M   11 الرداءة : الردأة M   12 الرداءة : الرَّداة M | after بفاسدة blank space of three letters M   15 قوّة : قوتِه M

corruption is linked to non-existence and non-existence is linked to
matter. Hence, badness is joined to matter. That is evident since wherever
there is no matter, there is no non-existence, and wherever there is no
non-existence, there is no corruption, and wherever there is no corrup-
tion, there is no badness. Matter is the source of badness, the wellspring
and root of evil from which it branches out. Opposite to this badness is
goodness, and goodness is linked to immortality, immortality is linked to
existence, and existence is the first form which the Creator, to Whom
belong might and majesty, has created. Therefore it is the pure good,
neither evil nor non-existence mixes with it and by it the active intellect is
specially distinguished. For the true existence in which there is no matter
at all nor any notion of passivity is the first intellect. The soul is a form
which completes the body through its existence. Therefore it is not matter 53b
and as we have also already made clear it is ⟨not⟩ a material form, i.e. [a
form which] needs matter for its existence. Thus there is no badness in the
soul and therefore the soul has neither corruption nor non-existence.
Therefore the soul is immortal. The course of the argument is [as follows]:
in the soul there is no badness. Nothing in which there is no badness is
corruptible. Therefore the soul is not corruptible.

As to the third argument, it is this: the soul is self-moved. Everything with
motion from itself is incorruptible. Therefore the soul is incorruptible.

Proclus makes the first proof clear by saying: everything that is opposed to
something which proceeds from some power is opposed to the power from
which that [other] thing proceeds. An example of this is the cold: for it is
opposed to heat which proceeds from fire and it is also opposed to that
from which heat proceeds, namely fire. So if it is like that, we say that if
death is opposed to the life which is in the body, then it is also opposed to
the life of the soul, from which the life of the body proceeds. Therefore the 54a
intellective soul is not receptive to death, which is the opposite of the life

غير قابلة للموت الذي هو ضدّ الحياة التي لها ، لأنّ المضادّ لحياة البدن هو مضادّ لحياتها أيضاً كما بيّنّا . فالنفس غير قابلة للموت المضادّ للحياة التي فيها ، فهي أذن غير مائتة ولا فائتة .

(92) وقال ارسطالس أترى النفس تعلم الأشياء المبسوطة والأشياء المركّبة بجزء واحد أم بأجزاء ؟ ثم أجاب فقال إنّها تعلم الجميع بجزء واحد وهو العقل إلا أنّها تعلم ذلك بنوع ونوع . ذلك أنّها إذا أرادت علم الصورة وحدها انبسط العقل ورجع إلى ذاته فيعرف تلك الصورة معرفةً صحيحةً ، وإذا أرادت النفس أن تعلم الشيء المركّب انحطّ العقل إلى الحسّ وانعرج إليه فيأخذ منه أوائل المعرفة ، كأنّه قال العقل يعلم الصورة العقليّة المفردة بلا توسّط ، أي بلا آلة ، بل يعرفها بنفسه ويعلم الأشياء المركّبة الجسمانيّة بتوسّط الحسّ . وذلك أنّه 54b
يستعمل الحواس كالآلة في معرفة الأشياء الجسمانيّة لأنّ الحواس تقبل الآثار الجسمانيّة وتؤدّيها إلى العقل ، والعقل يقبلها من الحواس ويؤدّيها إلى النفس فتعرفها . والدليل على ذلك أنّ الذي قد عدم البصر في أوّل مولده لا يعرف الألوان فإنّ البصر لا يؤدّي إليه آثار الألوان ، فإذا لم يقبل العقل آثار الألوان من البصر لم يقف على معرفة الألوان البتّة .

1 للموت الذي : للذي *K. al-Ḥikma* | لها : للبدن على ما تبين فيما مضى كانت أيضاً غير قابلة للموت الذي هو ضدّ الحياة التي لها *Fawz*   3 فائتة : فانية *K. al-Ḥikma, Fawz*   4 وقال . . . أترى : أفترى *K. al-Nafs* | والأشياء om. *K. al-Nafs*, MS A   5 أم بأجزاء : وإنما تعلم الأشياء المركّبة والمبسوطة *K. al-Nafs*, MS A : أو إنما تعلم الأشياء المبسوطة أعني الصور بجزء وتعلم الأشياء المركّبة بجزء آخر؟ ed. Arn | ثم . . . الجميع om. *K. al-Nafs*, MS A | إنّها : إنما *K. al-Nafs*, ed. Arn | الجميع : الأشياء المبسوطة والأشياء المركبة *K. al-Nafs*, ed. Arn   6 ونوع om. *K. al-Nafs*, MS A | ذلك[2] : وذلك *K. al-Nafs*   7 فيعرف : فتقرّب *K. al-Nafs*, MS A   8 وانعرج : انفرج *K. al-Nafs*, MS A   8–9 فيأخذ . . . المعرفة : وأخذ منها وأنا المعدود *K. al-Nafs*, MS A   9 يعلم : تعليم *K. al-Nafs*, MS A | الصورة : الصور *K. al-Nafs*, ed. Arn | المفردة om. *K. al-Nafs*, MS A : المجرّدة ed. Arn   10 بل . . . الأشياء : تعرفه وتعرف *K. al-Nafs*, MS A | ويعلم : ويعرف *K. al-Nafs*, ed. Arn   11 الحواس[1] : المشاعر *K. al-Nafs*, MS A   11–15 في . . . البتّة om. *K. al-Nafs*, MS A   11 لأنّ : وذلك أنّ *K. al-Nafs*, ed. Arn   12 إلى[1] . . . النفس : إليه *K. al-Nafs*, ed. Arn   13 أنّ . . . في : العادم لبصره من *K. al-Nafs*, ed. Arn | after مولده add. فإنما صار *K. al-Nafs*, ed. Arn   14 فإنّ : لأنّ *K. al-Nafs*, ed. Arn   15 يقف : يقو *K. al-Nafs*, ed. Arn

3 after فائتة blank space of more than half a line M   5 بأجزاء : باجزآ M   6 وحدها : وجدها M   7 تلك : ذلك M   8 وانعرج : واتعرج M   11 كالآلة : كَالَالةٍ M | الآثار : الاثار M   14 آثار[1] : اثار M | آثار[2] : اثار M   15 after البتّة blank space of ten letters M

which belongs to the soul. For what is opposed to the life of the body is also opposed to the life of the soul, as we have shown. So the soul is not receptive to death, which is opposed to the life in the soul. Therefore it is immortal and does not pass away.

(92) Aristotle says:[152] would you say the soul knows the simple and composite things through one or through [more] parts? Thereupon he answers, saying that it knows everything through one part which is intellect, yet it knows [all] that in different ways. For if it wants to have the knowledge of form alone, the intellect expands and returns to itself, so that it comes to know that form correctly. If the soul wants to know something composite, the intellect descends to sense perception and inclines to it to gather from it the first [principles] of [the wanted] knowledge. Like he says: the intellect knows the single intellectual form without mediation, i.e. without an instrument, but comes to know it through itself, whereas it knows composite corporeal things through the 54b
mediation of sense perception. For in order to come to know the corporeal things it employs the senses as instruments,[153] because the senses receive corporeal impressions and convey them to intellect. The intellect receives them from the senses and conveys them to the soul, so it comes to know them. The proof for this is that he who already lacks vision from the very moment of his birth does not come to know colours, for vision does not convey the impressions of colours to him. So if the intellect does not receive the impressions of colours from vision, it does not grasp the knowledge of colours at all.

[152] Cf. source in *DA*-paraphrase, ed. Arnzen, 311.8–313.5; and MS A, fol.95a4–15.

[153] The rest of the passage is not preserved in Aya Sofya 4156, maybe due to the loss of a folio.

(93) قال والعقل يفعل فعله دون البدن فإنّه يروي في الأشياء النائية عنه من غير أن يحضر بين يديه الشيء الذي يروي فيه ، ويروي في الجواهر العالية من غير حاجة منه إلى البدن فيه . والدليل على ذلك أنّ كلّ آلة هي ملائمة للشيء الذي هي له آلة لا يعوقه عن شيء من أفاعيله ، والبدن يعوق العقل عن أشياء كثيرة مخالفة ، وذلك أنّ البدن يمنع [افا] أفاعيل النفس العاقلة لأنّها 55a ربّما أرادت شيئاً فيخالفها البدن وتغلب شهواته المذمومة عليها فلا تدعها الفعل ، وربّما اشتاقت النفس الناطقة إلى عالمها فيجرّها البدن إلى عالمه الذي هو الواقع تحت الفساد . ولو أنّ قوام النفس بالبدن لكان إذا فسد البدن فسدت ولم تشتق إلى غير هذا العالم لأنّه ليس شيء من الأشياء يشتاق إلى غير عالمه .

(94) وقال أبو على مسكويه النفس تدرك الأمور البسائط والمركّبات وتدرك من المركّبات أنواعها وأشخاصها ، والموجودات منقسمة إلى هذه الأشياء وليس يفوت النفس منها شيء .

1 قال والعقل : والدليل على ذلك أن العقل K. al-Nafs, ed. Arn | قال ... البدن om. K. al-Nafs, MS A | فإنّه يروي : انه يُوَار K. al-Nafs, MS A : ويروي ed. Arn 2 يحضر : يكون يحصره K. al-Nafs, ed. Arn | يديه : بدنه K. al-Nafs, MS A | بين ... الشيء om. K. al-Nafs, MS A | ويروي ... الجواهر : وإذا رأى العقل أيضاً في الجواهر K. al-Nafs, MS A | يروي : يرى K. al-Nafs, MS A : وإذا روى العقل في الأشياء ed. Arn 2–3 من ... منه : لم يحتج إلى آلة بدنيّة لينال بها تلك الأشياء K. al-Nafs, MS A : ولا إلى آلة بدنيّة ينال بها تلك الأشياء فليس ed. Arn 3 فيه : ولا K. al-Nafs | والدليل ... ذلك : والدليل أيضاً على أن العقل لا يعوقه عن فعله شيء إذا كان مجرداً ed. Arn | أنّ : اي K. al-Nafs, MS A | هي ملائمة : يحتاج في معرفة الأشياء العالية إلى آلة K. al-Nafs | له ... يعوقه : آلة له ولا تعوقه K. al-Nafs 4 له ... لا : آلة له ولا K. al-Nafs, MS A | ملائمة : ملامسة K. al-Nafs, MS A | يعوق العقل : يفرق وأفاعيله K. al-Nafs, MS A | من : ثمّ K. al-Nafs, MS A | مخالفة : فيخالفه K. al-Nafs 5 العاقلة om. K. al-Nafs, MS A 6 وتغلب شهواته : ويبعد شهواتها K. al-Nafs, MS A | عليها : عنها K. al-Nafs, MS A 7 الفعل om. K. al-Nafs, MS A : تفعل فعلها ed. Arn | فيجرّها : فيردها K. al-Nafs, MS A : فحجزها ed. Arn | عالمه : غايته K. al-Nafs, MS A 7–8 الذي هو om. K. al-Nafs 8 الفساد : الفسق والبلى K. al-Nafs, MS A | أنّ ... لكان : كان K. al-Nafs, MS A | after النفس add. كان K. al-Nafs, ed. Arn 9 after فسدت add. النفس K. al-Nafs, ed. Arn | ولم : لم K. al-Nafs, MS A | لأنّه : لا انه K. al-Nafs, MS A | من الأشياء om. K. al-Nafs, MS A | after إلى[2] add. علـــم K. al-Nafs, MS A : عالم ed. Arn 11 وقال ... النفس : والنفس K. al-Ḥikma : إنّا نجد النفس Fawz

1 النائية : الثانية M 2 يحضر : يحصر M 3 آلة : الة M 4 آلة : آلَه M 5 أفاعيل the word was originally split over two pages, but completed on the second page by a later hand M 6 شهواته : شهواتها M 7 الفعل : العقل M 10 after عالمه break (∴) ind. M 13 after شيء blank space of six letters M

(93) He says:[154] the intellect performs its activity without the body, for it deliberates things which are distant from it without having the deliberated thing present. It deliberates the elevated substances without any need on its part of a body to do so. The proof for this is that every instrument is consistent with the thing for which it is an instrument [and] does not hold it back from any of its activities. [Yet] the body holds the intellect back from many different things. For the body prevents the activities of 55a
the intellective soul, because [the soul] may sometimes want something, [but] then the body opposes it and its blameworthy appetites prevail over [the soul] and do not let it act. Sometimes the rational soul may desire its world, then the body drags it towards its [own] world which is subject to corruption.[155] If the subsistence of the soul were due to the body, then when the body corrupted, [the soul] would corrupt [as well] and it would not desire anything else other than this world, for nothing desires anything but its own world.

(94) Abū ʿAlī Miskawayh says:[156] the soul perceives the simple and composite things, and of the composite ones, it perceives their species and individuals. These are the categories into which the existents are divided and none of them escapes the soul.

154 Cf. source in *DA*-paraphrase, ed. Arnzen, 317.16–319.9; and MS A, fol.97a1–b1.

155 The Aya Sofya has an interesting variant here, reading 'is subject to sinfulness and decline'.

156 Excerpted from Miskawayh, alternative *Fawz*; cf. parallels in *K. al-Ḥikma*, MS E, fols.64a14–65a13; and *Fawz*, 65.4–67.3.

أمّا الأمور البسيطة فمنها هيولانيّة ومنها غير هيولانيّة ، وغير الهيولانيّة منها
هي المعقولات ، أعني الموجودة بغير موادّ ، والهيولانيّة منها هي التي تعرى من
الموضوع وتوجد بالوهم ، وهي رسوم الجزئيّات كما يفعله أصحاب التعاليم ،
55b فإنهم يأخذون النقطة والخطّ والسطح والجسم التعليميّ ، أعني الأبعاد الثلاثة ،
في غير مادّة كأنّها أشياء موجودة بذواتها ، وكذلك يأخذون توابع الجسم
مفردة ، أعني الحركة والزمان والمكان والأشكال وبالجملة كلّ ما لا يوجد إلا
في الجسم وبه ، يفردونها عن موادّها ويلحظونها بأوهامهم بسائط مرّة ومركّبة
أخرى في غير حوامل . وربّما بلغ من قوّة أحدهم في هذا التوهّم أن يظنّ بهذه
الصور التي انتزعها من موادّها وجرّدها في وهمه أنّها موجودة من خارج الوهم
ولها حقائق في ذواتها من غير حوامل ولا موضوعات لها ، ويخلط بينها وبين
المعقولات حتّى لا تتميّز عنده بل يسمّيها كلّها معقولات . وهذه حال
موجودة للنفس ، أعني أنّها تدرك الأمور المركّبة ، ثم تحلّها إلى بسائطها ، ثم
تأخذ تلك البسائط في الوهم فتفردها تارةً وتركّبها أخرى ضروب التركيبات .
56a فربّما كانت لتلك التركيبات حقائق وربّما لم يكن لها حقائق كما يتوهّم عنقاء
مغرب وإنسان يطير وشخص خارج العالم وحيوان مركّب من حمار ونعجة ،
فهذه لا حقائق لها ولا وجود خارج الوهم ، وقد يجوز أن يركّب من البسائط
في الوهم ما له حقيقة ووجود من خارج . فهذه حال البسائط ما كان منها
هيولانيّاً وما كان غير هيولانيّ .

فأمّا المركّبات فمنها استقصات أوَل ومنها مركّبات من الاستقصات ،
والمركّبات حيوان ونبات وجماد ، ثم ينقسم كلّ واحد منها بضروب التركيبات
وأنواع المزاجات إلى أنواع كثيرة جدّاً ، وينقسم أيضاً أنواعها إلى أشخاص لا
تحصى والنفس تدرك جميع ذلك . ولما كانت الاستقصات أربعة ومزاجها
كثيرة بالأقلّ والأكثر والأشدّ والأضعف صار لها بالأمزجة توابع من الكيفيّات

1 أمّا : وأمّا *K. al-Ḥikma* 3 بالوهم : في الوهم *Fawz* 5 في : من *K. al-Ḥikma* 7 يفردونها : فيفردونها *Fawz* 13 تأخذ : تأخذها *K. al-Ḥikma* 15 حمار : جواد *K. al-Ḥikma* 16–17 من ... الوهم : في الوهم من البسائط *Fawz* 17 من خارج : خارج وأمثلته كثيرة *Fawz* 19 فمنها : فهي *K. al-Ḥikma* 20 after والمركّبات add. منها *Fawz* 22 ومزاجها : ومزاجاتها *K. al-Ḥikma, Fawz*

4 الثلاثة : الثلثة M 14 لتلك : التلك M 18 after هيولانيّ blank space of five letters M 22 تدرك : مُدرك M | جميع : جهعًا M

The simple things are material and immaterial. The immaterial ones are
the intelligibles, i.e. the existents without matter, and the material ones
are those stripped of their substratum and found in the imagination. They
are traces of the particulars, like the constructs of the mathematicians. For
they take point, line, plane and mathematical body, i.e. the three dimen- 55b
sions without regard to matter, as if they were things existent by
themselves. In the same way they take as isolated what follows the body,
i.e. motion, time, place, shapes and, to sum up, everything which only
exists in and due to a body. They isolate them from their materials and
regard them in their imaginations as simple at times and as composite at
other times, yet without reference to [their] bearers. Sometimes one of
them attains such a powerful imagination of this kind that he comes to
think that these forms, which he has extracted from their materials and
abstracted in his imagination, are existent outside the imagination and
have a reality in themselves without having bearers or substrata. He
confuses them with the intelligibles so that they are, in his opinion,
indistinguishable, and he even calls all of them intelligibles. This is one
condition which exists for the soul, that is to say that it perceives the
composite things, then resolves them into their simple [components],
thereupon takes those simple [components] in the imagination and
isolates them at times and composes them in different ways at other times.
Sometimes those compositions have a reality and sometimes they do not 56a
have a reality, as a phoenix, a man who flies, an individual outside the
world and an animal composed of a donkey and an ewe can be imagined.
These have neither reality nor existence outside the imagination, but it is
certainly possible that what is composed from simple [components] in the
imagination has reality and outside existence. So this is the condition of
the simple [components], be they material or immaterial.

As to the composites, to them belong the first elements and the composites of the elements. The composites are animals, plants and inorganic bodies, and each one of them is furthermore divisible into a great many species due to the varieties of the compositions and the kinds of the mixtures. Their species are further divisible into innumerable individuals, but the soul perceives them all. Since the elements are four and their mixture manifold due to decrease, increase, strength and weakness [of each element], different qualities resulting from the[se] mixtures befall

56b مختلفة ، وليس تخلو هذه الاختلافات من أن تكون إمّا لأنّ أحد الاستقصات فيها أقوى من الآخر أو إثنين منها أو ثلاثة ، وإمّا لأنّها متساوية في القوّة إلا أنّ بعضها أكثر من بعض ، أعني أنّها تمتزج بعد أن تصير في أجسام طبيعيّة . وإذا كانت النفس تدرك جميع هذه الأقسام فيجب على ظاهر النظر أن تدركها بأربعة أنحاء وبأربع آلات لتتفرّد كلّ واحد منها باستقص ، فتدركه على تصرف أحواله من الشدّة والضعف والقلّة والكثرة إذا كانت في أجسام طبيعيّة .

(95) وليس للنفس أجزاء كأجزاء الجسم ، فلا ينبغي أن تكون المدرِكات بعدد المدرَكات . وهو ظاهر لأنّ الحاكم في جميعها واحد ، فإنّ شيئاً واحداً من الإنسان يحكم في الصغير أنّه صغير وفي الكبير أنّه كبير وهو الحاكم في الألوان والطعوم والروائح وفي أنّ الأشياء المتساوية بشيء واحد بعينه هي أيضاً 57a متساوية . ولو كان المدركون مختلفين لما صحّ أن يحكم واحد منها على ما أدركه الآخر .

فأمّا ظنّ من ظنّ أنّ النفس واحدة ولكنّها تدرك المدركات المختلفة بقوى كثيرة وبأنحاء مختلفة فهو موضع البحث ، ونقول إنّ بعض الناس لما نظر في الأمور الموجودة فرأى منها مركّبة ومنها بسيطة ونظر في الآلات والقوّة المدركة فوجد أيضاً بعضها مركّبة وبعضها بسيطة حكم بأنّ المركّبة منها تدرك المركّبة والبسيطة البسيطة . ومثّل ذلك بأن قال وجدت من المركّبات المدركات ما هو كالحواسّ الخمس لا يدرك إلا المركّبات ، فإنّ العين لما كانت مركّبة من قوّة باصرة في آلات وطبقات من العين لا تتمّ إلا باجتماعها أدركت من الأمور

5 وبأربع : أو باربع *K. al-Ḥikma*   6 تصرف : تعرف *Fawz* | من : في *Fawz*   8 وليس : أمّا أنّه ليس *Fawz* | after الجسم add. فهو بيّن ممّا قدّمنا وذلك أنّ التجزؤ والانقسام إنما يكون للجسم *Fawz* | فلا : وأمّا أنّه لا *Fawz*   9 وهو : فهو *Fawz* | ظاهر لأنّ : ظاهرة أيضاً وذلك أنّ *Fawz* | فإنّ : لأنّ *Fawz*   10 after الألوان add. وفي الأشكال *Fawz*   11 المتساوية : المساوية *K. al-Ḥikma, Fawz* | بشيء : لشيء *K. al-Ḥikma, Fawz*   15 وبأنحاء : أو بأنحاء *K. al-Ḥikma, Fawz* | ونقول : وسننظر فيه فنقول *Fawz*   18 after والبسيطة add. تدرك *Fawz*   19 إلا : أكثر *K. al-Ḥikma, Fawz*   20 من[2] om. *Fawz*

1 تخلو : يخلوا M   2 الآخر : اخر M | ثلاثة : ثلثه with blank space of five letters after it M   3 بعد أن : هذَان M   5 أنحاء : انحًا M   8 أجزاء كأجزاء : أجزآ كاجزا M   12 لما : كما M   14 بقوى : يقوى M   15 وبأنحاء : وبانحا M | ونقول : ويقول M   17 تدرك : مدرك a later hand dotted the *mīm* as a *yāʾ* M   20 آلات : الات M

them. These differences occur either because one of the elements is stronger in them than the other or two or three of them, or because they are equally powerful yet one of them is more abundant than the other[s]. That is to say they become mixed after they have come to be in the natural bodies. Since the soul perceives all these parts, it is obviously necessary that it perceives them in four ways and through four instruments so that each one of [the instruments] is dedicated to [the perception] of one element alone. Thus it perceives [each element] when it exists in the natural bodies according to the free fluctation of its conditions between strength, weakness, scarceness and abundance. 56b

(95) The soul does not have parts[157] like body parts, and thus [its] organs of perception do not have to equal the number of the things perceived. This is obvious, on the grounds that the judge of all of them is one. For one thing in the human being judges the small to be small and the big to be big, and is the judge of colours, tastes, odours and of the fact that all things which are equal to the same thing are also equal [to each other]. If the 57a perceiving organs were different, it would not be sound that one of them judged what the other perceived.

A subject of discussion is the assumption of someone who assumes that the soul is one, yet perceives the different perceptibles through multiple powers and different methods. We say that when some people examine the existents and see that some of them are composite and others simple, and then examine the perceiving instruments and power and again find some of them composite and others simple, they judge that the composite [organs of perception] perceive the composite [perceptibles] and the simple the simple. They may exemplify that by saying: I have found the composite organs of perception, like the five senses, to perceive only the composites. For the eye, which is composite of the visual power in the instruments and tunics of the eye and is only completed by their conjunction, perceives the things which are composite of the elements in different

[157] Excerpted from Miskawayh, alternative *Fawz*; cf. parallels in *K. al-Ḥikma,* MS E, fols.65a13–66a13, 60b4–61a15, 67a3–9; and *Fawz,* 67.9–71.12.

المركّبة من الاستقصات بالمزاجات المختلفة ، ووجدت أيضاً من المدركات ما
هو بسيط كالعقل والفكر والراي لا يدرك إلا الأمور البسيطة كالعلوم بحقائق 57b
الأشياء والآراء التي تستخرج بالأفكار في الأمور ، فإنّ هذه بسيطة تدرك أموراً
بسيطة . فكلّ واحد منها إنما يدرك ما لاءمه وأشبهه إن كان بسيطاً فبسيطاً
وإن كان مركّباً فمركّباً .

إلا أنّ ارسطوطالس يبحث في هذا الموضع ويقول للنفس قوّة بها تدرك الأمور الهيولانيّة المركّبة وبها تدرك الأمور غير الهيولانيّة البسيطة ولكن النحو الذي تدرك به الهيولانيّة المركّبة غير النحو الذي تدرك به الأمور البسيطة . ولو كانت النفس الناطقة تدرك المحسوسات بقوّة ما وتدرك المعقولات بقوّة أخرى لما جاز أن تردّ حكم الحسّ فيما يغلط فيه وتردّه إلى ما حكم به العقل ، كما لا يردّ ما حكمت به حاسّة إلى أخرى . ومثال ذلك أنّ الحسّ دائم الغلط في محسوسه كالعين إذا نظرت من بعيد إلى الشيء الكبير فتراه صغيراً ، حتّى ترى
الشمس وهي مثل الأرض مائة ونيفاً وستّين مرةً مثل المرآة التي قطرها فتر ، 58a
وتنظر إلى ما على شاطئ النهر إذا كانت هي في سفينة مصعدة فتراه كأنّه متحرّك منحدر وهو بالحقيقة ساكن ، وترى الشيء في الماء كبيراً وهو صغير ، ومعوجاً وهو مستقيم ، وترى الأشباح بحسب البخارات التي بينها وبينها مختلفة في الشكل واللون . وكذلك غلط الذوق فإنّ الصفراويّ يحسّ الحلو مرّاً ، وأغلاط الحسّ كثيرة ، فتعلم النفس الناطقة أنّها قد غلطت وأنّ الحقّ غير ما أحسّت فتردّ الجميع إلى حقائقها . فلو كانت النفس لا تعلم المعقول والمحسوس بقوّة واحدة لما علمت الفرق بينهما ولما ردّت الجميع إلى أمر واحد بجمعه وتحكم فيه بحكم واحد .

4 فكلّ : وكلّ *Fawz* 6 after ويقول add. إن *K. al-Ḥikma* : فيقول إن *Fawz* | after قوّة add. واحدة *Fawz* 8 after به[1] add. الأمور *Fawz* | after البسيطة add. وسنبيّن ذلك فيما بعد قال *Fawz* 11 أخرى : أخر *K. al-Ḥikma* : حاسّة أخرى *Fawz* | دائم : الدايم *K. al-Ḥikma* 12 حتّى : حين *Fawz* 17 غلط : غلظ *K. al-Ḥikma* 18 قد om. *K. al-Ḥikma* 21 بجمعه : تجمعه *Fawz* | بحكم واحد : حكماً واحداً *Fawz*

3 والآراء : والارا M 4 لاءمه : لأمه M | وأشبهه : وأشبههه M 13 وهي : وهو M | المرآة : المرأه M | فتر : قيرٌ M 14 سفينة : سَّفيةٍ M 15 الماء : المآ M 19 حقائقها : حقًا يقها word split over two lines M 20 الفرق : للفرق M

mixtures. And again I have found the simple organs of perception, like intellect, thought and opinion, to perceive only the simple things like the 57b
sciences of the true nature of things and the opinions which are deduced by means of thoughts on the[se] matters. For these are simple [organs of perception] which perceive simple things. So each of [the organs of perception] only perceives what agrees with and resembles it. If it is simple, [it will perceive the] simple, and if it is composite, [it will perceive the] composite.

Yet, Aristotle discusses this subject saying: the soul has one power by which it perceives the material, composite things and by which it perceives the immaterial, simple things, but the method by which it perceives the material composites is unlike the method by which it perceives the simple things. If the rational soul perceived the sensibles by some power and the intelligibles by another power, it would not be permissible to refuse the judgement of sense perception where it is wrong and correct it to the judgement of the intellect, just as the judgement of one sense is not [to be] corrected to the judgement of another. For example, sense perception is always erroneous with regard to its sensible just like the eye, gazing at something big from afar, sees it as something small. It even sees the sun, which is equivalent to some 160 times the 58a
earth, as equivalent to a mirror the diameter of which is a small span. If it gazes at what is on the riverbank from onboard a ship going upstream, it sees it as if it were moving downwards, whereas it is in reality at rest. It sees something in the water as big whereas it is small, and crooked whereas it is straight. It sees distant objects as different in shape and colour due to intermediate vapours. Similarly, the taste errs, for the person with a bilious mixture perceives the sweet as bitter. The errors of sense perception are many, but the rational soul knows that [the senses] have erred and that the reality is not as they have perceived it, and corrects everything to its reality. If the soul did not know the intelligible and the sensible by one power, it would not know the difference between them and would not reduce the entirety [of the intelligible and sensible data on something] to one thing as a whole on which it judges in a single judgement.

ثم يقول إنّ النفس الناطقة تدرك الأمور المعقولة بغير النحو الذي تدرك به 58b الأمور المحسوسة . وذاك أنّها إذا طلبت الأمور المعقولة انبسطت ورجعت إلى ذاتها كأنّها تطلب شيئاً هو عندها ، وإذا طلبت الأمور المحسوسة خرجت عن ذاتها كأنّها تلتمس شيئاً خارجاً عنها ، فتحتاج حينئذ إلى آلة موصلة بها إلى مطلوبها ، فإن وجدت الآلة صحيحة استعملتها فأدركت الأمر الخارج ، ثم حصلت صورتها عندها في الوهم ، وإن لم تجد الآلة فإنّها تعدم ذلك المطلوب . ومثال ذلك الأكمه ، فإنّه لا يمكنه أن يتصوّر الألوان لأنّه لم يجد آلتها ، وإذا لم يدركها من خارج لم يمكنه تحصيلها عنده وفي وهمه ، وليس ذلك حاله في المعقولات .

فأمّا المثال على ما ذكرناه من أنّ النفس إذا طلبت الأمور المعقولة رجعت إلى ذاتها فهو أنّ الإنسان إذا همّ بتحصيل رأي بديع أو فكّر في عاقبة أو أراد استخراج علم عويص ، خلا بنفسه وأبعد جميع المحسوسات عنه وكره أن 59a يشغله شيء من الحواسّ واجتهد في تعطيلها كلّها ، فيداخل نفسه حينئذ وينبسط انبساط الراجع إلى ذاته ، فيدرك ما يلتمسه من ذلك المعنى بحسب قوّتها في الانبساط وخلوها من عوارض الوهم الذي فيه صور المحسوسات ، فإنّها عائقة للنفس عن الرجوع إلى ذاتها والنظر فيما هو عندها . وهذه الحال من النفس هي حركة ما ، أعني الجولان في الطلب وهو الذي يسمّى رويّة وهو الالتجاء إلى العقل . والعقل فيه جميع الأشياء حاضرة موجودة لأنّه هو والمعقولات شيء واحد ألا يتكثّر بها . فإذا فعلت النفس ذلك فقد تحرّكت نحو تمامها إن تستكمل بالعلوم وبتأحّد بالعقل . والنفس الناطقة تدرك الأمور

1 يقول : نقول *K. al-Ḥikma* : نعود فنقول *Fawz*   4 after شيئاً add. آخر *K. al-Ḥikma* | عنها : عن ذاتها *K. al-Ḥikma* | موصلة : يتوصّل *K. al-Ḥikma, Fawz*   5 فأدركت : وأدركت *K. al-Ḥikma, Fawz*   6 صورتها : صورته *Fawz*   9 حاله : حالها *Fawz*   10 فأمّا : وأمّا *K. al-Ḥikma*   11 فهو : فهي *K. al-Ḥikma* | بديع : ينفع *K. al-Ḥikma*   13 كلّها om. *Fawz* | فيداخل : فتتداخل *K. al-Ḥikma* : فيتداخل *Fawz*   16 after عندها add. وفي خزانتها *Fawz*   19 ألا : لا *K. al-Ḥikma, Fawz*   20 after تمامها add. وتمامها *Fawz* | وبتأحّد : وتأخذ *K. al-Ḥikma* : وتتأحد *Fawz* | والنفس : فالنفس *Fawz*

4 موصلة : موصل M   5 الآلة : الالة M   7 after المطلوب blank space of three letters M   9 الأكمه : الاكهه M   9 after المعقولات blank space of three letters M   11 فهو : فهي M | بديع : يدفع M   16 فيما : فيها M | after عندها blank space of three letters M   18 الالتجاء : الالتجأ M | حاضرة : حاصرة M   19 ألا M : or read إلا ?   20 وبتأحّد :

Then he says that the rational soul perceives the intelligibles by a method unlike the one by which it perceives the sensibles. For when it seeks intelligibles, it expands and returns to itself as if seeking something which 58b
is with it, whereas when it seeks sensibles, it steps out of itself as if searching for something outside itself. So it then needs an instrument by which it is led to what it seeks. If it finds the instrument to be sound, it employs it and perceives the thing outside. Thereupon the form [of the sensibles] occurs to it in the imagination. If [the soul] does not find an instrument, it is deprived of what it seeks. An example of that is a person born blind: he cannot imagine colours, because he cannot not find their instrument, and if he does not perceive them from outside, they cannot occur to him and in his imagination. But with regard to the intelligibles [the blind man's] condition is not like that.

An example of what we have said about the soul returning to itself when it seeks intelligibles is [this]: when man intends attaining an unprecedented opinion, or considers the consequence [of something], or wants to deduce knowledge difficult to understand, he is by himself, removes all sensibles from himself, disapproves of something belonging to the senses occupying him and strives to neglect all of them. He consequently enters his self and 59a
expands in the way of someone returning to himself. So he perceives that notion he has searched for according to [his soul's] ability to expand and to free itself from the impediments of the imagination in which the forms of the sensibles are. For [these forms] hold the soul back from reversion to itself and examination of what is with it. This condition of the soul is a certain motion, i.e. revolving in pursuit, and it is what is called deliberation and the recourse to intellect. In the intellect, all things are present and existent, because it and the intelligibles are the same, lest it should be multiplied by them. In doing this the soul has certainly moved towards its perfection if it is completed through knowledge and becoming one with the intellect. The rational soul perceives the simple things without instru-

البسيطة بغير آلة بل بنفسها وتدرك الأمور المركّبة المحسوسة بتوسّط الحواسّ .
وهذا المذهب لارسطاطالس ويتبيّن منه رأيه في النفس الناطقة ، وأنّها تدرك
59b المعقولات والمحسوسات معاً ، وليس كما ظنّه قوم من أنّ الأشياء المحسوسة إنما
تدركها الحواسّ فقط والأشياء المعقولة يدركها العقل فقط ، وأنّ تلك تعلم
الجزئيّات حسب وهذا يعلم الكلّيّات فقط ، بل النفس الناطقة تدرك الجميع
بقوّة واحدة ، أعني قوّة العقل ، وإنّها وإن أدركت الجميع فإنّها تدركه بوجه
ووجه . وقد شبّه ارسطوطالس فعل النفس الناطقة في إدراكها الأشياء البسيطة
بالخطّ المستقيم وفي إدراكها الأشياء المركّبة بالخطّ المنعطف .

(96) وقال الشيخ اليونانيّ لو كانت النفس جرماً أو القوى أجراماً ، لكانت
القوى الشديدة ذوات جثث عظام ، ولكانت القوى الضعاف ذوات جثث
لطاف ، فأمّا الآن فإنّا ربّما رأيناها على خلاف هذه الصفة ، وذلك أنّه ربّما
كانت الجثّة عظيمة وكانت القوّة ضعيفة ، وربّما كانت الجثّة لطيفة وكانت
60a القوّة شديدة . فإن كان الأمر على هذا ، قلنا إنّه لا ينبغي لنا أن نضيف
القوّة إلى عظم الجثّة بل إلى شيء آخر لا جثّة له فلا عظم .
فإن قالوا الحيّ إذا مات نزف دمه وانفشّت الريح الغريزيّة التي فيه هلك ولم
يبق ، فلو كانت النفس جوهراً غير جوهر الدم والريح وسائر الأخلاط التي في
البدن ، ثم عدمها البدن ، لما مات الحيّ إذ كانت النفس غير ذي الأخلاط ،
قلنا إنّ الأشياء التي تقيم الحيّ ليست هي الأخلاط البدنيّة فقط لكن أشياء
أخر غيرها أيضاً ، وقد يحتاج الحيّ إليها في ثباته وقوامه . وإنّما هذه الأشياء

2 ويتبيّن : ونتبيّن *Fawz* 6 الجميع om. *Fawz* | تدركه : تدرك *Fawz* 9 وقال ... أجراماً : ونقول إنّه لو كانت القوى أجراماً *Uthūlūjiyā* 10 ذوات[1] : ذات *Uthūlūjiyā* | ذوات[2] : ذات *Uthūlūjiyā* 11 فإنّا : فإنّه *Uthūlūjiyā* 12 عظيمة[2] ... الجثّة om. *Uthūlūjiyā* 13 الأمر ... هذا : هذا هكذا *Uthūlūjiyā* 14 فلا : ولا *Uthūlūjiyā* 15 after قالوا add. إنّ *Uthūlūjiyā* | مات : ما برد *Uthūlūjiyā* 16 فلو : فإن *Uthūlūjiyā* 17 إذ : إذا *Uthūlūjiyā* | ذي : هذه *Uthūlūjiyā* | الأخلاط : الاختلاط *Uthūlūjiyā*, ed. Die 18 after لكن add. هي *Uthūlūjiyā* 19 وقد om. *Uthūlūjiyā*, ed. Die | ثباته وقوامه : قيامه وثباته *Uthūlūjiyā*

1 آلة : الةٍ M 2 لارسطاطالس : الارسطاطالِس M | وأنّها : وانما M | ويتاحّدُ M | after بالعقل blank space of six letters M 7 after ووجه blank space of five letters M 8 after المنعطف blank space of three letters M 11 الآن : الْان M 14 آخر : اخر M 19 هذه : هدر M

ment and by itself, whereas it perceives the composite sensibles by mediation of the senses.

This is the teaching of Aristotle and therefrom his opinion on the rational soul becomes clear: it perceives the intelligibles and the sensibles together, 59b
and it is not like some people assume, namely that only the senses perceive the sensibles and only the intellect perceives the intelligibles and that the former only know the particulars, whereas the latter only knows the universals. Rather the rational soul perceives everything by one power, i.e. the power of intellect. For although it perceives everything, it perceives [this] in one mode and [that] in another. Aristotle has already compared the activity of the rational soul while perceiving the simple things to a straight line and while perceiving the composite things to a bent line.

(96) The Greek Sage says:[158] if the soul were a body or [its] powers were bodies, the strong powers would possess large masses and the weak powers would possess small masses. Yet we sometimes see them contrary to this description, for at times the mass is large and the power weak, whereas at other times the mass is small and the power strong. Thus if 60a
matter is like that, we say that we must not relate power to the magnitude of the mass, but to something else which has no mass and thus no magnitude.

Thus if they say: when a living being dies, its blood is drawn off and [when] the innate breath in it comes forth, it perishes and does not remain. So if the soul were a substance other than the substance of blood, breath and the other humours in the body, and the body then lacked them, the living being would not die, since the soul is different from the humoral substance. We say that the things which sustain the living being are not solely the bodily humours, but also things other than these which living beings certainly need to remain and subsist. These things serve as matter

[158] Belonging to *GS*; cf. parallel in *ThA*, ed. Badawī, 46.7–11, 46.14–48.8, 49.11–14; and Dieterici, 33.17–34.2, 34.6–35.19, 37.5–9.

بمنزلة هيولى للبدن تأخذها النفس فتهيئها على صورة البدن لأنّ البدن سيّال ،
فلو لا أن النفس تمدّ جوهر البدن بهذه الأخلاط ، لما ثبت الحيّ كثير شتى .
فإذا فنيت هذه العناصر ولم تجد النفس عنصراً تمدّ به البدن فعند ذلك يهلك
60b الحيّ ويفسد . والأخلاط إنما هي علّة هيولانيّة للحيّ والنفس علّة فاعلة ،
والدليل على ذلك أنّا قد نجد بعض الحيوان لا دم له وبعضه لا ريح فيه
غريزيّة ، ولا يمكن أن يكون حيّ من الحيوان غير ذي نفس البتّة ، فليست
النفس إذن بجرم .

ويقول إن كانت النفس جرماً فلا بدّ لها من أن تنفذ في جميع البدن وتمتزج به
كامتزاج الأجرام إذا اتّصل بعضها ببعض ، وإنما تحتاج النفس إلى أن تنفذ في
جميع البدن لتنيل الأعضاء كلّها من قوّتها . فإن كانت النفس تمتزج بالبدن
كامتزاج بعض الأجرام ببعض لم تكن النفس نفساً بالفعل . وذلك أنّ الأجرام
إذا امتزج بعضها ببعض واختلط ، لم يبق واحد منها على حاله الأولى بالفعل
لكنّها تكون في الشيء بالقوّة ، فكذلك النفس إذا امتزجت بالبدن لم تكن
نفساً بالفعل بل إنما تكون بالقوّة فقط ، فتكون قد هلكت ذاتها كما تهلك
61a الحلاوة إذا امتزجت بالمرارة . فإن كان هذا هكذا وكان الجرم إذا امتزج بالجرم
لم يبق واحد منهما على حاله ، فكذلك النفس إذا امتزجت بالبدن لم تبق
على حالها الأولى ، فإذا لم تبق على حالها الأولى لم تكن نفساً .

ويقول إنّ الجرم إذا امتزج بجرم آخر احتاج إلى مكان أعظم من مكانه الأوّل ،
لا ينكر ذلك أحد ولا يدفعه . والنفس إذا صارت في البدن لم يحتج البدن إلى
مكان أعظم من مكانه الأوّل ، وكذلك إذا فارقته النفس لم يأخذ البدن مكاناً

1 هيولى : الهيولى *Uthūlūjiyā* | فتهيئها : وتهيؤها *Uthūlūjiyā* 2 شتى : شيء *Uthūlūjiyā* 5 قد om. *Uthūlūjiyā* | فيه : له *Uthūlūjiyā* 8 ويقول : ونقول *Uthūlūjiyā* | جميع : سائر *Uthūlūjiyā* 9 إلى om. *Uthūlūjiyā* 12 ببعض : بعض *Uthūlūjiyā* | واختلط : واختلطت *Uthūlūjiyā* 13 لكنّها : لكنّهما *Uthūlūjiyā* | الأولى : الأوّل *Uthūlūjiyā* | منها : منهما *Uthūlūjiyā*, ed. Die 14 هلكت : أهلكت *Uthūlūjiyā* 16–17 [1]لم ... الأولى[2] : فكذلك فلذلك *Uthūlūjiyā*, ed. Die 18 ويقول : ونقول *Uthūlūjiyā* 19–20 لا ... الأوّل : om. *Uthūlūjiyā*, واذا انفصل منه وفارقه يكفيه مكان أقلّ وليست النفس اذا اتصلت بالبدن يعظم البدن ويحتاج إلى مكان أوسع Lew 19 في : إلى *Uthūlūjiyā* | لم يحتج : لن يحتاج *Uthūlūjiyā* 20 فارقته : فارقت *Uthūlūjiyā* | after النفس add. البدن *Uthūlūjiyā*

16 منهما : منها M | حاله : حالةٍ M 17 after نفساً blank space of five letters M
18 آخر : اخرَ M

for the body, which the soul takes and prepares in accord with the form of the body, because the body is fluid, and unless the soul supplied the substance of the body with these humours, the living being would not remain throughout manifold diverse [conditions]. So if these elements perish and the soul does not find an element with which to supply the body, in that case the living being perishes and corrupts. The mixtures are only a material cause for the living being, whereas the soul is an efficient
cause. The proof for this is that we find some animals which have no blood 60b
and some in which there is no innate breath, whereas it is impossible that there is any living being which does not possess a soul at all. Therefore the soul is not a body.

He says: if the soul were a body, then it would inevitably permeate the whole body and mix with it like bodies mix when one of them is joined to another. However, the soul only needs to permeate the whole body so that all body parts obtain [something] from its power. If the soul mixed with the body like one body mixes with another, the soul would not be soul in actuality. For when a body mixes and mingles with another, neither of them remains in its previous condition in actuality, rather [the previous condition] exists in the [new mixed] thing only in potentiality. So if the soul mixed with the body, it would likewise not be soul in actuality, but it would rather be it only in potentiality, since its essence would have certainly perished, just as sweetness perishes when it mixes with bitter-
ness. If it is like that and a body when mixing with another body does not 61a
remain in its condition,[159] then the soul when mixing with the body would likewise not remain in its previous condition. Yet if it does not remain in its previous condition, it will not be soul.

He says that when a body mixes with another body, it needs a space larger than its previously [needed] space. Nobody denies or rejects that.[160] [Yet] when the soul comes to be in the body, the body does not need a space larger than its previously [needed] space and when the soul departs from it, the body does likewise not take less space than its previously [needed]

[159] Literally 'none of the two [bodies] remains in its condition'.

[160] Lewis accepts and translates the MS reading of the oldest Istanbul MS which he gives in the apparatus to his translation in *Re-examination*, p.67 (given in the apparatus as 'Lew'): 'and when it parts from it and leaves it a smaller space is enough for it, whereas when the soul joins the frame the frame does not enlarge or need more ample space' instead of 'Nobody ... than its previous one'.

أقلّ من مكانه الأوّل والنفس فيه .

ويقول أيضاً إذا صار الجرم في الجرم وامتزجا كبرت جثّتهما وعظمت ، والنفس
إذا صارت في البدن لم تكبر جثّة البدن بل هو أحرى أن يجتمع بعضه إلى
بعض ويصغر . والدليل على ذلك أنّ النفس إذا فارقت البدن انتفخ البدن
وعظم غير أنّه عظم فاسد ، فليست النفس بجرم إذن .

61b ويقول إنّ الجرم لا ينفذ في الجرم كلّه لأنّه لا يقطع جميع أجزائه ، والنفس تنفذ
في البدن كلّه وفي جميع أجزائه لأنّها لا تحتاج في نفاذها إلى أن تقطع الأجزاء
قطعاً جرميّاً ، بل تقطعها قطعاً عليّاً ، أي تحيط بجميع أجزاء الجرم ، لأنّها علّة
للجرم ، والعلّة أكثر من المعلول ولن تحتاج إلى أن تقطع معلولها بنوع المعلول
بل بنوع آخر وأعلى وأشرف .

(97) وقال افلاطن إنّ النفس ليست بأنّيّة شريرة ، وكيف تكون شريرة وهي
التي صوّرت البدن وأثرت فيه هذه الآثار العجيبة وصيّرته سعيداً شريفاً ، وهي
التي تدبّره تدبيراً محكماً متقناً ؟ فإن كانت النفس على هذه الحال التي
وصفناها فليس للقائل أن يقول إنّه يمكن أن تكون النفس شريرة ، وأن يقول
إنّ الله تعالى علّة شرّها ، وكيف تكون النفس شريرة والله خلقها خيّرة فتفيض
62a الخير على الأجرام وصيّر الخير طبيعة فيها ؟ ولم يخرج النفس من طبيعتها التي
هي الخير بل الخير فيها لم يزل ولا يزال دائماً ، فالشيء الذي هو فيها دائماً
لا يمكن أن يكون فيها عرضاً ، فإن كان الخير فيها طبيعة والطبيعة لازمة ،
فالخير لازم فيها أيضاً غير مفارق لها ، لم يبتدئ فيها بزمان ولا ينتهي إلى
زمان ما فيفارقها ، فإن كان الخير في النفس على هذه الصفة ، فالنفس إذن

1 والنفس فيه : ونقول *Uthūlūjiyā* 2 ويقول : ولا ينكر ذلك أحد ولا يدفعه *Uthūlūjiyā* 4 ويصغر : ويقل *Uthūlūjiyā* | البدن² om. *Uthūlūjiyā* 5 بجرم إذن : بجرم إذاً *Uthūlūjiyā* 6 ويقول إنّ : فإن *Uthūlūjiyā* | لأنّه ... أجزائه om. *Uthūlūjiyā* 7 لأنّها om. *Uthūlūjiyā* | نفاذها add. في الجرم *Uthūlūjiyā* | after الأجزاء : add. كلّها *Uthūlūjiyā* 8 جرميّاً : جزئيّاً *Uthūlūjiyā* | عليّاً : كلّيّاً *Uthūlūjiyā* 9 أكثر : أكبر *Uthūlūjiyā* 10 وأعلى : أعلى *Uthūlūjiyā*

3 أحرى : اخرى M 7 الأجزاء : الاجزآ M 8 تقطعها : يقطها M | عليّاً : or read علياً ? 10 آخر : اخر M | after وأشرف blank space of half a line M | أجزاء : اجْزآ M | أي : لتي M 11 بأنّيّة : بآنيه M : بآنية Zim 12 الآثار : الاثار M, Zim 14 وصفناها : وصفنا M, Zim 18 طبيعة M: الطبيعة Zim

space when the soul was in it.
He also says: when a body comes to be in a body and they mix, both of their masses become bigger and larger, whereas when the soul comes to be in the body, the mass of the body does not get bigger; rather it is more likely that part [of the body] coalesces with some other part and gets smaller. The proof for this is that when the soul departs from the body, the body swells and becomes larger, yet it is a corruptible largeness. Therefore the soul is no body.
He says that a body does not permeate an entire body, because it does not 61b
cut through all its parts, whereas the soul permeates the entire body and all its parts. For when it permeates it does not need to cut through the parts in a bodily way, but cuts through them in a causative way, i.e. it encompasses all parts of the body, because it is a cause for the body. The cause is more abundant than the effect and it does not need to cut through its effect in the manner of the effect, but in another manner which is more elevated and nobler.
(97) Plato says[161] that the soul is not an evil being. How could it be evil when it is [the soul] which has formed the body, produced in it these admirable effects and let it become happy and noble, and when it is [the soul] which directs [the body] masterfully and perfectly? So if the soul is as we have described it, then no one has the right to say that it is possible for the soul to be evil and to say that God the Sublime is the cause of its evil. How could the soul be evil when God has created it good, so that it pours the good over the bodies, and [when] He has let the good become natural
to [the soul]? The soul never leaves its nature, which is the good, rather 62a
the good has not ceased and will never cease to be in it. What is always in it cannot be in it accidentally. So if the good is naturally in it and nature is inherent, then the good is also inherent in and inseparable from [the soul], it neither originates in [the soul] at a certain time nor comes to a certain point in time when it departs from it. So if the good is in the soul according to this description, then the soul is a perpetual good. There is no

[161] According to Zimmermann ('Origins', p.221) belonging to *GS*, see his edition, pp.220–21.

خير دائم . وليس للشرّ فيها مساغ إلا أن تفسد وتهلك ، والنفس لا تفسد ولا تهلك . وذلك أنّه إن كان الخير في النفس طبيعة وكانت الطبيعة لازمة ، فلا محالة إذن أن الخير لا يفارق النفس إلا بفسادها والنفس لا تفسد ، فالنفس إذن خير دائم .

(98) وقال النفس في البدن كالجودة في السيف . وقال قبول البدن النفس كقبول الهواء حرارة النار .

62b (99) وقال إنّ النفس لا تفارق مكانها ولا تزايل مركزها ولا تبيد ولا تتجزأ . والطبيعة أداتها ، تصوّر في هيولاها صورها وتكملها بقوّتها .

ومثل لفعلها مثلاً طبيعيّاً .

(100) وقال مثل الطبيعة كالفخّار الذي يصوّر من الطينة المعجونة المهيّأة لقبول الصور أشياء كثيرة مختلفة الصور متباينة الأشكال ، ويتحرّك إلى صنعته ويصوّره بصورة مشاكلة في نفسه ولم يعطه التمام إلا بمعاونة شيء خارج من نفسه وصنعته وهي النار . ولا يقدر أن يكسو ما صنعه الكمال والتمام إلا بعد أن يجعله في الأتون ويوقّد فيه النيران ويطبخه ، فإذا يردّ فعله إلى النار يولي النار تمام فعله وإكمال نفسه ، وعلى حسب قوى تلك الأشياء ينطبخ وينضج ويقبل القوّة ويكسى البهجة ويكمل ويتمّ . فالطبيعة تتحرّك والهيولى تقبل ، وتفعل فيها الأفعال وتردّ الأفعال إلى النفس فتكمّلها وتتمّمها وتكسوها نوراً ونفساً وتظهرها وتوجدها وتجوهرها .

63a (101) وقال الدليل على بقاء النفس أنّها كلّما ازدادت بالشيء علماً كانت أقوى بجوهرها ، وغيرها كلّما ازداد علماً بمحسوسه كان أضرّ له وأقرب من فساده ، مثل البصر فإنّه كلّما ازداد استقصاء لعلم محسوسه ازداد استضراراً به.

(102) قال فوثاغورس إنّ في الإنسان جنساً إلاهيّاً ، فمتى استعمل التميّز

23 إنّ : كان فإنّه إذ كان ! يتشجّع أن لك ينبغي الإنسان أيّها انت لكنّك *W. Fūthāghūrus* | جنساً

1 مساغ Zim : مساع M 4 after دائم blank space of three letters M 5 كالجودة : or read كالحدّة or كالحدّ ? 6 after النار blank space of half a line M 7 تتجزأ : تتجزّى M 9 after طبيعيّاً blank space of three letters M 12 ويصوّره : or read وتَصَوُّره ? 13 يكسو : يكسوا M 14 يردّ : لردّ M 18 وتجوهرها : وبجوهرها with blank space of half a line after it M 19 كانت : كان M 20 بجوهرها : لجوهرها M | بمحسوسه : لمحسوسه M 21 استقصاء :

potential in it for evil unless it corrupts and perishes, but the soul neither corrupts nor perishes. For if the good is in the soul naturally and nature is inherent, then, most certainly, the good will only depart from the soul when it corrupts, but the soul does not corrupt. Therefore the soul is a perpetual good.

(98) He says: the soul in the body is like the excellence of a sword. And he says: the body's reception of the soul is like air's reception of fire's heat.

(99) He says that the soul neither departs from its place, nor leaves its
position, nor perishes nor is it divisible. Nature is its tool; it forms its forms 62b
in its matter and completes them through its power.

He exemplifies the [soul's] activity by an example [taken from] nature.[162]

(100) And he says: an example of nature is the potter who forms many things of different forms and varying shapes from a clay paste prepared to receive the[se] forms. He moves towards his product, gives it a form resembling [the form] in his soul, but he only gives it perfection by the help of something outside himself and his product, namely fire. He can only clothe what he has produced with completion and perfection after he has put it in the kiln, lit the fire in it and baked it. When he commits his work to the fire, he puts fire in charge of completing his work and perfecting himself. According to the powers of those things, [his work] gets baked, ripens, receives power, is clothed in beauty, and becomes complete and perfect. Thus nature moves and matter receives. In it [nature] performs the actions and commits them to the soul so that it completes them, brings them to perfection, clothes them in light and soul, makes them apparent, brings them about and substantiates them.

(101) He says: the proof for the immortality of the soul is that the more 63a
knowledgeable of something it becomes, the more powerful its substance becomes, whereas the more knowledgeable of its sensible something else [tries to] become, the more harmful it is for it and the closer to its corruption. An example is vision: the more it exerts itself in seeking knowledge, the more it is damaged by this.

(102) Pythagoras says[163] that in man there is an [element of] divine genus.

[162] It is not clear whether this sentence belongs to passage (99) or introduces passage (100) or if the two passages are, in fact, one. In the Marsh MS a new line is started with passage (100), but that may only be due to the fact that not much space is left in the previous line.

[163] Cf. source in Pythagoras' *Dicta aurea* as given in the lemmata of Iamblichus' commentary, pp. 108, 116, 118, which are closest to the text of the *PR*.

الصحيح ، فعند مفارقة البدن يصير مخلًّى في الجو سياحاً غير عائدا إلى الإنسانيّة ولا قابل للموت .

(103) وقال زنون النفس عالية رفيعة بعيدة من الموت لطيفة غير مجسّمة كالجسد ، وإنما تسكن الجسد فتربّيه بحكمتها ، فإذا ودّعته عند الموت رحلت عنه إلى مستقرّها الذي تحلّ فيه مع إخوانها .

(104) وقال افلاطن حياة النفس الناطقة أعمالها المحصّنة لها من آفات النفس السبعيّة ، فإنّ تلك الشهوات تطفئ نورها . فأمّا الموت فغير واقع عليها 63b للطفها وعلوّها .

(105) وقال أكثر الأنفس استعمالاً للعقل أبعدها من العود إلى الآلات الحسّيّة .

(106) وقال بعضهم النفس جوهر إلاهيّ مدبّرة للجسد والروح جسم لطيف منبثّ في الجسد . والإنسان يصير إنساناً بالنفس . فأمّا الروح فللحيوان كلّها . وكلّ ذي نفس فذو روح وليس كلّ ذي روح ذا نفس .

(107) وقال ارسطالس انفعالات الحيّ أكثرها مشتركة للنفس والبدن ، والذي يشبّه أن يخصّ النفس فهو التصوّر بالعقل .

ولا يقال للنفس وحدها حيوان إلا مع البدن ، كما أنّه لا يقال لصورة القدوم وحدها قدوم إلا مع الحديد ، ولا يقال للقوّة التي في العين إنّها بصر إلا مع العين.

... واجعل القيم المشرف على ذلك : فمتى استعمل | إلاهيّاً : جنس إلهيّ *W. Fūthāghūrus* *W. Fūthāghūrus*
1 فعند . . . البدن : فإنك عند ذلك إذا فارقت هذا البدن حتى ... *W. Fūthāghūrus* | سياحاً : تكون حينئذ سائحاً *W. Fūthāghūrus* | غير : لا *W. Fūthāghūrus* 6 افلاطن : إن *R. fī Ārā'* 9 after وقال add. أيضاً *R. fī Ārā'* | العود : العدد *R. fī Ārā'* 16 ولا . . . البدن : فكذلك النفس لا يقال لها حيوان إلا مع الجسم الطبيعيّ الآليّ *K. al-Nafs* | كما أنّه : وكذلك *K. al-Nafs* | القدوم : القدور *K. al-Nafs*, MS A 17 قدوم : قدور *K. al-Nafs*, MS A | للقوّة . . . العين : وحدها[1] *K. al-Nafs* | إنّها om. *K. al-Nafs*, MS A | بصر : تبصر *K. al-Nafs*, ed. Arn

استقصاً M 22 after به blank space of almost the entire line M
1 يصير : تصير M | مخلًّى : مُخلىً M | عائدا : عايد M 2 after للموت blank space of ten letters M 4 فتربّيه : فترنيه M 8 after وعلوّها blank space of half a line M 9 الآلات : الالات M 10 after الحسّيّة blank space of half a line M 16 لصورة : الصوة M 18 after العين blank space of half a line M

So when he employs correct discrimination, upon separation from the body he is released into the region between heaven and earth, travelling, not returning to mankind and not receptive to death.
(103) Zeno says: the soul is elevated, exalted, remote from death, fine and not embodied like the body. It only dwells in the body to educate it with its wisdom. When it then takes leave of it at death, it journeys from it to its abode in which it resides with its brethren.
(104) Plato says:[164] the life of the rational soul is its activities, which fortify it against the harms of the brute soul, for those appetites extinguish its light. As to death, it does not befall it because of its fineness and exalted- 63b
ness.
(105) He says:[165] the souls which employ the intellect more are further from reversion to the sensible instruments.
(106) One of them says: the soul is a divine substance which directs the body, whereas the spirit is a fine body which is spread in the body. Man becomes man through the soul. As to the spirit, all animals have [it]. Everything possessing a soul also possesses a spirit, but not everything possessing a spirit possesses a soul.
(107) Aristotle says: most of the affections of the living being are shared by the soul and the body. What seems to be specific to the soul is imagining through intellect.
'Living being'[166] is not said of the soul alone, but only of [the soul] together with the body, just as 'adze' is not said of the form of the adze alone, but only of it together with the iron. It is [also] not said of the power in the eye that it is vision, but only together with the eye.

[164] Cf. parallel in *R. fī Ārā*ʾ, 319.3–6.
[165] Cf. parallel in *R. fī Ārā*ʾ, 319.7–12, which also provides a lengthy explanation, for which see the commentary on the passage.
[166] From here onwards cf. *DA*-paraphrase, ed. Arnzen, 219.5–6, 4–5, 3–4; and MS A, fol.68b2–3, 1–2, 68a15–b1, which is parallel, but in reverse order.

(108) وقال الأشياء المعقولة في جواهرها لا تقبل الاستحالة والتغيّر . فالنفس
64a إذا اتّحدت بالبدن لا تفسد ولا تتغيّر ولا تختلط ، بل هي حافظة لمحض
طبيعتها معطية الحياة للأجساد التي تحلّ فيها .

وكما أنّ الشمس إذا ظهرت تغيّر الهواء إلى الضياء فتجعله مضيئاً واتّحد الضوء بالهواء وانبثّ فيه انبثاثاً على غير اختلاط ، كذلك النفس إذا اتّحدت بالبدن بقيت غير مختلطة .

(109) وقوى النفس تنقسم إلى ثلاثة أقسام ، القوّة الناطقة التي يكون بها الفكر والتميّز والنظر في حقائق الأمور ، وآلتها التي تستعملها من البدن الدماغ . والقوّة الغضبيّة التي تسمّى سبعيّة ، وبها يكون الغضب والنجدة والإقدام على الأهوال والشوق إلى التسلّط والترفّع وضروب الكرامات ، وآلتها التي تستعملها من البدن القلب . والقوّة الشهويّة التي تسمّى بهيميّة ، وبها تكون الشهوة وطلب الغذاء والشوق إلى الملاذّ وضروب اللذّات الحسّيّة ، وآلتها التي
64b تستعملها من البدن الكبد . وهذه الثلاث متباينة ويعلم ذلك من أنّ بعضها
إذا قوي أضرّ بالأخر وربّما أبطل أحدها فعل الأخر .

فلذلك وجب أن يكون عدد الفضائل بحسب أعداد هذه القوى وكذلك أضدادها التي هي رذائل . فمتى كانت حركة النفس الناطقة معتدلة وغير

1 وقال : وإذ قد تبيّن أنّ *Fī Ṭabīʿa* | after والتغيّر add. إذا أن تكون اضطراراً فقد يتبع ذلك
after | *Fī Ṭabīʿa* قد add. إذا 2 after *Fī Ṭabīʿa* اتّحدت إلا تتغيّر ولا تفسد لاتّحادها بما اتّحدت به
بالبدن add. باقية في ذلك وهي ... من غير اختلاط اتّحاداً *Fī Ṭabīʿa* | ولا تتغيّر om. *Fī Ṭabīʿa*
بل هي om. *Fī Ṭabīʿa* 3 فيها ... معطية : ولا لها التي الحياة إلى تحيله فهي فيه حلّت وكلّما
تستحيل إليه *Fī Ṭabīʿa* 4 فتجعله : وتجعله *Fī Ṭabīʿa* | واتّحد : فيتّحد *Fī Ṭabīʿa* 5 وانبثّ :
وينبثّ *Fī Ṭabīʿa* 6 after بقيت add. أيضا لا محالة *Fī Ṭabīʿa* 7 وقوى النفس : وقد تبيّن للناظر
انها وقواها النفس هذه أمر في *Tahdhīb* | after أقسام add. أعني *Tahdhīb* | الناطقة om.
Tahdhīb | يكون بها : بها يكون *Tahdhīb* 8 والتميّز : والتمييز *Tahdhīb* 9 after الغضبيّة add.
هي *Tahdhīb* | سبعيّة : السبعيّة *Tahdhīb* | وبها : والقوة التي بها *Tahdhīb* 11 after الشهويّة
add. هي *Tahdhīb* | بهيميّة : البهيميّة *Tahdhīb* | وبها : والقوة التي بها *Tahdhīb* 12 after
الملاذّ add. التي في المآكل والمشارب والمناكح *Tahdhīb* 14 بالأخر : بالآخر *Tahdhīb* | الأخر :
الآخر *Tahdhīb*

3 after فيها blank space of three letters M 4 فتجعله : فجعله M | مضيئاً : مضياً M
6 after مختلطة break (ه) ind. M 7 ثلاثة : ثلثه M 9 وبها : ونها M | والنجدة : والتحدة M
11 after القلب blank space of half a line M 12 الغذاء : الغذأ M | وآلتها : والتها M
13 الثلاث : الثلث M 14 أبطل : انطل M | أحدها : احدهما M | after الأخر blank space of three letters M

(108) He says:[167] The intelligibles receive neither change nor alteration in their substances. Thus when the soul is joined to the body, it neither corrupts nor alters nor mingles. Rather it is the guardian of the pureness 64a
of its nature and the lifegiver of the bodies in which it resides.
Just as the sun when manifesting itself alters air to light and makes it lit, the light is united with the air and is spread through it without mingling. In the same way the soul, when united with the body, remains unmingled.
(109) The powers of the soul are divided into three parts:[168] [There is] the rational power due to which there is thought, discernment and the examination of the reality of things. Its instrument, which is the body part it employs, is the brain. [Then there is] the irascible power, which is called brute, and due to which there is anger, bravery, fearlessness in terrifying situations, desire for dominion, disdain and varieties of honour. Its instrument, which is the body part it employs, is the heart. [There is finally] the appetitive power, which is called beastly, and due to which there is appetite, pursuit of food, desire for delights and sorts of sensual pleasures. Its instrument, which is the body part it employs, is the liver. These three are 64b
distinct from one another. This is known from the fact that when one of them gathers strength, it impairs the others and sometimes one of them counteracts the action of the others.
Therefore it is necessary that the number of virtues, and likewise the number of their opposites, namely the vices, accords with the number of these powers. When the motion of the rational soul is in balance, not out

[167] Cf. source in Nemesius, *K. al-Abwāb fī ṭabīʿat al-insān*, 80.2–5, 80.18–81.4.

[168] Cf. Miskawayh, *Tahdhīb*, 15.10–12, 16.2–3, 4–5, 15.12–14, 16.5, 3, 15.14–16, 16.4, 15.16–17, 16.6–19.

خارجة عن ذاتها وكان شوقها إلى المعارف الصحيحة ، حدثت فيها فضيلة العلم وتتبعها الحكمة . ومتّى كانت حركة النفس البهيميّة معتدلة منقادة للنفس العاملة غير متباينة عليها فيما تقسّطه لها ولا منهمكة في اتّباع هواها ، حدثت عنها فضيلة العفّة وتتبعها فضيلة السخاء . ومتّى كانت حركة النفس الغضبيّة معتدلة تطيع النفس العاقلة فيما تقسط لها فلا تهيج في غير حينها ولا تحمى أكثر مما ينبغي ، حدثت عنها فضيلة الحلم وتتبعها فضيلة الشجاعة .
65a ثم يحدث عن هذه الفضائل الثلاث باعتدالها ونسبة بعضها إلى بعض فضيلة
هي كمالها وتمامها ، وهي فضيلة العدالة . فلذلك أجمع الحكماء أنّ أجناس الفضائل أربع وهي الحكمة والعفّة والشجاعة والعدالة .

(110) وكلّ واحد من هذه الفضائل وسط بين رذيلتين . فالحكمة وسط بين السفه والبله ، أعني بالسفه استعمال القوّة الفكريّة فيما لا ينبغي ، وكما لا ينبغي وسمّاه قوم الجربزة ، وأعني بالبله تعطيل هذه القوّة واطّراحها بالإرادة ، ولست أريد بالبله هاهنا نقصان الخلقة بل ما ذكرنا . والعفّة وسط بين الشره وهو الانهماك في اللذات وبين خمود الشهوة . والشجاعة وسط بين الجبن والتهوّر . والسخاء وسط بين السرف والتبذير والبخل والتقتير . والعدالة وسط بين الظلم والاستكانة .

65b (111) وهذه القوى الثلاث إحداها كريمة أديبة بالطبع ، وهي النفس الناطقة .

1 الصحيحة add. لا المظنونة معارف وهي بالحقيقة جهالات *Tahdhīb* | فيها : عنها *Tahdhīb* 3 العاملة : العاقلة *Tahdhīb* | متباينة : متأبية *Tahdhīb* 6 ينبغي add. لها *Tahdhīb* 9 أربع : أربعة *Tahdhīb* 10 وكلّ...رذيلتين : وينبغي أن يفهم من قولنا انّ كلّ فضيلة فهي وسط بين رذائل *Tahdhīb* | فالحكمة : أمّا الحكمة فهي *Tahdhīb* 11 أعني : وأعني *Tahdhīb* | بالسفه add. ههنا *Tahdhib* 12 قوم : القوم *Tahdhīb* | بالإرادة om. *Tahdhīb* 13 ولست...هاهنا : وليس ينبغي أن يفهم أنّ البله ههنا *Tahdhīb* | ما ذكرنا : تعطيل هذه القوّة بالإرادة *Tahdhīb* | والعفّة : وأمّا العفّة فهي *Tahdhīb* | after بين add. رذيلتين *Tahdhīb* 13–14 الشره...الشهوة : وهما الشره وخمود الشهوة وأعني بالشره الانهماك في اللذّات *Tahdhīb* 14 والشجاعة : وأمّا الشجاعة فهي *Tahdhīb* | after بين add. رذيلتين أحداهما *Tahdhīb* 15 والتهوّر : والأخرى التهوّر *Tahdhīb* | والسخاء : وأمّا السخاء فهو *Tahdhīb* | after بين add. بين رذيلتين أحداهما *Tahdhīb* | والبخل : والأخرى البخل *Tahdhīb* | والعدالة : وأمّا العدالة فهي *Tahdhīb* 16 والاستكانة : والانظلام *Tahdhīb* 17 وهذه...إحداها : بعد أن تعلم أنّ بعض هذه ...

4 السخاء : السَخا M 7 الثلاث : الثلث M 13 after ذكرنا blank space of six letters M 14 الانهماك : الا نهماك word split over two lines M 15 والتبذير : والتبدير M 16 after والاستكانة break (∴) ind. M 17 الثلاث : الثلث M | إحداها : احديها M

of kilter with itself and it desires to come to know what is correct, the virtue of knowledge emerges in it followed by wisdom. When the motion of the brute soul is in balance, yielding to the knowing soul, not abandoning it with regard to what the latter assigns to it, and not absorbed in the pursuit of its passion, the virtue of moderation emerges from it followed by the virtue of liberality. When the motion of the irascible soul is in balance, and it obeys the intellective soul with regard to what the latter assigns to it so that it neither is stirred up at an inopportune time nor flares up more than necessary, the virtue of self-control emerges from it followed by the virtue of courage. Thereupon a virtue emerges from these three virtues due to their balance and mutual relation, and this is their 65a
completion and perfection, i.e. the virtue of justice. Therefore the sages have agreed that the genera of the virtues are four, namely wisdom, moderation, courage and justice.

(110) Each one of these virtues[169] is a mean between two vices. Wisdom is a mean between impudence and stupidity. By impudence I mean the employment of the cogitative faculty where and how it should not be [employed]. Some people have called it 'deceitfulness'. By stupidity I mean the obstruction of this faculty and its wilful rejection. By stupidity here I do not mean a shortcoming in the inborn disposition, but what we have mentioned. Moderation is a mean between greed, which is the absorption in pleasures, and the extinction of the appetite. Courage is a mean between cowardice and recklessness. Liberality is a mean between excess and dissipation and avarice and stinginess. Justice is a mean between tyranny and submission.

(111) One of these three powers[170] is excellent and well disciplined by nature, i.e. the rational soul. The second one is abject, lacks discipline by 65b

[169] Cf. Miskawayh, *Tahdhīb*, 25.5, 26.9–13, 27.11–12, 21–2, 28.2–3, 28.6.

[170] Cf. Miskawayh, *Tahdhīb*, 51.18–19, 51.21, 51.19–20, 52.1–2, 51.20–1, 52.2–3, 52.5–9, 52.13–22.

والثانية مهينة عادمة للأدب بالطبع ليس فيها استعداد لقبول الأدب ، وهي النفس الشهوانيّة البهيميّة . والثالثة عادمة للأدب إلا أنّها تقبل التأديب وتنقاد للأدبيّة ، وهي النفس الغضبيّة .

وقد شبّه القدماء الإنسان وحاله في هذه الأنفس الثلاث برجل راكب بهيمة قويّة يقود كلباً أو فهداً للقنص . فإن كان الرجل هو الذي يروض دابّته وكلبه فيصرّفهما ويطيعانه في سيره وتصيّده وسائر متصرفاته ، فلا شكّ في رغد العيش المشترك بين الثلاثة . وإذا كانت البهيمة هي الغالبة ساءت حال الثلاثة لأنّها لا تطيع فارسها ، فمتى رأت عشباً من بعيد عدت نحوه وتعسّفت في عدوها وعدلت عن الخدد فاعترضتها الأودية والوهاد والشوك والشجر وتورّطت فيها ولحق فارسها أنواع المكاره والأشراف على الهلكة . وكذلك إن [66a] قوي الكلب لم يطع صاحبه ، فإن رأى صيداً من بعيد أخذ نحوه وجذب الفرس وراكبه وألحق الجميع من الضرّ والضرر أضعاف ما ذكرناه .

(112) وقوى النفس تابعة إلى مزاج البدن . فمن كان مزاج بدنه مستوياً كان الروح في بدنه مستوياً وكانت أفعال النفس مستوية ، ومن قصر مزاج بدنه عن الاعتدال المخصوص به قصر الروح والأفعال عن الاستواء ، ولذلك صارت

وهي النفس : *Tahdhīb* | أمّا الكريمة الاديبة بالطبع فالنفس *Tahdhīb*
1 والثانية : وبعضها *Tahdhīb* | ليس : وليس *Tahdhīb* | وهي : وهي مع وأمّا العادمة للأدب
Tahdhīb وبعضها : والثالثة | om. *Tahdhīb* 2 الشهوانيّة *Tahdhīb* ذلك غير قابلة له فهي
3 للأدبيّة : للتي هي أديبة *Tahdhīb* | وهي : فهي وأمّا التي عدمت الأدب لكنها تقبله وتنقاد له
Tahdhīb 4 برجل : بإنسان *Tahdhīb* 5 الرجل : الإنسان من بينهم *Tahdhīb* 6 فيصرّفهما :
يصرّفهما *Tahdhīb* | متصرفاته : تصرفاته *Tahdhīb* 8 تطيع . . . لأنّها : وكان الإنسان كلّهم
Tahdhīb فإن : فمتى | *Tahdhīb* وغلبت .after فارسها add | *Tahdhīb* مضعوفاً عندها فلم تطع
فارسها 10 after *Tahdhīb* فتقحمتها .after والشجر add | *Tahdhīb* الطريق النهج : الخدد 9
ما لا خفاء .after الهلكة add | *Tahdhīb* ما يلحق مثله في هذه الأحوال فيصيبها جميعاً من .add
فجذب : وجذب | *Tahdhīb* من بعيد صيداً أو ما يظنّه صيداً : صيداً . . . بعيد 11 *Tahdhīb* به
ولذلك قالت : وقوى 13 *Tahdhīb* ولحق : وألحق | *Tahdhīb* وفارسه : وراكبه 12 *Tahdhīb*
R. fī في غاية الاستواء : مستوياً | *R. fī l-Farq* لمزاج : إلى مزاج | *R. fī l-Farq* الفلاسفة إنّ قوى
R. fī l-Farq كانت الروح التي في بدنه في غاية الاستواء وكانت : كان² . . . وكانت 13–14 *l-Farq*
أعني الأعضاء التي فيها الروح .after بدنه² add | *R. fī l-Farq* فيها في غاية الاستواء : مستوية 14
قصرت أيضاً الروح عمّا يجب لها من : قصر . . . الاستواء | *R. fī l-Farq* بها : به 15 *R. fī l-Farq*
R. fī l-Farq الدقّة واللطف وقصرت أفعال النفس فيها على تلك النسبة

M سات : ساءت | M الثلثة : الثلاثة 7 M للقب : للقنص 5 M الثلث : الثلاث 4
M إلا : إلى 13 M after ذكرناه blank space of six letters 12 M الثلثة : الثلاثة 8
14–15 عن . . . الاستواء add. in marg. M

nature and has no inclination to receive discipline, i.e. the appetitive brute soul. The third one lacks discipline, yet it is receptive to the teaching of discipline and yields to the well-disciplined [soul]. This is the irascible soul.

The Ancients have already compared man and his condition with regard to the three souls to a man riding a powerful beast who leads a dog or a lynx in a hunt. So if the man is someone who trains his riding animal and his dog so that he directs them and they obey him in his conduct, his hunting and his other acts of directing [them], there is no doubt about the agreeable life shared between the three. If the beast prevails, the state of the three is bad, for it does not obey its rider: so when it sees grass from afar, it dashes towards it, takes the course of its run at random and deviates from the obvious tracks of the road, so river beds, hollows, thorns and trees stand in its way, it falls into them and [all] kinds of calamities afflict its rider and he comes to the verge of perdition. In the same way if the dog gathers strength, it does not obey its master: so if it sees prey from afar, it 66a
rushes towards it, drags the horse and its rider [along], and they all are afflicted by damage and harm which are several times as bad as what we have mentioned.

(112) The powers of the soul follow the mixture of the body.[171] In the case of having a well-balanced body mixture, one will have well-balanced spirit in one's body and well-balanced activities of one's soul. In the case of having a body mixture which fails to achieve its correct equilibrium, one will have spirit and psychic activities which also fail to achieve a balance.

[171] Cf. source in Qusṭā ibn Lūqā, *R. fī l-Farq*, ed. Troupeau, lines 280–6.

قوى النفس في الصبيان ناقصة وفي النساء ضعيفة ، وكذلك في الأمم التي غلب على أمزجتها الحرّ أو البرد كالزنج والصقالبة ومن شابههم .

(113) وقال ارسطالس النفس تابعة للبدن والبدن تابع للنفس ، يألم احدهما بألم الآخر ويتغيّر سجيّة الواحد منهما بتغيّر الثاني . ويتبيّن ذلك في السرور والحزن وهما خلقان من أخلاق النفس ، فإنّ الوجه عند الحزن يرى كئيباً عبوساً وعند السرور يرى هشّاً طلقاً . ويتبيّن بأكثر من ذلك وهو أنّ الجنون أحد العوارض التي تعرض للنفس ، فمتى حدث هذا العارض واستعمل معه الأطبّاء
66b تنقية البدن والتدبير الملائم أزالوه وأشفوا الإنسان منه ، ومع بطلان تلك الحال التي للبدن في وقت الجنون يكون بطلان ذلك الجنون .

(114) وقال افلاطن قد يصل الضرر إلى البدن من قبل مرض النفس كما يصل إلى أفعال النفس الضرر بسوء مزاج البدن .

(115) وقال بعضهم الأمور الطبيعيّة التي بها قوام البدن ستّة أركان وأمزجة وأخلاط وأعضاء وقوى وأفعال .

والأركان هي المادّة البعيدة لأنّها هي الموضوع الأوّل لجميع ما يكوّن ويفسد ، وهي النار والهواء والماء والأرض ، والنار حارّة يابسة من دوام حركة الفلك

1 after التي add. قد *R. fī l-Farq* 3 قال : وقد أرى أنّ : وقال ارسطالس *K. fī l-Firāsa* النفس والبدن : النفس ... للنفس | ارسطوطاليس في كتاب الفراسة وقد أرى أنّ *R. fī l-Ṭibb* *K. fī l-Firāsa, R. fī l-Ṭibb* | before يألم add. متى *K. fī l-Firāsa* 4 يألم : بألم *K. fī l-Firāsa, R. fī l-Ṭibb* | ويتغيّر ... الثاني : فمتى تغيّرت سحنة النفس تغيّرت معها صورة البدن ومتى تغيّرت أيضاً صورة البدن تغيّرت معها سحنة النفس *K. fī l-Firāsa, R. fī l-Ṭibb* | ويتبيّن : وسنبيّن *K. fī l-Firāsa* : وبيّن *R. fī l-Ṭibb* 5 الوجه : النفس *K. fī l-Firāsa* | after يرى add. رؤية بيّنة *K. fī l-Firāsa, R. fī l-Ṭibb* | عبوساً : عبساً *K. fī l-Firāsa, R. fī l-Ṭibb* 6 وعند ... يرى : ويرى عند السرور *K. fī l-Firāsa, R. fī l-Ṭibb* | هشّاً : باشّاً *K. fī l-Firāsa, R. fī l-Ṭibb* | ويتبيّن : وقد تبيّن *K. fī l-Firāsa, R. fī l-Ṭibb* | بأكثر ... ذلك : ذلك بأكثر من هذا القول ممّا أصف *K. fī l-Firāsa, R. fī l-Ṭibb* | after الجنون add. كما تعلم *K. fī l-Firāsa* : add. كما قد تعلم *R. fī l-Ṭibb* 7 معه : فيه *R. fī l-Ṭibb* 8 أزالوه ... منه : شفوا بذلك من الجنون بعلاجهم للبدن *K. fī l-Firāsa, R. fī l-Ṭibb* 12–13 وقال ... وأفعال : كم هي الأمور الطبيعيّة ؟ سبعة أشياء . وما هي ؟ الأركان والأمزجة والأخلاط والأعضاء والقوى والأفعال والأرواح *Masāʾil* 14–15 والأركان ... والأرض : كم هي الأركان ؟ أربعة . وما هي؟ النار والهواء والماء والأرض *Masāʾil* 15–200.2 والنار ... الماء : ما قوّة النار ؟ حارّة ياسبة . ما قوّة الهواء ؟ حارّ رطب . ما قوّة الماء ؟ بارد رطب . ما قوّة الأرض ؟ باردة يابسة *Masāʾil*

1 النساء : النسا M 2 after شابههم blank space of almost an entire line M 4 الآخر : الْاَخِر M 4–6 ويتبيّن ... ذلك add. in marg. M 8 تنقية : بنقّيه M | وأشفوا : واشفو M
9 after الجنون[2] blank space of half a line M

Therefore the powers of the soul are deficient in young people and weak in women, and likewise in the peoples over whose mixtures the warm or the cold prevails, as the Black Africans, the Slavs and their likes.
(113) Aristotle says:[172] the soul follows the body and the body follows the soul. Each of them suffers from the pain of the other and the temper of one of them alters due to an alteration of the second. That becomes clear in joy and grief, which are both natural dispositions of the soul. So the face[173] looks dispirited and gloomy in grief, but cheerful and happy in joy. It becomes even clearer than that in madness which is one of the accidents which befall the soul. When this accident occurs and the physicians treat it by purging the body and an appropriate regimen, they make it stop and cure the person of it. Along with the abolition of the condition in which 66b
the body is when mad, that madness is abolished.
(114) Plato says: sometimes harm comes to the body from an illness of the soul, just as harm comes to the activities of the soul due to a bad mixture of the body.
(115) Someone says: the natural matters[174] through which the body subsists are six: elements, temperaments, humours, body parts, faculties and activities.
The elements are remote matter because they are the first substratum for all that is generated and corrupts. They are fire, air, water and earth. Fire is hot and dry due to the perpetuity of the sphere's motion above it. Air is

[172] Cf. source in pseudo-Aristotelian *K. fī l-Firāsa*, 29.8–12, 30.4–8; cf. also parallel in Ibn Bakhtīshūʿ, *R. fī l-Ṭibb*, 35.1–5, 35.10–13, which version is closer to the *K. fī l-Firāsa* than the text of the *PR*.

[173] Here the *PR* and the *R. fī l-Ṭibb* agree on the same reading *al-wajh* against the *K. fī l-Firāsa* which reads *al-nafs*. The reading of the former should be accepted, as the edition of the *K. fī l-Firāsa* is based on only one manuscript.

[174] Cf. Ḥunayn's *Masāʾil*, pp. 2–5, 9–11, 13–18, 255–8, on the natural matters, the elements, the humours, the organs and the powers.

عليها ، والهواء رطب حارّ من مماسته للنار ، والماء بارد رطب من مماسة للهواء ، والأرض باردة يابسة من مجاورتها الماء .

والأمزجة اختلاط هذه الأركان واستحالة بعضها إلى بعض ليتهذّب منها مادّة أخرى أقرب إلى قبول صورة النبات والحيوان والمعادن .

67a والأخلاط مادّة للحيوان الذي له دم ، وهي الصفراء والدم والبلغم والسوداء ، والصفراء حارّة يابسة كالنار ، والدم حارّ رطب كالهواء ، والبلغم بارد رطب كالماء والسوداء باردة يابسة كالأرض .

والأعضاء البسيطة كاللحم مادّة للأعضاء المركّبة التي تسمّى آليّة كاليد والرجل ، فإنّهما مادّة لجملة البدن ، وجميع ذلك أربعة أصناف . (1) أصول تنبعث منها جميع القوى وهي القلب والدماغ 〈والكبد〉 والأنثيان . (2) وما ينشو منها ليخدمها وينفذ فيه قواها إلى جميع البدن وهو الشرايين والأوردة والعصب وأوعية المنى ، فالشرايين تخدم القلب ، والعصب تخدم الدماغ ، والأوردة تخدم الكبد وأوعية المنى يخدمان الأنثيين . (3) وما ينفرد بقوّة طبيعيّة تعمّر جميع البدن وهو العظم والغضروف ونحوهما . (4) وما يختصّ بقوّة تجري

3-4 والأمزجة... والمعادن : كم هي أصناف المزاج ؟ تسعة . وما هي ؟ ثمانية منها غير معتدلة ، وواحد معتدل . ومن الثمانية الخارجة عن الاعتدال . أربعة مفردة ، وهي الحارّ ، والبارد ، والرطب ، واليابس . وأربعة مركّبة وهي الحارّ اليابس ، والحارّ الرطب ، والبارد اليابس ، والبارد الرطب *Masāʾil* 5 والأخلاط... والسوداء : كم هي الأخلاط ؟ أربعة . وما هي ؟ الدم ، والبلغم ، والمرّة الصفراء ، والمرّة السوداء *Masāʾil* 6-7 والصفراء... كالأرض : ما قوّة الدم ؟ حارّ رطب . ما قوّة البلغم ؟ بارد رطب . ما قوّة المرّة الصفراء ؟ حارّة يابسة . ما قوّة المرّة السوداء ؟ باردة يابسة ... *Masāʾil* 8-9 والأعضاء... البدن : كم هي أصناف الأعضاء إذا قسمت جواهرها ؟ صنفان ، وما هما ؟ ان منها ما بسيطة مفردة عنه الحيّ ويقال لها المتشابهة الأجزاء ومنها ما هي مركّبة وتسمّى أعضاء آليّة . ما مثال الأعضاء المتشابهة الأجزاء ؟ ... واللحم المفرد ... ما مثال الأعضاء الآليّة ؟ الرأس واليد والرجل ... *Masāʾil* 9-10 وجميع... والأنثيان : كم هي أصناف الأعضاء ؟ أربعة . وما هي ؟ منها ما هي رئيسية كالأصول والمعادن وهي أربعة ، أعني الدماغ والقلب والكبد والانثيين *Masāʾil* 10-13 وما... الأنثيين : ومنها ما يخدم تلك الأعضاء الرئيسيّة . أعني الدماغ يخدمه العصب ، والقلب تخدمه العروق الضوارب . والكبد تخدمه العروق غير الضوارب والأنثيان تخدمهما أوعية المنى *Masāʾil* 13-14 وما... ونحوهما : من الأعضاء أعضاء فيها قوى غريزيّة بها يكون تدبيرها وقوام أمرها مثل العظام والغضاريف والأغشية والرباطات والشحم واللحم *Masāʾil* 14-202.2 وما... : ومنها ماله قوى غريزيّة فيها وقوى أخرى تجرى إليها من تلك الأصول ، مثل المعدة والأمعاء والكلى والطحال وجميع العضل ، فإن هذه الأعضاء فيها قوى غريريّة بها تجذب الغذاء والعضلات

1 after للنار blank space of three letters M 4-5 أخرى... مادّة add. in marg. M 8 آليّة : الية M 10 والأنثيان : والانثيين M 11 ينشو : ينشوا M | فيه : فيها M 13 والأوردة : والا ‌ وردة word split over two lines M

moist and hot due to its contact with fire. Water is cold and moist due to its contact with air. Earth is cold and dry due to its vicinity to water.

The temperaments[175] are the mixture of these elements and their interchange with each other, so that another matter is shaped from them which is more proximate to receiving the form of plants, animals and minerals.

The humours are matter for animals which have blood. They are yellow bile, blood, phlegm and black bile. Yellow bile is hot and dry like fire, blood 67a
is hot and moist like air, phlegm is cold and moist like water, and black bile is cold and dry like earth.

The simple body parts,[176] like flesh, are matter for the composite body parts which are called instrumental, like hand and foot. For these two [kinds of body parts] are matter for the entire body. All of them are of four kinds: (1) Foundations from which all powers originate, namely the heart, the brain, ⟨the liver⟩ and the testicles. (2) What emerges from them to serve them and to distribute their powers through the entire body, namely the arteries, the veins,[177] the nerves and the spermatic ducts. For the arteries serve the heart, the nerves the brain, the veins the liver and the spermatic ducts the testicles. (3) What possesses one specific natural power supplied to the entire body, namely bone, cartilage and their like. (4) What is specially distinguished by power flowing towards it from the

[175] The passage on the temperaments is completely different in Ḥunayn's *Masāʾil* (3.10–4.6), where nine temperaments are distinguished. Only one of them is balanced, whereas there are eight unbalanced ones, four of them single (hot, cold, moist, dry) and four composite (hot-dry, hot-moist, cold-dry, cold-moist).

[176] The first division of body parts is into simple and composite ones which, due to the lack of any introductory statement, is not very obvious.

[177] An indication, among others (see commentary), that the *PR* is not using Ḥunayn directly, is the fact the *PR* uses the terms *shiryān* (artery) and *warīd* (vein), whereas the *Masāʾil* have *ʿirq (ghayr) ḍārib*, which is also the more common term in the other passages of the *PR* deriving from different sources.

إليه من الأصول يكون بها الحسّ والحركة الإراديّة وهو المعدة والأمعاء والكلى والعضلات .

والقوى هي الصورة الأولى وأصنافها ثلاثة طبيعيّة وحيوانيّة ونفسانيّة ، فالطبيعيّة 67b ثلاثة المولّدة والمربّية والغاذية ، أمّا المولّدة فهي القوّة التي تحيل النطفة وتغيّرها حتّى يعمل منها أعضاء متشابهة الأجزاء ، ثم يؤلّف من هذه الأعضاء أعضاء آليّة ويؤلّف من هذه الأعضاء الآليّة جملة البدن . والمربّية ⟨هي⟩ القوّة التي تمدّد الأعضاء طولاً وعرضاً وتنقلها من الصغر إلى الكبر. والغاذية هي القوّة التي تشبّه الغذاء بالمغتذي وتجعله خلف ما نقص منه . والحيوانيّة صنفان فاعلة ومنفعلة ، فالفاعلة القوّة التي تحدث انبساط القلب والعروق الضوارب ، والمنفعلة القوّة التي يكون بها الغضب والأنفة والمنازعة للغلبة والنباهة . والقوى النفسانيّة ثلاثة المدبّرة والمحرّكة بإرادة والحسّاسة ، فالقوى المدبّرة ⟨ثلاثة وهي القوة التي يكون بها التخيل ، والقوة يكون بها الفكر والقوة التي يكون بها الذكر ، والقوى المحركة بإرادة⟩ هي القوى التي تحرّك العضل فيتحرّك بها

وتغيّره وتفعل سائر أفعالها على الحالة الطبيعيّة ولها أيضاً قوى أخرى تجرى إليها تلك الأصول إمّا أن
يكون بها الحسّ والحياة فقط وإمّا أن يكون بها مع ذلك الحركة الإرادية *Masāʾil*
3 والقوى . . . الأولى om. *Masāʾil* | وأصنافها . . . ونفسانيّة : وما كم هي أصناف القوى ؟ ثلاثة. وما
3–4 فالطبيعيّة . . . *Masāʾil* هي ؟ منها ما هي طبيعيّة ومنها ما هي حيوانيّة ومنها ما هي نفسانيّة
كم هي أصناف القوى الطبيعيّة المخدومة ؟ ثلاثة . وما هي ؟ المولّدة والمربّية والغاذية : والغاذية
والمربّية هي : *Masāʾil* | 4 أمّا . . . فهي : أي القوى هي المولّدة ؟ *Masāʾil* 6 هذه om. *Masāʾil*
7 after وعرضاً add. وعمقاً *Masāʾil* | الكبر : العظم *Masāʾil* أي القوى هي القوّة المربّية ؟
8 خلف ما : خلفا مكان ما *Masāʾil* | والغاذية هي : أي القوى هي القوّة الغاذية ؟ *Masāʾil*
8–9 والحيوانيّة . . . ومنفعلة : كم هي أصناف القوى الحيوانيّة : صنفان . وما هما ؟ ان *Masāʾil*
9 فالفاعلة : أمّا الفاعلة *Masāʾil* | after فالفاعلة *Masāʾil* منها ما هي فاعلة ومنها ما هي منفعلة
10 والمنفعلة : فالقوّة المنفعلة وأمّا القوّة المنفعلة فالقوّة *Masāʾil* والقوّة التي تحدث انقباضها add. الضوارب
والأنفة : والقوّة التي تكون بها الأنفة *Masāʾil* | الغضب : والقوّة التي تكون بها الغضب *Masāʾil*
والمنازعة للغلبة : والقوّة التي تكون بها المنازعة للغلبة والترأس *Masāʾil* | after والنباهة *Masāʾil*
10–11 والقوى . . . والحسّاسة : كم هي أصناف القوى والأحداث النفسانيّة *Masāʾil* add.
النفسانية ؟ ثلاثة . وما هي ؟ منها ما هي مدبّرة سياسة ، ومنها ما تحرّك بإرادة ومنها حسّاسة
Masāʾil 11 فالقوى : والقوى *Masāʾil*

1 إليه : اليهَا M 3 ثلاثة : ثلثَه M 4 ثلاثة : ثلثه M 6 آليّة : اليه M | الآليّة : الاليه M
after البدن blank space of three letters M 7 after الكبر blank space of ten letters M
8 الغذاء : الغدآ M | بالمغتذي : بالمعتَدي M 9 after الضوارب blank والغاذية : والعَاذيه M
space of three letters M 10 الغضب : العَضبُ M 11 ثلاثة[1] : ثلثه M | والمحرّكة : والمتحركة M
13 العضل : الفضل M

foundations and providing sense perception and voluntary movement, namely the stomach, the intestines, the kidneys and the muscles.

The powers are the first form and their kinds are three: natural, vital and 67b
psychic. The natural are three:[178] the reproductive, augmentative and nutritive faculties. The reproductive faculty transforms the sperm and alters it to produce homoeomerous organs from it. From these organs it then combines instrumental organs and from these instrumental organs the entire body. The augmentative faculty extends the organs lengthwise and breadthwise and turns them from small to big. The nutritive faculty assimilates the nutriment to what is nourished and uses it to replace what is depleted from it. The vital powers are two kinds: acting and being acted upon. The acting power causes the heart and the arteries to expand, whereas the power which is acted upon causes anger, pride, contentiousness and alertness. The psychic powers are three: the directive, voluntary locomotive and sensitive faculties. The directive faculties ⟨are three, namely the faculty which causes imagination, the one which causes thought and the one which causes memory. The voluntary locomotive faculties⟩[179] move the muscles by which then the voluntarily moving limbs

[178] Ḥunayn (p.12) distinguishes two sets of natural faculties: serving and being served. The first are the attractive, retentive, digestive and repellent faculties, which all serve the nutritive faculty, which, in turn, serves the augmentative faculty, which serves the reproductive power.

[179] I have completed the obviously incomplete text of the *PR* using Ḥunayn's *Masāʾil*.

68a الأعضاء المتحرّكة بإرادة . والحسّاسة هي قوّة البصر والسمع والمذاق والشمّ واللمس .

والأفعال ⟨هي⟩ الصورة الأخيرة ، أعني العلّة التامّيّة الصادرة عن القوى والأرواح بمنزلة الآلة الطبيعيّة ، والطبيعة هي القوّة المدبّرة للأبدان .

(116) قال افلاطن الأعضاء الرئيسيّة ثلاثة الدماغ والقلب والكبد ، فالدماغ مسكن النفس الناطقة وله الحسّ والحركة والفكر والرويّة . والقلب مسكن النفس المحسّة ومنه الغضب والشجاعة . والكبد مسكن النفس الغاذية وله التربية والنمو ومنه الشهوة .

(117) الدماغ أصل الحسّ والحركة ويخدمه العصب وهو يقبل ما يؤدّي إليه جميع الحواس ويتخيل الأشياء ويتوهّم المعقولات .

(118) والنفس الناطقة تسكن الدماغ ولهذا يكون الفكر فيه ، وهو يحوي روحاً
68b نفسانيّاً كثيراً ، والروح النفسانيّ يتولّد من الروح الذي يكون به الحياة ويصعد في العروق الضوارب التي ترتقي من القلب إلى الدماغ ، وذلك بمنزلة العنصر والمادّة المشاكلة .

(119) والدماغ بارد رطب بالطبع ، أمّا برده فلحكمتين ، إحداهما أنّ منه الحركات الكثيرة مثل حركة التخيّل وحركة الفكر وحركات الحسّ وكلّ حركة تتبعها حرارة ، فلو لم يكن الدماغ بارداً وعرضت له هذه الحركات التهب

وقوّة السمع وقوّة الشمّ : والسمع ... واللمس 1–2 Masāʾil والقوى الحسّاسة خمس : والحسّاسة 1
: والنفس 11 om. Masāʾil والأفعال ... للأبدان 3–4 Masāʾil وقوة المذاق وقوّة اللمس . ....
: ولهذا ... وهو | K. fī Manāfiʿ وقد أتيت في كتاب آراء بقراط وفلاطن بالبرهان على أنّ النفس
وإذ كان ذلك كذلك فالذي بيناه في كتاب : والروح 12 K. fī Manāfiʿ وإنما نتفكّر بالدماغ وأنّه
: ويصعد | K. fī Manāfiʿ به يكون : يكون به | K. fī Manāfiʿ آراء بقراط وفلاطن من أنّ الروح
K. fī صواب add. 14 after المشاكلة K. fī Manāfiʿ وهو : وذلك 13 K. fī Manāfiʿ الذي يصعد
برده ... 15–16 Masāʾil جعل الدماغ بالطبع بارداً رطباً ... : والدماغ ... بالطبع 15 Manāfiʿ
Masāʾil برد الدماغ بالطبع فجعل الشيئين أحدهما لأنّ الحركات فيه ومنه حركات كثيرة : الكثيرة
Masāʾil والحركات الإراديّة add. after الحسّ | Masāʾil وحركة الذكر add. 16 after الفكر
فجعل الدماغ لهذا السبب : فلو ... واحترق 17–206.1 Masāʾil تتصل بها وتتبعها : تتبعها 17

1 after بإرادة blank space of three letters M 4 الآلة : اَلاَلةِ M | after للأبدان blank space of more than half a line M 5 ثلاثة : ثلثه M 7 الغضب : الغَضبُ M 8 after الشهوة blank space of more than half a line M 9 العصب : الغضب M 10 after المعقولات blank space of three letters M 11 تسكن : تمسكن M 11–12 روحاً ... كثيراً : روح نفساني كبير M 12 يكون به : يكون ~~بها~~ به M 15 إحداهما : احديهما M

are moved. The sensitive faculties are vision, hearing, taste, smell and touch. 68a

The activities[180] are the last form, i.e. the final cause proceeding from the powers and spirits as a natural instrument. Nature is the power directing the bodies.

(116) Plato says: the principal organs are three, namely the brain, the heart and the liver. The brain is the dwelling place of the rational soul and due to it there are sensation, movement, thought and consideration. The heart is the dwelling place of the sensible soul and belonging to it are anger and courage. The liver is the dwelling place of the nutritive soul and due to it there is augmentation and growth, and belonging to it is appetite.

(117) The brain is the source of sensation and movement, and the nerves serve it. It receives what all senses convey to it, forms images of the things and imagines the intelligibles.

(118) The rational soul dwells in the brain[181] and therefore the brain thinks. It comprises abundant psychic spirit. The psychic spirit is brought forth by the spirit due to which life exists and [which] rises in the arteries 68b
ascending from the heart to the brain. That [spirit, i.e. the vital spirit] serves as matter and proper material.

(119) The brain is cold and moist by nature.[182] It is cold for two good reasons: one of them is that it causes many motions, for example the movements of imagination, thought and sensation, each of which is followed by heat. So if the brain were not cold and these motions befell it, it would be inflamed and burned. The second [reason] is that it is the

[180] Ḥunayn does not call the psychic activities the 'last form', but instead describes the powers as the 'first form' which is taken up in the *PR*. However, his discussion of the activities and the spirits (see commentary) is not quoted by the compiler of the *PR*.

[181] Cf. source in Galen, *K. fī Manāfiʿ*, appendix 1, excerpt 1.

[182] Cf. Ḥunayn, *Masāʾil*, 258.1–259.9, although the *PR* passage is considerably rephrased and incomplete at the end.

واحترق ، والثانية أنّه معدن الفكر والذكر الذين يصحّان بالثبات والرفق ، فلما كان بارداً صار ثابتاً متمكّناً . وأمّا رطوبته فلسببين ، أحدهما أن لا يجفّ بكثرة حركاته إذ كانت الحركة إذا كثرت تبعها اليبس والجفاف ، فلمّا كان رطباً صبر على الحركات ولم تجفّ جفافاً مفرطاً ⟨...⟩ .

(120) وقال جالينوس الدماغ والنخاع هما مبدأ العصب كلّه والدماغ مبدأ النخاع ، والقلب مبدأ العروق الضوارب كلّها ، والكبد مبدأ العروق غير 69a الضوارب كلّها . والعصب يؤدّي القوّة النفسانيّة من الدماغ والقوّة النفسانيّة هي التي يكون بها الحسّ والحركة ، والعروق الضوارب تؤدّي القوّة الحيوانيّة وهي النبض من القلب ، والعروق غير الضوارب تؤدّي القوّة الطبيعيّة من الكبد . فليكن مما تقدّم أنّ منفعة العصب تأدية قوّة الحسّ والحركة من الدماغ إلى جميع أجزاء البدن ، ومنفعة ⟨العروق⟩ الضوارب حفظ الحرارة الطبيعيّة وغذاء الروح النفسانيّ ، ومنفعة العروق غير الضوارب العون على كون الدم وتأديته من الكبد إلى جميع الأعضاء .

(121) وقال الدماغ مبدأ الحسّ والحركة ينشو منه الأعصاب ، والكبد مبدأ الاغتذاء وينبوع الدم ينشو منه العروق غير الضوارب التي هي أوعية للدم ،

Masāʾil باردا كي لا يعرض له عندما يتحرّك هذه الحركات كلّها أن يلتهب ويحترق
1–2 والثانية... متمكّناً : والسبب الآخر أنّ الدماغ هو أصل ومبدأ للفكر يحتاج إلى تأييد وثبات
وكما أن الحرارة من شأنها سرعة الحركة وسرعة التنقّل ، كذلك من شأن البرودة الثبات وطول اللبث
ولذلك جعل الدماغ بارداً ليصير للفكر بسبب برودته تمكن وثبات Masāʾil 2 رطوبته فلسببين :
إذا... 3–4 Masāʾil بسبب كثرة : بكثرة 3 Masāʾil رطوبة الدماغ فجعلت أيضاً بالطبع لسببين
يلزمها توليد الحرارة وكانت الحرارة متى كثرت فاليبس والجفاف تابع لها ولذلك جعل الدماغ : مفرطاً
رطباً ليحتمل الحركات ويصبر عليها ولا يجفّ جفافاً مفرطاً والسبب الثاني أنّه احتيج إلى أن يكون
ليّناً لسببين أحدهما ليكون سهل التغيّر سريع الحركة والثاني لتنبت منه أعصاب ليّنة تصلح للحسّ
فنقول إنّا قد بيّنا في الكتاب الذي قد وضعناه في آراء ابقراط : وقال جالينوس 5 Masāʾil
وفلاطون أن K. fī Manāfiʿ 6 النخاع والقلب : النخاع أيضا وأن القلب K. fī Manāfiʿ
K. fī التي بها يكون الحسّ والحركة .add النفسانيّة[1] after | K. fī Manāfiʿ وإن العصب : والعصب 7
Manāfiʿ 7–8 والحركة... والقوّة om. K. fī Manāfiʿ 9 النبض : الفاعلة للنبض K. fī Manāfiʿ
14 وقال الدماغ : وهو الثلاثة المبادئ المدبّرة للحيوان فإنّ الدماغ وهو Fī Ṭabīʿa | after والكبد add.
وهو Fī Ṭabīʿa 15 للدم : الدم Fī Ṭabīʿa

1 أنّه : أن M | والرفق : M or read والوقف ? 2 after متمكّناً blank space of three letters M | يجفّ : يخفَّ M 4 after مفرطاً blank space of six letters M 5 والنخاع : والشجاع M | العصب : الغضب M 6 النخاع : الشجاع M 7 والعصب : الغضَبُ M 10 العصب : الغَضب M 11–12 ومنفعة... النفسانيّ add. in marg. M 11 وغذاء : وغدا M 13 after الأعضاء blank space of ten letters M 14 ينشو : ينشوا M 15 الاغتذاء : الاعتذا M | ينشو : ينشوا M

source of thought and memory, which are both sound due to stability and softness. Thus when [the brain] is cold, it is stable and consolidating. There are also two reasons for [the brain's] moisture: one of them is so that it does not dry out by its many motions since whenever movement increases, dryness and desiccation follow. So since [the brain] is moist, it endures the motions and does not dry out excessively.[183]

(120) Galen says:[184] the brain and the spinal cord are the starting point for all nerves, the brain being the starting point for the spinal cord. The heart is the starting point for all arteries and the liver for all veins. Starting from 69a
the brain the nerves convey the psychic power which allows for sensation and movement. Starting from the heart the arteries convey the vital power, which is pulsation,[185] and starting from the liver the veins convey the natural power. According to what has been mentioned, the usefulness of the nerves is to convey the power of sensation and movement from the brain to all parts of the body, the usefulness of the arteries to preserve the natural heat and to nurture the psychic spirit, and the usefulness of the veins to help with generating blood and conveying it from the liver to all body parts.

(121) He says:[186] the brain is the principle of sensation and movement and the nerves grow out of it. The liver is the principle of nutrition and the

183 The second reason for the brain's moisture is missing. Ḥunayn explains it as follows: moisture ensures softness, which allows the brain to change quickly and easily and to grow soft nerves, which convey sensation.

184 Cf. source in Galen, *K. fī Manāfiʿ*.

185 Although the Arabic Galen reads 'which produces pulsation', the reading of the *PR* 'which is pulsation' may be intended by the compiler as he may have been influenced by Nemesius' *Fī Ṭabīʿa*, 159.10, which has 'the pulsating motion is called vital power'.

186 Cf. source in Nemesius, *Fī Ṭabīʿa*, 160.3–17.

والقلب مبدأ الروح الحيوانيّ ينشو منه العروق الضوارب التي هي أوعية الروح . وهذه الثلاثة بمجاورة بعضها بعضاً يكتسب المنافع ، فإنّ العروق غير الضوارب 69b تعطي العروق الضوارب والأعصاب الغذاء ، والعروق الضوارب تعطي العروق غير الضوارب حرارة طبيعيّة وروحاً حيوانيّة ، ولذلك لا يوجد عرق ضارب إلا وفيه دم رقيق ولا يوجد عرق غير ضارب إلا وفيه روح بخاريّة ، والعروق الضوارب تنقبض وتنبسط وانقباضها وانبساطها بتأليف ومناسبة ومبدأ حركتها من القلب ، فهي لشدّة انبساطها يجذب الدم الرقيق من العروق غير الضوارب التي تجاورها قسراً ، وهذا الدم إذا صار في حدّ البخار كان غذاء للروح النفسانيّة ، وبانقباضها يخرج عنها البخارات الغليظة من جميع البدن بالمنافذ الخفيّة كما أنّ القلب يخرج عنه بالتنفّس إلى خارج البخار الدخانيّ .

(122) ومبدأ الحركة الإراديّة هو الدماغ ، وآلتها الأعصاب المنبعثة من الدماغ والنخاع والرباطات والعضل مؤلّف من لحم وليف عصبيّ ووترات مضفورة بالليف العصبيّ ، ولذلك ظنّ قوم أنّ الأوتار حسّاسة بسبب الحسّ الذي 70a يكون من قبل العصب الذي قد ضفر بها . والوترة مركّبة من رباطات ومن أعصاب دقاق ، والفصل بين العصبة والوترة [و]أنّ كلّ عصبة حسّاسة مستديرة وأنّها ألين ومبدأها من الدماغ فأمّا الوترة فإنّها صلبة وهي في نفسها لا حسّ لها وربّما كانت عريضة لا مستديرة كالعصب .

(123) والحيوان محتاج إلى التنفّس لمكان القلب ، لأنّ القلب يحتاج إلى الهواء

1 after والقلب add. وهو *Fī Ṭabīʿa* 2 بعضاً : ببعض *Fī Ṭabīʿa* | المنافع : بعضها منافع بعض *Fī Ṭabīʿa* 6 after وانبساطها add. وذلك شديد جدّاً *Fī Ṭabīʿa* 7 لشدّة : بشدّة *Fī Ṭabīʿa* 9 بالمنافذ : المنافس *Fī Ṭabīʿa* 11 ومبدأ : ان مبدأ *Fī Ṭabīʿa* | after الإراديّة add. والحركة بالاختيار *Fī Ṭabīʿa* | after الدماغ[1] add. والنخاع أيضاً هو جزء من الدماغ *Fī Ṭabīʿa* | وآلتها : وآلة هذه الحركة *Fī Ṭabīʿa* 12 والرباطات om. *Fī Ṭabīʿa* 14 ضفر : ظفر *Fī Ṭabīʿa* | والوترة : والوتر *Fī Ṭabīʿa* 15 والوترة : وبين الوترة *Fī Ṭabīʿa* 16 فأمّا : وأما *Fī Ṭabīʿa* 18 والحيوان محتاج : أقول إنّي قد بيّنت في كتاب غير هذا أنّ الحيوان احتاج *K. fī Manāfiʿ* | الهواء : جوهر الهواء *K. fī*

1 ينشو : ينشوا M | أوعية : the original reading ادَعِيَه has been corrected by adding a squiggel on top of the *dāl* M 2 الثلاثة : الثلثه M | بمجاورة : the original reading مجاورة has been corrected by adding a squiggel above of the *mīm* M 3 العروق . . . الضوارب[1] [2] add. in marg. M | الغذاء : الغدا M 6 حركتها : حركها M 8 غذاء : غدا M 9 بالمنافذ : بالمنافس M 10 after الدخانيّ blank space of ten letters M 11 وآلتها : والَتُها M 12 والنخاع : والثجاع M | ووترات مضفورة : ووتراب مصفورة M 16 ومبدأها : وَمبدأوهَا M 17 عريضة : عريضته M

wellspring of blood, and the veins, which are vessels for the blood, grow out of it. The heart is the principle of the vital spirit and the arteries, which are vessels for the spirit, grow out of it. These three benefit from their vicinity to each other. For the veins give nutriment to the arteries 69b
and the nerves, and the arteries give natural heat and vital spirit to the veins. Therefore, there is no artery without fine blood and no vein without vaporous pneuma. The arteries contract and expand, their contraction and expansion being in sequence and proportion, its motion starting from the heart. They forcibly draw the fine blood from the veins in their immediate vicinity due to the intensity of their expansion, and when this blood is on the point of vapourising, it is nutriment for the psychic pneuma. Through their contraction the thick vapours exude from them to the entire body through invisible passages, just as the smoky vapour exudes from the heart to the outside by respiration.

(122) The starting point for voluntary movement[187] is the brain. Its instruments are the nerves, which originate in the brain and in the spinal cord, the ligaments and the muscles which are combined from flesh, sinewy fibre and tendons interwoven with neural fibre.[188] Therefore some assume that the tendons are sensitive because of the sensation caused by the 70a
nerves with which they are certainly interwoven. The tendon is composite of ligaments and thin nerves. The difference between a nerve and a tendon is that every nerve is sensitive and circular, and that it is softer and starts from the brain, whereas the tendon is hard, has no sensation in itself and is sometimes broad, not circular like nerves.

(123) The living being needs respiration[189] for the sake of[190] the heart. For

[187] Cf. source in Nemesius, *Fī Ṭabīʿa*, 166.2–11.

[188] The reason for translating *ʿaṣabī* first as 'sinewy', then as 'neural' is based on the Greek text, 88.5–9: οἱ δὲ μύες συνεστήκασιν ἔκ τε σαρκῶν καὶ νευρωδῶν ἰνῶν καὶ τενόντων συμπεπλεγμένων ταῖς τῶν νεύρων ἴνεσιν.

[189] Cf. source in Galen, *K. fī Manāfiʿ*.

[190] The translation of διὰ 'for the sake of' by *li-makān* is attested by Ullmann, *WGAÜ*, p.193, and *WGAÜ Suppl. I*, p.268.

والتروّح به لتلهّب الحرارة وغليانها فيه . واستنشاق الهواء يروّح القلب بما يوصله إليه من الهواء البارد ، وإخراج الهواء يروّحه بما يدفع عنه من البخار المحترق الذي هو بمنزلة الدخان . ومن أجل ذلك يتحرّك حركتين متضادّتين ، فإذا انبسط جذب الهواء البارد ، وإذا انقبض دفع وأخرج عنه ما يتولّد فيه من الفضل .

70b (124) ولو عدم الحيوان التنفّس ساعةً لعدم حياته جملة . وذلك أنّ الحرارة التي فيه كانت تختنق بالبخار الغليظ ، فيطفئ وبانطفائها ذهاب الروح ، كما أنّك لو غطيت ناراً بآلة لا متنفّس فيها طفئت النار بالدخان . ولهذا السبب الضروريّ صارت النفس في حال النوم أيضاً تفعل بهذا الجزء مثل فعلها في اليقظة ، لعلمها أنّها إن توانت في ذلك بطلت الحياة .

وللتنفّس ثلاث علل ، منفعته والقوّة الفاعلة له والآلات التي تفعل بها . فأمّا منفعته فحفظ الحرارة الغريزيّة بإدخال التنفّس وإخراجه ، وذلك أنّ بإدخال التنفّس تستروح الحرارة وتبرد برودة معتدلة وبإخراج التنفّس يصفو القلب من البخارات الدخانيّة ، ثم تغذية الروح النفسانيّ ويكون بإدخال التنفّس وحده ، لأنّه يتجذّب بانبساط القلب جزء من الهواء إليه . فأمّا القوّة الفاعلة للتنفّس
71a فهي القوّة النفسانيّة ، وهذه القوّة هي التي تحرّك آلات التنفّس بالعضل ، وأوّل

Manāfiʿ

1 والتروّح : وهو مع ذلك يحتاج إلى التروّح *K. fī Manāfiʿ* 2–3 من²... الدخان : مما يحترق فيه *K. fī Manāfiʿ* 3 يتحرّك : صار يتحرّك *K. fī Manāfiʿ* 6 ولو ... جملة : فأمّا فيصير بمنزلة الدخان *K. fī Manāfiʿ*
كانت | فينا *Fī Ṭabīʿa* 7 فيه : التنفّس فإنّا إن عدمناه ولو عشر ساعة لم يمكن أن نحيا *Fī Ṭabīʿa*
Fī Ṭabīʿa فينطفي : فيطفئ | بالبخارات الغليظة : بالبخار الغليظ *Fī Ṭabīʿa* | om. *Fī Ṭabīʿa*
7–8 وبانطفائها ... فيها : ويتبع ذلك على المكان الموت فإنّك إن غطيت لهيب نار بآلة ما لا
وذلك : يتنفّس لها *Fī Ṭabīʿa* 9 مثل فعلها : ليس بدون فعلها به *Fī Ṭabīʿa* 10 لعلمها ... إن
Fī أقلّ تواني بطل الحيوان : بطلت الحياة | عن : في *Fī Ṭabīʿa* | لأنّها تعلم أيضاً أن
Ṭabīʿa 11 after علل add. وهي *Fī Ṭabīʿa* 12 after منفعته add. فإنّها شيئان احدهما حفظ
after | فأمّا الحفظ : فحفظ *Fī Ṭabīʿa* | الحرارة الغريزيّة والأخرى تغذية الروح النفسانيّ
Fī وأمّا غذاء : ثم تغذية 14 *Fī Ṭabīʿa* يتصفي : يصفو 13 *Fī Ṭabīʿa* فيكون add. الغريزيّة
Fī Ṭabīʿa وأما : فأمّا | *Fī Ṭabīʿa* ينجذب : يتجذّب 15 *Fī Ṭabīʿa* فإنّه يكون : ويكون | *Ṭabīʿa*
16 القوّة : فهي وهي *Fī Ṭabīʿa* | وهذه : فإنّ هذه *Fī Ṭabīʿa* | هي om. *Fī Ṭabīʿa*

5 after الفضل blank space of half a line M 7 الغليظ : بالغليظ M 8 طفئت : طقيت M
11 ثلاث : ثلث M | والآلات : والآلاتُ M | after بها blank space of five letters M
12 فحفظ : محيط M 13 يصفو : تصفوا M 14 تغذية : تَعديه M 15 جزء : جزو M | after
إليه blank space of six letters M 16 آلات : الات M | بالعضل : بالفضل M

the heart needs air and to be aerated by it, because heat burns and boils in it. Inhalation aerates the heart by bringing it cold air and exhalation by removing burning smoke-like vapour from it. Therefore it moves in two contrary motions. For whenever it expands, it draws in cold air, and whenever it contracts, it removes and exhales waste brought forth in it.

(124) If the living being lacks respiration[191] for a while, it will lose its life
entirely. For the heat in it is choked by thick vapour and thus extinguished, 70b
and with its extinction the spirit vanishes. It is as, if you cover fire with a device which does not allow it to breathe, the fire is extinguished by the smoke. For that inevitable reason the soul also plays this part in the state of sleep as it does while awake. For it knows that if it slackened therein, life would perish.

There are three causes for respiration: its usefulness, the power performing it and the instruments through which it is performed. Its usefulness lies in preserving the innate heat by breathing in and out. For by breathing in the heat is refreshed and cools off in a well-balanced way, and by breathing out the heart is cleared of smoky vapours. Further [usefulness lies] in nourishing the psychic spirit, which happens by breathing in alone. For a part of the air is drawn into the heart through its expansion. The power which performs respiration is the psychic power. This power is the one which moves the instruments of respiration through the muscles.
What it moves first is the thorax and by the thorax's movement the lungs 71a

[191] Cf. source in Nemesius, *Fī Ṭabīʿa,* 169.17–170.6, 170.13–171.5.

ما تحرّكه الصدر وبحركة الصدر تتحرّك الرئة وقصبتها وهي جزء من الرئة .

(125) والقلب ليس يجتذب الهواء من خارج بالحنجرة ، بل جعلت الرئة واسطة فيما بينه وبينها ، فصيّرت بمنزلة الخزانة تجتمع فيها الهواء ليكون عدّة للنفس والصوت . والرئة للقلب بمنزلة الفراش الليّن والصدر بمنزلة السور الحصين لحفظ القلب والرئة . ووضع القلب في الموضع الوسط من تجويف الصدر لأنّه كان أصلح المواضع وأوفقها لحفظه .

والرئة وإن لم يكن لها في نفسها شيء من الحركة ، فإنّها تتحرّك بحركة الصدر دائماً ، فإذا انقبض الصدر انقبضت معه لأنّه يزحمها ويضغطها من جميع النواحي ، وذلك يكون في وقت خروج النفس والصوت ، وإذا انبسط تبعته أيضاً وامتدّت وانبسطت وذلك في وقت استنشاق الهواء .

71b (126) وقصبة الرئة ما كان منها غضروفيّاً فهو آلة للصوت ، وما كان من رباطها من جنس الأغشية فهو آلة للتنفّس ، والمجتمع من هذين جميعاً هي القصبة نفسها التي هي آلة للصوت وآلة للتنفّس .

وقصبة الرئة تتّصل بها الحنجرة وهي مؤلفة من ثلاثة غضاريف عظام ، والحنجرة تتّصل بالحلق ثم يتّصل بذلك الفم والأنف ، فإنّا نجذب الهواء من خارج بهما جميعاً . والهواء يدخل من الأنف فينفذ في العظم الشبيه بالمصفى الإسفنجي ، وإنما جعلت فيه المجاري على هذه الصفة لئلا ينال الدماغ ضرر من قبل إفراط يكون في كيفيّة الهواء بمفاجأته إيّاه دفعة .

2 والقلب ... يجتذب : فأقول إن أوّل الأمور المحمودة التي يحدّها في خلقة آلات التنفّس أن القلب لم يجعل ليجتذب *K. fī Manāfiʿ* | بالحنجرة : بلا متوسّط فيما بينه وبينها *K. fī Manāfiʿ* 3 للنفس : للتنفّس *K. fī Manāfiʿ* 4 والصوت : وللصوت *K. fī Manāfiʿ* | والرئة : كما قالنا . فإنّ الرئة صارت *K. fī Manāfiʿ* | after الليّن add. كما قال فلاطن *K. fī Manāfiʿ* | after والصدر add. صار *K. fī Manāfiʿ* | لحفظ : يحوط *K. fī Manāfiʿ* 5 after والرئة add. ويهمهما جميعاً *K. fī Manāfiʿ* | لأنّه : لأنّ هذا الموضع *K. fī Manāfiʿ* 6 after لحفظه add. وتروّحه من الرئة كلّها بالسوا *K. fī Manāfiʿ* 7 والرئة : وقد بيّنا في كتاب حركة الصدر والرئة أن الرئه *K. fī Manāfiʿ* | لم يكن : كان ليس *K. fī Manāfiʿ* 10 after وانبسطت add. مثل انبساطه *K. fī Manāfiʿ* 13 نفسها om. *Fī Ṭabīʿa* للتنفّس : التنفّس *Fī Ṭabīʿa* 17 وإنما : فإنّه إنما *Fī Ṭabīʿa*

1 وبحركة ... تتحرّك : وتحركه الصَّدر بتحرّك M | after الرئة[2] blank space of half a line M 10 استنشاق : اشتياق M | after الهواء blank space of five letters M 11 آلة : الة M 12 آلة : الةٌ M | after للتنفّس blank space of five letters M 12–13 والمجتمع ... للتنفّس add. in marg. M 14 ثلاثة : ثلثه M 16 الشبيه : الشَبْه M 18 after دفعة blank space of six letters M

and its windpipe, which is a part of the lungs, are moved.
(125) The heart does not draw air[192] from outside through the larynx, rather the lungs are put in between [the heart] and [the larynx] and serve as a storage in which air is gathered in order to be ready for breath and voice. The lungs serve as a soft cushion for the heart, whereas the thorax is like a fortified wall protecting heart and lungs. The heart is placed in the middle position of the thorax cavity, because that is the best and most suitable position for its protection.
Even if the lungs do not have any motion on their own, they are continuously moved by the motion of the thorax. So whenever the thorax contracts, they contract with it, for it presses and squeezes them from all directions, which happens when breath and voice emerge. Whenever [the thorax] expands, [the lungs] follow it as well, extending and expanding, which happens when air is inhaled.
(126) The cartilaginous part of the windpipe[193] is the instrument of voice, 71b
and its membranous ligament is the instrument of breath. These two combined together are the windpipe itself which is an instrument of voice and respiration.
The larynx, which is composed of three large cartilages, is joined to the windpipe. The larynx is also joined to the throat, to which mouth and nose are joined next. For we draw air from outside through both of them together. The air enters through the nose[194] and then penetrates the bone which resembles a spongy sieve. The channels are only put into [the bone] according to this description so that the brain does not suffer any harm from an excess in the quality of air, were it to enter suddenly at a stroke.

[192] Cf. source in Galen, *K. fī Manāfiʿ*.
[193] Cf. source in Nemesius, *Fī Ṭabīʿa,* 171.5–8, 171.16–172.4.
[194] The Greek has 'mouth and nose'.

(127) وقصبة الرئة تعدّ الهواء الذي يخرج فيها لقبول الصوت في الحنجرة ، ثم إنّ الصوت إذا تمّ واستحكم في الحنجرة زاد فيه أعلى الحنك واللهاة ، وذلك أنّ أعلى الحنك قدام الحنجرة بمنزلة النحاس الموصول برأس المزمار واللهاة بمنزلة المضراب .

72a وجرم طبق الحنجرة من طبيعة الأغشية ، لكيما لا يتمزّق إذا امتلأ من الهواء ، ولا ينداه شيء من التخرّق عندما يضطرّه الحنجرة إلى أن يتحرّك بحركاتها المتضادّة ، إذ كانت الحنجرة تتّسع وتنبسط مرّةً وتنقبض وتضيق أخرى . وجعل هذا الجرم رطباً ولم يقتصر به على الرطوبة وحدها دون أن صيّر لزجاً دسماً لكيما تندى وتبلّ الحنجرة برطوبته الطبيعيّة ولا تحتاج إلى رطوبة تستعين بها من خارج ، كما يحتاج إلى ذلك لسان المزمار الذي يجفّ دائماً فيحتاج إلى رطوبة من خارج . ولو لا هذه الرطوبة لكان الصوت يفسد من قبل سرعة جفاف طبق الحنجرة ، كما يعرض في بعض الأوقات واتفقت أسباب قويّة يفسد بها مجري أفعال الطبيعة .

(128) والحيوان ذوات الرئة يستنشق الهواء ويوصله إلى الرئة بهذه القصبة ويخرجه أيضاً منها ، ولذلك يكون الحيوان العديم للرئة عديم الصوت .

72b (129) وأوّل آلات القوّة الغاذية الفم فإنّه يعدّ الطعام للمعدة بأن يقطعه قطعاً

1 وقصبة : أنّ قصبة الصوت في كتاب أيضا وبيّنا ... *K. fī Manāfiʿ* | تعدّ : تهيّئ وتعدّ *K. fī Manāfiʿ* 5 وجرم : وأمّا جرم *K. fī Manāfiʿ* | after الحنجرة add. فإنّه جعل *K. fī Manāfiʿ* 8 هذا الجرم : جرم هذا الطبق *K. fī Manāfiʿ* | صيّر : جعل *K. fī Manāfiʿ* 11 ولو . . . يفسد : فجعلت رطوبته كما قلت لزجة دسمة لكيما تبقي ولا تنحلل سريعاً ولا تنفرق وذاك أنّ الرطوبة الرقيقة المائية تنحلل وتصير بخاراً فتبعثر وتنحلّ سريعاً وهي مع هذا مجرى وينفرق أيضاً ولا تلبث كلبث الرطوبة اللزجة الدسمة وسيما إذا كان المجرى الذي هي مصبوبة فيه قائماً منضباً فأمّا الرطوبة اللزجة الدسمة فإنّها تمكث زماناً طويلاً من غير أن يجري وينفرق ولا يجفّ سريعاً فلو أنّه كان قد احتيط هيئة الحنجرة غاية الاحتياط في سائر حالاتها واغفلت هذه الرطوبة وبسبب لكان في ذلك فساد الصوت *K. fī Manāfiʿ* 12 جفاف : جفوف *K. fī Manāfiʿ* | after الحنجرة add. وسائر أجزاء الحنجرة *K. fī Manāfiʿ* | واتفقت : متى حدثت *K. fī Manāfiʿ* | after كما add. يجد ذلك *K. fī Manāfiʿ* 13 يفسد : ففسد *K. fī Manāfiʿ* 14 والحيوان : جميع الحيوان *K. fī Manāfiʿ* 15 after منها add. والنفع أيضاً الذي بينا أن به يكون الصوت وهو كالعنصر له هو من فعل هذه القصبة والصوت أيضاً لا يكون خلو منها وأوّل آلات الصوت وأشرفها وأولاها به أعني الحنجرة هي طرف هذه القصبة *K. fī Manāfiʿ* | يكون : صار جميع *K. fī Manāfiʿ* | للرئة : الرئة *K. fī Manāfiʿ* 16 وأوّل وآلات : آلات *Fī Ṭabīʿa* | الفم : هي الفم والمري والبطن والكبد والعروق كلّها والأمعاء والشريان

2 زاد : راد M 9 برطوبته : برطوبة M 13 after الطبيعة blank space of five letters M
15 after الصوت blank space of half a line M 16 آلات : اَلَات M | يعدّ : يغد M

(127) The windpipe adapts the air[195] which comes through it, for receiving the voice in the larynx. When the voice is then fully developed and strong in the larynx, the uppermost part of the palate and the uvula intensify[196] it. For the uppermost part of the palate, i.e. the front part of the larynx, is similar to the metallic sounding-plate linked to the head of a musical reed and the uvula is similar to a plectrum.

The body of the glottis is of membranous nature so it does not get torn 72a
whenever it is filled with air, nor is it affected by any piercing whenever the larynx obliges it to move in its contrary motions. For the larynx widens and expands at times and contracts and tightens at other times. This body is made moist, yet not only moist alone, but also adhesive and greasy so that the larynx is moistened and wet by its natural moisture and does not need moisture from outside to use, as the tongue of a musical reed, which is permanently dry and thus needs moisture from outside. If it were not for this moisture, the voice would fail due to the rapidity with which the glottis would dry out, just as it happens at times when strong causes coincide which corrupt the course of nature's activities.

(128) Animals possessing lungs[197] inhale air, bring it to the lungs through this windpipe, and exhale it through it as well. Therefore animals lacking lungs lack voice.

(129) The first of the instruments of the nutritive faculty[198] is the mouth, 72b
as it prepares the food for the stomach by cutting it small with teeth and

[195] Cf. source in Galen, *K. fī Manāfiʿ*.

[196] I follow the reading *zāda*, which is attested in the Arabic version of Galen and renders the Greek ἐπαυξάνω. However, the reading of the *PR*, *rādda* 'to return', may also be correct.

[197] Cf. source in Galen, *K. fī Manāfiʿ*.

[198] Cf. source in Nemesius, *Fī Ṭabiʿa,* 157.3–158.4, 158.10–159.3.

صغاراً بالأسنان واللسان . فإنّ اللسان قد ينفع في المضغ منفعة عظيمة لأنّه يجمع الطعام ويوقعه تحت الأسنان ، كمن يطحن برحى اليد في جمع ما يطحن وإلقائه إلى الرحى . ثم ينفده إلى المعدة في المرىء وهو طريق ينحدر فيه الغذاء من الفم إليها . فإنّ المرىء عند الابتلاع يتصاعد إلى فوق نحو الفم وتجذب الغذاء ويؤديه إلى المعدة . فإذا تسلّمته ميّزته وأفردت عنه الأخلاط التي لا غذاء لها ودفعتها إلى أسفل ، وغيّرت ما كان غاذياً موافقاً وأنفذته إلى الكبد في العروق التي تمتصّ الغذاء في المعدة . وهذه العروق تقوم للكبد مقام العروق التي تنشعب من أصول الأشجار ، فتجذب الغذاء من المعدة كما
73a تجذب تلك العروق الغذاء من الأرض . فإنّ المعدة تقوم للحيوان مقام الأرض للنبات لأنّ الأرض هي التي تعطي النبات غذاءه .

والكبد إذا تسلّمت هذا الطعام من المعدة طبخته وهضمته بالحرارة الغريزيّة التي فيه وشبّهته بجوهرها وهو الدم ، لأنّ لحمية الكبد في غاية المشابهة للدم . وهذا الدم يصفّى ويتهذب بالطحال والمرارة والكلى ، وذلك أنّ الطحال يجذب ما كان منه غليظاً شبيهاً بالدردي فيجعله غذاءً له ، والمرارة تجذب منه ما كان

فإنّ الفم أوّلاً *Fī Ṭabīʿa* : فإنّه | جميعاً والكلى *Fī Ṭabīʿa*
1 ينفع : ينتفع *Fī Ṭabīʿa* 2 كمن : الطحان الذي : وفعله هذا مثل الفعل الذي يكون من الطحان الذي *Fī*
Ṭabīʿa فإنّ اللسان في المضغ : ثم ينفده | بيده *Fī Ṭabīʿa* 3 وإلقائه : جمعه *Fī Ṭabīʿa* | جمع : جمع
يقوم مقام اليد الذي يجمع ما تطحن إلى الرحا حتى يستحكم طحنه فإذا أعد الطعام هذه الأعداد
فإنّ المري مع أن كانت منفعته انّه : وهو . . . إليها 3–4 البطن *Fī Ṭabīʿa* | المعدة : نفذ *Fī Ṭabīʿa*
Fī Ṭabīʿa وذلك أنّه : فإنّ المرىء 4 آلة الحسّ بالجوع قد ينفع أيضاً في أنّه طريق الطعام
المعدة : ميّزته . . . موافقاً 5–6 تسلمت *Fī Ṭabīʿa* | تسلّمته : ويجتذب *Fī Ṭabīʿa* 5 وتجذب :
الطعام من المرىء ميّزت ما كان منه غاذياً موافقاً مما كان منه حجرياً أو خشبياً أو لا غذاء فيه
: في² 7 وينفده *Fī Ṭabīʿa* 6 وأنفذته : فالغاذي منه النافع بغيره ويجعله امشاجاً وأخلاطاً
Fī Ṭabīʿa فهذه : وهذه | ويصير له طريقاً إلى الكبد *Fī Ṭabīʿa* add. after المعدة | من *Fī Ṭabīʿa*
Fī Ṭabīʿa الامشاج والأخلاط : هذا الطعام 11 تتشعب *Fī Ṭabīʿa* : تنشعب 8
وهو | وشبهتها *Fī Ṭabīʿa* 12 وشبّهته : طبختها وهضمتها *Fī Ṭabīʿa* طبخته . . . فيه 11–12
: يصفّى 13 فإنّه تحيل ذلك الغذاء إلى الدم *Fī Ṭabīʿa* add. after للدم | الدم om. *Fī Ṭabīʿa*
منه ما : ما . . . غليظاً 13–14 وبالمرارة وبالكي *Fī Ṭabīʿa* : والمرارة والكلى | يتصفي *Fī Ṭabīʿa*
شبيهاً بالدردي om. *Fī Ṭabīʿa* 14 كان غليظاً رديّاً *Fī Ṭabīʿa*

2 برحى : برحَا M 3 الرحى : الرّحا with blank space of five letters after it M 5 الغذاء :
الغدآ M | after المعدة الغدآ M 6 ودفعتها : ودفعته M | وأنفذته : انفديه M 7 الغذاء : الغدآ M
blank space of six letters M 10 غذاءه : غذاه with blank space of ten letters after it M
11 طبخته : طحنته M 12 after للدم blank space of six letters M 13 يصفّى : يتصفّي M
14 after له blank space of five letters M

the tongue. For the tongue is certainly of great use for chewing, because it gathers the food and places it under the teeth just as the one who grinds with the hand mill does by gathering what he is about to grind and throwing it in the mill. Then it sends [the food] to the stomach via the oesophagus, which is the route by which the nutriment slides down to it from the mouth. For at swallowing the oesophagus rises upwards towards the mouth, draws the nutriment and conveys it to the stomach. When [the stomach] has taken [the nutriment], it separates it, isolates the non-nutritious components from it and pushes them downwards. It transforms what is nutritious and suitable, and sends it on to the liver via the veins, which absorb the nutriment in the stomach. For the liver these veins take the place of roots which branch out from the stems of trees. They draw nutriment from the stomach, just as those roots draw nutriment from the
soil. In the animal the stomach takes the place that soil has for plants, as 73a
soil is what gives the plants their nutriment.

When the liver takes this food from the stomach, it concocts[199] and digests it by the innate heat in it and assimilates it to its substance, i.e. blood. For the flesh of the liver is of utmost similarity to blood. This blood is cleared and refined by the spleen, the gall bladder and the kidneys. For the spleen draws out of it the thick part with the apparence of dregs and then makes it into nutriment for itself. The gall bladder draws out of it the sharp and

[199] The Arabic of Nemesius as well as of the *PR* reads *taḥana* ‘grind’, which can be emended to *tabakha* ‘cook, concoct’ on the basis of the Greek 84.8–9: τὸ γὰρ ἧπαρ δεξάμενον ἐκ τῆς κοιλίας τὸν χυμὸν συμπέσσει τε καὶ ἐξομοιοῖ ἑαυτῷ.

بقي فيه حاداً حريقاً من جوهر الأطعمة التي تولّد منها وذلك الكيموس ، والكلى تجذب منه ⟨ما⟩ كان مائعاً دقيقاً . فإذا صار الدم نقياً صافياً يصلح لأن يغتذي به البدن ، توزعته سائر أعضاء البدن في العروق المنشعبة فيها . فيجذب كلّ واحد من الأعضاء قسطه من هذا الدم ويحيله وينقله إلى طبيعته التي تخصّه .

73b (130) والطحال قد يعين أيضاً على الهضم معونة ليست باليسيرة . وذلك أنّه لما كان في طبيعته ذا حموضة وقبض وكان يصبّ فضل الكيموس الأسود إلى المعدة صار يعيدها قبضاً وقوّة ويبعثها على الهضم .

والكلّيتان وإن كانتا أحد أسباب تصفية الدم فإنّها سبب شهوة الجماع ، لأنّ العروق التي تصير إلى الأنثيين تنفذ في الكلى فتفيدهما حدّة تهيج الشهوة مثل الحكة التي يفعلها الشيء الحادّ إذا كان تحت الجلد .

(131) والأمعاء منظومة هذا النظام ، أوّلها مما يلي أسفل المعدة المعاء ، المسمّى أوّل نبات الأمعاء ، وهو ممتدّ مستقيم على الصلب ، فإذا انتهى هذا المعاء المستوي التف ما يليه من الأمعاء وانثنى وانطوى وصارت له استدارات .

والثاني المعاء الصائم وإنما سمّى كذلك لأنّه يوجد فارغاً أبداً ، ليس فيه من
74a الغذاء قليل ولا كثير . والثالث المعاء الدقيق ، والرابع المعاء المعور ، والخامس

1 بقي : مايياً بقي *Fī Ṭabīʿa* | وذلك . . . تولّد : كان عنها ذلك *Fī Ṭabīʿa* 2 تجذب : تحدث *Fī Ṭabīʿa* تهذب | صار : مايياً رقيقاً معما بقي فيه من الحدّة : مائعاً دقيقاً | *Fī Ṭabīʿa* 3 after البدن[2] add. كلّها *Fī Ṭabīʿa* 4 الدم . . . قسطه : من هذا الدم بقسطه *Fī Ṭabīʿa* 6 after والطحال add. أيضاً *Fī Ṭabīʿa* | أيضاً om. *Fī Ṭabīʿa* 8 يعيدها : يفيدها *Fī Ṭabīʿa* | ويبعثها : فإنّها *Fī Ṭabīʿa* | فإنّها : وهما أيضاً *Fī Ṭabīʿa* | أيضاً هما أيضاً *Fī Ṭabīʿa* 9 وإن كانتا : يعينها *Fī Ṭabīʿa* | سبب : سبباً وهما مع ذلك *Fī Ṭabīʿa* | شهوة : الشهوة *Fī Ṭabīʿa* | لأنّ : فان *Fī Ṭabīʿa* | after حدّة add. ما *Fī Ṭabīʿa* | فتفيدهما : وينفذ منها *Fī Ṭabīʿa* | قلنا إنّها *Fī Ṭabīʿa* 10 after التي add. وهذه الحدّة هي التي *Fī Ṭabīʿa* 12 والأمعاء : وهي الأمعاء … *K. fī Manāfiʿ* | after الصلب add. بقدر … *K. fī Manāfiʿ* 13 وهو . . . مستقيم : بل جعله ممتداً مستقيماً … *K. fī Manāfiʿ* 14 المستوي : المتوالي *K. fī Manāfiʿ* الحاجة كانت إلى توسيع الموضع وإخلائه للأجرام التي ذكرتها *K. fī Manāfiʿ* 15 كذلك : بذلك *K. fī Manāfiʿ*

1 after الكيموس blank space of three letters M 4 ويحيله : تخيله M 5 after تخصّه blank space of half a line M 8 after الهضم blank space of half a line M 10 التي : اِلَى M 11 الحكة : الحكمة M | كان : كانت M | after الجلد blank space of ten letters M 13 المعاء : المعاد M 16 after كثير break (∴) ind. and blank space of more than half a line M | الدقيق : الرقيق M

burning remnants of the food's substance.[200] This is the chyle. The kidneys draw from it what is semiliquid and thin. So when the blood becomes pure and unmixed, it is suitable for feeding the body. The remaining parts of the body divide it among themselves via the veins which branch out in them. Each one of the body parts draws its share from this blood, changes it and transfers it to its own specific nature.

(130) The spleen[201] also makes a not insignificant contribution to digestion. 73b
Since it is, by nature, acidic and gripping[202] and empties the surplus of the black chyle into the stomach, it restores[203] contraction and power to it and stimulates it to digest.

The kidneys, among [other body parts], cause the cleansing of the blood, but also the appetite for sexual intercourse. For the veins which get to the testicles penetrate the kidneys and give them a sharpness which stirs up the appetite like the itching something sharp causes under the skin.

(131) The intestines[204] are organised in this order: the first of them belongs to what follows the lower part of the stomach. It is called the 'first outgrowth [into] the intestine', and extends and stretches along the spinal column. When this straight intestine ends, the following part of the intestines winds, bends, folds and has coils. The second is the fasting intestine (*jejunum*), which is called that because it is always found empty with no nutriment in it, neither little nor much. The third is the small intestine, 74a
the fourth is the blind gut (*caecum*), the fifth is the intestine called 'colon',

[200] Literally 'the sharp and burning remnants of the substance of the food from which they were generated'.

[201] Cf. source in Nemesius, *Fī Ṭabīʿa*, 173.5–8, 173.9–174.3.

[202] The pair of Greek terms is ὀξώδης 'acidic' and στρυφνός 'astringent'. The latter can be rendered by different variants of the root *qbḍ* (see Ullmann, *WGAÜ*, p. 644).

[203] Nemesius provides the reading *yufīdu*, 'gives it'.

[204] Cf. source in Galen, *K. fī Manāfiʿ*, appendix 1, excerpt 2, but the *PR* passage is slightly rearranged: the Galenic text starts with some information on the first two intestines and then lists the six intestines. The compiler inserts some of this earlier information directly into his list.

المعاء المسمّى قولون ، والسادس المعاء المستقيم الذي في طرفه العضل الذي يضمّه ويمنع الثقل ويحبسه .

(132) قال ابقراط إذا قوي زرع المرأة والرجل جميعاً كان الولد ذكراً ، وإن رقّ زرعهما وضعف كان أنثى ، فإنّ في زرعهما جميعاً الذكران والإناث .

(133) وقال ارسطاطالس إذا غلبت على الزرع الحرارة كان الولد ذكراً ، وإذا [74b] غلبت عليه البرودة كان الولد أنثى ، وكذلك صار الذكر أسرع حركة وأجهر كلاماً وصار ذكره بائناً متدلّياً ، فهو يقذف الزرع بحرارته إلى داخل قذفاً قويّاً . فأمّا الأنثى فهي أبطأ حركة وأرطب نعمة وقبلها غائر منقبض 〈إلى〉 داخل ، وينصبّ كذلك زرعها في الرحم انصباباً لكثرة رطوبتها . ولذلك [75a] النساء صرن أسرع إدراكاً في الرحم وأسرع انقطاع ولادة ، لأنّ الشيء الضعيف الناقص أسرع إدراكاً ونَهاءً من التامّ القويّ .

(134) قال ومن علل الذكر والأنثى هبوب الرياح ، لأنّ ريح الجنوب يرخي الأبدان ويذيب الزرع فيخرج رفيعاً نِيءاً غير نضيج ، وريح الشمال تصلّب البدن وتمنع الحرارة من الانتشار فيخرج الزرع وقد أنضجته الحرارة . وذكر أنّ الرعاة يعرفون ذلك من فعل الرياح في نتاج غنمهم . وكذلك صار المشائخ والغلمان أكثر ولدهم الإناث ، وأكثر ولد الشباب الذكور لقوّة حرارة الشباب

3 before قال add. فصل *K. al-Ḥikma*   4 فإنّ : وإنّ *Firdaws*   5 وقال ارسطاطالس : وقال ارسطوطيلس إنّه لو لا تضادّ العنصرين الفاعلين لكان الولد كلّه ذكراناً أو أناثاً وانّه *Firdaws* : قال ارسطو *K. al-Ḥikma* | وإذا : وإن *Firdaws*   6 وكذلك : ولذلك *Firdaws, K. al-Ḥikma*   7 بائناً : بارداً *Firdaws* : نايتاً *K. al-Ḥikma* | بحرارته : لحرارته *Firdaws, K. al-Ḥikma*   8 فأمّا : وأمّا *K. al-Ḥikma*   9 كذلك : لذلك *Firdaws, K. al-Ḥikma* | الرحم : الزرع *K. al-Ḥikma*   10 النساء صرن : صرن النساء *Firdaws*   11 ونَهاءً : وانتهاءً *Firdaws* : om. *K. al-Ḥikma*   12 قال : وقال ارسطوطيلس *Firdaws* | after والأنثى add. أيضاً *Firdaws, K. al-Ḥikma* | ريح om. *Firdaws, K. al-Ḥikma* | وريح الشمال : والشمال *Firdaws, K. al-Ḥikma*   13 رفيعاً : رقيقاً *Firdaws, K. al-Ḥikma*   15 وكذلك : ولذلك *Firdaws, K. al-Ḥikma*   16 الشباب[1] : الشبّان *K. al-Ḥikma* | الذكور : الذكورة *Firdaws* | الشباب[2] : الشبّان *K. al-Ḥikma*

1 قولون : قولى M | والسادس : والشادي M   2 الثقل : الثِفلهِ M | after ويحبسه blank space of more than half a line M   3 والرجل : وللرجل M | كان : ~~ان~~ كان M   4 after والإناث blank space of more than half a line M   8 after قويّاً blank space of half a line M وأرطب : وابطبَ M   9 الرحم : الزرع M   11 ونَهاءً : وبهَاا M | after القويّ blank space of three letters M   12 ريح add . in marg. M   13 نيءاً : نِياً M | وريح الشمال : ريح add. in marg. M   14 after الحرارة[2] blank space of three letters M   15 غنمهم : غمهم M

and the sixth is the straight intestine (*rectum*) which has at its end the muscle which closes it, stops and confines the waste.

(132) Hippocrates says:[205] if the woman's and the man's semen together are strong, the offspring will be male. If their semen is thin and weak, it will be female. For in the combination of their semen lies the male and female [gender of the offspring].

(133) Aristotle says:[206] if heat prevails over the semen, the offspring will be
male. If coldness prevails, the offspring will be female. So the male is faster 74b
in motion, louder in speech and his penis is evident and hanging down. By
means of his heat he powerfully ejects the semen inwards. The female is
slower in motion, moister in softness and her anterior pudendum is hollow
and contracted towards the interior. By means of her abundant moisture
her semen thus flows into the womb. Therefore women reach maturity
with regard to their womb faster and stop childbearing sooner. For 75a
something deficient and weak is faster to reach maturity and cessation
than something perfect and strong.

(134) He says:[207] among the causes for the male and female [gender] is the blowing of the winds. For the south wind slackens the bodies and dissolves the semen, so that it emerges fine, unripe and immature; whereas the north wind hardens the bodies and prevents the heat from spreading out, so that the semen emerges already ripened by the heat. He mentions that the shepherds know that due to the effect of the winds on the breeding of their sheep. Equally, most of the offspring of old men and lads is female, whereas most of the offspring of young men is male due to the young

[205] Cf. source in ʿAlī ibn Rabban, *Firdaws*, ed. al-Ṣiddīqī, 34.12–14; parallel in *K. al-Ḥikma*, MS E, fol. 101b12–14.

[206] Cf. source in ʿAlī ibn Rabban, *Firdaws*, ed. al-Ṣiddīqī, 34.17–35.1; parallel in *K. al-Ḥikma*, MS E, fols. 101b14–102a1.

[207] Cf. source in ʿAlī ibn Rabban, *Firdaws*, ed. al-Ṣiddīqī, 35.1–9; parallel in *K. al-Ḥikma*, MS E, fol. 102a2–9.

وضعف حرارة أولئك ، وملاك ذلك كلّه بالاعتدال ، فإنّ الحرارة الشديدة تحرق الزرع والحرارة الضعيفة تعجز عن إنضاجه .

(135) وقال إن السمان من الناس وسائر الحيوان يقلّ زروعهم فيقلّ لذلك 75b ولدهم . وكذلك عظام الشجر يقلّ ثمرها لأن أغذيتها تتفرّق وتذهب في تربية أبدانها وأغصانها ، ولذلك يكسح الفلاحون أغصان الشجر لتصير أغذيتها زيادة في ثمارها دون الأغصان . وهذا بيّن في الفيلة فإنّها تلد في إثنتا عشرة سنة مرّة ولداً واحداً ، وتلد السنانير والكلاب والجرذان في السنة مراراً وفي كلّ مرّة عدّة أولاد .

(136) وقال بقراط السمان من الناس أقلّ عمراً من المهازيل ، معناه رقيق البدن ، أنّ السمن يسدّ مجاري أبدانهم فتختنق الحرارة الغريزيّة فيها وتنطفئ من أدنى علّة ، فإمّا المهازيل فإنّ مجاري أبدانهم واسعة وحرارتهم أقوى .

(137) وقال إنّ الزرع إذا جرى من يمين الرجل إلى يمين رحم المرأة كان الولد ذكراً ، وإذا جرى من يسار الرجل إلى يسار الرحم كان الولد أنثى ، وإن 76a جرى من يسار الرجل إلى يمين الرحم كان الولد أنثى مذكراً ، وإن جرى من يمين الرجل إلى يسار الرحم كان الولد ذكراً مؤنثاً .

وقد يكون من جماع واحد عدّة أولاد مثل ما يكون من جماع الكلب والخنزير .

فأمّا زيادة العضو ونقصانه فيكون مما يجري إلى ذلك العضو من كثرة مادّة

2 إنضاجه : الضاجه *Firdaws* 3 الناس : الإنسان *K. al-Ḥikma* | زروعهم : زرعهم *Firdaws* 4 after وكذلك add. أيضاً أمر *Firdaws* | تتفرّق وتذهب : تذهب وتفرّق *Firdaws* : تتفرّق وتذهب ثمرتها *K. al-Ḥikma* 5 يكسح : نشرت *K. al-Ḥikma* | الشجر : الشجرة *Firdaws* 6 دون : وورق *K. al-Ḥikma* | إثنتا عشرة : إثنا عشر *Firdaws, K. al-Ḥikma* 9 بقراط : ابقراط إنّ *Firdaws* 9–10 رقيق البدن om. *K. al-Ḥikma* : في هذا الأمر *Firdaws* 10–11 فتختنق . . . فإنّ : والمهازيل *K. al-Ḥikma* 10 وتنطفئ : فتنطفيء *Firdaws* 12 after وقال add. غيرهما *Firdaws* | من : عن *Firdaws* | رحم om. *Firdaws, K. al-Ḥikma* 13 وإذا : وإن *Firdaws* | after جرى add. الزرع *Firdaws* | وإن : وإذا *K. al-Ḥikma* 15 يسار الرحم : يسارها *K. al-Ḥikma* 16 جماع² om. *Firdaws* | الكلب : الكلاب *Firdaws, K. al-Ḥikma* 17 والخنزير : والخنازير *Firdaws* 18 فأمّا : وأمّا *K. al-Ḥikma*

1 أولئك : آوليك M 2 after إنضاجه blank space of ten letters M 6 إثنتا عشرة : اثنى عشره M 8 after أولاد blank space of half a line M 9–10 رقيق البدن add. in marg. M 10 أنّ M : or read فإنّ or لأنّ ? 11 after أقوى blank space of ten letters M 13–14 وإن . . . أنثى add. in marg. M

men's strong heat and the weak heat of the others. All of that is based on equilibrium. For strong heat burns the semen whereas weak heat is unable to ripen it.[208]

(135) He says:[209] fat people and fat animals have little semen and therefore few offspring. Equally, large trees have little fruit, because their 75b nutriment[s are] dispersed and go into growing their trunks and branches. Therefore farmers clear the branches of the trees so that their nutriments increase their fruits, not the branches. This is evident in elephants, for they give birth to a single offspring once every twelve years whereas cats, dogs and rats [give birth] several times a year and each time to a number of offspring.

(136) Hippocrates says:[210] fat people have a shorter lifespan than slim people, i.e. the ones with a slender body. The fat clogs their body channels, so the innate heat is choked in them and extinguished by the slightest cause. As for the slim, their body channels are wide and their heat is stronger.

(137) He says[211] that if the semen flows from the right side of the man to the right side of the woman's womb, the offspring will be male and if it flows from the left side of the man to the left side of the womb, the offspring will be female. If it flows from the left side of the man to the right side of the womb, the offspring will be female with male characteristics, whereas if it flows from the right side of the man to the left side of 76a the womb, the offspring will be male with female characteristics.

There may be a number of offspring from one single copulation as in the case of dogs and swine.

Excess and deficiency in a body part are caused by the abundance or shortage of the semen's matter which flows to that part.

[208] Siggel (*Propädeutische Kapitel*, p.400, n. 3) points out that the reading *al-ḍājih*, edited by al-Ṣiddīqī [and al-Jundī], should be emended to *inḍājihī*.

[209] Cf. source in ʿAlī ibn Rabban, *Firdaws*, ed. al-Ṣiddīqī, 35.9–15; parallel in *K. al-Ḥikma*, MS E, fol. 102a9–15.

[210] Cf. source in ʿAlī ibn Rabban, *Firdaws*, ed. al-Ṣiddīqī, 35.16–19; parallel in *K. al-Ḥikma*, MS E, fol. 102a15–18.

[211] Cf. source in ʿAlī ibn Rabban, *Firdaws*, ed. al-Ṣiddīqī, 35.21–36.13; parallel in *K. al-Ḥikma*, MS E, fol. 102a18–b15.

الزرع وقلّتها .
وأمّا قصر البدن وصغر المولود فإنّه يكون من ضيق الرحم وقلّة غذاء الجنين . وذلك مثل الأترجّة التي يدخّلها في قنينة ضيقة وهي على غصنها ، فلا تقدر أن تكبر فيها ، وإن كانت القنينة واسعة وللأترجّ غذاء كاف عظمت الأترجّة واتّسعت فيها . وكذلك الشجرة إذا نبتت بين أحجار معطشة لم تقدر أن تعظم ، فأمّا ما ينبت منها على الأرض الطيّبة فإنّها تعظم مثل أشجار بلاد طبرستان ، فإنّها لسهولة أرضها وكثرة أمطارها تعظم جدّاً وتلتف التفافاً
76b وتذهب في السماء طولاً ، وربّما رأيت في غياضها جنساً من شجر قد اتّصل أغصان بعضها ببعض والتحمت حتّى لا يعرف الناظر غصن هذه الشجرة من غصن تلك الشجرة ، ويسمّى هذا الجنس الجولق .
(138) وقال في ذكر علّة استدارة الرأس وما فيه من الدليل إن المزاجات التي تكون في زرع الأبوين إذا صارت مضغة وتحرّكت الحرارة ، صعد ما كان في الزرع من جوهر العنصرين الخفيفين إلى فوق لخفتها وانحدر ما كان فيه من جوهر العنصرين الثقيلين إلى أسفل ، وترفع الحرارة معها مادّة ، وإذا انتهى البدن منتهاه من الامتداد والطول في الرحم ، دارت في أعلاه الهوائيّة الصاعدة مع الناريّة ، فيستدير لذلك تلك المادّة هناك ويكوّن منه الرأس ، كما يستدير
77a الزجاج إذا نفخ فيه ، وكالنفاخات التي تظهر من المطر مستديرة . فالرأس الفاضل ما كان فلكيّاً معتدل المزاج . وأمّا ما صغر جدّاً فإنّه يدلّ على خفّة

2 وأمّا : فإمّا *Firdaws* | وقلّة : أو قلّة *Firdaws* 3 غصنها : غضها *Firdaws* 4 وإن : فإن *K. al-Ḥikma* | وللأترجّ : وللاترجة *K. al-Ḥikma* | الأترجّة om. *Firdaws* 5 معطشة om. *Firdaws* : وأرض صلبة غير طيبة *K. al-Ḥikma* 6 فأمّا ما : وما *K. al-Ḥikma* | ينبت : نبت *Firdaws* | منها om. *K. al-Ḥikma* | بلاد om. *Firdaws* 8 after جنساً add. ما *Firdaws* | شجر : الشجر *Firdaws* 9 أغصان om. *Firdaws* | بعضها : بعضه *K. al-Ḥikma* | الشجرة : الشجر *Firdaws* 10 الشجرة om. *Firdaws, K. al-Ḥikma* | الجولق : الجولي *Firdaws* : الحيوان *K. al-Ḥikma* 11 وقال ... ذكر : في *Firdaws* | علّة : قلّة *K. al-Ḥikma* | الدليل : الدلائل *Firdaws, K. al-Ḥikma* 12 تكون om. *Firdaws* 13 لخفتها : لخفتهما *Firdaws* 14 وترفع : وتدفع *Firdaws* | وإذا : فإذا *Firdaws, K. al-Ḥikma* 16 ويكوّن : ويتكوّن *Firdaws* 17 وكالنفاخات : وكنفاخات *Firdaws* : وكالنفحات *K. al-Ḥikma* | التي ... من om. *Firdaws* | مستديرة : المستديرة *Firdaws, K. al-Ḥikma* 18 المزاج : الدماغ *Firdaws, K. al-Ḥikma* | وأمّا : فأمّا *Firdaws* | خفّة : قلّة *Firdaws*

1 after وقلّتها blank space of six letters M 5 واتّسعت : واشبعت M 8 السماء : السمآ M 9 الشجرة : الشجر M 10 الجولق : الحيوان M 17 وكالنفاخات : كالنَّفحات M | مستديرة : المستدير M

The shortness of the body and the smallness of the newborn are caused by a constriction of the womb and a shortage of nutriment for the embryo. That is like the citron which one puts in a narrow bottle when still on its branch, and which can then not get big in it. If the bottle were wide and the citron had enough nutriment, the citron would get large and expand in it. Likewise when trees grow between stones in which there is no water, they cannot get large, whereas trees growing in good soil get large like the trees of the land Tabaristan. For due to the smoothness of its soil and the abundance of rainfall [the trees there] get very large, intertwine and disappear lengthwise into the sky. Often I have seen in the thicket [of 76b
Tabaristan] a kind of tree the branches of which had already fused with those of others and clumped together so that who gazed [at them] did not know the branch of this tree from the branch of that tree. This kind of tree is called *Jawlaq*.[212]

(138) He says[213] in an account of the reason for the head's roundness and of the evidence thereof that when the mixtures which are in the parents' semen become a foetus and the heat is in motion, the substance of the two light elements in the semen rises due to their lightness whereas the substance of the two heavy elements in [the semen] descends. Together with them the heat lifts up the matter and when the body has reached its final extension and length within the womb, the airy [element] rising with the fiery begins to rotate in [the body's] most elevated part. That matter there, therefore, becomes round and the head is made from it, just as glass becomes round when one blows into it and just as the bubbles which emerge from the rain are round. Thus the ideal head is spherical and of 77a
well-balanced mixture. A very small [head] indicates a light brain and a

[212] Following the emendation *al-jawlaq* suggested by Siggel (*Propädeutische Kapitel*, p.401, n.4), which he identifies with the *Euphorbia antiquorum*.

[213] Cf. source in ʿAlī ibn Rabban, *Firdaws*, ed. al-Ṣiddīqī, 45.6–20; parallel in *K. al-Ḥikma*, MS E, fols.102b15–103a10.

الدماغ وفساد الذهن ، وما عظم جدّاً دلّ على الخبل والبله ، وما اعتدل منه وشاكل البدن دلّ على الذكاء والوقار . وكذلك إفراط طول البدن أو قصره وإفراط عظم العين وجحوظها وإفراط صغرها وغؤورها دليل على فساد الذهن والاعتدال . وكما أنّ الفلك جعل مسكناً للنيرات وكذلك جعل الرأس مسكناً للنفس الناطقة ولجميع الحواسّ .

(139) وقال بقراط من العلامات الدالّة على أنّ المرأة حامل ذكراً أن يكون لونها حسناً وحركتها خفيفة وثديها الأيمن أكبر من الأيسر والنبض من اليد اليمين عظيماً سريعاً صلباً ، وإذا كانت حاملاً بأنثى فإنّ العلامات بالضدّ مما ذكر .

(140) وقال إنّ الذكورة من الأجنّة تحملون في الجانب الأيمن أكثر والإناث في 77b
الجانب الأيسر أكثر .

(141) وقال بقراط المني يقوم للجنين مقام المادّة ومقام الفاعل المصوّر ، ودم الطمث يقوم مقام المادّة فقط . والجنين يتمّ بامتزاج مني الذكر بمني الأنثى ، والسبب فيه أنّ مني الذكر غليظ حارّ المزاج لا يمكنه أن يتمدّد وينبسط جيّداً ، ومني الأنثى رقيق بارد المزاج فيعدّل غلظ ذاك وحرارته . ويكوّن منهما غشاء يحيط بالجنين . كما يتكوّن الخبز المختبز من النشا‹ستج› على الطابق ، ويكون هذا الغشاء غير لاصق به على مثال بيضة لم تبلغ في رحم

والتوقير : Firdaws التوقد : والوقار 2 Firdaws, K. al-Ḥikma البله والخبل : الخبل والبله 1
الذهن 3–4 K. al-Ḥikma وقصره : أو قصره | K. al-Ḥikma كذلك : وكذلك | K. al-Ḥikma
فكذلك : Firdaws كذلك : وكذلك 4 Firdaws, K. al-Ḥikma الاعتدال والذهن : والاعتدال
Kāmil والعلامات : وقال ... العلامات 6 K. al-Ḥikma وبجميع : ولجميع 5 K. al-Ḥikma
Kāmil في : من[2] | Kāmil وحلمته أكبر add. الأيسر after 7 Kāmil حبلى بذكر : حامل ذكراً
هذه add. فإنّ after | Kāmil حبلى : حاملاً | Kāmil فأمّا متى : وإذا | Kāmil ممتلئاً : صلباً 8
om. مما ذكر | Kāmil على الضدّ : بالضدّ | Kāmil تكون منها add. العلامات after | Kāmil
وقال 11 K. fī Manāfiʿ وكذلك قال بقراط . وهذا القول موافق لقوله حيث يقول : وقال 9 Kāmil
: مقام ... المصوّر | om. Kāmil للجنين | Kāmil فنقول إن جالينوس وابقراط يعتقدان أنّ : بقراط
: والسبب فيه 13 Kāmil وأنّ الجنين نما : والجنين 12 Kāmil مقام الفاعل والمادّة في كون الجنين
: جيّداً 14 Kāmil ومنى الأنثى رقيق بارد المزاج فمنى الذكر لغلظه add. المزاج after | Kāmil وذلك
om. Kāmil ومنى ... المزاج | Kāmil جداً ولحرارته تفسد مادة الجنين فاحتاج إلى منى الأنثى
: النشاستج 15 Kāmil فيكون : ويكوّن | Kāmil غلظه : غلظ ذاك | Kāmil لتعديل : فيعدّل
وذلك أنّك ترى ذلك الغشاء لاصقاً بالرحم في مواضع : ويكون ... الغشاء 16 Kāmil النشاشيج
Kāmil البيضة التي : بيضة | Kāmil افواه العروق المعروفة بالنقر وترى سائره متبريا عن الرحم

2 الذكاء : الذكا M 3 وغؤورها : وعوورها M 10 after أكثر blank space of half a line M
11 المصوّر : والمصور M

damaged mind, whereas a very large [head] indicates insanity and stupidity. [A head] which is well-balanced and fits the proportion[s] of the body indicates intelligence and seriousness. Equally an excessively long or short body and excessively large and bulging eyes or excessively small and sunken eyes are evidence for corruption of the mind and the equilibrium. As the sphere is made the dwelling place for the luminaries, so the head is made the dwelling place for the rational soul and all senses.

(139) Hippocrates says:[214] among the signs which indicate that a woman is with a male child are her good complexion, the lightness of her movements, her right breast being bigger than the left, and the pulse of her right hand being strong, fast and hard. When she is with a female child the signs are the opposite of the ones [just] mentioned.

(140) He says:[215] the male embryos are borne more often on the right side whereas the female ones [are borne] more often on the left side. 77b

(141) Hippocrates says:[216] the sperm serves as the embryo's matter and the embryo's formative agent, whereas the menstrual blood solely serves as matter. The embryo becomes perfect through the mixing of the male with the female semen. The reason thereof is that the male semen is thick, of hot mixture and cannot extend and expand well, whereas the female semen is fine, of cold mixture and thus balances the thickness and heat of the former. From both of them the membrane surrounding the embryo is generated. Just as the baked bread containing starch[217] is generated along [its] baking dish, this membrane is non-adhesive in the manner of an egg

[214] Cf. parallel in al-Majūsī, *Kāmil*, I 121.18–20.

[215] Cf. Galen, *K. fī Manāfiʿ*.

[216] Cf. parallel in al-Majūsī, *Kāmil*, I 117.19–21 (partially repeated in 118.21–2), 117.33–118.1, 118.4, 118.6–7, 118.10–12, 118.15–17, 118.18–19, 118.23–7, 119.9–11, 119.12–16, 119.19–20, 119.23–4, 119.26, 119.30, 120.14–16, 121.10–11.

[217] I suggest the emended reading *nashāstaj* (Persian for 'starch') based on the *rasm* of Majūsī's reading.

الدجاجة ولم تصلب قشرها . ثم يصير إليه دم الطمث وروح حيوانيّ في
الشرايين التي تصير إلى الرحم ، فينفذان جميعاً في جوهر الغشاء ، وتصير من
ذلك في الغشاء ثقب ومجار ، ولا تزال تلك المجاري تتّسع ولا تلتحم لاتّصال
78a الجريان فيها ، فإنّ المنى لا يتقطع اجتذابه للدم بما فيه من القوّة الجاذبة . ثم
يصلب الغشاء ويشتدّ ويتولّد فيه عند المنافذ التي يجري فيها الدم إلى الجنين
عروق وشرايين . ثم تجتمع وتشتبك ويصير منها المشيمة . ثم يحدث القوّة
المصوّرة من المنى والدم الأعضاء والعروق والغضاريف . وتبدأ بالأعضاء الثلاثة
التي هي الأصول ، فيحدث الدماغ من نفس المنى والقلب من دم الشرايين
والكبد من دم العروق الصائرة من المشيمة إلى بدن الجنين . وتكون هذه
الأعضاء الثلاثة أوّلاً يقرب بعضها من بعض ، ثم تنفرق وتتباعد ويتفرّع منها
ثلاثة فروع ، فيفرّع من الدماغ العصب والنخاع ومن القلب الشريان العظيم
ومن الكبد العرق الأجوف . ثم يكون له بعد هذه الأصول والفروع العظام
78b المحيطة بها لتكون جنّة لها ، وبعدها الأعضاء الباقية ، وعند ذلك يتبدئ الجنين
يتحرّك . فإن كان ذكراً فصورته تتمّ في أربعين يوماً وحركته في ثمانين وتمامه في
مائتين وأربعين يوماً ، وإن كانت أنثى فصورتها تتمّ في خمسة وأربعين يوماً

رحم : من الرحم *Kāmil*
لطيف : الطمث | *Kāmil* ويصير أيضاً إليه : ثم ... إليه | *Kāmil* الخارج add. قشرها 1 after
om. *Kāmil* تلك | *Kāmil* فلا : ولا[1] 3 *Kāmil* فيصير : وتصير | *Kāmil* في : إلى 2 *Kāmil*
ان : يصلب الغشاء 5 *Kāmil* روح حيوانيّ وروح طبيعيّ add. المنى after | *Kāmil* لأنّ : فإنّ 4
6 after في الغشاء : فيه | *Kāmil* من المنى add. ويتولّد | *Kāmil* ذلك الغشاء يصلب
أفواهها متصلة بأفواه العروق والشرايين التي تصير إلى الرحم فيتصل العرق منها بفم add. وشرايين
أنّ هذه العروق والشرايين المتولّدة تشتبك : تجتمع ... المشيمة | *Kāmil* العرق الشريان بفم الشرايين
Kāmil ان القوى المصوّرة تحدث : يحدث ... المصوّرة 6–7 *Kāmil* وتنتسج وتستدير معاً على الغشاء
أعضاء الجنين فيحدث من المنى : الأعضاء ... والغضاريف | *Kāmil* هذين أعني add. من 7 after
فانّه يحدث من دم الشرايين وأوّل الشيء تبدأ : وتبدأ ... الثلاثة | *Kāmil* نفسه الأعضاء البيض
لأكثر الأعضاء التي في البدن وهي الدماغ add. الأصول 8 after *Kāmil* القوّة المصوّرة فبالاعضاء
إلى بدن الجنين من المشيمة : من[2] ... الجنين 9 *Kāmil* والقلب والكبد وسائر الأعضاء اللحمية
: منها | *Kāmil* ثم يتّفرق من هذه الأصول : ثم ... منها | *Kāmil* بالقرب : يقرب 10 *Kāmil*
12 after *Kāmil* أزواج add. الدماغ after | *Kāmil* فيتفرع : فيفرّع 11 *Kāmil* من هذه الأصول
يتكوّن مع كون هذه الأصول والفروع والعظام المحيطة : يكون ... بها 12–13 *Kāmil* انه add. ثم
... وأمّا add. يتحرّك 14 after *Kāmil* ... ثم يتكون من بعده هذه : وبعدها 13 *Kāmil* بها
om. *Kāmil* تتمّ 15 *Kāmil* يوماً add. ثمانين after | *Kāmil* المولود لتسعة أشهر

وبمجار : ومجارى M 4 بما : ولما M | الجاذبة : الجاذثه M 7 الثلاثة : الثلثة M 10 الثلاثة :
M ثلثه 11 ثلاثة : M الثَلثه

which neither matures nor does its shell harden in the womb of the hen. Then the menstrual blood and the vital spirit in the arteries coming to the womb reach [the embryo]. They both penetrate the substance of the membrane together which causes perforations and channels in the membrane. Those channels do not stop widening and do not close because of the continuous flow in them. For there is no disruption in the semen drawing blood by means of its attractive power. Next the membrane 78a
hardens, becomes firm and veins and arteries are generated in it along the passages in which the blood flows to the embryo. They then join, form a net and from that the placenta results. Thereupon the forming power brings forth organs, veins and cartilages from semen and blood. It starts with the three organs which are the foundations. It brings forth the brain from the semen itself, the heart from the blood of the arteries and the liver from the blood of the veins going from the placenta to the embryo's body. At first these three organs are close to each other, then they become separate, distant and three branches branch out from them: the nerves and the spinal cord branch out from the brain, the large artery (*aorta*) from the heart and the hollow vein (*vena cava*) from the liver. After these foundations and branches [the embryo] gets bones which encompass those so that they have protection, and afterwards [the embryo gets] the remaining body parts. At that [stage] the embryo begins to move. If it is male, its form 78b
will be achieved in 40 days, its motion in 80 and its completion in 240 days.

وحركتها في تسعين يوماً وتمامها في مائتين وسبعين يوماً . والجنين الذكر تولّده يكون في الجانب الأيمن لأنّه أسحن بمجاورة الكبد وحركته تبيّن في هذا الجانب ، والأنثى فإن تولّدها في الجانب الأيسر لأنّه أبرد مزاجاً بمجاورته الطحال وحركتها تبيّن في هذا الجانب .

(142) قالت الحكماء إنّ للنفس النباتيّة سبع قوى فعّالة ، وهي الجاذبة والماسكة والهاضمة والدافعة والغاذية والنامية والمصوّرة . وإنّ أوّل فعلها عند استقرار النطفة في الرحم هو جذبها دم الطمث إلى الرحم وإمساكها له هناك وهضمها . فإذا جذبت هذه القوّة الدم إلى هناك جفّفته حول النطفة وأدارته عليها كما يدور بياض البيضة حول محّها ، فيكون عند ذلك النطفة كالمحّة 79a
ودم الطمث حولها كالبياض . ثم إنّ حرارة النطفة تسخّن رطوبة الدم وتنضجها فتثخن وتنعقد تلك الرطوبة وتصير علقة كما ينعقد اللبن الحليب من الإنفحة .

ويستولي عند ذلك على تلك الجملة قوى روحانيّات زحل وتبقى في تدبيرها شهراً ، وإنّما صار ابتداء تدبير النطفة من زحل لأنّه أعلى الكواكب السيّارة فلكاً مما يلي فلك المكوكب الذي هو مكان الجواهر الشريفة ومصبّ القوى الروحانيّة . وما دام التدبير إلى زحل فإنّ تلك العلقة تكون بحالها غير مختلطة ولا ممتزجة ، بل جامدة ممتسكة جاذبة إليها الموادّ لغلبة برد زحل وسكونه وثقل طبيعته إلى أن يدخل الشهر الثاني ويصير التدبير للمشتري الذي فلكه يتلو

1 والجنين الذكر : ينبغي أن تعلم ان كلّ جنين ذكر Kāmil 1–2 تولّده يكون : يكون تولده Kāmil 2 لأنّه ... الكبد om. Kāmil | تبيّن : تتبين Kāmil 3 والأنثى : وكلّ أنثى Kāmil | فإن تولّدها : فتولدها Kāmil 3–4 لأنّه ... الطحال om. Kāmil 4 تبيّن : تتبين Kāmil 5 قالت ... إنّ : ثم إعلم يا أخي أن Rasāʾil 7 له : لها Rasāʾil 8 فإذا : ثم إعلم يا أخي بأنّه إذا Rasāʾil | جفّفته : أخفته Rasāʾil 9 البيضة : البيض Rasāʾil | after ذلك add. حول Rasāʾil 11 وتنضجها فتثخن : فتنضجها فتسخن Rasāʾil | وتصير : فتصير Rasāʾil 13–14 تدبيرها شهراً : تدبيراتها بمشاركة قوى روحانيّات سائر الكواكب شهراً واحداً ثلاثين يوماً ، سبع مائة وعشرين ساعة ، كما ذكر ذلك في كتب أحكام النجوم بشرح طويل ونريد أن نذكر من ذلك طرفاً ليكون دستوراً لما أن نتكلّم فيما بعد Rasāʾil 14 وإنّما صار : وإعلم يا أخي بأنّ Rasāʾil | من ... لأنّه : إنما صار لزحل من أجل أنّه Rasāʾil 15 المكوكب : الكواكب Rasāʾil | ومصبّ : منصب Rasāʾil 16 وما : ثم إعلم يا أخي بأنّه ما Rasāʾil | إلى زحل : لزحل إلى تمام شهر ثلاثين يوماً Rasāʾil | after تكون add. ياقية Rasāʾil 17 ممتسكة : متمسكة Rasāʾil | جاذبة : جارية Rasāʾil

3 مزاجاً : مراحاً M 4 after الجانب blank space of six letters M 7 وإمساكها : اوامساكها M 8 القوّة M : or read القوى ? 9 كالمحّة : كالمخة M | محّها : مُخهَا M

If it is female, its form will be achieved in 45 days, its motion in 90 days and its completion in 270 days. The development of the male embryo takes place on the right side, as it is hotter due to the liver's vicinity and its motion is apparent on this side. The development of the female takes place on the left side, as it is colder in mixture due to its vicinity to the spleen and its motion is apparent on this side.

(142) The wise men say[218] that the vegetative soul has seven active faculties: the attractive, retentive, digestive, repellent, nutritive, augmentative and formative. Their first activity upon insemination is to draw menstrual blood to the womb, keep it there and digest it. So when these power[s] draw the menstrual blood there, they cause it to dry and move around the sperm just as the egg white surrounds its yolk. In doing that 79a
the sperm is like the yolk and the menstrual blood around it like the egg white. Then the heat of the sperm warms and ripens the moisture of the blood and so that moisture thickens, coagulates and becomes a blood clot just as cow milk is coagulated by rennet.

At that [time] the spiritual powers of Saturn master all of that and remain in charge for a month. Saturn is the first to take charge of the sperm, for it is the most elevated of the planets and a sphere next to the sphere of the stars, which is the position of the noble substances and the place from where the spiritual powers pour forth. As long as Saturn is in charge, the blood clot is in a pure and unmixed condition. Moreover it is congealing, retentive and attracts matter to itself due to the prevailing coldness of Saturn, its tranquillity and heavy nature until the second month sets in and Jupiter, the sphere of which follows the sphere of Saturn, takes charge.

218 Taken from the 25th Epistle of the *Rasāʾil Ikhwān al-Ṣafāʾ*, II 421.2–17, 422.8–11, 434.19–20, 422.11–16, 436.6–7, 422.16–423.11, 423.22–425.4, 425.15, 439.24–440.2, 425.15–20, 441.7–10, 425.20–426.22, 451.20–452.1, 427.1, 446.2–8, 446.14–447.10, 447.15–18, 447.22–448.4, 448.7–10, 449.3–11, 448.11–19.

فلك زحل .

79b وهو كوكب الاعتدال وعلّة صحّة المزاج . فحينئذ يستولي عليها قوى روحانيّة فيتولّد منها في تلك العلقة حرارة تسخنها وتعدل مزاجها ، ويختلط الماءأن ويعرض لتلك الجملة حركة مثل الاختلاج والارتعاش والهضم والنضج ، فلا تزال تلك حالها ما دام في تدبير المشتري إلى شهرين .

ثم يصير التدبير إلى المرّيخ ، وهو ينبوع الحرارة والنضج ، فيستولي على العلقة قوى روحانيّة . فيشدّ اختلاجها وارتعاشها ويتولّد فيها فضل حرارة وسخونة وتصير تلك الجملة مضغة حمراء .

فإذا دخل الشهر الرابع وصار التدبير للشمس ، التي هى رئيس الكواكب وقلب العالم ، استولى على المضغة قوى روحانيّاتها ونفخ فيها روح الحياة وسرت فيها النفس الحيوانيّة . وذلك أنّ الشمس هي المستولية على الكائنات
80a التي دون فلك القمر وخصوصاً على مواليد الحيوانات ذوي الرحم . وجرمها في هذا العالم بمنزلة جرم القلب في البدن وسائر أجرام الكواكب والأفلاك بمنزلة أعضاء البدن ومفاصل الجسد ، وقوى روحانيّاتها في العالم كسريان الحرارة الغريزيّة المنبثّة من القلب السارية في جميع أعضاء البدن . فإذا بلغت بمسيرها

2 فحينئذ يستولي : وتستولي *Rasāʾil* | روحانيّة : روحانيّته *Rasāʾil* 3 فيتولّد : فيولد *Rasāʾil* | منها : عند ذلك *Rasāʾil* | تسخنها : تسخن *Rasāʾil* | وتعدل : يعتدل *Rasāʾil* | after الماءأن add. ويمتزج الخلطان *Rasāʾil* 5 تلك : هذا *Rasāʾil* | دام : دامت *Rasāʾil* | after إلى add. تمام *Rasāʾil* 6 يصير : يدخل الشهر الثالث ويصير *Rasāʾil* | إلى المرّيخ : للمرّيخ الذي يلي المشتري في الفلك *Rasāʾil* | after الحرارة add. والأسخان *Rasāʾil* | after والنضج add. في الكائنات *Rasāʾil* | فيستولي : وتستولي *Rasāʾil* | after على add. تلك *Rasāʾil* 7 روحانيّة : روحانيّته *Rasāʾil* | فيشدّ : ويشتد *Rasāʾil* 8 الجملة : العلقة *Rasāʾil* | after حمراء add. فلا تزال تتقلّب حالاً بعد حال من النضج والاستحكام بمشاركة قوى روحانيّات سائر الكواكب للمرّيخ إلى تمام ثلاثة أشهر *Rasāʾil* 9 فإذا دخل : ثم يدخل *Rasāʾil* | وصار : ويصير *Rasāʾil* | التي . . . رئيس : رئيسة *Rasāʾil* | after الكواكب add. وملكة الفلك *Rasāʾil* 10 after العالم add. بإذن البارئ جلّ ثناؤه *Rasāʾil* | استولى : وإعلم يا أخي بأنّه إذا دخل الشهر الرابع من مسقط النطفة وصار التدبير للشمس واستولت *Rasāʾil* | ونفخ : ونفخت *Rasāʾil* 11 أنّ : لأنّ *Rasāʾil* | after هي add. رئيسة الكواكب في الفلك ونفسها هي روح العالم بأسره وهي *Rasāʾil* 12 وخصوصاً : وخاصّة *Rasāʾil* | وجرمها : وأشدّ اختصاصاً بمواليد الإنس وذلك أنّ جرمها *Rasāʾil* 13 هذا om. *Rasāʾil* 14 وقوى : وسريان قوى *Rasāʾil* 15 جميع om. *Rasāʾil*

3 الماءأن : المااَن M 4 لتلك : له ذلك M | والنضج : والنّصح M 6 العلقة : العقله written on the line M : العلقة written above the line M 7 فضل : فصل M 8 مضغة : مُصْيغه M after حمراء blank space of six letters M 10 after استولى blank space of six letters M الحياة : الحيوَة M 14 البدن : القلب M | كسريان : لَشريان M

Jupiter is the planet of balance and the cause of sound mixture. At that time [Jupiter's] spiritual powers master the [blood clot] and due to them 79b
heat is generated in it which warms the blood clot and balances its mixture. The two fluids [i.e. sperm and blood] mix and their entire [mixture] is exposed to motion like convulsion, tremor, digestion and ripening. The [blood clot] continues in that condition as long as it is in Jupiter's charge until two months [have passed since the insemination].

Then Mars, the wellspring of heat and ripening, takes charge and [its] spiritual powers master the blood clot. So its convulsion and tremor become strong, a surplus of heat and warmth is generated in it, and that entire [hot and moving blood clot] becomes a red foetus.

When the fourth month sets in and the sun, which is the leader of the planets and heart of the world, takes charge, the powers of its spirits master the foetus and breathe the spirit of life into it, and the animal soul pervades it. For the sun masters the beings under the sphere of the moon, especially the births of animals possessing a womb. The [sun's] body is in 80a
this world like the heart in a body, and the bodies of the other planets and spheres are like the body parts and the joints of the body. The powers of its spirits are in the world like penetrating innate heat spreading from the heart which penetrates all body parts. When, four months after insemination, the sun has reached the end of the fourth zodiac sign and has covered

بعد أربعة أشهر من مسقط النطفة إلى آخر البرج الرابع وقد قطعت من الفلك ثُلث الدور ، وهو من المسافة مقدار ما بين شرفها إلى بيتها ، تكون المضغة قد استوفت طبائع البروج من المثلّثات الناريّة والترابيّة والهوائيّة والمائيّة ، واختلطت الطبائع الأربع من الأركان الأربعة في تركيب بنية الجنين ، اعتدل المزاج وانتقشت الصورة واستبانت الخلقة وظهرت أشكال العظام وتركيب المفاصل.

80b وإذا سارت الشمس إلى البرج الخامس المسمّى بيت الولد الموافق طبيعته للبرج الذي كانت فيه يوم مسقط النطفة ، وصار التدبير للزهرة صاحبة النقش والتصاوير ، استتمّت الخلقة واستكملت البنية وظهرت صورة الأعضاء واستبانت رسوم العينين وأنشقت المنخران وانفتح الفم وتميّزت المفاصل ، ولكن الجنين يكون مجموعاً منضماً منقبضاً كأنّه في صرّة ، ركبتاه مجموعتان إلى صدره ومرفقاه منضمان إلى حقويه ، وهو منكّس رأسه وذقنه على ركبتيه وكفاه على خدّيه كنائم محروز ، وسرّته متّصلة بسرّة أمّه يمتصّ الغذاء منها إلى يوم ولادته ، ويكون وجهه مما يلي ظهر أمّه إن كان ذكراً وإن كانت أنثى فبعكس ذلك .

ثم يدخل الشهر السادس ويصير التدبير لعطارد ، صاحب العلوم والآداب

1 بعد om. *Rasāʾil* 2 مقدار : بمقدار *Rasāʾil* | المضغة om. *Rasāʾil* 3 من المثلّثات om. *Rasāʾil*
4 واختلطت : وعند ذلك تكون قد اختلطت *Rasāʾil* | الأربع om. *Rasāʾil* | اعتدل : واعتدل
Rasāʾil 5 واستبانت : وأنشئت *Rasāʾil* | وتركيب : وركّبت *Rasāʾil* 6 after المفاصل add.
وتهندم التركيب والتفّت الأعصاب على المفاصل وامتدت العروق في خلل اللحم وظهرت البنية مخلّقة
غير مخلّقة *Rasāʾil* 7 وإذا سارت : إعلم يا أخي بأنّه إذا دخل الشهر الخامس وسارت *Rasāʾil*
8 صاحبة : وصاحبة الأصغر الساعد *Rasāʾil* 9 after والتصاوير add. واستولى على المخلّقة قوى
روحانيّاتها *Rasāʾil* 10 واستبانت رسوم : واستبان رسم *Rasāʾil* | وأنشقت : وأنشق *Rasāʾil*
after الفم add. وثقب الأذنين ومجرى السبيلين *Rasāʾil* 11 منضماً : منظماً *Rasāʾil* | after
كأنّه add. مصرور *Rasāʾil* 12 وذقنه : على دقّته *Rasāʾil* | على : وعلى *Rasāʾil* 13 كنائم...
وهو شبه نائم محزون فلو رأيته يا أخي لرحمته لضيق مكانه وضعف أحواله ولكنّه لا يحسّ : وسرّته
الولادة : ولادته 14 *Rasāʾil* بما هو فيه رفقاً من الله تعالى بخلقه ولطفاً بهم وتكون سرّته *Rasāʾil*
: فبعكس 15 *Rasāʾil* كان : كانت | *Rasāʾil* إن كان ذكراً مما يلي ظهر أمه : مما... ذكراً
after | *Rasāʾil* إعلم أنّه عند دخول الشهر السادس : يدخل... السادس 16 *Rasāʾil* فعكس
العلوم add. والشعور والحسن والمعارف والحكم والحركات *Rasāʾil* | after والآداب add. *Rasāʾil*

1 آخر : اخر M 2 ما بين : مَاتين M 13 خدّيه : خدّمْه M | محروز : مَحرونٍ M | الغذاء :
الغذآ M 15 فبعكس : فيعكس M 16 والآداب : والادَاب M

a third of the sphere's circumference, which equals the distance between its exaltation and its house,[219] the foetus has already received its full share of the natures of the zodiac signs belonging to triplicities[220] of fire, earth, air and water. The four natures of the four elements mix during the composition of the embryo's natural constitution. The mixture is well-balanced, the form engraved, the inborn disposition evident, and the shape of the bones and the composition of the joints become visible.

When the sun approaches the fifth zodiac sign, which is called the house 80b
of the offspring,[221] and which accords in its nature with the zodiac sign where [the sun] was on the day of insemination, and when Venus the mistress of engraving and images takes charge, the [embryo's] inborn disposition becomes perfect, the natural constitution complete, the form of the body parts visible, the traces of the eyes become evident, the nostrils inhale, the mouth opens and the joints are set apart. However, the embryo is drawn together, closely packed and contracted as if it were in a bundle, its two knees drawn together towards its chest and its two elbows towards its loins. It is turned upside down, its head and its chin on its two knees, the palms of its two hands on its two cheeks like a protected sleeper, its navel joined to the navel of its mother, and it sucks the nutriment from her until the day of its birth. Its face is close to the back of its mother if it is male, and if it is female, then it is the reverse of that.

Then the sixth month sets in and Mercury, the master of the sciences,

[219] According to al-Qabīṣī (Alcabitius), *The Introduction to Astrology* (pp.24–5), the exaltation of the sun, i.e. the position in which it exerts the strongest influence, is in the ninteenth degree of Aries and its house is Leo.

[220] Al-Qabīṣī defines the term 'triplicity' saying 'When three of these [zodiac – EW] signs agree in one nature [being either fiery or earthy or airy or watery – EW], they are called a triplicity' (tr. Burnett et al., p.25).

[221] The Arabic does not differentiate between the house as the zodiac sign which is understood to be the 'domicile (οἶκος) of a planet' (as above), and the house which indicates 'an astrological place (τόπος)' (as here). These latter are also twelve starting with the ascendant and defined in astrological tables. See al-Qabīṣī's *Introduction*, p.47 and n.31.
About the fifth place or house al-Qabīṣī says: 'The fifth is the place of children [*bayt al-walad* – EW], joy, eating and drinking, messengers and gifts; of life it indicates what kind of reputation – beautiful or ugly – comes after death' (tr. Burnett et al., pp.51–3).

والصنائع والنطق ، ويستولي عليه قوى روحانيّاته فيتحرّك عند ذلك الجنين في 81a الرحم ويمدّ يديه ورجليه وينبسط جوارحه ويفتح فاه وتحرّك شفتيه ويتنفّس من منخريه ويدير لسانه في فمه ، تارة يتحرّك وتارة يسكن ، ولا يزال كذلك إلى أن يتمّ الشهر السادس ويدخل الشهر السابع ويصير التدبير للقمر .

وهو النيّر الأصغر القابل طبائع الكواكب والمؤدي الفيض من العالم العلويّ إلى العالم السفليّ . وتستولي عليه قوى روحانيّته فتستوي قامته وتشتدّ أعضاؤه وتقوى حركته ويحسّ بضيق مكانه ويطلب التنقّل والخروج ، فإن قدّر له ذاك وولد عاش وتربى .

وإن بقي هناك إلى أن يدخل الشهر الثامن وتحصل الشمس في بيت الموت ويرجع التدبير ثانياً إلى زحل ، عرض للجنين ثقل وسكون وغلب عليه البرد والنوم وقلّة الحركة ، فإنّ ولد في هذا الشهر كان بطيء النشوء ثقيل الحركة قليل العمر وربّما كان ميتاً .

وإذا دخل الشهر التاسع وانتقلت الشمس إلى هذا البرج ، وهو بيت النقلة 81b والأسفار ، ورجع التدبير إلى المشتري ، ظهرت أفعال النفس الحيوانيّة من الجسد ، لأنّ الشمس قد استوفت طبائع البروج مرّتين في هذه الأشهر الثمانية وسارت من فلك البروج مائتين وأربعين درجة ، وهذه المسافة ما بين بيتها إلى

1 after والنطق add. والبيان والكلام والفصاحة والتمييز والفطنة والقراءة والنغمة والرياضات والحكمة *Rasāʾil* 2 ويمدّ ... وينبسط : ويركض برجليه ويمد يديه ويبسط *Rasāʾil* | after جوارحه add. ويضطرب ويحسّ بمكانه *Rasāʾil* 3 فمه : فيه *Rasāʾil* | تارة يتحرّك : فيكون تارة متحرّكاً *Rasāʾil* after يسكن add. وتارة ينام وتارة يستيقظ *Rasāʾil* | ولا : فلا *Rasāʾil* | كذلك : ذلك دأبه *Rasāʾil* 5 وهو om. *Rasāʾil* | القابل : نظير الشمس في المنظر والمخالف في المخبر المتوسّط بين العالمين الآخذ من *Rasāʾil* | after الكواكب add. فيضها من العالم العلوي *Rasāʾil* | والمؤدي : الفائض المؤدي *Rasāʾil* | الفيض : تلك الفيضان والخيرات *Rasāʾil* | من ... العلويّ om. *Rasāʾil* 6 روحانيّته : روحانيّاته فيربو لحم الجنين حينئذ وتسمن جثته *Rasāʾil* | فتستوي : وتنتصب *Rasāʾil* after أعضاؤه add. وتصلب مفاصله *Rasāʾil* 7 ذاك : ذلك بما يوجد أحكام النجوم بأسباب يطول شرحها وخروجها على المجرى الطبيعيّ *Rasāʾil* 8 وولد ... وتربى : وكان الجنين كاملاً عاش وتربى وعمّر *Rasāʾil* 9 وتحصل : وتدخل *Rasāʾil* | في om. *Rasāʾil* 10 ثانياً om. *Rasāʾil* | after زحل add. من الرأس فتستولي عليه قوى روحانيّاته *Rasāʾil* 13 هذا البرج : البرج التاسع *Rasāʾil* وهو om. *Rasāʾil* 14 after المشتري add. السعد الأكبر واستولت عليه قوى روحانيّاته واعتدل المزاج وقويت روح الحياة *Rasāʾil* 15 after الشمس add. تكون *Rasāʾil* | after البروج add. المثلّثات الناريّة والمائيّة والهوائيّة والترابيّة *Rasāʾil* | هذه ... الثمانية : الثمانية الأشهر *Rasāʾil* 16 وسارت من : وقد سارت الشمس في *Rasāʾil* | after المسافة add. مقدار *Rasāʾil*

2 شفتيه : شغته M 8 وتربى : وَنزَباً M 11 بطيء النشوء : يَطئ الشُو M 16 مائتين : مايتى M

culture, skills and rationality takes charge. The powers of its spirits master [the embryo] and in doing so the embryo is set in motion in the womb, extends its hands and legs, expands its extremities, opens its mouth, 81a
moves its lips, breathes through its nostrils, and turns its tongue in its mouth; it is in motion at times and at rest at other times. It does not stop being like that until the sixth month ends, the seventh month sets in and the moon takes charge.

[The moon] is the smallest luminary, which receives the natures of the planets and conveys the emanation from the upper to the lower world. The powers of its spirituality master [the embryo], so its structure straightens up, its body parts become firm and its motion strong, it perceives the constriction of its place, and seeks a change of locality and an exit. If that is ordained for it and it is born, it lives and is raised.

But if it remains there until the eight month sets in, the sun occurs in the house of death and Saturn takes charge for a second time, the embryo is beset by heaviness and tranquillity, and coldness, sleep and rarity of motion prevail over it. If it is born in that month, it is slow in developing, heavy in motion, of limited lifespan and sometimes already dead.

When the ninth month sets in, the sun wanders to the zodiac sign which is the house of transposition and journeys,[222] and Jupiter returns to take charge, the activities of the animal soul arise from the body. For the sun 81b
has already received its full share of the natures of the zodiac signs twice in these eight months and journeyed 240 degrees of the sphere of the zodiac. This distance [equals] what is between its house and its exaltation,

[222] About this ninth house al-Qabīṣī has to say: 'The ninth is the place of the journey [*bayt al-safar* - EW], absence, roads (caravan routes), religion, religious observance, sciences, philosophy, books, messengers, messages and visions; of life it indicates the beginning of the middle of life' (tr. Burnett et al., p.53).

شرفها التاسع من بيتها الموافقَين في طبيعة واحدة . وطبيعة الجنين قبلت قوى روحانيّات الكواكب المنحطّة من الفلك مرّتين بمسير الشمس في المثلثات مرّةً إلى البرج الخامس ومرّةً إلى البرج التاسع ، ويكون الذي بقي للشمس إلى أن تعود إلى الدرجة التي كانت فيها وقت مسقط النطفة أربعة أبراج درجها مائة وعشرون درجة .

فإذا خرج الجنين بعد ثمانية أشهر استأنف العمر في الدنيا لكلّ درجة سنة وهو العمر الطبيعيّ ، فإن مكث تسعة أشهر فالذي يبقى لها ثلاثة أبراج تسعون درجة ، فيستأنف العمر تسعين سنة ، وإن مكث عشرة أشهر بقي ستون 82a درجة وعاش ستون سنةً . فأمّا الذي يزيد وينقص عن هذا المقدار فلأسباب وعلل يطول شرحها .

ولكلّ مولود من الحيوان أبوان في الفلك كما له والدان في الأرض ، أحدهما دليل عمره يسمّى كدخداه وهو فارسيّ معرّب أي ربّ البيت ، والآخر هيلاج وهو فهلويّ وأصله هيل أي ربّة البيت . فإن كانا مسعودين عند ولادته عاش المولود بخير وعمر ، وإن كانا منحوسين فالعكس من ذلك ، وإن كان الكدخداه مسعوداً والهيلاج منحوساً كان المولود طويل العمر سيّء الحال ، وإن كان الهيلاج مسعوداً والكدخداه منحوساً كان المولود حسن

1 الموافقَين : المتفقين *Rasāʾil* | وطبيعة ... قبلت : وتكون أيضاً في هذه المدّة قد قبلت طبيعة الجنين *Rasāʾil* 2 after في add. البروج *Rasāʾil* 3 after التاسع add. وكما تقدّم ذكرها ويبقى مرة أخرى كما نبيّن بعد هذا الفصل *Rasāʾil* | بقي : يبقى *Rasāʾil* 4 درجها مائة : ومائة *Rasāʾil* 5 وعشرون درجة : وعشرين درجة إلى تمام الدور *Rasāʾil* 6 سنة وهو : سنة الذي هو *Rasāʾil* 7 لها : له *Rasāʾil* 8 فيستأنف : ويستأنف المولود *Rasāʾil* | وإن : فإن *Rasāʾil* | after أشهر add. فالذي *Rasāʾil* | بقي : يبقى له برجان *Rasāʾil* 9 وعاش : فيستأنف المولود العمر *Rasāʾil* ستون : ستين *Rasāʾil* | فأمّا : وأمّا *Rasāʾil* 11 ولكلّ : وإعلم يا أخي بأنّ لكلّ *Rasāʾil* أبوان : أبوين *Rasāʾil* | after كما add. أن *Rasāʾil* | والدان : والدين *Rasāʾil* 12 كدخداه : كدَخداي *Rasāʾil* | وهو ... معرّب om. *Rasāʾil* | after والآخر add. يسمّى *Rasāʾil* 13 وهو ... هيل om. *Rasāʾil* 14 وعمر : عمراً طويلاً *Rasāʾil* | فالعكس : فبالعكس *Rasāʾil* 15 الكدخداه : الكدخداي *Rasāʾil* | after العمر add. فقيراً *Rasāʾil* 16 والكدخداه : الكدخداي *Rasāʾil*

6 ثمانية : ثمنية M 7 ثلاثة : ثلثه M 10 after شرحها blank space of half a line M 12 كدخداه : كذَخُداي M | والآخر : والاخر M 13 مسعودين : معودين written in the text, corrected to مسعودين in the marg. by a later hand M 14 ولادته : ولادتها M 15 الكدخداه : الكذحداي M | سيّء : سيي M 16 والكدخداه : واللدخاى M

which is the ninth after its house[223] and they both accord in one single nature. The nature of the embryo has received the powers of the planets' spirits which have descended from the sphere twice during the sun's journey through the triplicities [of the twelve zodiac signs], one time towards the fifth zodiac sign and another time towards the ninth zodiac sign. There remain four zodiac signs and their 120 degrees until the sun returns to the degree in which it was at time of insemination.

If the embryo emerges after eight months, it anticipates a lifetime in this world, [which amounts to] a year for every [remaining] degree, and this is a natural lifetime. If the embryo stays [in the womb] for nine months, then three zodiac signs of 90 degrees remain [until the sun's return] and it thus anticipates a lifetime of 90 years. If it stays ten months, 60 degrees remain and it lives 60 years. The [lifetime] which exceeds or falls short of this 82a
amount is due to causes and reasons the explanation of which would take too long.

Every newborn living being has two parents in the sphere just as it has two parents on earth. One of these two [spherical parents] is the guide of its life and is called *kadkhudāh* (owner of a domicile), which is Persian and Arabisized as 'master of the house'. The other one is *haylāj* (prorogator), which is Pahlevi and its root is *hayl*, i.e. 'mistress of the house'.[224] If they are both beneficent at the time of birth, the newborn will live well and long. If they are both maleficent, then it is the opposite of that. If the *kadkhudāh* is beneficent, but the *haylāj* maleficent, the newborn will have a long life, but in a bad condition. If the *haylāj* is beneficent, but the *kadkhudāh* maleficent, the newborn will have a good condition and be rich,

[223] See notes 219 and 221 above.

[224] On the *haylāj* (ἀφέτης) and the *kadkhudāh* (οἰκοδεσπότης) and how to define them, see Qabīṣī's *Introduction*, pp.110–17, and Kunitzsch, *Glossare*, pp.35–6 and 49–50.

الحال غنياً قصير العمر .

والمولود من يوم الولادة إلى تمام أربع سنين شمسيّة فى تدبير القمر ، صاحب
النموّ والزيادة والنشوء ، ومشاركه في التدبير كلّ واحد من الكواكب سبع تلك
المدّة . فتتصرّف الأحوالُ بالطفل من النموّ والزيادة والصحّة والسلامة وغيرها 82b
بحسب ما توجبه أحوال تلك المدبّرات .

ثم يصير في تدبير عطارد ثلاث عشرة سنة وهو صاحب النطق والحركة والآداب والفهم ، ويشاركه في التدبير سائر الكواكب كلّ واحد سبع هذه المدّة ، وكلّ ما انتهى التدبير إلى واحد منها ظهر من المولود الأخلاق والأفعال المشاكلة لتلك القوى التي انعجنت وامتزجت وانغرست في جبلته وهو جنين في الرحم ، كما يظهر من زهر النبات حبوبها ومن الشجر ثمارها وروائحها وألوانها وطعومها عند بلوغها وتمامها بحسب ما في طباعها وأشباحها .

ثم يصير في تدبير الزهرة ثمان سنين وهي صاحبة الحسن والزينة والشهوة واللذّة والرغبة في النكاح ، ويشاركها في التدبير سائر الكواكب سبع هذه المدّة ، فيظهر من المولود الرغبة في الأحوال المذكورة .

83a ثم يصير في تدبير الشمس صاحبة العزّ والرئاسة والتدبير والسياسة عشر

Rasāʾil ثم إعلم يا أخي بأنّه متّفق بين أهل صناعة التنجيم في أحكام المواليد أنّه : والمولود 2
after شمسيّة add. يكون الطفل Rasāʾil 3 ومشاركه : وتشاركه سائر الكواكب Rasāʾil | من
الكواكب om. Rasāʾil 4 after المدّة add. التي تسمّى سني التربية Rasāʾil | من النموّ : من التربية
Rasāʾil والعزّ والكرامة والأعلال والأمراض والبؤس والهوان واللذّة والألم : وغيرها | Rasāʾil والنموّ
في هذه السنين مذكور شرح ذلك في كتب .after المدبّرات add | Rasāʾil توجب : توجبه أحوال 5
7 after والآداب add. والتعاليم Rasāʾil 6 after والحركة add. تحاويل سني المواليد Rasāʾil
: وهو ... الرحم 9–10 Rasāʾil ظهرت في : ظهر من | Rasāʾil وكلّما : وكلّ 8 Rasāʾil والتمييز
ونور : ومن | Rasāʾil وحبوبها : حبوبها | om. Rasāʾil من 10 Rasāʾil في الرحم وهو جنين
12 after Rasāʾil وكمالها ونضجها .add وتمامها 11 after Rasāʾil وثمارها : ثمارها | Rasāʾil
Rasāʾil والشهوات : والشهوة | Rasāʾil ثماني سنوات : ثمان سنين | Rasāʾil المولود .add يصير
13 after النكاح add. والحرص على السفاح Rasāʾil | after الكواكب add. كلّ واحد منها Rasāʾil
14 after المولود add. في هذه المدّة Rasāʾil | الرغبة ... المذكورة : الرغبة في التزوج في هذه المدّة
Rasāʾil والنكاح

1 غنياً : عينا M | after العمر blank space of six letters M 2 سنين : سيتن M | تدبير :
M ثلثة عشر : ثلاث عشرة 6 blank space of five letters M after المدبّرات 5 M تربيه
انعجنت ... وانغرست : انعجن وامترجَ وانغرسَّ M | تلتلك : لتلك 9 M والآداب : والاْدَاب 7
M 14 after المذكورة blank space of three letters M 15 صاحبة : صاحب M

but during a short life.
From the day of its birth to the completion of four solar years the newborn is in the charge of the moon, the master of growth, increase and development, and each of the planets shares being in charge with [the moon] for one-seventh of this period. The conditions of the child with regard to 82b growth, increase in size, health, well-being and the like are managed according to what is imposed by the conditions of these directive [powers].
Next Mercury, the master of rationality, motion, culture and understanding takes charge for thirteen years. Each one of the other planets shares being in charge with Mercury for one-seventh of this period. Whenever one of them finally takes charge, in the offspring the character traits and activities arise which correspond to those powers which have been kneaded, mixed and implanted into the natural disposition of the embryo while it was in the womb. In this manner seeds arise from the plants' flowers and fruits, odours, colours and tastes arise from trees when they have ripened and reached perfection according to their natures and bodily forms.
Then Venus, the mistress of beauty, grace, passion, pleasure and longing for marriage takes charge for eight years. The other planets share being in charge with [Venus] for one-seventh of this period and the longing for the mentioned conditions arises in the offspring.
Then the sun, the mistress of might, leadership, direction and ruling takes 83a

سنين ، ويظهر من المولود الكدخدائيّة في المنزل وتربية الأولاد وتأديب الأهل والخدم وطلب العزّة والرفعة .

ثم يصير في تدبير المرّيخ سبع سنين وهو صاحب الحزم والعزم والشجاعة والصرامة والبذل في السخاء والحميّة والعزّة ، ويشاركه سائر الكواكب في التدبير كلّ كوكب سبع المدّة .

ثم يصير في تدبير المشتري إثنتي عشرة سنة وهو صاحب الدين والورع والزهد والعبادة ، وتشاركه سائر الكواكب في التدبير .

ثم يصير التدبير إلى زحل إحدى عشرة سنة ، وهو صاحب السكون والهدوء وخمود نيّران الشهوة الجسمانيّة وذهاب القوى الحيوانيّة ووقوف الحواس عن مباشرة المحسوسات وتناول اللذّات . فعند ذلك يقلّ رغبه في الدنيا وحظّه منها
83b ويجيئه الموت الطبيعيّ على التدريج إذا انطفأت الحرارة الغريزيّة من البدن وانسلّت الروح الجسمانيّة من الجسد ، كما ينطفئ السراج ويذهب الضوء إذا فني الدهن واحترقت الفتيلة .

وربّما ظهرت أفعال هذه الكواكب متنافصة من أجل القوى المتضادّة لأنّ الإنسان ربّما حصل في هذه المدّة من أمرين متضادّين ، وذلك أنّ الزهرة إذا

1 سنين : سنوات *Rasāʾil* 2 والخدم : والجيران ومراعاة أمر الأقرباء والإخوان *Rasāʾil* | العزّة : العزّ والسلطان *Rasāʾil* | after والرفعة add. والعلو والشرف في المنزلة وما شاكل ذلك *Rasāʾil* 3 سنين : سنوات *Rasāʾil* 4 والصرامة ... السخاء : والإقدام والعطاء والطلب والمواهب *Rasāʾil* | after والعزّة add. وبالجملة كلّ خصلة وخلق وسجية لا بدّ منها لساسة الأمور وقادة الجيوش ورعاة الجماعات ومدبّري الملك والناموس جميعاً *Rasāʾil* | after والحميّة add. والإنصاف *Rasāʾil* 5 كوكب سبع : هذه سبع واحد *Rasāʾil* 6 after يصير add. المولود *Rasāʾil* | after والورع add. والصدقة والصلاة والصوم بالصوم والصلاة والصدقة add. after والعبادة 7 *Rasāʾil* والتوبة والندامة والرجوع إلى الله جل ثناؤه بالصوم والصلاة والصدقة والاستغفار وطلب الآخرة والرغبة فيها والتزود للرحلة من هذه الدار الفانية إلى دار القرار الباقية *Rasāʾil* | after زحل add. بعد *Rasāʾil* 8 ثم : إلى أن *Rasāʾil* | في التدبير om. *Rasāʾil* | after والهدوء add. والكسل *Rasāʾil* 9 وخمود : وجمود *Rasāʾil* | الشهوة : الشهوات *Rasāʾil* | after الحيوانيّة add. والآلات الجسدانيّة واسترخاء الأعصاب وذبول *Rasāʾil* 10 after المحسوسات add. ثم لا يمكن للنفس إظهار الأفعال *Rasāʾil* | وتناول : ولا تناول *Rasāʾil* | رغبه : رغبته *Rasāʾil* 10–11 وحظّه ... ويجيئه : وينقطع طمعه في المقام في عالم الكون *Rasāʾil* | after في add. هذا *Rasāʾil* 12 الجسمانيّة : الحيوانيّة *Rasāʾil* 14 أفعال ... متنافصة : أفعالها والفساد ثم تجيئه *Rasāʾil* 14–15 لأنّ الإنسان : وذلك أن الإنسان العاقل *Rasāʾil* 15 من أمرين : متناقضة متجاذباً بين أمرين إثنين *Rasāʾil*

1 الكدخدائيّة : الْكذخذانية M 4 السخاء : السّخا M 6 إثنتي عشرة : اثنى عشر M 8 عشرة : عشر M 11 ويجيئه : وتجيه M | انطفأت : انطفت M 12 وانسلّت : وانسلب M | السراج : السِّراح M

charge for ten years and the offspring shows the ability to be master of the house, educate children, cultivate family and servants, and [shows] the craving for honour and high standing.

Thereupon Mars, the master of determination, zeal, courage, rigor, generous spending, zeal and honour takes charge for seven years and each of the other planets shares being in charge with [Mars] for one-seventh of the period.

Next Jupiter, the master of religion, piety, asceticism and devotion takes charge for twelve years and the other planets share being in charge with it.

Then Saturn takes charge for eleven years. It is the master of tranquillity, calmness, extinction of the fires of bodily passion, cessation of the vital powers, and prevention of the senses from pursuing sensibles and obtaining pleasures. At that time the [person's] longing for the world and his
share in it subsides, and natural death comes to him gradually when the 83b
innate heat is extinguished in his body and the bodily spirit slips away from the corpse. Likewise the lamp is extinguished and the light disappears when the oil has vanished and the wick has been consumed by fire.

Sometimes the activities of these planets arise incompletely because of powers which are opposed to them, as the human being is brought about by two opposite causes in this period. For when Venus masters the

استولت بشركة المرّيخ على أحوال المولود دلّت على الرغبة في الدنيا والحرص على شهواتها ، ويزيده المرّيخ قوّة ونشاطاً وعطارد لطفاً ورفقاً وزحل ثباتاً وصبراً والقمر زيادة ونموّاً والشمس عزّاً ورفعةً ، والمشتري إذا استولى على المولود بشركة زحل كان بالضدّ .

(143) قال ولما كانت المادّة التي خلق منها الحيوان غير ممكن فيها البقاء ، إذ
84a كان الشيء المخلوق المركّب من عروق ضوارب وغير ضوارب وعصب وعظام ولحم لا يمكن أن يكون غير قابل للفساد والغير ، لطف بحكمته بالتناسل الذي يقوّم به بدل الشخص الهالك شخص آخر ، وجعل في الأعضاء التي يتناسل بها قوّة غريزيّة لها خاصّة بها تكوّن اللذّة وحبّب إلى النفس استعمالها . فصار الحيوان إذا بعثته هذه المحبّة وحرّكته الشهوة انبعث على طلب التناسل واستبقاء جنسه ألا يبيد .

ولما كانت العناية بأمر الخلقة ألا يبيد شيء من أجناس الحيوان ، تلطّف لما كان منها قليل البقاء قصير العمر إمّا لضعف بدنه وإمّا لأنّه طعمة الحيوان أقوى منه ، فجعل تلافي تواتر فساده وتوالي تلفه بكثرة توالده . وأعجب من

1 استولت : استوت بدلالتها Rasāʾil | after دلّت add. له Rasāʾil 2 after شهواتها add. ولذّاتها ووقوفاً Rasāʾil | after ثباتاً add. وحيلة Rasāʾil | after ورفقاً add. فيزيده Rasāʾil | ويزيده : Rasāʾil 3 after ورفعةً add. وبالضدّ من هذه كلّها Rasāʾil | والمشتري : أمّا المشتري وطباعه على أحوال المولود دلّ له Rasāʾil 4 كان بالضدّ : الإنسان العاقل بدلالته Rasāʾil | المولود : منها خُلقوا على الزهد في الدنيا Rasāʾil 5 قال ولما : ولكن لما K. fī Manāfiʿ | خلق … الحيوان : التي فيها ينفذ K. fī Manāfiʿ 6 after وعظام add. ذلك K. fī Manāfiʿ | البقاء : فأرفق خليفته … وستبقى كذلك بما لطف K. fī Manāfiʿ 7 والغير : التغيّر K. fī Manāfiʿ | after بحكمته add. من الحيوان K. fī Manāfiʿ 8 after الهالك add. لها بحكمته K. fī Manāfiʿ | بالتناسل : من التناسل K. fī Manāfiʿ | after آخر add. وقد قصدنا في كلامنا هذا إلى تعليمك آثار هذه الحكمة في الإنسان … فأقول إنّه خلق لجميع الحيوان أعضاء يتناسل بها K. fī Manāfiʿ 8–9 الأعضاء … بها : المستعملة K. fī Manāfiʿ | after النفس add. تلك الأعضاء K. fī Manāfiʿ 9 وحبّب : وحبّبت K. fī Manāfiʿ | after استعمالها add. وشوّقت إليه شوقاً عجيباً لا يوصف K. fī Manāfiʿ لتلك الأعضاء إلى K. fī Manāfiʿ | after الشهوة add. وهذا الشوق المطبوع فيه K. fī Manāfiʿ 10 after المحبّة add. مباطشة ذلك بحيوان هو K. fī Manāfiʿ 12 العناية … ولما : إنّما عنى أنّه عدد الأرحام في اختلاف والسبب K. fī Manāfiʿ | الحيوان : من الحيوان K. fī Manāfiʿ 13 منها : كيلا K. fī Manāfiʿ | ألا : وذلك أيضاً من K. fī Manāfiʿ 14 تلافي : بتلافي K. fī Manāfiʿ 14–246.1 وأعجب … مساواة : الأشياء العجيبة في الخلقة الا أنّ هاهنا شيء لا أشكّ أنّك تقرّ بأنّه أعجوبة تفوق كلّ عجب

4 after بالضدّ break (∴) ind. and blank space of almost an entire line M 5 إذ : اذا M 8 آخر : اخر M 9 تكوّن : يكُون M 11 after يبيد blank space of three letters M 13 لضعف : الصعف M

offspring's conditions with the participation of Mars, this indicates longing for the world and wishing for its passions. Mars increases the [offspring's] power and energy, Mercury [his] kindness and friendliness, Saturn [his] stability and patience, the moon [his] increase and growth and the sun [his] might and high standing. Whereas when Jupiter masters the offspring with the participation of Saturn, the opposite happens.

(143) He says:[225] since the matter from which an animal is created does not allow for permanence – as it is impossible that something created and composite of arteries, veins, nerves, bones and flesh is incorruptible and 84a
unchangeable – He in His wisdom has benevolently provided procreation by means of which a deceased individual is replaced with another one. He has put into the reproductive organs an innate power specific to them, which produces pleasure, and in the soul He has evoked a love for using them. So when this love rouses the animal and passion moves it, it is incited to desire procreation and ensure the continuance of its kind so that it may not perish.

Since providence with regard to creation is [to ensure] that no kind of animal perishes, He has acted benevolently towards animals which are transient and short-lived because of either the weakness of their bodies or their being food for animals stronger than them. So He has prepared compensation for their constant corruption and continuous destruction by means of abundance in their reproduction. More admirable than that is

[225] Cf. source in Galen, *K. fī Manāfi*ʿ, appendix 1, excerpt 3.

ذلك مساواة عدد أولاد الحيوان الأرحام لعدد الثدى ، فيكون اللبن في الثدى
بعد تمام الجنين وقرب حاجته إلى الخروج بالولاد ، فإنّه لما كان الحيوان في أوّل 84b
ميلاده ليّن البدن ضعيف النحيزة لم يمكنه استمراء الغذاء الصلب ، فجعل له
غذاء مهيئاً من والدته مثل الغذاء الذي كان يغذّى به منها وهو فى الرحم .
(144) ولو كانت الأعضاء تبقى دائمة على حالة واحدة لاستغنى الإنسان عن
الغذاء ولما كان للهرم والشيخوخة والفناء موضع ، ولكن لما كانت الأعضاء
تذوب وتنحلّ وتستفرغ احتجنا إلى الغذاء المساوي لمقدار الجوهر الذي يتهيّأ
ويذوب من البدن ، والذي يذوب من الأجرام الحارّة الليّنة الدائمة الحركة أو
القويّة الحركة مقدار كثير ، والذي يذوب من الأجرام الباردة الصلبة اليسيرة
الحركة مقدار يسير ، وذلك أنّ البرودة تكثف وتجمع الأجسام وتسدّها وتمنعها
من التحلّل ، والحرارة بخلاف ذلك تذيّب وتلطّف وتحلّل . وكذلك أيضاً ما 85a
كان من الجواهر أكثر صلابة وأكثر قرباً من الحجارة كالعظام فإنّه يكون عسر
الاستفراغ كثير البقاء ، وما كان منها رطباً ليّناً فإنّه يتحلّل من الحرارة ويذوب
ويتفرّق ويهلك كالرئة .

1 الأرحام ... عدد : عدد لعدد أولاده K. fī Manāfiʿ | الثدى[1] : وكذلك ... أثدائه K. fī Manāfiʿ مساواة ... فإنّ الذي يجده في سائر الحيوان من مساواة K. fī Manāfiʿ
2 فيكون : من كون يجده ما K. fī Manāfiʿ 2 فإنّه : من فعل أنّه يظنّ الناس من أحداً أحسب لا مما K. fī Manāfiʿ | after كان add. جميع البخت عن غيرمعرفة ... وتصدفهم عن رأيهم الباطل وإنه K. fī Manāfiʿ
3 ضعيف : ضعيف ورخوة K. fī Manāfiʿ | لم : لا صار K. fī Manāfiʿ 4 يغذّى : K. fī Manāfiʿ
5 الأعضاء ... ولو : شيء منها ويستفرغ يذوب يكن لم لو أنّه وذلك K. fī Manāfiʿ يغتذي
6 كان ولما : K. fī Manāfiʿ لقد كان في ذلك غنى : لاستغنى الإنسان | K. fī Manāfiʿ وكانت والفناء | K. fī Manāfiʿ للهدم : للهرم | K. fī Manāfiʿ ودفع الحاجة إليه وكان مع ذلك لا يكون K. fī Manāfiʿ وكان لا يحدث الموت ولا يتخوّف ناحيته .add موضع after | om. K. fī Manāfiʿ
6–7 المساوي ... كانت : الغذاء يكون أن ينبغي قد نستفرغ أنّا قبل من نغتذي أن نحتاج كنا K. fī Manāfiʿ مساوياً 7 يتهيّأ : يتهيّى K. fī Manāfiʿ 8 يذوب : يتهيّى ويذوب K. fī Manāfiʿ
8–9 أو القويّة : والقويّة K. fī Manāfiʿ 9 after يذوب add. ويتهى K. fī Manāfiʿ 10 after وذلك add. من قبل K. fī Manāfiʿ | وتمنعها : وتمنع K. fī Manāfiʿ 11 التحلّل : التحليل K. fī Manāfiʿ باقياً .add يكون after | om. K. fī Manāfiʿ كالعظام 12 K. fī Manāfiʿ توسّع وتذيب : تذيّب
13 كثير البقاء om. K. fī Manāfiʿ | ويذوب : ويذوب سريعاً يتبخّر K. fī Manāfiʿ
14 كالرئة : ... فالرئة وشكياً K. fī Manāfiʿ

4 مهيئاً : مهيا M | after الرحم blank space of almost an entire line M 7 الغذاء : الاعبدا M
9 القويّة : اليقّوية M 10 تكثف : يتكاثف M 14 after كالرئة break (∴) ind. and blank space of half a line M

that in mammals the number of offspring[226] equals the number of teats.
Then there is milk in the teats after the embryo has reached completion
and is close to its necessary emergence through delivery. For when at the 84b
first time after its birth the animal is soft in body and weak in nature, it is
not able to digest solid nutriment. So He has made nutriment for it, which
is prepared by its mother like the nutriment which it has been fed by her
while in the womb.

(144) If the body parts always remained in the same condition,[227] man
would be able to spare nutriment and there would be no place for decrep-
itude, old age and death. However, as the body parts dissolve, disintegrate
and decompose, we need nutriment equal to the amount of the substance
which is disposed of and dissolved by the body. An abundant amount is
dissolved by hot, soft bodies which are either always or vigorously in
motion, whereas an insignificant amount is dissolved by cold, hard bodies
moving a little. For coldness thickens, contracts, clogs and prevents the
bodies from dissolving. Contrary to that, heat dissolves, diminishes and 85a
decomposes. Likewise the substances which are hardest and most similar
to stone, such as bones, are also difficult to disengage and are long-lasting,
whereas moist and soft [substances], such as the lungs, are decomposed by
heat, dissolved, dispersed and wiped out.

[226] The *PR* omits the first Galenic example (the equality of the number of 'sinuses' (κόλπος – *tajwīf*) to the number of teats) and quotes only the second, which Galen seems to consider as even more convincing than the first.

[227] Cf. source in Galen, *K. fī Manāfiʿ*, ed. Savage-Smith, 103.3–14, 104.4–6, which the *PR* quotes with some omissions.

والقلب صلب الجوهر إلا أنّه بكثرة حرارته يفني غذاءً كثيراً ، والدماغ جوهره كثير الليّن إلا أنّه لقلّة حرارته لا يتحلّل منه إلا اليسير .

(145) قال قسطا بن لوقا الروح في البدن الإنسانيّ روحان وهما جسمان لطيفان ، أحدهما يسمّى الروح الحيوانيّ مادّته الهواء وينبوعه القلب ، وينبعث بالشريانات إلى سائر البدن فيفعل الحياة والتنفّس والنبض . وجعل النبض
85b والتنفّس لصلاحه ، أعني لترويحه بالهواء الوارد عليه من خارج ولإخراج الأبخرة الدخانيّة عنه . ويسمّى الآخر الروح النفسانيّ ومادّته الروح الحيوانيّ وينبوعه الدماغ ، يفعل في الدماغ نفسه الفكر والذكر والرويّة ، وينبعث منه في الأعصاب إلى سائر الأعضاء فيفعل الحسّ والحركة . ويفعل هذا الروح الذي في تجويفات الدماغ أفعالاً مختلفة ، أمّا الذي في التجويفين المقدّمين فيفعل الحسّ السمعيّ والبصريّ والذوقيّ والشمّيّ وبعض اللمس . وينبعث من هذين التجويفين سبعة أزواج عصب فيها روح ، الأوّل منها يتّصل بالعين فيكون به البصر ، والثاني يتّصل باللسان فيكون به حسّ الذوق ، والثالث يتّصل بالمنخر فيكون به حسّ الشمّ ، والرابع يتّصل بالحنك ، والخامس يتّصل بالصماخين فيكون به حسّ السمع ، والسادس يتّصل بالأحشاء فينيلها الحسّ ويرجع
86a شيء منه بعد إلى الحنجرة فيحرّكها لانضمام فوهتها ، والسابع يتّصل باللسان

1 والقلب ... الجوهر : ومنها ما جوهره صلب K. fī Manāfiʿ | يفني : يغذى K. fī Manāfiʿ 2 والدماغ : ومنها ما مثل القلب K. fī Manāfiʿ 2 after اليسير add. مثل الدماغ K. fī Manāfiʿ 3 قال ... لوقا : فقد حصل إذن ممّا قلنا إنّ R. fī l-Farq | البدن الإنسانيّ : بدن الإنسان R. fī l-Farq 3–4 وهما ... لطيفان om. R. fī l-Farq 4 يسمّى : يقال لها R. fī l-Farq | الروح om. R. fī l-Farq 5 الحيوانيّ : الحيوانيّة R. fī l-Farq 5 والتنفّس والنبض : والنبض والتنفّس R. fī l-Farq | وجعل : وأنّ R. fī l-Farq 6 لصلاحه : لصلاح هذه الروح R. fī l-Farq 7 ويسمّى ... الروح[1] : والأخرى يقال لها R. fī l-Farq | النفسانيّ : النفسانيّة R. fī l-Farq | الحيوانيّ : الحيوانيّة R. fī l-Farq | وينبوعه : وينبوعها R. fī l-Farq 9 ويفعل ... الذي : فقد صحّ مما قلنا أنّ الروح التي R. fī l-Farq 10 after الدماغ add. جزء الدماغ : تفعل R. fī l-Farq | الذي : التي R. fī l-Farq 11–12 هذين التجويفين : من التجويفين المقدمين R. fī l-Farq 12 after روح add. المقدّم R. fī l-Farq 13 باللسان ... الذوق : باللسان فيكون به حسّ الدوق R. fī l-Farq 13–14 بالمنخر ... الشمّ : بعضل العين فيحرّكها R. fī l-Farq 14 after بالحنك add. فيؤدي إليه حسّ الطعم . R. fī l-Farq

2 after اليسير break (∴) ind. and blank space of half a line M 5 الحياة : الحيوة M 6 ولإخراج : والاخراج M 7 الآخر : الاحر M 6 النبض : للنَبضُ M | والتنفّس : النَّفسُ M 12 أزواج : ارواح M | after روح break (∴) ind. and blank space of ten letters M 13 after البصر blank space of five letters M 14 after بالحنك blank space of three letters M 16 لانضمام : لانضِام M

The heart is hard in substance, yet it consumes much nutriment due to its abundant heat; and the brain is extremely soft in substance, yet only an insignificant part of it becomes decomposed due to its little heat.

(145) Qusṭā ibn Lūqā says:[228] The spirit in the human body is of two kinds, both of which are fine bodies. One of them is called the vital spirit, its matter is air and its source the heart. It is sent out through the arteries to the rest of the body to enable life, breathing and pulsation. Pulsation and breathing are set up for [the body's] health, i.e. to aerate it with air entering from outside and to rid it of smoky vapours. The other is called 85b
the psychic spirit, its matter is the vital spirit and its source the brain itself where it enables thought, memory and consideration, and from where it is sent out through the nerves to the remaining body parts to enable sense perception and movement. This spirit in the ventricles of the brain performs different activities. In the two anterior ventricles it enables the senses of hearing, sight, taste, smell and, in part, touch. Seven pairs of nerves with spirit in them proceed from these two ventricles. The first [pair] of them is joined to the eye, so vision exists through it. The second is joined to the tongue, so taste exists through it. The third is joined to the nose, so smell exists through it. The fourth is joined to the palate and the fifth to the auditory canals, so hearing exists through it. The sixth is joined to the bowels, so sense perception reaches it and then part of [the sixth pair of nerves] turns to the larynx, so it moves it to close its orifice. The 86a
seventh is joined to the tongue, so its movement exists through it. These

[228] Cf. source in Qusṭā, *R. fī l-Farq*, ed. Troupeau, lines 146–8, 54–6, 148–50, 141–3, 99–100, 102–14, 144–5.

فيكون به حركته . وكلّ ذلك تفعله هذه الأعصاب بالروح التي تنفذ فيها من الدماغ إلى هذه الأعضاء ، والدليل على ذلك أنّه متى عرض عارض فسدّ مجرى الروح التي في عصبة من هذه الأعصاب ومنعه أن يصل إلى العضو ، فيبطل ذلك العضو ، كالماء المجتمع في العين فإنّه يحول بين الروح التي في العصبة وبين الناظر فيفعل عمى ، وكالأخلاط والأبخرة التي تسدّ وتحول بين الروح الذي في الصماخين وبين ⟨آلة السمع أو⟩ آلة الشمّ أو آلة الذوق فيضعف السمع أو الشمّ ، فإذا انفتحت تلك المجاري إمّا بعلاج أو بمقاومة الطبيعة للعلّة ، عاد العضو إلى فعله فصار صحيحاً مستوياً . والذي في التجويف الأوسط فيفعل الفكر والتمييز والرويّة ، والذي في التجويف المؤخّر يفعل الذكر والحركة .

86b (146) وقال جالينوس الروح النفسانيّ في بطون الدماغ وهو أوّل آلات النفس الناطقة التي مسكنها جرم الدماغ لا بطونه .

والقلب بمنزلة المستوقد للحرارة التي في أبدان الحيوان ، وفي النبات جنس آخر من الحرارة تشركه فيها أبداننا أيضاً ومن هذا الوجه تشركه في القوّة التي تسمّى طبيعيّة وغاذية . والحرارة التي تنبعث من القلب إلى جميع البدن ، فتحسّه بسمتها حاراً باللمس أيضاً غير هذه الحرارة ، فإنّه تبلغ من قلّة الحرارة الطبيعيّة ألا تدرك بالحسّ ، ولهذا نقول في النبات إذا قسناه ببدن الحيوان وجعلنا امتحان الحرارة بالحسّ إنّه لا حرارة أصلاً .

والكبد للقوى الطبيعيّة بمنزلة الأصل والينبوع كالأصل من الشجر ، فإنّ الشجرة ما دام أصلها حافظاً لقوّته فهي سليمة وإن جفّ بعض اغصانها أو

3 ومنعه : ومنعها من *R. fī l-Farq* 4 after فيبطل add. فعل *R. fī l-Farq* 5 عمى : العمى *R. fī l-Farq* 6 وبين : أو بينها وبين الصماخين فيفعل الصمم *R. fī l-Farq* 5–6 الصماخين . . . بين : وإمّا *R. fī l-Farq* | أو[2] : فيبطل المذاقة واللمس والشمّ *R. fī l-Farq* 7 فإذا . . . السمع : بين وأنّ الروح *R. fī l-Farq* | والذي : تفعل *R. fī l-Farq* 9 فيفعل : وأنّ الروح التي *R. fī l-Farq* 8 والذي : التي *R. fī l-Farq*

4 كالماء : كمالما M 5 والأبخرة : وللابخرة M 6 آلة[2] : اله M | آلة[3] : الة M 8 after مستوياً break (∴) ind. M 9 الفكر : الذكر M | والتمييز : والتميز M 10 after والحركة blank space of ten letters M 11 آلات : الأت M 13 آخر : اخر M 14 تشركه فيها : بشركة فيه M 16 بسمتها : يسميهَا M 17 نقول : يقول M 18 after أصلاً blank space of ten letters M 19 والينبوع كالأصل add. in marg. M

nerves perform all that by means of the spirit which passes in them from the brain to these body parts. The proof for this is that if there is an accident and the flow of the spirit in one of these nerves is blocked and prevented from reaching the body part, then that body part becomes ineffective. [It is] like water gathering in the eye, for it obstructs the course of the spirit in the nerve to the pupil and causes blindness; and like humours and vapours which clog and obstruct the course of the spirit in the auditory canals to the instrument of ⟨hearing or⟩ smelling or tasting so that hearing or smelling becomes impaired. If those passages are opened [again], either by treatment or by nature resisting the disease, the body part returns to its activity and becomes [once more] sound and well balanced. [The psychic spirit] in the middle ventricle enables thought, discernment and consideration, and the one in the posterior ventricle enables memory and motion.

(146) Galen says:[229] the psychic spirit [exists] in the ventricles of the brain 86b
and is the first of the instruments of the rational soul which dwells in the mass of the brain, not in its ventricles.

The heart is like a hearth for the heat which is in the animals' bodies. In plants there is another kind of heat in which our bodies also share with them, and in this way [our bodies] share with [plants] in the power which is called natural and nutritive. The heat which is sent out from the heart to the entire body, and by the characteristic of which you perceive [the body] as warm to touch, is unlike this heat [of plants]. For [the heat of plants] attains such a low degree of natural heat that it is not perceived by sense perception. Therefore when we compare [plants] with the body of animals and test the heat by sense perception, we say about plants that there is no heat [to perceive in them] at all.

The liver is for the natural powers like a root and a source like the stem of a tree. For as long as the tree's stem keeps its power, the tree is healthy even if some of its boughs or some of the branches which spread out from

[229] Most probably excerpted from the Arabic version of Galen's *De propriis placitis*.

87a بعض الشعب المتفرّعة من ذلك الأصل . وهو مبدأ الحرارة الغريزيّة الذي من الحيوان بمنزلة الشيء الملائم لجوهر النبات الذي هو له في الموضع الثابت فيه أصله .

(147) قال ارسطالس النفس المحسّة هي التي تدرك الأشياء بقبول صورها دون أجسامها ، وإدراكها إيّاها أن تتشبّه بها وتكون مثلها ، وذلك أنّه إذا كانت الأشياء المحسوسة بائنة عن النفس الحسّيّة فهي مثلها بالقوّة ، وإذا حضرت كانت مثلها بالفعل .

(148) وقال البصر هو الذي يدرك الألوان بأن يقبل آثارها ويتشبّه بها بتوسّط الهواء بلا حركة ولا زمان ولا انفعال ولا استحالة زمانيّة .

(149) والصوت هو قرع هواء بتنسّم في آلات متهيّئة لقبوله مع وهم يدلّ على
87b شيء ما ، والسمع مثل الصوت بالقوّة إذا كان الصوت عنه نائياً . فإذا حضر كان مثله بالفعل .

(150) والقرع يأتي إلى السمع بتوسّط الهواء ، والدليل على ذلك أنّك إن أدنيت شيئاً من مجرى السمع ثم قرعتَه بشيء آخر ، لم تسمع له صوتاً البتّة لأنّ الهواء

4 قال ارسطالس : فلما فرغ الفيلسوف من تلخيص الأشياء المحسوسة وبيّنها وصف عند ذلك : K. al-Nafs, MS A | النفس المحسّة : ان الحسية K. al-Nafs, MS A | النفس الحسّيّة وهي الحيوانيّة فقال K. al-Nafs | النفس الحسّيّة ed. Arn | 5 وإدراكها : فدركها K. al-Nafs, MS | صورها : صورتها K. al-Nafs, MS A | وتكون . . . وذلك : ويكون ذلك K. al-Nafs, MS A | تتشبّه بها : تنسه فيها K. al-Nafs, MS A | ودركها ed. Arn | فإذا : وإذا A: K. al-Nafs, MS A | الحسّيّة om. K. al-Nafs, ed. Arn | 6 بائنة : نائية K. al-Nafs | 8 وقال البصر : فلما بيّن الفيلسوف ما اللون وما الضوء انصرف إلى صفة K. al-Nafs, ed. Arn البصر فقال : ولما بيّن الفيلسوف ما اللون وما الضوء صفة البصر فقال: K. al-Nafs, MS A | بها om. K. al-Nafs, MS A | آثارها : أثره K. al-Nafs, MS A | الألوان : اللون K. al-Nafs, MS A | 9 ولا انفعال om. K. al-Nafs, ed. Arn | 10 والصوت هو : ثم انصرف إلى صفة الصوت فقال هو K. al-Nafs, MS A : ووصف الصوت فقال الصوت ed. Arn | هواء om. K. al-Nafs, MS A | بتنسّم . . . متهيّئة : ينسم <ينبث> إلى أنبوبة الرئة والحلقوم المتهيّئة K. al-Nafs, MS A | بتنسّم K. al-Nafs, ed. Arn | 11 ما om. K. al-Nafs, MS A | والسمع : فلما وصف الصوت وصف... السمع فقال هو K. al-Nafs | عنه نائياً om. K. al-Nafs, MS A | فإذا حضر : وإذا كان حضره K. al-Nafs, MS A | 13 والقرع : فقال القرع K. al-Nafs | يأتي : اتى K. al-Nafs, MS A | after بتوسّط add. في K. al-Nafs, MS A | إن : لو K. al-Nafs, MS A : إذا ed. Arn | 14 له صوتاً om. K. al-Nafs, MS A | البتّة om. K. al-Nafs, MS A | لأنّ الهواء : لم تشمّ وكذلك لو ألصقت شيئاً ذا رائحة K. al-Nafs

1 الذي : التي M 2 الملائم : الملاوم M | الذي : الثى M 3 after أصله break (∴) ind. and blank space of more than half a line M 7 after بالفعل break (∴) ind. and blank space of half a line M 8 آثارها : اثارهَا M 9 after زمانيّة break (∴) ind. and blank space of half a line M 10 آلات : الاتٍ M | متهيّئة : متهية M 14 آخر : أخَرَ M

that stem dry out. [The liver] is the principle of the innate heat [and it is the liver] which is, with regard to animals, like the thing suitable for the plants' substance which [the plants] have in the firm place where their stems are.[230] 87a

(147) Aristotle says:[231] it is the sensible soul which perceives things through receiving their forms without their bodies. Its perception of them is to imitate them and to be like them. For when the sensibles are separate from the sensible soul, it is like them in potentiality, whereas when they are present, it is like them in actuality.

(148) He says:[232] it is vision which perceives the colours by receiving their impressions and imitating them, [which happens] through the mediation of air, [but] without motion, time, passivity or temporal change.

(149) Voice is the stroking of air[233] through exhaling within the [bodily] instruments receptive to it together with [an act of] imagination indicating something. Hearing is like sound in potentiality when the sound is distant from it, for when it is present, [hearing] is like [sound] in actuality. 87b

(150) The stroke comes to hearing[234] through the mediation of air. The proof for this is that if you bring something close to the auditory canal, then strike it with something else, you will not hear any sound from it at

[230] Cf. *De propriis placitis*, 186.1–3 for this difficult sentence: εἶναί τε καὶ τὸ ἧπαρ ἤτοι αὐτὴν μὲν ἀρχὴν τῆς φυσικῆς θερμασίας οἵαν ἔχει καὶ τὰ φυτὰ κατὰ τὴν καλουμένην ῥίζωσιν οἰκεῖαν τῆς οὐσίας ἑαυτῶν. The liver of the animals is likened to the 'taking roots' (ῥίζωσις) of the plants.

[231] Cf. source in *DA*-paraphrase, ed. Arnzen, 293.5–9; and MS A, fol.89a1–8.

[232] Cf. source in *DA*-paraphrase, ed. Arnzen, 265.11–13; and MS A, fol.78a6–10.

[233] Cf. source in *DA*-paraphrase, ed. Arnzen, 269.16–271.1, 271.5–6; and MS A, fols.81b8–10, 81b14–82a2.

[234] Cf. source in *DA*-paraphrase, ed. Arnzen, 267.1–4; and MS A, fol.78a13–b3; the sentence omitted here is taken up later on, passage (169).

هو الذي يؤدّي القرع إلى السمع والرائحة إلى الخياشيم .

(151) وقال الحسّ الذي به يشمّ هو مثل الشيء ذي الرائحة بالقوّة إذا كان عنه نائياً ، فإذا حضر كان مثله بالفعل بلا حركة ولا زمان ، وذلك أنّه إذا حضرت الرائحة كانت القوّة الشامّة مثلها سواء .

(152) قال وحاسّة الذوق تخالف سائر الحواسّ فإنّها تعرف طعم الشيء بلا توسّط الهواء ، وذلك أنّا إن لم نضع الشيء ذا الطعم على اللسان لم نعرف طعمه ما هو ، وليس سائر الحواسّ كذلك فإنّها لا تعرف محسوساتها إلا
بتوسّط الهواء ، مثل [مثل] البصر والسمع والشمّ . وذلك أنّ الإنسان إذا أدنى 88a
الشيء من عينه حتى تلاصقه لم يعرف لونه ، وكذلك إن صاح إنسان في داخل أذنه حيث حسّ السمع لم يسمع صوته ، وإن جعل الشيء ذا الرائحة في داخل الخياشيم لم يشمّه .

(153) ثم وصفه فقال الذوق قوّة حسّيّة هي مثل الطعم بالقوّة ، فإذا لامست

وكذلك لو ألصقت شيئاً ذا رائحة بالخياشيم لم تشمّ رائحة البتة : K. al-Nafs, MS A له رائحة فالهواء ed. Arn فالهواء
1 after يؤدّي add. إليه K. al-Nafs, MS A 2 وقال الحسّ : فلما وصف الرائحة وكيف تصل إلينا K. al-Nafs, MS A : فلما وصف الرائحة وكيف تصل إلينا وصف الحاس انصرف الى صفة الحاسّ ed. Arn | after يشمّ add. فقال K. al-Nafs, ed. Arn : يشمّ : نشم K. al-Nafs, MS A | به : فيه K. al-Nafs, MS A | الشيء om. K. al-Nafs, MS A | هو : هذا K. al-Nafs, MS A | إذا : اذ K. al-Nafs, MS A 3 عنه om. K. al-Nafs, ed. Arn 5 قال وحاسّة : فقال حاسّة K. al-Nafs | فإنّها : لأنّها K. al-Nafs, MS A : بأنّها ed. Arn | تعرف : تتعرف K. al-Nafs, ed. Arn 5–7 بلا ... ما : بما K. al-Nafs, MS A 7 طعمه : طعم الشيء K. al-Nafs, ed. Arn | فإنّها : لأنها K. al-Nafs, MS A | محسوساتها : محسوسها K. al-Nafs, MS A 8 after والسمع add. واللمس K. al-Nafs, MS A 9 الشيء ... عينه : الشيء إلى عينه جدّاً K. al-Nafs, MS A : الشيء ⟨ذا اللون⟩ جدّاً من عينه ed. Arn | حتى ... لونه : لم يعرف لونه K. al-Nafs, MS A : لم يبصره ed. Arn 9–10 إن ... أذنه : لو ان رجلاً صاح رجل K. al-Nafs, MS A : لو جعل إنسان ⟨الشيء ذا الصوت⟩ في داخل أذن صاحبه ed. Arn 10 حسّ السمع : جسم السمع K. al-Nafs, MS A : عصب الحسّ ed. Arn 11 في داخل : دخل K. al-Nafs, MS A | يشمّه : يشمّ رائحة الرائحة K. al-Nafs, MS A 12 ثم وصفه : ولما فلما أخبر أن الذوق يعرف الطعوم بالملامسة انصرف إلى صفته K. al-Nafs, MS A : وصفه أخبر أن الذوق يعرف الطعوم بالملامسة وصفه ed. Arn | after الذوق add. هو K. al-Nafs, MS A | هي om. K. al-Nafs, MS A

1 after الخياشيم break (∴) ind. and blank space of half a line M 2 الشيء : الشم M 4 after سواء break (∴) ind. and blank space of ten letters M 6 نضع : يضع M | ذا : اذا M 7 ما : با M | نعرف : يعرف M 9 وكذلك : ولذلك M 11 يشمّه : يشبهه with blank space of half a line M

all, for it is the air which conveys the stroke to the [organ of] hearing and the odour to the nostrils.

(151) He says:[235] the sense through which one smells is like the fragrant thing in potentiality when it is distant from it, for when it is present, [the sense of smell] is like it in actuality without motion or time. For when the odour is present, the power to smell is completely like it.

(152) He says:[236] the sense of taste is different from the other senses for it gets to know the taste of something without the mediation of air. For if we do not place the thing possessing taste on the tongue, we do not know what taste it has. The other senses – vision, hearing and smell – are not like that for they only get to know their sensibles through the mediation of air. For if man brings something close to his eye so that [his eye] touches 88a
it, he does not come to know its colour. Likewise if someone yells in the inside of [someone else's] ear where the sense of hearing is, [the latter] does not hear the voice [of the former]; and if he places the fragrant thing in the inside of the nose, he does not smell it.

(153) Next he describes it saying:[237] the sense of taste is a sensible power which is like flavour in potentiality, and when it is in contact with flavour

[235] Cf. source in *DA*-paraphrase, ed. Arnzen, 273.15–275.2; and MS A, fol.83a6–11.

[236] Cf. source in *DA*-paraphrase, ed. Arnzen, 275.4–12; and MS A, fol.83a13–b6.

[237] Cf. source in *DA*-paraphrase, ed. Arnzen, 277.9–11; and MS A, fol.79b11–13.

الطعم صارت مثله سواء .

(154) وقال اللمس يدرك الملموس بتوسّط الهواء غير أنّه خفي . وضرب لذلك مثلاً فقال لو أن إنساناً غمر يده في ماء ثم أمسك حجراً لم يكن بدّ من أن يكون بين الحجر وبين اليد ماء إلا أن ذلك يخفى علينا للطافته ، فبالحرى أن يكون الهواء يخفى علينا لأنّه ألطف من الماء كثيراً .

88b (155) وقال ما بال الحواسّ تفسد من محسوساتها إذا أفرطت ، مثل الضوء الساطع فإنّه يفسد البصر ، والصوت الشديد فإنّه يفسد السمع ، والرائحة المفرطة تفسد المشمّ ، والشيء المرّ الطعم جدّاً يفسد الذوق ، والحارّ المفرط يفسد اللمس ؟ ثم أجاب فقال لأنّ جميع الحواسّ تثبت على الاعتدال والشيء المعتدل لا يفسد من شيء معتدل آخر مثله ، بل إنما يفسده الشيء الخارج من الاعتدال . فلذلك صارت للأشياء المحسوسة إذا زالت عن اعتدالها وأفرطت أفسدت الحواسّ .

ومن أجل ذلك أنّه لا يكون للنبات حسّ لأنّ الأرضيّة غالبة عليه ، وكذلك أعضاء الإنسان التي غلبت عليها الأرضيّة مثل العظام والشعر والأظفار فإنّها لا تحسّ .

1 after مثله add. بالفعل *K. al-Nafs* | سواء om. *K. al-Nafs*, MS A 2 وقال : فقال أترى *K. al-Nafs*, MS A : وقال إن ed. Arn | أنّه خفي : ان ذلك يخفى *K. al-Nafs*, MS A : ان ذلك خفي ed. Arn 3 غمر : صيّر *K. al-Nafs*, MS A : ألقى ed. Arn | ثم ... حجراً : وقبض على حجر ed. Arn 4 بين ... اليد : بين اليد وبين الحجر ed. Arn فأمسك حجراً أو غيره : *K. al-Nafs*, MS A أو غيره *K. al-Nafs* | علينا om. *K. al-Nafs* | after للطافته add. فإن ذلك توسّط الماء بين اللمس ed. Arn فإن كان توسّط الماء بين اللمس والملموس يخفى : *K. al-Nafs*, MS A والملموس يخفى علينا 5 لأنّه : الا ان *K. al-Nafs*, MS A | كثيراً om. *K. al-Nafs*, MS A 7 فإنّه[2] om. *K. al-Nafs* 8 after المفرطة add. في النتن *K. al-Nafs*, MS A : add. الطيب ed. Arn | المشمّ : الشم *K. al-Nafs* الطعم جدّاً om. *K. al-Nafs*, MS A 9 ثم أجاب om. *K. al-Nafs*, MS A : ثم أطلق المسألة ed. Arn | لأنّ : ذلك لأنّ *K. al-Nafs*, MS A : إن ed. Arn | تثبت : تثبته *K. al-Nafs*, MS A : بنيت ed. Arn 10 شيء om. *K. al-Nafs* | مثله om. *K. al-Nafs*, MS A | الشيء : المفرط *K. al-Nafs*, MS A : الشيء المفرط ed. Arn 11 من : عن *K. al-Nafs*, ed. Arn | زالت : زاغت *K. al-Nafs* 13 ومن ... الأرضيّة : فإن قال قائل وما بال النبات ليس بذي حسّ قلنا الحسّ إنما يكون من اعتدال أستقسات المركّب الأربعة والأرضيّة *K. al-Nafs*, ed. Arn | عليه : على النبات *K. al-Nafs*, ed. Arn | وكذلك : فلذلك صار لا يحسّ والدليل عليه *K. al-Nafs*, ed. Arn 14 after التي add. قد *K. al-Nafs*, ed. Arn | مثل العظام : كالعظام *K. al-Nafs*, ed. Arn

1 after سواء break (∴) ind. and blank space of half a line M 3 ماء : ما M | بدّ : بداً M 4 ماء : مآ M 5 الماء : المآ M | after كثيراً break (∴) ind. and blank space of ten letters M 10 آخر : اخر M

it becomes like it completely.

(154) He says:[238] the sense of touch perceives the touchable through the mediation of air, yet the air is hidden. He gives an example of that by saying: if somebody submerges his hand in water, then grabs a stone, it is inevitable that there is water between the stone and the hand, yet that is hidden from us due to its fineness. How much more must the air thus be hidden from us as it is much finer than the water.

(155) He says:[239] why is it that the senses are corrupted by their sensibles if 88b
they are excessive, like radiant light, for it corrupts vision; and powerful sound, for it corrupts hearing; excessive odour corrupts [the organ of] smell; something of very bitter taste corrupts [the organ of] taste; and something excessively hot corrupts touch? Then he answers saying: for all senses adhere to balance and the well-balanced thing is not corrupted by another well-balanced thing like itself. Rather only something out of balance corrupts it. Therefore, it happens to the sensibles that they corrupt the senses when they leave their balance and become excessive.

On that account plants do not have any sense, for the earthy [element] prevails over them; and in the same way the body parts of man over which the earthy [element] prevails such as bones, hair and nails do not perceive.

[238] Cf. source in *DA*-paraphrase, ed. Arnzen, 281.12–17; and MS A, fol.84b6–13.

[239] Cf. source in *DA*-paraphrase, ed. Arnzen, 285.16–287.4; 345.16–347.1; and MS A, fol.86a15–86b10, but without parallel to the second part of the *PR* passage.

(156) وقال الحواسّ إذا انفعلت من الأشياء المحسوسة فإنّها تقبل صورها ولا [89a] تقبل هيولاها ، كالبصر فإنّه يقبل لون الشيء ولا يقبل جسمه ، وكذلك جميع الحواسّ . وما ليس بحيوان مثل النبات والحجارة فإنّه يقبل صورة المحسوس وجسمه جميعاً ، وذلك أن النبات والحجارة تنفعل من وصول الماء إليها بصورته وجرمه ، فإنّ الماء رطب الشجر بوصول جرمه إليها مع رطوبته . فعلى هذا تفعل الأشياء المحسوسة في الشجر والحجارة من غير أن يكون للشجر والحجارة حاسّة يحسّ بها ذلك الفعل .

(157) وقال كلّ واحد من المحسوسات يعرف بالحاسّة المجانسة له . فإنّ الإنسان يحسّ بالحاسّة الجسمانيّة الأرضيّة التي هي اللمس الطبيعة الأرضيّة ، وبالحاسّة النوريّة التي هي البصر الطبائع المضيئة ، وبالحاسّة الهوائيّة انفعالات [89b] الهواء فإنّ جوهر الصوت هو الهواء أو قرع الهواء ، وبالحاسّة المائيّة الطعوم . فأمّا الحاسّة الخامسة التي هي الشمّ فإنها تدرك البخارات والروائح بطبيعتها التي بين الهواء والماء لأنّها أغلظ من الهواء وألطف من الماء .

(158) قال ويحدّ الحسُّ بأنّه قوّة للنفس تدرك المحسوسات ، والحاسّة آلة تدرك بها المحسوسات .

1 وقال : فإن قال قائل إن K. al-Nafs, MS A : قلنا إن ed. Arn | الحواسّ : الحيوان K. al-Nafs, MS
A | فإنّها : فإنما K. al-Nafs, ed. Arn | after صورها add. فقط K. al-Nafs 2 جسمه : شأنها
K. al-Nafs, MS A 2–3 وكذلك ... الحواسّ : وكذلك ولا يقبل صور أجسامها K. al-Nafs, MS A : وكذلك
السمع وسائر الحواس تقبل صور محسوساتها ولا تقبل أجسامها ed. Arn 3 وما : فأمّا K. al-Nafs,
MS A : وأما ما ed. Arn | مثل : من K. al-Nafs, MS A 3–4 النبات ... والحجارة om. K. al-Nafs,
MS A 5 فإنّ : وذلك أن K. al-Nafs | رطب : يرطب K. al-Nafs 6 تفعل : شغل K. al-Nafs, MS
A | للشجر : الشجر K. al-Nafs, MS A 7 والحجارة : أو للحجارة K. al-Nafs, ed. Arn | حاسّة :
حاسّ K. al-Nafs, MS A | يحسّ بها : تعرف لها K. al-Nafs, ed. Arn | بها om. K. al-Nafs, MS A
8 وقال : فإنّ Fī Ṭabīʿa | after المحسوسات add. إنما من شأنه أن Fī Ṭabīʿa 8–9 فإنّ ... يحسّ :
فتحسّ Fī Ṭabīʿa 9 الجسمانيّة الأرضيّة : الأرضيّة الجسمانيّة Fī Ṭabīʿa 10 وبالحاسّة¹ : والحاسّة Fī
Ṭabīʿa | وبالحاسّة² : وكذلك يحسّ بالحاسّة Fī Ṭabīʿa 11 after المائيّة add. اختلاف
الكيموسات التي هي Fī Ṭabīʿa 12 فأمّا ... تدرك : وقد كان يلزم على هذا المثال القياس إذ
كانت العناصر أربعة أن تكون الحواس أيضاً أربعة ولكن من أجل أنّ Fī Ṭabīʿa | والروائح² ... التي :
في جنسها وطبيعتها Fī Ṭabīʿa 14 قال ... بأنّه : وقد نحدّ ايضاً بحدّ آخر وهو أنّه Fī Ṭabīʿa
تدرك المحسوسات : مدركة للمحسوسات Fī Ṭabīʿa | after والحاسّة add. هي Fī Ṭabīʿa

3 after الحواسّ blank space of six letters M 4 الماء : المآ M 7 after الفعل break (.˙.) ind.
and blank space of the entire next line M 10 المضيئة : المصغيه M 11 after الهواء³ break
(.˙.) ind. and blank space of three letters M | after الطعوم blank space of five letters M
13 أغلظ : اغلط M | الماء : المآ followed by blank space of ten letters M 14 آلة : الة M

(156) He says:[240] when the senses are acted upon by the sensibles, they receive their forms, but not their matter, like vision. For it receives the colour of something but not its body, and the same [is true for] all the 89a
senses. What is not an animal, such as plants and stones, receives the form and the mass of the sensible together. For when water reaches plants and stones, they are acted upon by its form and mass, because water reaching the trees moistens them by its mass and moisture together. So in accord with this the sensibles act on trees and stones, without trees and stones having any sense organ by which they perceive that action.

(157) He says:[241] each of the sensibles is known by the sense organ which is of the same genus as it. For man perceives through the bodily, earthy sense organ, namely touch, the earthy nature; through the luminary sense organ, namely vision, the luminous natures; through the airy sense organ affections of the air, because the substance of sound is air or the stroke of air; and through the watery sense organ the tastes. As to the fifth sense 89b
organ, namely smell, it grasps the vapours and odours through their nature, which is between air and water as they are thicker than air but finer than water.

(158) He says:[242] sense is defined as a power of the soul, which perceives the sensibles whereas the sense organ is the instrument by which the sensibles are perceived.

[240] Cf. source in *DA*-paraphrase, ed. Arnzen, 289.13–291.4; and MS A, fols.87b11–88a5.
[241] Cf. source in Nemesius, *Fī Ṭabīʿa*, 102.15–16, 10–15, 102.16–19.
[242] Cf. source *ibid.*, 103.11–13.

(159) وقال الناظر من ماء والسمع من هواء . ولم تفعل الطبيعة حاسّة من نار لأنّ الحرارة موجودة في مزاج الأعضاء .

E68b 〈(160) والأجسام الطبيعيّة أقرب الأشياء التي نبحث عنها إلينا لأنّا بعضها ومناسبون لها ، ولذلك نحسّها بالحواسّ الخمس . وذلك أنّ كلّ حاسّة إنما تحسّ من الأمور ما لائمها لأنّ لكلّ حاسّة اعتدالاً موضوعاً لها . فإذا أورد عليه من جنسه ما يخالفه بكيفيّة ما أحسّ به ، مثال ذلك أن الذوق حسّ بالرطوبة للرطوبة المخالفة ، والسمع حسّ بالهواء للهواء المخالف ، واللمس حسّ بالأرض للأرض ، والبصر حسّ بشعاع ناريّ لشعاع ناريّ . فأمّا الشمّ وهو الخامس فإنّه مركّب لأنّه إدراك بالبخار للبخار والبخار مركّب من الهواء والماء . وينبغي أن نذكر حال واحدة منها ليستدلّ بها على أحوال الباقيات ، فأقول إنّ الهواء الموضوع لتجويف الأذن له اعتدال موافق له ، فإذا تغيّر بهواء آخر يطرقه مما فيه حركة واقتراع أحسّ به الإنسان ، وكذلك حال الرطوبة الموضوعة للسان.

ولكلّ جسم طبيعيّ حركة تخصّه ، وذلك أن الجسم ما كان منه موجوداً وما كان منه متكوّناً فإنما قوامه بصورته الخاصّة به ، وصورته الخاصّة به هي المقوّمة لذاته ، والصورة المقوّمة لذاته هي طبيعته ، وطبيعته مبدأ حركته الخاصّة به ، E69a وحركته | الخاصّة به هي التي تحركه إلى تمامه ، وتمام كلّ شيء هو ما لائمه ووافقه . ولذلك كلّ متحرّك يتحرّك إلى تمامه وإلى ما يوافقه ، ومن ههنا قيل إنّ كلّ متحرّك يتحرّك إلى تمامه فهو بالشوق ، والذي يشتاق فهو معلول بما يشتاق إليه . والعلّة تتقدّم على المعلول بالطبع ، فلذلك صار الاستدلال بالحركة أظهر الأشياء وأولاها بالدلالة على الصانع عزّ وجلّ .

3 والأجسام : قد قلنا إن الأجسام *Fawz*   5 أورد : ورد *Fawz*   11 لتجويف : بتجويف or لتحاويف in different MSS of *Fawz*   14 ولكلّ : فأقول الآن إن لكلّ *Fawz*   16 طبيعته : طبيعية *Fawz*   18 after متحرّك add. فإنما *Fawz* | ومن ههنا : ولهذا *Fawz*   19 بما : مما *Fawz*   21 عزّ وجلّ : جلّ وجهه *Fawz*

1 ماء : مآ M   2 after الأعضاء blank space of five letters in this line after which the page is empty M   3 الأشياء : الاءشيَآ E | نبحث : يبحث with *yā*ʾ and *bā*ʾ undotted in E   5 لائمها : لآءَمهَا E   6 مثال ذلك rubricated in E   7 بالهواء للهواء : بالهوَآءِ للهوآءِ E   9 الهواء : الهوآءِ E   10 والماء : وَالمآءِ E | وينبغي *waw* rubricated in E   11 الهواء : الهوَآء E بهواء : بهوَآءٍ E   12 آخر : اخر E   14 ولكلّ... طبيعيّ rubricated in E   17 لائمه : لاءَمهُ E

(159) He says: the pupil is of water and the [organ of] hearing of air. Nature does not make any sense organ from fire, because of the heat existing in the mixture of the body parts.

<(160) ...[243] Among the things which we are examining the natural bodies are closest to us, for we are part of and related to them. Therefore we perceive them with the five senses. For every sense organ only perceives what is consistent with it, because every sense organ has its underlying balance. For when something of its genus, but different from it in some quality, is conveyed to it, it perceives it. An example of that is taste perceiving a different moisture by means of moisture, hearing perceiving a different air by means of air, touch perceiving earth by means of earth, and vision perceiving fiery rays by means of fiery rays. As for smell, which is the fifth [sense], it is composite, for it grasps vapour by means of vapour, and vapour is composite of air and water. We must mention one of their conditions so as to infer from it the conditions of the others. So I say that the air which is placed in the cavity of the ear has a balance which is accordant with it. When it is altered by some other air, which strikes it by what motion and stroke it contains, man perceives it. Likewise is the condition of the moisture placed on the tongue.

Every natural body has a motion proper to it, for the body, be it a body already existent or a body in the process of generation, subsists through its proper form. Its proper form causes the subsistence of its essence, and the form which causes the subsistence of its essence is its nature, and its nature is the starting point of its proper motion. It is its proper motion which moves it towards its perfection. The perfection of everything is what is consistent and accordant with it. Therefore everything moved is moved towards its perfection and towards what is accordant with it. For this reason it is said that everything moved moves towards its perfection and [this happens] due to desire. The desiring one is caused by what it desires. The cause precedes the effect by nature. So therefore the inference from motion is the most manifest thing and the best suited for indicating the Creator to Whom belong might and majesty.

[243] Here (89b8–92b7) the manuscript is deliberately left blank. As the text is resumed (92b8, on the assumption that the entire folio should contain thirteen lines and six lines are there) in the middle of a sentence paralleled in the *K. al-Ḥikma* and Miskawayh's *Fawz*, it seems plausible that the entire gap is meant to be filled with textual material from the alternative *Fawz*, i.e. MS Esad Efendi, fols.68b5–69b5, and paralleled in the standard *Fawz*, 42.14–44.2, 74.10–13, 141.11–142.10. Therefore I have inserted passage (160) and the beginning of passage (161) from the *K. al-Ḥikma*.

ومن شأن الحسّ أن يفسد من غلبة المحسوس القويّ ، كالعين فإنها تكلّ وتضعف من الضوء القويّ والأشياء النيّرة هي التي تفوق قوّتها ، وكالسمع فإنّه يكلّ ويضعف من الأصوات الهائلة القويّة ، وكذلك باقي الحسّ .

ومن هذه القوى التي هي الحواسّ ما هو في أفق النبات ومنها ما هو في أفق الحيوان البهيميّ ومنها ما هو في أفق الإنسان ، أعني حسّ السمع والبصر . وذلك أنّ أوّل ما يقبله الحيوان من أثر النفس ويتميّز به عن النبات هو حسّ اللمس الذي يوجد في أنواع الصدف ، ثم حسّ الذوق والشمّ اللذين في أصناف الدود وكثير من الفراش ، ثم إذا قبل صورة السمع والبصر صار شريفاً في الحيوان . وإنما صار هذان الحسّان شريفين لأنّهما أبسط وأقلّ مخالطة للهيولى ، وذلك أنّهما يقبلان صورة الأمور من غير استحالة إليها . فأمّا تلك الحواسّ الأخر فإنّها لا تقبل الأثر إلا بمخالطة وممازجة واستحالة هيولانيّة . وإذا كانت صور الحقائق التي تأتي النفس من فوق غير ملائمة لشيء من الهيولى ، لم تتجاوز حسّ السمع والبصر ، لأنّه ليس في طاقة الحواسّ الأخر أن تقبلها بنوع | من الأنواع ولا بجهة من الجهات ، وعلى أن تلك المعاني البسيطة E69b الشريفة إذا انتهت إلى السمع والبصر صار فيها ظلّ للهيولى . ولذلك تظهر في معرض منها ولم يمكن بعد ذلك أن تتجاوزها إلى كثافة أخرى لأنّ في ذلك خروجاً عن ذواتها ، وهذا محال .

(161) وللحواسّ الخمس حسّ مشترك جامع يجمعها ويؤلّفها في ذاته ، ولو لا

1 ومن : فقد فرغنا من ذكر الأشياء التي تشترك فيها جهتا العقل والحسّ فأمّا الأشياء التي تتباين فيها فهي هذه من *Fawz* 3 القويّة : التي تفوق قوته *Fawz* | الحسّ : الحواس *Fawz* 4 ومن . . . وأعلاها مرتبة .add after الإنسان 5 *Fawz* فمنها .add after الحواسّ | *Fawz* فأمّا قواها : القوى أنّا .add after وذلك 6 *Fawz* البصر والسمع : السمع والبصر | *Fawz* ما كان في أفق الإنسان منه الحيوان 9–8 شريفاً . . . الحيوان : *Fawz* بآخره .add after ثم 8 *Fawz* قد بيّنا فيما تقدّم *Fawz* الشريف الذي شرحنا من أموره ما شرحنا فيما سلف 12 ملائمة : ملابسة *Fawz* 15 للهيولى : الهيولى *Fawz* 16 تتجاوزها : يتجاوزهما *Fawz* | في[2] om. *Fawz* 18 وللحواسّ : في كيفيّة ارتقاء الحواسّ الخمس إلى القوّة المشتركة ومنها إلى ما فوقها قد قلنا فيما تقدّم أنّ للحواسّ *Fawz* | حسّ . . . جامع : حسّاً مشتركاً جامعاً *Fawz* 18–264.1 ولو . . . هو : ولولاه *Fawz*

2 والأشياء : وَالاشيآء E 3 الهائلة : الهآيلة E 4 ومن . . . القوى rubricated in E | ومنها *waw* rubricated in E 5 ومنها *waw* rubricated in E 7 ثم rubricated in E 9 وإنما *waw* rubricated in E 10 فأمّا rubricated in E 11 وإذا rubricated in E 12 الحقائق : الحقآيق E 18 وللحواسّ الخمس rubricated in E 21 الأشياء : الاشيآءِ E

It is of the nature of a sense that it is corrupted due to the predominance of a strong sensible [object], like the eye, for it becomes weary and weak due to strong light and luminous things which outbalance its power; and like hearing, for it becomes weary and weak due to terrible strong voices. The same applies to the remaining senses.

Among these powers which are the senses, some occur in the domain of plants, some in the domain of brute animals and some in the domain of man, namely hearing and vision. For the impression of the soul which the animal receives first and which distinguishes it from the plants is touch, which is [already] found in species of shellfish. Next are taste and smell which are [already] found in species of worms and many [species] of moths. If [living beings] have then received the form of hearing and vision, they have become noble among living beings. These two senses are noble, because they are simpler and less mingled with matter. For they receive the form of things without change in them. As to the other senses, they only receive an impression by mingling, mixing and materially changing. If the forms of the realities which come to the soul from above are not suited to anything material, they will not go beyond the senses of hearing and vision, for it is not possible for the other senses to receive them in any way and any respect. Although, when those noble, simple notions finally get to hearing and vision, there is a shadow of matter in them. Therefore they manifest themselves in the way of [matter], but after that they are not able to go beyond [these two senses] towards the coarseness of the other [senses], for that [would be] leaving their essences, and that is unthinkable.

(161) The five senses[244] have a common sense collecting them all and

[244] The first part of this passage is only preserved in the *K. al-Ḥikma*, MS Esad Efendi, fols.69b5–15; and paralleled in *Fawz*, 123.2–13, whereas the second part (starting in the mid-sentence) also occurs in the *PR* as preserved in the Marsh MS, immediately after the deliberate gap.

هو لتفرّق علم الحواسّ ولم يكن لها ما يؤلّفها ولا ما يحفظها بعد أن تزول آثارها . والنفس لما تحرّكت الحركة المنسوبة إلى أسفل على ما كنّا بيّنّاه لم يكن ممكناً في الجسم المركّب على جفائه وغلظه أن يتّصل بالنفس على لطفها وبعدها من الجوهر الجسميّ إلا بوسائط يلطف فيها الجسم أوّلاً حتّى ينتهي إلى غاية ما يمكنه أن ينتهي إليه . فحينئذ يمكن أن يقع بينهما الاتّصال الذي يصيّر أحدهما قابلاً أثراً من الآخر ، مثال ذلك أن المعدة إذا لطفت الغذاء بالهضم وحصل منه في القلب دم رقيق ألطف ما يمكن ، عادت الحرارة التي في القلب عليه فزادته تلطيفاً وأجرته في العروق الجوف التي تسمّى شريانات ، وهو ألطف ما يكون من الدم وحصل منه في العرق الأجوف〉

92b التي ترتقي إلى الدماغ ، فيجري فيه جريان الماء في الأنابيب ، أعني أنّه يبقي فيه فضاء ما فلا يختنق فيه بأن يملأه . وذلك الدم حارّ قريب العهد بالقلب فيرتفع منه بخار لطيف يحصل منه في فضاء العرق الأجوف الخالي من الدم ، وكلّما ارتفع لطف هذا البخار حتّى يحصل في الدماغ ، فيتشعّب إلى عروق

93a دقاق كثيرة شبيهة بالشعر في الدقّة ، ويتفرّق في الدماغ فيعدّل برده بحرّه ويعتدل أيضاً هو ببرد ذاك ، ويصير منه ما يسمّى روحاً وبحسب صفاء هذا الروح وتهذّبه في آلاته يكون صدر قوى النفس عنه واستعداده لقبول آثارها من الحسّ والفهم . وتنشئ الطبيعة حينئذ من الدماغ أعصاباً يكون بها الحسّ والحركة الإراديّة في جميع البدن وبها يتميّز الحيوان من النبات . فمنها العصبة الجوفاء التي تنقسم إلى ثقبتي العين ، وينفذ فيها ذلك الروح وقد تهذّب فيها غاية تهذّبه ولطف جداً فيكون به البصر ، ومنها التي تأتي الأذن فيكون بها

1 لتفرّق : لتفرقت *Fawz* | علم : علوم *Fawz* 2 والنفس : ونقول الآن إن النفس *Fawz* 4 after أوّلاً add. فأوّلاً *Fawz* 7 القلب ... منه[1] : في القلب منه *Fawz* | after يمكن add. من الغذاء *Fawz* 10 التي : الذي *K. al-Ḥikma, Fawz* 11 ما om. *K. al-Ḥikma* | فيه[2] om. *K. al-Ḥikma* | بأن : أن *Fawz* 12 منه[2] om. *Fawz* 14 دقاق : دقائق *Fawz* 15 أيضاً هو : هو أيضاً *Fawz* | ببرد : برد *K. al-Ḥikma* 16 صدر : صدور *K. al-Ḥikma, Fawz* 19 ثقبتي العين : ثقبي العين *Fawz* | فيها[1] : فيهما *Fawz* | فيها[2] om. *K. al-Ḥikma, Fawz*

2 آثارها : اثارهَا E | والنفس rubricated in E 4 بوسائط : بوَسآيَط E 5 يمكنه : يمكنها E 6 الآخر : الاخرِ E | مثال ذلك rubricated in E | الغذاء : الغِذآءَ E 10 الماء : المآ M 11 فضاء : فضآ M 12 فضاء : فضآ M 13 وكلّما : كما M 16 آلاته : اَلاتِه M | after عنه نفس M | آثارها : انارهَا M 19 الجوفاء : الجوفآ M | ثقبتي : blank space of three letters M

combining them in itself. If it were not for this, the information of the senses would be disunited and they would have nothing to combine and nothing to preserve [this information] after its impressions have ceased. When the soul moves in a motion towards the lower [world], as we have made clear, the composite body in its roughness and thickness can only be joined to the soul in its fineness and remoteness from the bodily substance through mediums, in which the body is firstly refined until it reaches the peak of what it can reach. Then they can be joined in a way which lets one of them become receptive to the impression of the other. An example of this is that when the stomach refines the food through digestion and from this fine blood, the finest possible, is produced in the heart,[245] the heat in the heart goes over it for a second time, increases it in refinement, and emits it into the hollow veins which are called arteries. It is the finest blood existent. Some of it goes to the hollow vein⟩
which ascends to the brain. It flows in it like water in tubes, that is it leaves 92b8
some empty space in [the hollow vein] and does not become congested in
it while filling it up. That blood is hot, in close contact with the heart, and
from it a fine vapour rises which is produced from it in the empty space of
the hollow vein which is free from blood. The more this vapour rises, the
more it is refined until it reaches the brain, branches out towards the
many thin veins resembling hair in their thinness, is dispersed in the 93a
brain, and balances the coldness [of the brain] with its heat and is, in turn, balanced by the coldness of the latter. From [this] what is called the 'spirit' results, and according to the clarity of this spirit and its refinement in its instruments the powers of the soul proceed from it and it is prepared to receive impressions of these [powers], for example, sense perception and understanding. Nature then causes nerves to grow out from the brain, by means of which there is sense perception and voluntary movement in the entire body and by which the animals are set apart from the plants. Among them is the hollow nerve which separates into both eye sockets. That [above-mentioned] spirit, being already extremely refined in it and having become very delicate, permeates it and produces vision. Among

245 Sweetman has some additional textual material, translating (*Fawz*, p.168): 'When the food reaches the stomach, that organ first refines it by means of digestion and then sends it to the liver, whereafter, making it finer still, the liver turns it into blood and sends it to the heart.'

السـمع وكـذلك الباقيات . فإذا حصل في كلّ واحـد مـن الحواسّ أثـر مـن
93b المحسوس تأدّى منه إلى الحسّ المشترك وهو قوّة من قوى النفس في أفق هـذا
الجوهر اللطيف من الجسم يقبل هذه الآثار كلّها . وكما أنّ كلّ حسّ من
الحواسّ الخمس يختصّ بنوع من المحسوسات فيقبل آثاره ثم يميّز بين أشخاصه ،
فكذلك الحسّ الجامع المشترك يقبل الآثار من الحواسّ كلّها ، ثم يميّز بينها ،
إلا أنّ الفرق بينهما أن الحواسّ الخمس إنّما تقبل الصور بأن يحصل فيها آثار
الجزئيّات من المحسوس شيئاً بعد شيء ، وأمّا الحسّ المشترك فإنّه يقبل الصور
من الحواسّ في دفعة واحدة من غير أن يتأثّر منها بما يحصل فيه من تلك
الصور ، لأنّه في نفسه صورة ، والصورة لا تقبل الصورة على طريق التأثر ، بل
على طريق آخر وبنحو أعلى وأشرف ، وكذلك يدرك الجميع بلا زمان ولا
94a تجزئة ولا انقسام ولا تختلط الصور هناك ولا تتزاحم كما تتزاحم في الأجسام .
(162) التخيّل أفضل من الحسّ دون التميّز وكأنّه بينهما . وهو يتلو الحسّ
والعقل يحتاج ضرورة إلى أن يتقدّم التخيّل . وكثيراً ما يسمّى العقل تخيّلاً
والتخيّل تصوّراً بالعقل . نحن نتخيّل في النوم ولا نحسّ فيه ، فليس الحسّ هو
التخيّل ، التخيّل هو قوّة تتمسّك برسوم الأشياء التي يسلّمها الحسّ إليها ،
وتلحقها الخطأ والصواب من جهة الحسّ وهو حركة للنفس تحدث عن الحسّ
الذي بالفعل .

(163) قال الشيخ اليونانيّ البصر وسائر الحسائس لا تنال المحسوس إلا أن
يكون بينها وبينه جرم متوسّط ، وذلك أنّ الحسّ إنما هو من حيّز النفس ،
والنفس إذا كانت في المكان العقليّ علمت الأشياء العقليّة من غير أن تحتاج
94b إلى شيء من الأجرام ، وإذا كانت في العالم الحسّيّ احتاجت في معرفة

3 من الجسم om. *Fawz* 4 المحسوسات : المحسوس *K. al-Ḥikma, Fawz* | فيقبل : فقبل *K. al-Ḥikma* 9 after والصورة add. والصورة *K. al-Ḥikma* 10 وكذلك : ولذلك *K. al-Ḥikma, Fawz*

3 الآثار : الاثار M | after كلّها blank space of three letters M 4 آثاره : اثاره M 5 الآثار : الاثار M 6 آثار : اثار M 7 الجزئيّات : الجزيات M | after شيء break (∴) ind. and blank space of half a line M 10 آخر : اخر M | وبنحو : وتنحوٍ M 11 after الأجسام blank space of more than half a line M 12 يتلو : يتلوا M 14 نتخيّل : يتخيل M | نحسّ : يحسّ M 15 يسلّمها : يسلّمه M 17 after بالفعل break (∴) ind. and blank space of half a line M

[the nerves] there is [further] the one which come to the ear and through this there is hearing. The same applies to all other [nerves]. So when the impression of a sensible occurs to any of the senses, [the nerves] convey [it] from there [i.e. the sense organ] to the common sense which is one of the powers of the soul in the domain of this fine bodily substance which 93b
receives all these impressions. As each of the five senses is specially distinguished by one kind of sensible, receives its impressions and then discriminates between its individual instances, in the same way the collecting common sense receives the impressions from all senses and then discriminates between them. Yet the difference between the two is that the five senses only receive the forms by the impressions of the sensible particulars arising in them one after the other. As to the common sense it receives the forms from the senses at one stroke without being impressed by them through these forms which arise in it. For it is in itself a form, and a form does not receive a form by being impressed, but by another way, namely a more elevated and nobler method. Similarly, it receives everything without time, partition or division, and there [i.e. in the common sense] the forms do not mingle or crowd together as they would crowd together in the bodies. 94a

(162) Imagination is better than sense perception, but below discrimination as if between the two. It succeeds sense-perception, and intellect, in turn, precedes imagination by need and necessity. Often intellect is called imagination and imagination is called conception [of forms] through intellect. We imagine during sleep, but we do not perceive during it. Sense perception is not imagination, and imagination is a power which retains the traces of the things sense perception delivers to it. It is subject to error and truth from sense perception and is a motion of the soul brought forth from the sense which is in actuality.

(163) The Greek Sage says:[246] vision and the other senses only attain the sensible if there is an intermediary body between them and it. For sense perception rather belongs to the realm of the soul. While the soul is in the intellectual place, it knows the intelligibles without needing any body, whereas while it is in the sensible world, it needs the body to know the 94b

[246] Belonging to *GS*; see ed. Rosenthal, pp.370–3; and ed. Badawī, 189.18–190.12.

المحسوسات إلى الجسم . والنفس إذا اتّحدت بالأشياء المحسوسة فإنما تنال
معرفتها بالأشياء الشبيهة بها القويّة على قبول الآثار المحسوسة ، فتحسّ النفس
الأشياء المحسوسة وتعرفها بتوسّط .

والبصر يدرك الألوان بمتوسّط يقبل أثر اللون أوّلاً ، ثم يؤدّيه إلى البصر ، وإلا
لم يقو البصر على نيل اللون ، وذلك أنّ البصر ألطف من المحسوس ذي اللون
كثيراً والهواء ألطف من اللون قليلاً ، أعني أن الهواء أغلظ من البصر قليلاً
وألطف من المحسوس ، فلمّا كان الهواء متوسّطاً على هذه الجهة قبل أثر
المحسوس فأدّاه إلى الحسّ . لأنّ الشيء إنما يؤثّر في الشيء الشبيه به الذي لم
ينائيه منائية بعيدة ولا يؤثّر في الشيء الذي كان مبايناً . فالهواء إذن يقبل أثر
95a لون الجرم فيؤدّيه إلى البصر ، فإذا أدّاه إليه تشبّه به البصر أيضاً وصار مثله
يحسّ حسّاً صحيحاً .

(164) فإن قال قائل فإنّ المبصر يحتاج إلى شيء متوسّط يكون بينه وبين المبصر
إليه وإلا لم يبصره ، لأنّ المحسوس لا يصل إلى الحسّ إلا أن يسلك في الهواء
فيغيّره ، ثم يصل بذلك التغيير إلى البصر ، قلنا فلسنا نرى كلّ سالك في الهواء
يغيّر الهواء لكنّه ربّما فرج الهواء فقط من غير أن يغيّره ، كالحجر فإنّه إذا سقط
من فوق وسلك في الهواء فإنما يفرج الهواء فقط ولا يغيّره . فإن قال قائل إنّ
الحجر إذا سقط من العلو وانفرج له الهواء فإنما ذلك الانفراج أثر من الآثار ،
لأنّ الحجر يدفع الهواء دفعاً قويّاً والهواء يندفع والاندفاع أثر ، فالهواء إذن يتأثّر
من الحجر ، وكذلك النار إذا سلكت من أسفل إلى فوق ، فإنما تدفع الهواء
95b فينفرج لها إلا أنّ انفراج الهواء من النار أسرع من انفراجه من الحجر ، قلنا إنّ
سلوك النار إلى فوق وسلوك الحجر إلى أسفل إنما يكون بانفراج الهواء ، وذلك
أنّ الحجر يشقّ الهواء ويمرّ فيه سفلاً ، فكذلك النار تشقّ الهواء وتمرّ فيه علواً ،
والشقّ لا يكون [إلا] بالدفع لكنّه يكون بأنّ الهواء ينفرج لجرم السالك فيه ،

2 الآثار Ros, Bad : الاثار M | فتحسّ Ros, Bad : فيحسن M   6 أغلظ Ros, Bad : اغلط M   7 متوسّطاً Ros, Bad : متوسط M   9 ينائيه : ينائه Ros : يناييه M : يباينه Bad | منائية Ros : منايية M : مباينة Bad   11 يحسّ Bad : بحس M : ⟨و⟩يحسّ Ros | حسّاً : إحساساً Bad   12 يكون Ros, Bad : يلون M   16 وسلك M, Bad : سلك Ros   16–17 وانفرج . . . الهواء[2] add. in marg. M   17 الآثار Ros, Bad : الاثار M   19 أسفل M, Ros : الثقل Bad   22 فكذلك M : وكذلك Ros, Bad

sensibles. When the soul is united with the sensibles, it attains knowledge of them through things which resemble them[247] and which are capable of receiving the impressions of the sensible. So the soul perceives and gets to know the sensibles through mediation.

Vision perceives colours through an intermediary which receives the impression of colour first and then conveys it to vision. Otherwise, vision would not be able to attain colour. For vision is much more fine than the coloured sensible, whereas air is [only] a little finer than colour, that is air is a little thicker than vision, but finer than the sensible. Thus when the air is intermediary in this respect, it receives the impression of the sensible and then conveys it to the sense. For something rather produces an impression on something which resembles it and is not at a great distance, and not on something which is different. Therefore air receives the impression of a body's colour and then conveys it to vision. So when [air] 95a
has conveyed it to [vision], vision also imitates it and becomes like it perceiving with correct sense perception.

(164) Someone may say:[248] he who sees needs something intermediary between him and what he sees or else he would not see it, because the sensible only reaches the sense by passing through the air, altering it and then reaching vision through this alteration. We say: we are not of the opinion that everything passing through the air alters the air, but sometimes it only splits the air without altering it. For example when a stone falls from above and passes through the air, then it only splits the air, but does not alter it. So someone may say that when the stone falls from an altitude and the air splits apart for it, that splitting apart is a kind of impression. For the stone projects the air forcefully and the air is being projected, projection being an impression. Therefore the stone makes an impression on the air and similarly, when fire rises upwards from below, it rather projects the air, which then splits apart for it, except that the air splitting apart by fire is faster than its splitting apart by a stone. We say 95b
that fire's passing upwards and the stone's passing downwards [through the air] exist only by the splitting of the air. For the stone cuts through the air and goes through it downwards, and in the same way fire cuts through the air and goes through it upwards. The cutting does not result from projecting,[249] but through the splitting apart of the air for the body

[247] The Arabic text could also be translated as: '[the soul] only obtains its knowledge through things which resemble it [i.e. the soul]', but the translation given above is the one supported by the Greek and adopted by Rosenthal and Lewis.

[248] Belonging to *GS*; see ed. Badawī, 190.13–192.2; and partial edn Rosenthal, pp.372–7.

[249] The deletion of *illā* is accepted by all editors and translators.

فإذا انفـرج وسـلك السالك ، امتـلأ أيضاً ورجع إلى الحال الـتي كان عليـها بدءاً . فإن كان هكذا وكان الهواء أنما ينفرج للأجرام التي تمرّ به فقط من غير أن يقبل شيئاً من آثارها ، فما الذي يمنعنا أن نقول إنّ صور الأشياء المحسوسة تسلك إلى البصر من غير أن ينفرج الهواء ، إلا إنّها روحانيّة تسلك ولا تحتاج إلى فرجة الهواء .

فإن قال إنّ صـور الأشـياء المحسوسـة إن لم تكـن معلّقـة محمولـة في الهواء اضمحلّت وبادت قبل أن تأتي البصر ، فلا يقع عليها البصر البتّة ، فلذلك صـار الهواء يقبل صـور الأشـياء المحسوسـة أوّلاً ، ثم يقبلـها بعـد ذلك البصـر 96a بتوسّط الهواء ، قلنا إن كنّا نحن لا نحسّ المحسوس حتّى ينفعل الهواء أوّلاً، لزم من ذلك أن نكون نبصر الشيء من غير أن نلقى أبصارنا عليه ، لأنّا إنما حسسنا بالهواء الذي يلينا كحسّنا بحرارة النار . فإنّ النار إذا كانت بعيدة منّا ، لم تسخن أبداننا بل الهواء الذي يسخن أبداننا الهواء المحيط بنا القريب منّا ، لأنّ الحرارة إنما تسخن الشيء بالملامسة ، فإن لم تلامسه لم تسخنه . فأمّا المبصر فلا يكون بالملامسة ، والدليل على ذلك أنّه إذا وضع المحسوس على العين ولامسـها ، لم تبصره لكن ينبغي أن يكون الشـيء الواقع تحت البصر بعيداً من البصر بعداً معتدلاً وأن كان المتوسّط بين البصر وبين المبصر نيّراً مضيئاً . وإنما احتاج البصر إلى الضياء لأنّ الهواء المتوسّط مظلم ، فلو أنّ 96b الهواء لم يكن مظلماً لم يحتج البصر إلى الضوء ، لأنّ الظلمة إنما هي حجاب بين المبصِر والمبصَر إليه ، فينبغي أن يزاح المانع بالضوء لينال البصر محسوسه . وإنما صار البصر لا يبصر الشيء إذا وُضع عليه لأنّ البصر والشيء المحسوس ظلالان ، فإذا وضع المحسوس على العين لم تره من أجل الظلّ الحادث منهما . والدليل على أن البصر يقع على المبصر إليه من غير أن ينفعل الهواء

2 بدءاً Bad : بدياً M: بدئا Ros 4 تسلك[1] Ros, Bad : فيسلك M | إلى[2] . . . تسلك[1] add. in marg. M | الهواء Ros, Bad : الهوآ M 9 كنّا Ros, Bad : لنا M 10 نكون نبصر : يكون يُنضر M : نكون ننظر Bad | نلقى Bad : تلقى M 11 كحسّنا M : لحسنا Bad 16 كان M : يكون Bad 17 مضيئاً Bad : مُضياً M 21 ظلالان : ظلالين M : ظلان Bad | الظلّ M : المضلّ Bad

passing through it. So when it has split apart and the thing passing through it has passed, [the air] fills up again and returns to its original condition. If it is like that, and the air only splits apart for the bodies which go through it without receiving any kind of impression from them, then what prevents us from saying that the forms of the sensibles pass to vision without the air splitting apart? They are certainly spiritual and do not need the air to split while passing [it].

Then someone might say that if the forms of sensibles were not attached to and borne in the air,[250] they would disappear and perish before they came to vision, so that vision would not meet them at all. Therefore, the 96a
air starts to receive the forms of the sensibles first, then, after that, vision receives them through the mediation of air. We say: if we did not perceive the sensible until the air has been acted upon first, it would follow from that that we saw something without casting our glances on it, as we would perceive the air adjacent to us, like our perception of fire's heat. For when the fire is remote from us, it does not warm our bodies, rather the air which warms our bodies is the air surrounding and near to us, because heat only warms something by contact. So if it is not in contact with [something], it does not warm it. As to vision, it is not by contact. The proof for this is that when the sensible is placed on the eye and touches it, [the eye] does not see it. The thing which can be seen must rather be moderately remote from [the organ of] vision, and the intermediary between [the organ of] vision and the visible must be luminous and bright. Vision only needs light, because the intermediary air is dark. If the air 96b
were not dark, vision would not need light, as darkness is a barrier between the one who sees and the object seen. So the obstacle must be removed by light so that vision attains its sensible. It only happens that vision does not see something when [the thing] is placed on [the organ of] vision, because [the organ of] vision and the sensible thing both cast a shadow.[251] So when the sensible is placed on the eye, it does not see it on account of the shadow occurring from both of them. The proof that vision meets the seen object without the air being acted upon by the form of the

[250] Lewis ('Plotiniana', p.157) translates: 'If the forms of the objects of sense-perception are not attached to the substratum of the air', which is almost to Rosenthal's translation ('aš-Šayẖ', p.377). See also Hansberger, 'Mediating', p. 69, n.25.

[251] The Greek text is much clearer than the Arabic and talks about the shadow of air and the visible object, cf. *Enn.* IV 5, 2, 60–1: 'But perhaps when the object is applied to the eye it is not seen because it brings with it the shadow of the air and its own shadow' (tr. Armstrong). Lewis ('Plotiniana', p.159) translates: 'The reason why the eye does not see a thing when it is placed upon it is that the eye and the thing to be seen are both in shadow.'

من صورة المحسوس الأشياء التي يراها بالليل ، كالنار والكواكب ، فإنّا قد نرى النار وضوء الكواكب بالليل ، ولا يقدر أحد أن يقول إنّ ضوء النار والكواكب يؤثّر في الهواء فيأتي أبصارنا ، فإنّه لو كان ذلك كذلك ، لما كان الهواء مظلماً لأنّ النار كانت مضيئةً ونيّرة بصورتها التي صارت فيه . وربّما استبانت النار بالليل من ثقب برج الديدبان والكواكب مستترة بالغيم . 97a

(165) قال والبصر لا يحتاج في نيله المحسوس إلى الهواء المتوسّط بينه وبين المحسوس بل إنما يحتاج إلى الضوء فقط . فلما صار الضوء إنما قوامه بالهواء إلا أنّه تمامه ، صار الهواء متوسّطاً بين البصر وبين المحسوس بعرض .

(166) وقال ارسطاطالس السمع والبصر أخصّ بالنفس من الأحساس الباقية . وأشكال المسموع مركّبة في بسيط ، وأشكال المبصر مبسوطة في مركّب .

(167) وقال بعضهم السمع أفضل من البصر ، لأنّ البصر يذهب في طلب محسوساته ويخدمها حتّى يدركها مثل العبد ، والسمع يحمل إليه محسوساته حتّى تخدمه مثل الملك .

وقال آخر البصر لا يدرك محسوساته إلا على خطوط مستقيمة ، والسمع يدركها من محيط الدائرة .

97b وقال آخر محسوسات البصر أكثرها جسمانيّة ، ومحسوسات السمع أكثرها روحانيّة .

وقال السمع أدقّ تميّزاً من البصر ، أذ كان يعرف بجودة الذوق الموزون والنغمات المتناسبة والفرق بين الصحيح والمنزحف والخروج من الإيقاع واستواء اللحن ، والبصر يخطئ في أكثر مدركاته فإنّه ربّما يرى الكبير صغيراً والصغير

11 بعضهم : آخر لا بل *Rasāʾil* 12 العبد : العبيد *Rasāʾil* 13 الملك : الملوك *Rasāʾil* 14 after آخر add. إن *Rasāʾil* | محسوساته : المحسوسات *Rasāʾil* 16 أكثرها[2] : كلّها *Rasāʾil* 18 after وقال add. آخر *Rasāʾil* | تميّزاً : تمييزاً *Rasāʾil* | after الذوق add. الكلام *Rasāʾil*

1 يراها M : نراها Bad 4 مضيئةً Bad : مُضيةً M 5 مستترة : مستنيرة M, Bad | after بالغيم break (∴) ind. and blank space of more than half a line M 8 after بعرض break (∴) ind. and blank space of half a line M 10 after مـركّب blank space of five letters M 12–14 والسمع ... يحمل add. in marg. M 13 الملك : الملوك with break (∴) ind. and blank space of three letters after it M 14 آخر : اخر M 15 after الدائرة break (∴) ind. and blank space of half a line M 16 آخر : اخر M 17 after روحانيّة break (∴) ind. and blank space of half a line M

sensible is the things which one sees at night like fire and the stars. For we certainly see fire and the light of the stars at night and nobody can say that the light of fire and the stars produces an impression on the air so that it comes to our eyes. If that were like that, the air would not be dark, because fire is bright and luminous by its form which would [then] come to be in [the air].[252] Sometimes fire is apparent at night from the windows 97a
of a watch-tower while the light of the stars is hidden by the clouds.[253]

(165) He says:[254] vision does not need the intermediary air between it and the sensible in order to attain the sensible, but it needs only light. So since light subsists through air, yet is the perfection [of air], air becomes the intermediary between the eye and the sensible accidentally.

(166) Aristotle says: hearing and vision are more proper to the soul than the other sensations. The shapes of the audible are composite in something simple and the shapes of the visible are simple in a composite.

(167) Someone says:[255] hearing is nobler than vision, for vision travels in pursuit of its sensibles and serves them in order to perceive them, like a servant, whereas with hearing its sensibles are brought to it so that they serve it like a sovereign.

Someone else says: vision only perceives its sensibles along straight lines, whereas hearing perceives them from all around.

Someone else says: most of the sensibles of vision are corporeal, whereas most of the sensibles of hearing are spiritual. 97b

He says: hearing is finer in discrimination than vision, since it recognises the excellence of a tasteful [musical] metre, proportionate tones, the difference between what is sound and what is discordant, dissonant rhythm and harmonious melody, whereas vision errs about most of its perceptibles. For it may see what is big as small and what is small as big,

[252] Lewis apparently changes the Arabic text to read *nayyarathu* instead of *nayyira* as he translates ('Plotiniana', p.159): 'If that were so, the air would not be dark, since the fire, being bright, would illuminate it with the form it contains.' The Greek does not necessarily support that; cf. *Enn.* IV 5, 3, 5–6: 'or there would not be a darkness, as the fire would radiate its own form' (tr. Armstrong).

[253] The suggested emendation, '*mustatira* (hidden)' instead of '*mustanīra* (apparent)', is based on the Greek and on the argument itself (see *Enn.* IV 5, 3, 1–15).

[254] Belonging to *GS*; see ed. Rosenthal , pp.376–7; and ed. Badawī, 192.3–5.

[255] Cf. *Rasāʾil Ikhwān al-Ṣafāʾ*, I, 236.15–20, 236.22–237.2.

كبيراً ، والقريب بعيداً والبعيد قريباً ، والمتحرّك ساكناً والساكن متحرّكاً ، والمستوي معوجّاً والمعوجّ مستوياً .

(168) وقال ارسطالس حدوث الصوت يتعلّق بصلابة الأجسام وملاستها وعمقها وشدّة القرعة وسرعتها ، أمّا الملاسة فلأن تلقى الهواء جملةً من غير أن ينفرق ، فلا يحرك السمع ، وأمّا التقعير فليكون بعد القرعة الأولى قرعات حتّى

98a تحرّك الهواء ويتردّد مدةً طويلةً ، كما يكون في اواني النحاس ، وشدّة القرعة وسرعتها لئلا يسبق الهواء فيتفرّق فيفوت حركة الضارب لأنّه سريع التشتّت . والهواء أعون على حدوث المصوت من الماء .

(169) والهواء هو الذي يؤدّي القرع إلى السمع والرائحة إلى الخياشيم . ولذلك إن أدنيت شيئاً من مجرى السمع ثم قرعته ، لم تسمع البتّة ، وكذلك لو ألصقت بالخياشيم شيئاً ، لم تشمّ رائحته .

(170) وقال السمع يحتاج إلى توسّط الهواء في نيل الصوت . ولو لم يكن الهواء ثم اصطكّ الأجرام ، لما حدث من اصطكاكها صوت البتّة ، وإنما يكون الصوت بأنّ الهواء يصطكّ باصطكاك الأجرام ، ثم يندفع من بينهما فيصل الهواء الذي يليه ، فلا يزال بعض الهواء يصل بعضاً إلى أن يأتي السمع فيحسّ به .

98b (171) وقال السمع يحدث من قرع هواء واحد متّصل إلى أن يبلغ السمع شبه الأمواج ، لأنّ في الأذنين أبداً هواء متّصل لأمّ الدماغ ، فإذا تحرّك الهواء من خارج تأدّى إلى ذلك الهواء فيقرعه ويصيران واحداً لا بمماسة .

(172) وقال الهواء الذي في الأذنين يتحرّك دائماً بأجزائه حركة ملائمة لها هادئة

9 والهواء : فالهواء *K. al-Nafs* | after يؤدّي add. إليه *K. al-Nafs*, MS A | ولذلك : والدليل على ذلك *K. al-Nafs* 10 إن : أنك لو *K. al-Nafs*, MS A : أنك إذا ed. Arn | after قرعته add. بشيء آخر *K. al-Nafs* | تسمع البتّة : يسمع *K. al-Nafs*, MS A 11 after ألصقت add. شيئاً ذا رائحة *K. al-Nafs* | بالخياشيم om. *K. al-Nafs*, MS A | شيئاً om. *K. al-Nafs* | رائحته : له رائحة *K. al-Nafs*, MS A : add. البتّة ed. Arn

1 والقريب : والتقريب M 2 after مستوياً break (∴) ind. and blank space of more than half a line M 5 after السمع blank space of three letters M 11 ألصقت : الصفت M | after رائحته break (∴) ind. and blank space of ten letters M 16 after به break (∴) ind. and blank space of six letters M 19 فيقرعه : فيقزغه M | after بمماسة break (∴) ind. and blank space of more than half a line M

what is near as remote and what is remote as near, what is moving as being at rest and what is at rest as moving, the straight as crooked and the crooked as straight.

(168) Aristotle says: the emergence of sound depends on the solidity of bodies, their smoothness, their depth, and the force and rapidity of the stroke: smoothness so that it meets the air entirely without [the air] being dissipated, [which would result in] hearing not being set in motion; hollowness so that after the intial stroke there are [more] strokes to move the air and be reflected over a long period of time, just as happens in 98a
copper vessels; the force and rapidity of the stroke so that the air neither gets ahead nor is dissipated nor escapes the motion of the [body] striking [it], as it is rapidly dispersed. Air is of greater help in the emergence of sound than water.

(169) It is air[256] which conveys the stroke to hearing and odour to the nostrils. Therefore if you bring something close to the auditory canal and then strike it, you will not hear [anything] at all. In the same way, if you press something against the nostrils, you will not smell its odour.

(170) He says: hearing needs the mediation of air to attain sound. If there were no air and then bodies were knocked together, no sound would emerge from their being knocked together at all. Sound occurs only through air being knocked by bodies being knocked together. Then [the sound] is expelled from between the two [bodies] and reaches the air adjacent to it. No part of the air ceases to transmit [it] to [the next] part until it reaches hearing and is perceived.

(171) He says: hearing emerges from striking one single continuum of air 98b
until it reaches hearing, which resembles waves. For in the ears there is always air contiguous to the [tympanic] membrane.[257] So when the air is moved from outside, it is led to that air, then strikes it and they two become one [although] not by contact.

(172) He says: the air in the ears is perpetually moved by its parts in a movement which is consistent with them, calm and without any disturb-

[256] This passage, which in the Marsh MS is not set apart from the previous one, is paralleled in the *DA*-paraphrase, ed. Arnzen, 267.4, 267.1–3; and MS A, fols.78b2–3, 78a14–b2, and partly in passage (150) above.

[257] The Greek term translated by *umm al-dimāgh* was probably μῆνιγξ, which means 'membrane' in general, as well as 'eardrum', but mostly the 'cerebral or tympanic membrane', the 'meninx' (see Liddell and Scott).

لا اضطراب فيها ، فأمّا بكلّيّته فلا يتنقّل ولا يتغيّر أصلاً ، وهو داخل لقّيه مواصل لأمّ الدماغ والسمع أبداً ، ولذلك يحسّ في الأذنين بِدويّ دائماً .

(173) وقال ثقب الأذنين لولبان كيما يعسّر دخول مجرى السمع الماء والهواء ، فكما أنّ الماء الذي على الناظر يقبل رسوم الألوان ويوصّلها إلى البصر ، 99a كذلك الهوا الذي في الأذنين يقبل الأصوات ويوصّلها إلى السمع . وتلك الرطوبة قد كنفتها الطبيعة وسترتها بأغشية حذراً من تشدّدها ، وكذلك الهواء الذي في الأذنين تحتوي عليه أمّ الدماغ وتقوم لها مقام الوقاء .

(174) وقال في طبع الهواء كلّه أن يتحرّك ، ويعلم ذلك من الهباء لأنّه لا تسكن أصلاً في وقت . ولذلك أن ضممنا إلى الأذنين مرنّاً سمعنا صوتاً ، لأنّ الهواء الذي فيه يتحرّك حركته التي تخصّه .

(175) وقال الصدى لقاء الهواء الذي يقرعه الصوت جسماً صلداً أملس ، فيحويه المكان الذي يقرعه ويمنعه من التشتّت ، فيعود إلى ورائيه كالكرة ، ولهذا إذا تكلّمنا في مكان أجوف عميق سمعنا رجع كلامنا .

99b (176) قال افلاطن الروائح تكون من بخارات لطيفة تخرج من المشموم وتصل إلى المشامّ ، ويعلم ذلك من ذبول النفاحة ، ومن أنّا متّى وضعنا على أنف الشامّ خرقة ، فإنّ الهواء يصل إليه والرائحة لا تصل إليه ، لأنّ موضوعها أغلظ من الهواء .

(177) وقال ارسطالس يتولّد في الهواء من المشموم رائحة مشابهة للرائحة الموجودة فيه من غير أن تنتقل عن المشموم ، كما يتولّد في الحديد حرارة مشابهة لحرارة النار . وهذه الرائحة يتولّد فيه من غير أن يستحيل ، وذلك أنّ الشيء إنما يستحيل من ضدّه وليست الرائحة الموجودة في المشموم مضادّة للهواء ، بل هي ملائمة لأنّها هوائيّة ومن طبيعته . والدليل على أنّ الهواء لا يستحيل هو أنّ الرائحة تمتدّ و توجد على مسافة بعيدة ، حتّى أنّ الرغامى

2 after دائماً break (∴) ind. and blank space of more than half a line M 3 لولبان : لَولَبتَين M 6 تشدّدها : تشدُدُهَا M 7 after الوقاء break (∴) ind. and blank space of six letters M
10 تخصّه : تخصها with break (∴) ind. and blank space of more than half a line after it M
12 فيحويه : فتجوية M 13 after كلامنا break (∴) ind. and blank space of three letters M
17 أغلـظ : اغلـط M | after الهواء break (∴) ind. and blank space of ten letters M
23 الرغامى : الرغم M

ance. In its totality it neither shifts nor alters at all. It is always inside its two coils, contiguous to the [tympanic] membrane and [the organ of] hearing, and therefore one perpetually perceives a hum in the ears.

(173) He says: the perforations of the ears are two spirals in order to make it difficult for water and air to enter the auditory canal. Just as the water on the pupil receives the traces of colours and transmits them to vision, in the same way the air in the ears receives the sounds and transmits them to 99a
hearing. Nature has certainly protected the humour [of the pupil] and covered it with a membrane to guard it against hardening. In the same way the air in the ears is enclosed by the [tympanic] membranes and they serve as [the ears'] protection.

(174) He says: [it is] in the nature of air to be altogether in motion. This is known from dust particles floating in the air for they are never at rest at all. Therefore, when we cover the ears with something resonating we hear some sound, because the air which is in there moves in its proper motion.

(175) He says: an echo [occurs when] air struck by sound meets with a hard and smooth body. So the place which strikes the air and prevents it from dissipation contains it so that it comes back like a ball. Therefore, when we talk in a deep hollow place we hear our talk rebounding.

(176) Plato says: odours are caused by fine vapours which are emitted from 99b
the object being smelt and reach the nose. That is known from the fading of fragrances and from [the fact] that if we place a rag on the nose of someone who can smell, although the air [still] reaches [his nose], the odour does not reach it [any longer], because its substratum is thicker than air.

(177) Aristotle says: the object being smelt generates an odour in the air which is similar to the odour existent within it, without anything being transferred from the object being smelt. In the same way heat which is similar to the heat of fire is generated in iron. This odour is generated in [the air] without [the air] being changed. For something only changes through its opposite whereas the odour existent in the object being smelt is not opposed to the air, but consistent [with it], as it is airy and of [air's] nature. The proof that the air does not change is that the odour extends

100a يشمّ رائحة الفتلي من الفراسخ الكثيرة . ولو كان يستحيل لم يوجد الرائحة على هذه المسافة ، إذ ليس فيها قوّة على أن تحيل هذا المقدار الكبير من الهواء .

(178) الضوء فعلٌ ما به يتمّ الجسم الصافي القابل اللون ، لأنّه لا يتلوّن إلا بتوسّط الضياء .

(179) وقال بعضهم الضياء جسم ، واستدلّوا عليه بانعكاس شعاع الشمس . فإنّه لو لم يكن جسماً ، لنفذ في الجسم ولم ينعكس ، وذلك أنّ الذي ليس بجسم لا يمتنع من أن ينفذ في الأجسام ، فالضوء إذن جسم . وقالوا لو لم يكن جسماً لم يسخن شعاع الشمس الهواء ، فإنّه إنما يسخن الهواء باحتكاكه به ، والاحتكاك يكون بالأجسام . وقالوا إن لم يكن الضوء جسماً ، فما بال شعاع الشمس إذا دخل في كوة من بيت ، لم ينبثّ في جميع البيت لكنّه يكون مقابل الكوة ، وهذا هو من خاصّة صفة الأجسام ، فالضوء إذن جسم .

(180) وقال آخرون ليس بجسم ، واحتجّوا بأنّ كلّ جسم إذا تحرّك فإنما يتحرّك

4 الضوء : ثم انصرف إلى الضوء فقال الضوء *K. al-Nafs*, MS A : ووصف الضوء فقال الضوء ed. Arn | ما ... يتمّ : فإنه يتمّ *K. al-Nafs*, MS A | لأنّه : الا انه *K. al-Nafs*, MS A 6 وقال ... الضياء : وقال الذين يزعمون أن الضوء *K. al-Nafs* | واستدلّوا ... بانعكاس : الدليل على ذلك انعكاس *K. al-Nafs*, MS A : الدليل على أن الضوء جسم انعكاس ed. Arn 7 فإنّه : لانه *K. al-Nafs*, MS A | الجسم : الأجسام *K. al-Nafs*, MS A | after أنّ add. الشيء *K. al-Nafs*, ed. Arn 8 من om. *K. al-Nafs*, MS A | after الأجسام add. ولا تمتنع الأجسام من نفاذه فيها *K. al-Nafs*, ed. Arn | after وقالوا add. أيضاً *K. al-Nafs*, ed. Arn 8–9 لم ... جسماً : كان الضوء ليس بجسم *K. al-Nafs*, MS A : لم يكن الضوء جسماً ed. Arn 9 يسخن[1] ... الشمس : يكن يسخن شعاع *K. al-Nafs*, MS A : يكن شعاع الشمس يسخن ed. Arn | فإنّه : لانه *K. al-Nafs*, MS A | فإنّه[2] ... الهواء om. *K. al-Nafs*, ed. Arn 10 به om. *K. al-Nafs*, MS A | والاحتكاك add. ⟨لا⟩ *K. al-Nafs*, ed. Arn | يكون om. *K. al-Nafs*, MS A : add. ⟨إلا⟩ ed. Arn | بالأجسام : للأجسام فالضوء إذاً جسم *K. al-Nafs* 11 في[1] : من *K. al-Nafs* | من : في *K. al-Nafs* | لم ينبثّ : لا يثبت *K. al-Nafs*, MS A 12 وهذا هو : والمقابلة *K. al-Nafs*, ed. Arn | وهذا ... صفة : وهذه هي صفة خاصّة *K. al-Nafs*, MS A | صفة : وصف *K. al-Nafs*, ed. Arn 14 وقال آخرون : وقال قد اختلف الأولون في الضياء فقال بعضهم بجسم فأمّا الذين قالوا إنه *K. al-Nafs*, MS A : وقال قد اختلف الأولون في الضياء فقال بعضهم إنه جسم وقال بعضهم إنه ليس بجسم فأمّا الذين قالوا

2 المسافة : لمسافة M | فيها : فيه M 3 after الهواء break (∴) ind. and blank space of ten letters M 5 after الضياء blank space of three letters M 7 لنفذ : ليغَدَ M 7–8 ولم ... الأجسام add. in marg. M 13 after جسم break (∴) ind. and blank space of half a line M 14 آخرون : اخرونَ M

and continues for a long distance, so that the nose smells the odour of some blossom from many parasangs. If [the air] were changed, the odour 100a
would not continue for this distance, since it has no power to transform this large amount of air.

(178) Light[258] is some activity by which the transparent body receptive to colour is perfected, as it only colours by the mediation of light.

(179) Some say:[259] light is a body, and they infer that from the reflection of the rays of the sun. For if it were not a body it would permeate the body and not be reflected. For what is not a body is not hindered from permeating bodies. Therefore light is a body. They say: if it were not a body, the rays of the sun would not warm the air. For they only warm the air by rubbing against it and rubbing occurs in bodies. They say: If light were not a body, how about the rays of the sun that enter through a window of a house: they are not spread through the entire house, but are opposite the window. This belongs to the specific attribute[s] of the bodies, therefore light is a body.

(180) Others say:[260] [light] is not a body, and they advance as an argument

[258] Cf. source in *DA*-paraphrase, ed. Arnzen, 259.2–3; and MS A, fol.76a4–7.

[259] Cf. source in *DA*-paraphrase, ed. Arnzen, 261.11–263.8; and MS A, fol.77a1–13.

[260] Cf. source in *DA*-paraphrase, ed. Arnzen, 259.4–14, 261.5–10, 263.9–11; and MS A, fols.76a7–b6, 76b10–77a1, 77a13–b1.

100b في زمان ، والضياء يتحرّك بلا زمان . وذلك أنّه مع طلوع الشمس يضيء الأفق كلّه ، ومع دخول المصباح البيت المظلم يضيء كلّ البيت بلا زمان ، فليس الضياء إذن بجسم . وقالوا كلّ جسم إمّا أن يكون مبسوطاً أو مركّباً ، فإن كان مبسوطاً أو مركّباً فلا يعدو أن يتحرّك حركة مستوية ، مثل النار والهواء وسائر الأجسام الأولى ، وإمّا أن يتحرّك حركة مستديرة ، مثل السماء ، ولسنا نرى الضياء يتحرّك حركة مستوية ولا حركة مستديرة ، بل يتحرّك يمنة ويسرة وفوق ، وأسفل فليس الضياء إذن بجسم .

وقالوا لو كان الضوء جسماً لكان إذا اختلط بالهواء كثّف الهواء وأظلم ، ومثال ذلك أنّا لو أخذنا صفيحة من جسم نيّر ، ثم ضممنا إليها أخرى لكثفت الصفيحة وغلظت وصارت مظلمة ، ولسنا نرى الهواء يكثف إذا كان 101a الضوء فيه ، بل يستنير ويلطف ، فليس الضوء إذن بجسم .

وقد وافقهم ارسطالس على هذا القول وذكر أنّه لو كان الضوء جسماً وهو

ed. Arn فاحتجوا فقالوا : K. al-Nafs, MS A فاحتجوا أن : واحتجّوا بأنّ | ed. Arn إنه
فإنما . . . إذا om. K. al-Nafs, MS A
1 after زمان[1] add. ما K. al-Nafs | بلا يتحرّك : بل لي نتج K. al-Nafs, MS A 2 ومع : وموضع
K. al-Nafs, MS A | after المصباح add. في K. al-Nafs, ed. Arn | المظلم : الذي K. al-Nafs, MS A
: om. ed. Arn | كلّ البيت : البيت كلّه K. al-Nafs 3 إذن om. K. al-Nafs, MS A | after
وقالوا add. إن K. al-Nafs | إمّا أن : أنما K. al-Nafs, MS A | أو مركّباً : وإمّا أن يكون مركباً
K. al-Nafs, MS A 4 فإن . . . مركّباً : فإنه ايضاً K. al-Nafs, MS A : ⟨والجسم المبسوط والمركّب⟩ ed.
Arn | فلا : لا K. al-Nafs | أن : إما أن ⟨إذا تحرك⟩ K. al-Nafs, ed. Arn 5 الأولى : الأول
K. al-Nafs 6 ولسنا : ونحن K. al-Nafs, ed. Arn | after يتحرّك[1] add. لا K. al-Nafs, ed. Arn
مستوية : مستديرة K. al-Nafs, MS A | حركة مستديرة : مستوية K. al-Nafs, MS A | بل : وذلك
أنه K. al-Nafs 6–7 وأسفل . . . يتحرّك[2] : يتحرّك بغير زمان معاً وخلف وأمام وأسفل وفوق ويسرة يمنة نفع
: بجسم 7 ed. Arn يتحرّك فوق وأسفل ويمنة ويسرة وأمام وخلف بلا زمان : K. al-Nafs, MS A
جسماً K. al-Nafs, MS A 8 وقالوا : فقالوا K. al-Nafs, ed. Arn | لكان : كان K. al-Nafs, MS A
الهواء om. K. al-Nafs, MS A | وأظلم : واظلام K. al-Nafs, MS A 9 ومثال : مثل K. al-Nafs
نيّر : ينير K. al-Nafs, MS A | ثم . . . أخرى : ⟨مثلها⟩ فألقينا عليها صفيحة أخرى K. al-Nafs, ed.
Arn | after إليها add. صفيحة K. al-Nafs, MS A 10 لكثفت : كثقت K. al-Nafs, MS A
after الصفيحة add. الأولى K. al-Nafs, ed. Arn | وغلظت om. K. al-Nafs, MS A | الهواء :
الضوء K. al-Nafs, MS A | كان : صار K. al-Nafs 11 يستنير ويلطف : ويتبيّن and corrected
above as وينير K. al-Nafs, MS A : يلطف ويستنير ed. Arn 12–282.1 وقد . . . إلينا : فردّ على
وقد وافق : K. al-Nafs, MS A هولاء ارسطاطاليس بأن قال إن كان جسماً وإنما يصل إليه
الفيلسوف الذين قالوا إن الضوء ليس بجسم واحتج له ولهم فقال إن كان الضوء جسماً وإنما يصل

3 after بجسم blank space of ten letters M 4 يعدو : يَعْدوا M 7 after بجسم blank space of ten letters M 11 after بجسم blank space of ten letters M

that every body while moving only moves in time, whereas light moves without time. For at sunrise it lights the entire horizon and when one 100b
brings a lamp into a dark house it lights the entire house without time. Therefore light is not a body. They say:[261] every body is either simple or composite. If it is simple or composite, it will inevitably move either in a linear motion as fire, air and the other first bodies, or in a circular motion as heaven. We do not see light move either in a linear or in a circular motion, but right and left and up and down. Therefore light is not a body. They say: if light were a body, then when it mingled with air, it would thicken and darken the air. An example of that is that if we take a plane of a luminous body, then attach another one closely to it, the plane thickens, coarsens and becomes dark. Yet we do not see air thickening when light comes to be in it; rather it is lit and fine. Therefore light is not a body. 101a
Already Aristotle has agreed with them with regard to this statement and mentioned that if light were a body and came to us by the mediation of air,

[261] See Cottrell's edition, p.438, of the following part of passage (180) and passages (181–2).

يصل إلينا بتوسّط الهواء ، لوجب أن يكون جسم ينفذ في جسم ، وهذا محال .

(181) وقال الشيخ اليونانيّ الضوء كيفيّ . والكيفيّات لا تقوم بأنفسها لكن في غيرها ، لأنّ الكيفيّات محمولة . فإن كان الضوء كيفيّاً والكيفيّة محمولة ، اضطرّ الضوء إلى أن يطلب له حاملاً جرميّاً يكون فيه . فأمّا فعل الضوء فإنّما يكون من شيء آخر لا من حامل الضوء ، وفعل الشمس ، أعني الضوء إنّما هو حياة للجرم الصافي المستشفّ وهو ينبوع الضوء وبدؤه . وأمّا الضوء الكائن في ظاهر أفق الجرم الصافي فإنما هو صنم للضوء الباطن ، وهو فعل ثان غير مفارق للفعل الأوّل .

(182) قال أبو زيد اللون هو كيفيّة السطح الأعلى من كلّ جسم محسوس
101b بالبصر ، والملوّن هو الجسم الذي هذه الكيفيّة محسوسة فيه . وأكثر الألوان وأشدّها ظهوراً إنّما هو في الأجسام المركّبة لأنّ سبل الألوان قريبة من سبل الأراييح والطعوم ، وهذه صفات يوجد أكثرها في الأجسام المركّبة ، فأما العناصر الأربعة الأولى فهي فيها قليلة .

وقد اختلف في الملوّن منها ، اتّفق رأي القدماء من أصحاب الطبائع أن الملوّن في الحقيقة منها الأرض ، وذلك لحصانتها وتماسك أجزائها وأنّها في طباعها باردة يابسة ، وهاتان الطبيعتان توجبان الانحصار . وإذا انحصر الجسم غاية الانحصار ولم يتموّج ولم يتبدّل أجزاؤه عن ملاقاة ، أردك البصر كيفيّته . فأمّا الماء فقد أوجبوا أن له لوناً ضعيفاً شديد الاستحالة ، وذلك أنّه وإن كان محسوساً بالبصر فإنّه لملاقاة الأرض وما تحيط به من الأماكن ، وعلى قدر
102a كميّته وهيئة انحصاره تختلف الألوان المحسوسة فيه ، فإنّه إذا كان منبسطاً

إلينا ed. Arn
1 لوجب : فقد وجب *K. al-Nafs*, MS A : وجب من ذلك ed. Arn | after جسم[1] add. ما *K. al-Nafs*, MS A | وهذا : وهو *K. al-Nafs*, ed. Arn 2 محال : ممتنع *K. al-Nafs*, MS A

2 after محال break (∴) ind. and blank space of half a line M 3–5 كيفيّ[1] ... الضوء add. in marg. M 5 after فيه blank space of ten letters M 7 وبدؤه Ros, Bad, Cot : ويدوه M 8 ثان Ros, Bad : تاتى M : ثاني Cot 9 after الأوّل break (∴) ind. and the following line entirely blank M 12 سبل[1] : سيل M : سبيل Cot | سبل[2] : سيل M : سبيل Cot 13 الأراييح : الارايح M, Cot 14 after قليلة blank space of five letters M 15 الملوّن[1] : المكون M 16 لحصانتها : لحصَا فيهَا M 18 كيفيّته : ليقيته M 21 وهيئة : وهيه M

it would have to be a body permeating a body, but this is unthinkable.
(181) The Greek Sage says:[262] Light is qualitative. Qualities do not subsist by themselves but in something else, because qualities are borne [by a substratum]. So if light is qualitative and quality is borne, light is obliged to seek a bodily bearer for itself to be in. As to light's activity, it is [caused] by something else, not by the bearer of light. The activity of the sun, i.e. the light, is life for the clear, transparent body, and it is the source of the light and its beginning. As to the light which exists on the outer surface of the clear body, it is only an image of the inner light, and it is a secondary activity inseparable from the first activity.
(182) Abū Zayd says:[263] colour is the quality of the uppermost surface of every body perceptible through vision. The coloured object is the body in which this quality is perceptible. The colours which appear most, as well as 101b
most strongly, are in composite bodies, as the ways of colours are similar to the ways of odours and tastes. Most of these attributes are found in composite bodies, whereas there are little of them in the first four elements.
There has certainly been disagreement on [which] of [the four elements] is the coloured one, yet the opinion of the ancient natural scientists has been consistent that the truly coloured among them is earth. That is due to [earth's] strength and the cohesion of its parts and [the fact] that [earth] is in its nature cold and dry. These two natures necessitate condensation. When a body is utterly condensed, does not undulate and its parts do not change upon contact, vision perceives its quality. As for water they have affirmed that it has a weak and extremely changeable colour. For even if it is perceived through vision, it is due to contact with earth and the places encompassing it. The colours perceptible in it vary according to the extent of its quantity and the form of its density. So when it expands over an 102a

[262] Belonging to *GS*; see ed. Rosenthal, pp.378–9; ed. Badawī, 192.6–13; and ed. Cottrell, pp.438–9.
[263] See Cottrell's edition, p.439, of the first paragraph of this passage.

على سطح الأرض مصحراً لوقوع الضياء عليه ، نرى لونه بهيّة الصفاء والبياض ، وإذا كان محصوراً في موضع عميق وكان كثير الكميّة ، نرى بهيئة الخضرة والسواد والألوان المختلفة ، وذلك لرقّته وتموّج أجزائه . وأمّا الهواء فقد أوجبوا أنّه لا لون له أصلاً ، وإن ظنّ ظانّ أن ما يرى راكداً فيه في أوقات من الضباب والبخار وما أشبههما إنما هو لون الهواء ، فقد أخطأ ، لأنّ ذلك المحسوس إنما هو أجزاء أرضيّة انحلّت وصعدت بها الحرارة تربعت تحت البصر . وأمّا النار فهي غير ذات لون وذلك لحرارتها وانبساط أجزائها ، فأمّا المحسوس الذي يسمّى ناريّاً فإنّما هو لهب ونار عاملة في جسم أرضيّ من بخار أو دخان قد التهب واستضاء لحركة النار فيه ، ومتّى فارقت تلك المادّة ، تمّ ملحقها 102b البصر وإن كان إحراقها قائماً .

وكذلك إذا صدمها الهواء والريح ففرق أجزآها وازالها عن سنن المادّة التي هي متعلّقة بها ، طفئت ، وظفوؤها إنّما هو تفرّق أجزائها ، كما تنفق أجزاء الماء المحصور في موضع إذا صدمه هواء قاهر له ، ففرّق أجزآه وأبطل صورته .

(183) وقال الشيخ القوّة النامية منبثّة في البدن كلّه ، فلذلك صارت أعضاء البدن كلّها تلتذّ وتألم ، والشهوة مثل ذلك لأنّها من حيّز القوّة النامية وليس عضو من أعضاء البدن إلا والشهوة راتبة فيه ، والدليل على ذلك الأوجاع واللذّات ، فإنّ كلّ عضو قد يألم ويلتذّ ، والألم واللذّة من باب الشهوة ، وذلك إذا نال العضو شهوته التذّ ، وإذا لم ينلها وجد ألماً لعدم تلك الشهوة. 103a فإن قال قائل إنّ الأوّلين جعلوا موضع الشهوة الكبد ، قلنا لهم إنّما جعلوا ذلك لا من أجل أنّ الشهوة هناك فقط ، لكنّه لما كانت القوّة النامية تبدأ بفعلها من ذلك الموضع ، قالوا إنّه موضع القوّة النامية والقوّة الشهوانيّة ، فالنفس إنّما تبتدئ في فعل النماء والشهوة من الكبد ، ثم تبثّه في جميع الأعضاء للبدن . وقد قلنا في قوى النفس إنّها ظلّ لها وظلّ النفس النامية

2 بهيئة : بهيه M 3 after أجزائه break (∴) ind. and blank space of six letters M 6 أجزاء : تلَحقُه M 9 ملحقها : ثم M | تمّ : فارق M | فارقت : تُوبِعَت M | تربعت : اجزاً M 10 إحراقها : احراقه M | after قائماً break (∴) ind. and blank space of six letters M 11 أجزآها : اجزاها M 12 وظفوؤها : وظفُوها with a tilde over the *hā*ʾ M | أجزاء : اجزا M 13 after صورته the following line entirely blank M 14 فلذلك M, Bad : ولذلك Ros 15 after وتألم blank space of five letters M 16 راتبة M, Bad : رابية Ros 18 ينلها Bad : ينله Ros : نيله M

earthy surface and is dust-coloured because light grasps it, we see its colour radiant with transparency and whiteness; whereas when it is condensed in a deep place and is abundant in quantity, we see it radiant with greenness, blackness and different colours, which is due to its thinness and the undulation of its parts. As for air, they have affirmed that it has no colour at all, even if someone assumes that what he sees suspended in it at times of mist, vapour and their like is the colour of air; he is certainly mistaken. For that [sort of] perceptible is only earthy parts which have disintegrated and which heat has made ascend, [so that] they have become subject to vision. As for fire, it does not possess colour which is due to its heat and the expansion of its parts. The perceptible which is called fiery is rather flame and fire active in an earthy body of vapour or smoke, which has already been lit and made to emit light due to the motion of the fire in it. When [the fire] departs from that material, its 102b
connection to vision comes to an end, even if it continues to burn.

Similarly, when air and wind break [fire] up, so that its parts are dispersed, and they cause [fire] to leave these scattered [parts] of the material to which it has been attached, it is extinguished. Its extinction is the dispersal of its parts. Likewise the parts of water condensed in a place dwindle away when air overpowering it breaks it up, then it disperses its parts and destroys its form.

(183) The Sage says:[264] the augmentative faculty is spread in the whole body and, therefore, all body parts come to feel pleasure and pain. Appetite is like that as it belongs to the realm of the augmentative faculty and none of the body parts is without appetite established in it. The proof for this is pains and pleasures. For each body part may feel pain and pleasure, and pain and pleasure are associated with appetite. For when the body 103a
part attains its appetite, it feels pleasure, and when it does not attain it, it experiences pain due to the lack of that appetite.

So if some said that the Ancients made the liver the seat of the appetite, we would say to them: they did not do that on account of appetite being only there; but since the augmentative faculty starts its activity from that site, they said that it was the site of the augmentative and the appetitive faculties. For the soul initiates the activity of growth and appetite from the liver and then spreads it to all body parts. We have already said about the faculties of the soul that they are a shadow of [the soul]. The shadow of the

[264] Belonging to *GS*; see ed. Badawī, 192.14–193.18; and partial edn Rosenthal, pp.378–81.

والشهوانيّة إنّما يبدأ من هذا العضو ، أعني الكبد ، ثم ينبسط في جميع البدن .
والغضب من حيّز النفس النامية ، وذلك أنّ النفس النامية هي التي تصيّر
البدن حسّاساً للأشياء المؤلمة وتجعل البدن ذا مرّة ودم ، وحيث ما كانت
الشهوة فهناك الغضب ، وحيث ما كان الغضب فهناك الشهوة ، والدليل
103b على أنّ ذلك كذلك أهل القوّة والقمع للشهوة المستعلين عليها ، فإنّهم أيضاً
أهل القوّة على الغضب ، فأمّا الذين استعبدتهم الشهوة وقهرتهم وغلبت
عليهم ، فلذلك يغلب عليهم الغضب أيضاً ، فيكونون في سرعة الغضب
وشدّته عليهم كأنّهم السباع . وبدء الغضب الجسمانيّ إنّما يكون من قبل المرّة
والدم بسرعة حركتهما ، وذلك أنّه إذا ألم البدن الكثير الدم والمرّة ألماً هاج
لذلك الدم والمرّة ، فإذا هاجا حسّ الوهم بهيجانهما فأدّى ذلك إلى النفس
وأعلمها ما نال البدن من ذلك الأذى ، وإنّما يؤدّي الوهم إلى النفس الألم
الداخل على البدن ، لتنهض النفس فتدفع عن البدن ذلك الشيء المؤلم .
وأوّل الغضب بدؤه من القوّة البهيميّة المرّية ومنتهاه إلى النفس الناطقة .
وأصحاب الدم الحارّ المغليّ والمرّة اللهيبة سريع غضبهم ، والذين قليلة مررهم
104a فغائرة دماؤهم ، والذين طباعهم ...
107a 〈...〉

(184) وقال جالينوس العشق من فعل النفس وهي كامنة في الدماغ والقلب والكبد . وفي الدماغ ثلاثة مساكن ، التخيّل وهو في مقدّم الرأس ، والفكر وهو في وسطه ، والذكر وهو في مؤخره . وليس يخلو عاشق إذا فارقه من يحبّه

17 وقال : قال *K. al-Zahra, Dhamm* | وقال جالينوس : من كلامه في العشق ... آداب جالينوس *M. al-Ḥikam, ʿUyūn* وقال : *Ādāb* 18 ثلاثة مساكن : ثلاث قوى *M. al-Ḥikam, ʿUyūn* | مقدّم الرأس : om. *Ādāb* | وهو : *Dhamm* مسكن للتخيّل : التخييل *Ādāb, K. al-Zahra* التخيّل : مقدّمه *Ādāb* : مقدم الدماغ *M. al-Ḥikam* | والفكر : ومسكن للفكر *Dhamm* 19 وهو[1] om. *Ādāb* | والذكر : ومسكن للذكر *Dhamm* | وهو[2] om. *Ādāb* | وليس ... عاشق : وليس يكمل لأحد اسم العاشق حتّى يكون *Ādāb* : وليس يكمل لأحد اسم عاشق إلا حتّى *K. al-Zahra* : وليس يكمل لأحد معنى عاشق حتّى يكون *M. al-Ḥikam* : وليس يكمل لأحد اسم عاشق حتّى يكون *ʿUyūn* : ولا يسمّى عاشقاً إلا من فارق من يعشقه *Dhamm* | فارقه ... يحبّه : لم يخل *Ādāb,*

2 التي : الى Bad: M 6 استعبدتهم : استعبد بهم Bad : M 7–8 الغضب[1] ... عليهم add. in marg. M 8 وبدء : وبَدْو Bad : M 11 وأعلمها : واعلمهما M, Bad 12 after المؤلم break (∴) ind. and blank space of half a line M 14 مررهم : مرّتهم Bad : M 18 ثلاثة : ثلثه M 19 يخلو : يخلوا M

augmentative and the appetitive soul starts from this organ, i.e. the liver,
and then it expands throughout the body.
Anger belongs to the realm of the augmentative soul, for it is the augment-
ative soul which makes the body sensitive to painful things and lets the
body have bile and blood. Wherever appetite is, there is anger, and
wherever anger exists, there is appetite. The proof that it is thus are the 103b
people who have power and control over their appetite, and have
mastered it, because they are also people with power over anger. As for
those whom appetite enslaves, overpowers and dominates, anger therefore
dominates them as well. With regard to the rapidity of anger [rising in
them] and its force over them, they are like brutes. The beginning of anger
in the body is bile and blood owing to the rapidity of their movement. For
when the body which is rich in blood and bile suffers pain, blood and bile
are stirred up due to that. Thereupon the two stir up the sense of imagina-
tion by their agitation, so it conveys that [agitation] to the soul and lets it
know of the harm the body has suffered. The imagination only conveys to
the soul the pain which has entered the body, so that the soul rises to repel
that painful thing from the body.
The first [flaring up] of anger begins in the brute and bilious faculty and
ends in the rational soul. Those who have hot boiling blood and flaming
bile are quick to become angry whereas those who have little bile, their
blood has receded deep [into their bodies], and those whose nature[265] ... 104a
⟨...⟩ 107a
(184) Galen says:[266] passionate love belongs to the activity of the soul
which has its seat in the brain, the heart and the liver. In the brain there
are three dwelling places: imagination in the front of the head, thought in
its middle and memory in its rear part. When the beloved has departed
from the lover, the lover is not free from imagining and remembering him,

265 This is, in the current state of the MS, the beginning of the second gap left deliberately in the Marsh MS. The two words (*alladhīna ṭabāʿihim*) which must have started off fol. 104a can be retrieved from the catchword at the bottom left corner of fol. 103b. The Greek text of the *Enn.* IV 4, 28, 29–32, which is the model for the last incomplete sentence, reads: 'and when those whose blood and bile are boiling are prompt to fly into a rage, and those who are said to be "without bile" and "chilly" are easy-going and slow to anger' (tr. Armstrong).

266 The text of the Marsh MS is resumed in line 7 of fol. 107a. Cf. parallels in al-Anṣārī, *Ādāb*, 123.9–15; Ibn Dāwūd, *K. al-Zahra*, 18.14–19.2; Mubashshir, *M. al-Ḥikam*, 294.3–7; Ibn al-Jawzī, *Dhamm*, 247.4–8, and Ibn Abī Uṣaybiʿa, *ʿUyūn*, I 88.27–32.

من تخيّله وذكره وقلبه وكبده ، فيمتنع من الطعام والشراب باشتغال الكبد ، ومن النوم باشتغال الدماغ بالتخييل والذكر له والفكر فيه ، فيكون جميع 107b مساكن النفس اشتغلت به .

(185) وقال الحياء هو الخوف من إتيان القبائح والحذر من الذم الصادق ، وهو أوّل ما يحمد من الصبيّ ويستدلّ منه على عقله ونجابته . والكرم إنفاق الكثير بسهولة من النفس في الأمور الجليلة القدر الكثيرة النفع كما ينبغي . والنبل سرور النفس بالأفعال الجميلة وابتهاجها بلزوم السيرة الحميدة . والورع لزوم الأحوال الحسنة التي فيها كمال النفس . والحرّيّة اكتساب المال في وجهه وإخراجه في وجهه . والخجل جزع النفس من أن يظهر منها قبيح .

(186) والضحك اعتدال دم القلب وانبساط النفس حتّى يظهر سرورها ، 108a والبكاء حزق من النفس يجمّد الرطوبات التي في السبل اللطاف ، فتحللها الحرارة الغريزيّة بأنواع مختلفة لتريح الدماغ والقلب .

(187) وقال افلاطن الضحك غليان الدم الغريزيّ ، إذا رأى الإنسان أو سمع

K. al-Zahra, M. al-Ḥikam, ʿUyūn : لم يخل فارق معشوقه Dhamm
1 تخيّله : تخييله K. al-Zahra, Ādāb | وذكره : وفكره وذكره Ādāb, K. al-Zahra, M. al-Ḥikam, ʿUyūn | وكبده ... وذكره om. Dhamm 1–2 وقلبه ... فيه om. M. al-Ḥikam 1 من[2] : عن Dhamm 2 بالتخييل : والتخييل K. al-Zahra : بالتخيّل Dhamm | والذكر ... فيه : والفكر فيه Dhamm 3 after النفس add. قد Ādāb, K. al-Zahra, M. al-Ḥikam, ʿUyūn, Dhamm | والذكر له : والفكر والذكر Ādāb | after به add. فإذا لم يكن عاشقاً فمتّى لم تشتغل به وقت الفراق لقيه خلت هذه المساكن Ādāb, K. al-Zahra, ʿUyūn 4 وقال[1] ... من : أمّا الحياء فهو انحصار النفس خوف Tahdhīb | after الذم add. والسبّ Tahdhīb 4–5 وهو ... والكرم om. Tahdhīb 5 والكرم إنفاق : أمّا الكرم فهو إنفاق المال Tahdhīb 6 after ينبغي add. وباقي الشروط التي ذكرناها في السخاء Tahdhīb | والنبل : وأمّا النبل فهو Tahdhīb 7 الجميلة : العظام Tahdhīb | السيرة الحميدة : هذه السيرة Tahdhīb | والورع : وأمّا الورع فهو Tahdhīb 8 الأحوال الحسنة : الأعمال الجميلة Tahdhīb | والحرّيّة اكتساب : وأمّا الحرّيّة فهي فضيلة للنفس بها يكتسب Tahdhīb | في : من Tahdhīb 9 وإخراجه : ويعطى ما يجب Tahdhīb | after وجهه add. ويمتنع من اكتساب المال من غير وجهه Tahdhīb 13 وقال ... الضحك : فأمّا الضحك فإنّه Firdaws | رأى ... سمع : ما رأى أو سمع الإنسان Firdaws

3 after به break (∴) ind. M 5 after ونجابته break (∴) ind. and blank space of half a line M 6 after ينبغي break (∴) ind. and blank space of five letters M 7 after الحميدة break (∴) ind. and blank space of ten letters M 8 after النفس blank space of ten letters M 9 after وجهه break (∴) ind. and blank space of five letters M | after قبيح break (∴) ind. and blank space of five letters M 11 حزق : حُرق M | فتحللها : فيحلّلُه M 12 after والقلب break (∴) ind. and blank space of half a line M

and from [having] his heart and liver [occupied with] him. So he abstains from eating and drinking, because the liver is occupied [with the beloved], and from sleep because the brain is occupied by imagining, remembering and thinking of him. All the dwelling places of the soul are thus occupied 107b
with [the beloved].

(185) He says:[267] shame is the fear of performing shameful actions and caution against justified blame. It is immediately praised in a youth, indicating his intellect and his nobleness. Generosity is light-heartedly spending much on matters of great worth and abundant benefit just as one should. Nobility is the delight of the soul in good actions and its rejoicing in adherence to praiseworthy conduct. Piety is the adherence to good attitudes in which the perfection of the soul [lies]. Liberality is the acquisition and spending of possession in its appropriate way. Embarrassment is the soul's worry about revealing something disgraceful.

(186) Laughter is the balancing of the blood in the heart and its becoming 108a
dilated[268] in the soul so that its delight manifests itself, whereas weeping is a compression [caused] by the soul which congeals the moistures in fine pathways so that the innate heat [has to] dissolve them in different ways to relieve the brain and the heart.

(187) Plato says:[269] laughter is the natural boiling of blood whenever man

[267] Cf. Miskawayh, *Tahdhīb*, 20.4–5, 22.11–13, 22.16–17, 21.7, 20.12–14, which is ordered differently, and the *PR* passage has some additional material.

[268] *Inbisāṭ* can mean both 'happiness, joy' and 'expansion', and may have been chosen here in order to contrast the latter meaning to the term 'compression' (*ḥazq* – if this is the correct reading) in what follows.

[269] Cf. source in ʿAlī ibn Rabban, *Firdaws*, ed. al-Ṣiddīqī, 88.9–11, 88.13–15, 88.19–21.

شيئاً يلهيه فيهزّه ، فإذا لم يستعمل في ذلك الفكرة غلبه الضحك . فأمّا البكاء فإنّ الإنسان إذا حزن ، انعصر الدماغ ببرد الحزن حتّى تسيل رطوباته من العين . وحدّ الفرح غليان الدم وانتشاره إلى ظاهر البدن لملاقاة الشيء الذي فرحت النفس به ، ولذلك يسخن البدن عند الفرح وتدّر العروق .

(188) الغريزة طبيعة حالّة في القلب أعدّت فيه لينال منه الحياة .

108b (189) قال والسبب في أنّ الإنسان لا يجدّر إلا دفعة واحدة أنّ الجدري ربّما يكون من فضل الغذاء الذي يغتذي به الطفل وهو في بطن أمّه من دم الحيضة ، فإذا اتّفق له هواء مثل الهواء الذي اغتذى به الطفل ففصّته الطبيعة إلى خارج دفعة واحدة من غير أن يبقى منه شيء .

(190) والسبب في ارتعاش الفزع واصفرار لونه أنّ الحرارة الغريزيّة ، التي هي منتشرة في ظاهر جلده ومفيدة طلاقة الوجه وقوّة الحركة في الأعضاء ، تغور بالخوف إلى غمق الجسد وباطنه ، فإذا خلت الأعضاء من الحرارة التي تمسكها والدم الذي يملأها ، وقعت عليها الرعشة والرعدة وغلب على اللون من الصفرة ما يعتري مثله المنزوف ، فالحرارة الغريزيّة في تلك الحال شبيهة بالنار التي مالت إلى أن تطفأ ، فيحتاج إلى تذكيتها بتحريكها والنفخ فيها لتعاود الأبعاد والاشتغال .

109a (191) وقال الفرق ينقسم إلى ستّة أقسام ، وهي النكول والحياء والخجل والذعر والحيرة والمجاهدة . فأمّا النكول فهو فرق من عمل يحتاج إلى عمله ، وأمّا الحيرة فهو فرق من تخيّل أمر عظيم ، والذعر فرق من تخيّل أمر غير مألوف ، والمجاهدة فرق من خيبة أمل ، والحياء فرق من ذمّ يتوقّع ، والخجل فرق يحدث

1 فيهزّه : فيستفزه ويهزه *Firdaws* | غلبه : غلب عليه *Firdaws* 2 after حزن add. من أمر ما *Firdaws* | ببرد : لرد *Firdaws* | رطوباته : رطوبته *Firdaws* 3 وحدّ الفرح : إن حدّ الفرح أنّه *Firdaws* | لملاقاة : لملاقات *Firdaws* 4 وتدّر : وتدور *Firdaws* 17 وقال om. *Fī Ṭabīʿa* | النكول : النكل *Fī Ṭabīʿa* 18 النكول : النكل *Fī Ṭabīʿa* | إلى عمله : أن بعمله *Fī Ṭabīʿa* 19 والذعر : وأمّا الذعر فهو *Fī Ṭabīʿa* 20 والمجاهدة : وأمّا المجاهدة فهي *Fī Ṭabīʿa* | من[1]... أمل : أن يخيب الرجا *Fī Ṭabīʿa* | والحياء : وأمّا الحياء فهو *Fī Ṭabīʿa* | after يتوقّع add.

1 after الضحك blank space of three letters M 2 ببرد : ليرد M 3 after العين blank space of half a line M 4 after العروق break (∴) ind. and blank space of five letters M 5 after الحياة break (∴) ind. M 7 يغتذي به : يَغتذِيه M 16 after والاشتغال break (∴) ind. and blank space of five letters M 18 after عمله blank space of ten letters M 19 after عظيم blank space of three letters M

sees or hears something which amuses him and thus stirs him. When he does not employ thought in such [a situation], laughter overpowers him. As for weeping, whenever man is sad the brain is squeezed by cold sadness so that its moistures flow from the eye. He defines joy as the boiling of blood and its spreading out towards the surface of the body because it has encountered something at which the soul rejoices. Therefore at times of joy the body warms up and the veins pulsate uninterruptedly.

(188) The innate disposition is a nature residing in the heart which has been prepared in it so that life can be obtained from it.

(189) He says: the reason why man is attacked by smallpox only at once is 108b
that smallpox may be caused by a surplus of nutriment the child is fed while being in its mother's womb through the menstrual blood. When a [draught of] air like the air the child has been fed, then hits it, nature displaces [this surplus] outside at once without any of it remaining.

(190) The reason for the tremor of someone frightened and the paleness of his complexion is that the innate heat spreading on the surface of his skin, which causes the cheerfulness of the face and the power of motion in the body parts, withdraws to the innermost depths of the body because of fear. So when the body parts are devoid of heat taking hold of them and devoid of blood filling them up, tremor and shivering befall them and the paleness which afflicts, for example, the one exhausted through loss of blood, overpowers the complexion. The innate heat in that condition resembles fire which is close to dying down, and thus needs to be rekindled by poking and blowing onto it to resume its proper dimension and get going again.

(191) He says:[270] fear is divided into six kinds. They are dread, shame, 109a
embarrassment, terror, confusion and distress. Dread is the fear of doing what needs to be done. Confusion is the fear caused by imagining something overwhelming. Terror is the fear caused by imagining something unusual.[271] Distress is the fear of failing expectations. Shame is the fear of anticipated blame. Embarrassment is the fear caused by an evil

[270] Cf. source in Nemesius, *Fī Ṭabīʿa*, 153.2–4, 153.7–9, 153.5–7 and 154.4–7.

[271] In the Greek the explanations for confusion and terror are inverted.

عن شرّ قد فعل . والفرق يحدث عن برودة ، وذلك لغور الحرارة كلّها إلى القلب الذي هو أصل وينبوع لها بمنزلة الجند الذين إذا فرقوا فزعوا إلى قوادهم .

(192) وقال افلاطن الخوف انقباض الحرارة الغريزيّة إلى داخل البدن هرباً من المكروه ، ولذلك يخضر وجه من يخاف خوفاً شديداً . وقد يكون نوعاً من الموت سببه انتشار الحرارة من القلب بالإفراط عند الفرح ، وموتاً سببه احتقان الحرارة في القلب بالإفراط عند الخوف .

109b (193) سئل بلينوس عن العشق فقيل له ما بال العاشق إذا نظر إلى معشوقه اصفرّ لونه واحمرّ لون المعشوق ؟ قال لأنّ مع الزمع الفزع ومع الفزع الجزع ، والدم لا يأوي حيث يستحكم الفزع بل يفرّ منه ، فإذا فرّ الدم استحال اللون واصفرّ . وربّما نظر العاشق إلى المعشوق فجأةً ، فيضطرب قلبه ويشتعل بالحرارة ثم يخمد ، فإذا خمد القلب برد التأمور وإذا برد التأمور جمد الدم وإذا جمد الدم استحال اللون إلى الخضرة ثم يصفرّ . وأمّا ظهور الحمرة في وجه المعشوق فمن الحياء والخجل . والخجل حركة من تأمور القلب ترتفع منها حرارة صاعدة إلى الرأس ، فتحيل الدم وتلطّفه فيظهر في أرقّ موضع من

Fī Ṭabīʿa وأمّا الخجل فهو : والخجل | Fī Ṭabīʿa وهذا الفعال محمود جدّاً
1 after فعل add. وصاحب هذا الانفعال ليس بمؤيّس من صلاحه Fī Ṭabīʿa 3 وقال . . . الخوف :
وحدّ الخوف أنّه مرّةً ويبيض أخرى ويبرد البدن ويصغر النبض Firdaws 4 after شديداً add.
الغريزيّة Firdaws 5 after الحرارة add. فإن اشتدّ الفرح وانتشار : وقد . . . انتشار 4–5 Firdaws
انتشاراً : بالإفراط . . . الفرح | Firdaws في البدن add. القلب | Firdaws عن : من | Firdaws
فإن دام الخوف وتراجعت : وموتاً . . . الخوف 5–6 Firdaws مفرطاً برد القلب ومات الإنسان
: سئل . . . له 7 Firdaws الحرارة إلى القلب تراجعاً مفرطاً اختنقت الحرارة في القلب ومات الإنسان
: واحمرّ . . . المعشوق 8 Malady of Love المعشوق : معشوقه | Malady of Love وقيل لبلينوس
إنّ الفرح يكون مع الدم : [1]لأنّ . . . الفزع | Malady of Love وليس منهما أحد إلا وفؤاده يخفق
وإذا : فإذا 9 Malady of Love والحزن : الجزع | Malady of Love مع : ومع | Malady of Love
فإذا : [1]وإذا | Malady of Love جمد ما اشتعل في القلب بالحرارة : خمد القلب 11 Malady of Love
Malady يشقر ويصفرّ : يصفرّ | Malady of Love والسواد .add الخضرة 12 after Malady of Love
عرض عن : [2]والخجل 13 Malady of Love الدم : الحمرة | Malady of Love فأمّا : وأمّا | of Love
فتحُلّ : فتحيل 14 Malady of Love منه : منها | Malady of Love .om من | Malady of Love
Malady of Love مكان في : موضع من | Malady of Love

1 after فعل blank space of three letters M | لغوره : لغور M 2 after قوادهم break (∴) ind. and blank space of more than half a line M 4 after شديداً blank space of five letters M 6 after الخوف blank space of six letters M 12 after يصفرّ break (∴) ind. and blank space of more than half a line M 13 مرتفع : ترتفع M

already done. Fear is caused by coldness, because all heat withdraws to the heart, which is its source and well-spring. It resembles an army, which, when afraid, takes refuge to its leaders.
(192) Plato says:[272] fear is the contraction of innate heat towards the inside of the body in flight from something unpleasant. Therefore someone who is in great fear has a greenish complexion. There certainly exists a kind of death caused by the excessive spreading of heat from the heart at times of joy and a kind of death caused by the excessive concentration of heat in the heart at times of fear.
(193) Apollonius was asked about passionate love and it was put to him:[273] 109b
why is it that, whenever the lover gazes at his beloved, the complexion of the lover pales, whereas the complexion of the beloved reddens? He said: it is because with fear there is anxiety and with anxiety uneasiness,[274] and the blood does not lodge where anxiety takes root, but flees from it. So when the blood flees, the complexion changes and pales. Sometimes the lover gazes at the beloved suddenly, then his heart becomes agitated and flares up with heat and thereupon calms down. When the heart calms down, the pericardium cools off; when the pericardium becomes cool, the blood clots; and when the blood clots the colour changes to greenness and then pales. As to the appearance of redness in the face of the beloved, it is caused by shame and embarrassment. Embarrassment is a motion of the pericardium of the heart due to which the heat ascends rising to the head. It transforms the blood and refines it, so that it manifests itself in the most

[272] Cf. source in ʿAlī ibn Rabban, *Firdaws*, ed. al-Ṣiddīqī, 89.2–5, 88.22–89.1, 89.5–6.
[273] Cf. parallel in long version of 'Malady of Love' as reconstructed by Biesterfeldt and Gutas ('Malady of Love', pp.37–9).
[274] This sentence is not attested in the reconstruction of the long version as the textual witnesses used by Biesterfeldt and Gutas are corrupted ('Malady of Love', p.44, n.55).

الوجه .

(194) قال افلاطن الشعر بخارات ترتفع إلى الرأس والجلد من فضول الأغذية ،
110a فما دامت تلك البخارات دسمة غليظة قويّة كان ما ينبت منها شعر أسود ،
فإذا بردت البخارات ورقّت ابيضّ الشعر .

(195) قال بقراط الشعر ينبت للرجال أكثر مما ينبت للنساء لسببين ، أحدهما أنّ الحرارة الغريزيّة في أبدان الرجال أقوى منها في أبدان النساء ، والبخارات الحارّة الدخانيّة التي هي مادّة الشعر يكون في الرجال أكثر . فليست تكتفي الطبيعة أن تصرفها في وجه واحد فتصرفها في وجهين ، أحدهما في شعر الرأس والآخر في الشعر اللحية ، ولذلك قد نجد كثيراً من النساء اللواتي مزاجهنّ حارّ ينبت لهنّ في موضع الذقن شعر ، وكثيراً من الرجال الذين مزاجهم بارد لا ينبت لهم لحىً ، ولذلك صار الخصيان لا ينبت لهم لحية لأنّ مزاجهم بارد ، إذ قد نقص منهم عضو غريزيّ الحرارة وهو الأنثيان . والثاني أنّ النساء
110b لما كنّ مستترات في البيوت وليس لهنّ أن يبرزن ولحياهن مكشوفان ، استغنين عن شعر يغطّي لحييهن ، وكان ذلك بهنّ أزين وأوفق .

(196) قال والعضو إذا كان حاراً رطباً تولّد فيه بخار دخانيّ كثير ، فدفعته الطبيعة إلى خارج وكان منه الشعر في ذلك العضو . ولذلك أكثر ما ينبت من الشعر في العانة لقرب هذا العضو من موضع الأنثيين اللذين مزاجهما حارّ

2 قال ... الشعر : وإنّما ينعقد الشعر ويتكوّن من *Firdaws* | ترتفع : تندفع *Firdaws* 3 منها : منه *Firdaws* | شعر أسود : أسود أعني من الشعر *Firdaws* 4 after بردت add. تلك *Firdaws* 5 قال ... مما : فصارت اللحية تنبت للرجال ولا *Kāmil* 7 يكون om. *Kāmil* 8 فتصرفها : فهي تصرفها *Kāmil* 9 after مزاجهنّ add. مزاج *Kāmil* 11 لحية : اللحى *Kāmil* 12 after إذ add. كان *Kāmil* | غريزيّ : غزير *Kāmil* | وهو : وهي *Kāmil* | والثاني : والسبب الثاني *Kāmil* 13 ولحياهن : لحاهن *Kāmil* | مكشوفان : مكشوفات *Kāmil* 14 لحييهن : اللحيين *Kāmil* 15 قال والعضو : وذلك أنّ العضو *Kāmil* | فدفعته : تدفعه *Kāmil* 16 وكان منه : فيكون معه *Kāmil* | أكثر : تجد كثيراً *Kāmil* | من : هذا *Kāmil*

1 after الوجه break (∴) ind. and blank space of half a line M 3 غليظة : غَليطه M 4 بردت : تردت M | after الشعر blank space of three letters M 7 فليست تكتفي : فليس يكتفي M 9 والآخر : والاخر M 10 الذقن : الذقى M | after شعر break (∴) ind. and blank space of ten letters M | مزاجهم : مزاجهن M 12 غريزيّ : غريز M | after الأنثيان break (∴) ind. M | النساء : النّسآ M 13 after البيوت blank space of six letters M 14 after وأوفق break (∴) ind. M 15 فدفعته : قد فعنه M 17 اللذين : اللّين M

delicate part of the face.

(194) Plato says:[275] hair exists through vapours which rise towards the head and the skin due to a surplus of nutriments. So as long as these 110a vapours are rich, thick and powerful black hair grows from them, whereas when the vapours cool off and become thin, the hair becomes white.

(195) Hippocrates says:[276] men grow more hair than women for two reasons. One of the two is that the innate heat in the bodies of men is stronger than in the bodies of women. The hot and smoky vapours which are the material for hair are more abundant in men. Nature is not content with distributing [the material of hair] in only one guise, but distributes it in two. One of them [consists] in the hair of the head and the other in the hair of the beard. Therefore we may find many women whose mixture is hot growing hair on the chin, whereas many men whose mixture is cold do not grow beards. Therefore eunuchs do not come to grow beards for their mixture is cold, since they lack the innate organ of heat which is the testicles. The second [reason for men growing more hair than women] is that since women are secluded in the houses and must not be seen with 110b their cheeks uncovered, they can do without hair to cover their cheeks. That is more beautiful and appropriate for them.

(196) He says:[277] when a body part is hot and moist, an abundant smoky vapour is generated in it. So nature pushes it outside and because of it hair emerges on that body part. Therefore most hair grows in the pubic region because that body part is close to the position of the testicles, the mixture

[275] Cf. source in ʿAlī ibn Rabban, *Firdaws*, ed. al-Ṣiddīqī, 52.11–15.
[276] Cf. parallel in al-Majūsī, *Kāmil* I, 81.29–82.4.
[277] Cf. parallel *ibid.*, 82.6–10.

رطب وفي البطن والصدر والإبطين لحرارة مزاج القلب والكبد الموضوعين بالقرب من هذه المواضع.

(197) قال جالينوس الجلد في البلاد المعتدلة يدلّ على طبيعة الأعضاء التي هي دونه . فأمّا في البلاد الواغلة في الجنوب فإنّ الحرارة التي في الأبدان قد مالت إلى الجلد ، فصارت طبيعته لاحتراقه من الحرارة الغريزيّة التي تتحرّك منهم إلى خارج صلبة سوداءاً ، وصار البدن قليل الحرارة الغريزيّة لكنّه يكون حاراً بحرارة 111a غريبة مستفادة . والبلاد الباردة مثل الترك والصقالبة فإنّ الحرارة الغريزيّة منهم قد غارت مع الدم إلى الأحشاء ، والجلد من أهلها يكون بارداً رطباً ولذلك يكون ناعماً أبيض .

(198) قال واللحم الصلب الكثير في الأبدان يدلّ على قلّة الحسّ ، والأزعر يدلّ على جودة الطبع وقلّة الثبات . وأمّا الحركات فالبطؤ منها يدلّ على البلادة والسريعة على الطيش . والصوت الغليظ المرتفع يدلّ على الشجاعة ، والدقيق الهادئ يدلّ على الجبن .

1 وفي : ومن بعد ذلك في *Kāmil* | الموضوعين : اللذين هما موضوعان *Kāmil* 3 قال . . . الجلد : قال حنين ليس هكذا يعلم جالينوس من أمر هؤلاء وما يشهد الحقّ على صحته ولكن ضدّه فإنّ جالينوس يقول في هؤلاء القوم في المقالة الثانية من كتابه في المزاج ... والجلد *K. fī l-Firāsa* | البلاد المعتدلة : في بلدنا هذا وهو العراق لاعتداله قد *K. fī l-Firāsa* 4 فأمّا . . . البلاد : وأمّا البلدان التي هي في الشمال أو في الجنوب واغلة من قبل أنّ الحار في البلدان *K. fī l-Firāsa* 4–5 فإنّ . . . الجلد : قد مال الى الجلد بجذب الحرارة المحيطة به له *K. fī l-Firāsa* 5 فصارت . . . لاحتراقه : فإنّ طبيعة الجلد منهم لاحتراقها *K. fī l-Firāsa* | after الغريزيّة add. المحيطة بأبدانهم من الحرارة الغريزيّة *K. fī l-Firāsa* | منهم : منها *K. fī l-Firāsa* 6 صلبة : تصير حاصّة صلبة *K. fī l-Firāsa* | وصار . . . الغريزيّة : ويصير البدن منهم كلّه لما فيه من الحرارة الغريزيّة أقلّ القليل *K. fī l-Firāsa* 7 والبلاد . . . والصقالبة : وذلك أنّ الجلد من أهل البلدان الباردة مثل الأتراك وأهل برجان وسائر الأجناس التي تأوي بلاد تراقى وبلاد الصقالبة يكون بارداً رطباً ولذلك يكون ليّناً ناعماً أبيض أزعر *K. fī l-Firāsa* 8–9 والجلد . . . أبيض om. *K. fī l-Firāsa* 10 قال . . . الصلب : وأمّا اللحم فالصلب *K. fī l-Firāsa* | في الأبدان : منه *K. fī l-Firāsa* 10–11 والأزعر يدلّ : وأمّا الأزعر فيدلّ *K. fī l-Firāsa* | وأمّا : فإنّ وإذ كانت *K. fī l-Firāsa* 11 فالبطؤ : فالبطيئة *K. fī l-Firāsa* 12 after والسريعة add. تدلّ *K. fī l-Firāsa* | والصوت . . . المرتفع : وأمّا الصوت فالغليظ المرتفع منه *K. fī l-Firāsa* 13 والدقيق . . . يدلّ : وأمّا الرقيق الهادئ فيدلّ *K. fī l-Firāsa*

2 after المواضع break (∴) ind. and blank space of half a line M 5 الجلد[2] . . . إلى add. in marg. M | فصارت : فصار M | طبيعته : الطبيعية M 6 صلبة : صلته M | سوداءاً : سوَدآ M 9 after أبيض break (∴) ind. and blank space of six letters M 11 جودة : حودة M | after الثبات blank space of three letters M | فالبطؤ : فلَبطيهُ M 12 after الطيش blank space of six letters M 13 after الجبن break (∴) ind. and blank space of six letters M

of which is hot and moist, and on the stomach, the chest and the armpits because of the heat [arising from] the mixture of the heart and the liver, which are both situated close to these positions.

(197) Galen says:[278] in temperate regions the skin indicates the nature of the organs beneath it. In regions of the deep south the heat in the bodies is turned towards the skin, so its nature has become hard and black due to [the skin] being burnt by the innate heat which moves from [inside the bodies] outwards. The body has been left with little innate heat, yet it is 111a
hot through the extraneous heat it has acquired. The innate heat of the inhabitants of the cold regions, for example the Turks and Slavs, has already sunken with the blood towards the bowels and their skin is cold and moist, and therefore soft and white.

(198) He says:[279] firm abundant flesh in the bodies indicates little sensitivity, and [flesh] with scanty hair indicates an excellent natural disposition, but little stability. As to movements their slowness indicates apathy and [their] rapidity unsteadiness. The deep loud voice indicates courage and the thin quiet one cowardice.

[278] Cf. Galenic gloss in Ps.-Aristotle, *K. al-Firāsa*, (14.18–15.3), 16.5–6, 16.8–10, 17.5–9, 16.13–17.3.

[279] Cf. source in Ps.-Aristotle, *K. al-Firāsa*, 18.1–3, 18.13–15.

(199) ما يعرض للبدن من قبل صورته يسمّى خاصّة ، وما كان من قبل هيولاه يسمّى عرضاً . والخاصّة منها ما يتبع صورة الشيء التي بها هو ما هو ، ومنها صورة موضوعه القريب ، ومنها صورة موضوعه البعيد ، كالضحك فإنّه من حيث استطراف وتعجّب يعرض للإنسان من قبل صورته التي هي النفس 111b الناطقة ، ويتبعها التعجّب لأنّه طلب وجود الأمر المتطرّف ، ومن حيث هو قهقهة فعارض ومن قبل موضوعه القريب الذي هو البدن ، عندما تتردّد النفس بين الإيجاب والسلب في طلب السبب للمبدأ الذي منه تنشؤ الحركة في الحيوان وهو مبدأ الغضب ، فيحدث حركتان متجاذبتان في البدن ويظهر أكثره في الوجه لأنّه مجمع الحواس .

(200) قال افلاطن الخير المحض هو غاية متوخّاة لذاتها ، أي هو الأمر المطلوب نفسه المقصود لذاته ، الذي يتشوقه الكلّ .

فالخير في الطبّ الصحّة وفي تدبير الحرب الظفر وفي صناعة البناء البيت ، وكذلك في كلّ شيء أي الغاية المقصودة في كلّ فعل واختيار .

(201) وقال ارسطالس الخيرات منها ما هي شريفة ، ومنها ما هي ممدوحة ، 112a ومنها ما هي بالقوّة كذلك ، ومنها ما هي نافعة . فالشريفة منها هي التي شرفها من ذاتها وتجعل من اقتناها شريفاً ، وهي الحكمة والعقل ، والممدوحة مثل الفضائل والأفعال الجميلة الإراديّة ، والتي هي بالقوّة مثل التهيّؤ والاستعداد لنيل الأشياء التي تقدّمت ، والنافعة هي جميع الأشياء التي تطلب لا لذاتها بل ليوصل بها إلى الخيرات . ولله تعالى هو الخير الأوّل فإنّ جميع الأشياء يتحرّك نحوه بالشوق إليه وأن ينال الخيرات الإلهيّة من البقاء والسرمديّة

10 قال افلاطن : وذلك أنّ *Tahdhīb* 11 الذي . . . الكلّ om. *Tahdhīb*, but part of *K. al-Akhlāq* 12 فالخير : وهو *K. al-Akhlāq* 13 وكذلك : وهو *K. al-Akhlāq* | شيء أي : واحد من الأشياء شيء غيره في آخر وهو *K. al-Akhlāq* 14 وقال ارسطالس : الخير على ما قسمه أرسطوطاليس وحكاه عنه فرفوريوس وغيره هكذا قال *Tahdhīb* 15 after نافعة add. فيها *Tahdhīb* 16 after اقتناها add. أيضاً *Tahdhīb* 19 ليوصل : ليتوصل *Tahdhīb* | ولله تعالى : فالله تعالى تقدّس *Tahdhīb* 20 وأن : لأن *Tahdhīb*

2 منها : منه M 3 ومنها[1] : ومنه M | موضوعه[1] : مَوضوعة M | ومنها[2] : وَمنه M | موضوعه[2] : موضوعَة M 4 استطراف M : or read استظراف ? | وتعجّب : وبعجبٍ M 5 المتطرّف M : or read المتظرّف ? 9 after الحواس break (∴) ind. and blank space of more than half a line M 13 after واختيار break (∴) ind. and blank space of half a line M 20 والسرمديّة : وللسرمديه M

(199) What comes to the body on the part of its form is called property and what is on the part of its matter is called accident. To the property [of the body] belong what is subordinate to the form of a thing by which it is what it is, [and further belong] the form of [the thing's] proximate substratum and the form of its remote substratum. Like laughter, for with regard to finding something unusual[280] and being astonished, [laughter] befalls man 111b
on the part of his form which is the rational soul, [since] being astonished belongs to [the soul], because it is the desire to find something unusual. With regard to being a guffaw, [laughter] is accidental and on the part of [man's] proximate substratum which is the body. When the soul wavers between affirmation and negation searching the reason for the principle from which motion in the living being arises and which is the principle of anger, two contending motions emerge in the body and most of [this] manifests itself on the face, for this is where the senses meet.

(200) Plato says:[281] the pure good is the end which is aimed at for itself, i.e. it is the matter sought for itself and pursued in itself which everything desires.

The good in medicine is health, in warfare victory, in architecture the building, and similarly with regard to everything [there is] the end pursued in each action and choice.

(201) Aristotle says:[282] among the goods are those which are noble, those 112a
which are praiseworthy, those which are potentially like that and those which are beneficial. The noble goods are the ones the nobleness of which is caused by themselves, and which render him who acquires them noble. They are wisdom and intellect. The praiseworthy [goods] are like the virtues and the good voluntary actions. The potential [goods] are like the preparedness and aptitude to attain the things mentioned. The beneficial [goods] are all things which are sought not for themselves, but in order to reach [other] goods through them. God the Sublime is the First Good, because all things move towards Him in desire for Him and to attain the

280 The Marsh MS reads 'unusual', but a single added dot would allow one to read 'witty'.

281 Cf. partly Miskawayh, *Tahdhīb*, 88.7–8; and partly Aristotle, *al-Akhlāq*, 113.5 and 131.4–6.

282 Cf. source in Miskawayh, *Tahdhīb*, 76.17–77.2, 78.3–5.

والتمام منه .

(202) وقال في كتاب الأخلاق (a) كلّ صناعة وكلّ مذهب وكلّ فعل وكلّ اختيار فقد يعلم أنّه إنّما يتشوّق به خير ما ، وقد أجاد من حكم على الخير أنّه الشيء الذي يتشوّقه الكلّ .

(b) وكما أنّ للمطبوعات تماماً تقصده الطبيعة بذاتها ، وللمعقولات تماماً 112b
يقصده العقل بذاته ، كذلك للمصنوعات تمام تقصده المهنة الإنسيّة بذاتها ، وذلك التمام هو الخير المقصود إليه وهو أفضل من سائر الأشياء التي تراد له وتسبّبه .

(c) والخير على ضربين ، أحدهما خير بذاته ، والآخر خير من أجل غيره ، والمطلوب لذاته آثَرُ من المطلوب لغيره .

(d) ونحن نريد السعادة القصوى لذاتها ، إذ هي غايتنا وغرضنا ، ونريد الكرامة والفضائل من أجل السعادة لأنّا بهما نصير إليها . (e) والسعادة هي الشيء الكافي التي يستغني بها في تصيّر سيرة الإنسان محمودة . والسعادة أقصى المختارات وأفضل الخيرات ، إذ هي تمام المفعولات . (f) والسعيد هو الذي يفعل أفعاله على ما توجبه الفضيلة الكاملة ، وتساعده الأمور التي تردّ عليه
من خارج ، لا زماناً يسيراً ولكن في عمره بأسره . 113a

(g) والخير ينقسم على ثلاث جهات ، خير في البدن ، وخير في النفس ،

2 وقال . . . الأخلاق : قال إنّ K. al-Akhlāq : من المقالة الأولى Ikhtiṣār 2–4 كلّ . . . الشيء : كلّ صناعة ومذهب وهِمّة وفعل واختيار فإنّه يظنّ به أنّه يقصد إلى خير ما فالخير هو خيراً Ikhtiṣār 3 خير : واختيار K. al-Akhlāq 2–3 وكلّ اختيار : وكذلك كل K. al-Akhlāq 2 وكلّ[2] : وهو الخير المقصود Ikhtiṣār 7–8 وذلك . . . وتسبّبه : ولذلك K. al-Akhlāq | وقد : K. al-Akhlāq 9 أحدهما om. Ikhtiṣār | على ضربين : ضربان Ikhtiṣār | والخير : وقال الخير M. al-Ḥikam : الخير Ikhtiṣār 10 لذاته : بذاته Ikhtiṣār | والآخر خير : وخير Ikhtiṣār | خير[1] om. M. al-Ḥikam | غايتنا وغرضنا : غاية غرضنا Ikhtiṣār 11 ونحن : فنحن Ikhtiṣār | آثَرُ : أفضل Ikhtiṣār | M. al-Ḥikam 11–12 الكرامة والفضائل : الفضائل Ikhtiṣār 12 لأنّا . . . إليها om. Ikhtiṣār 14 والسعيد : وقال السعيد M, fol. 115b12 14–15 هو . . . يفعل : الذي هو يعفل M, fol. 115b12 15 وتساعده : ويساعد عليه M, fol. 115b13 | عليه : عليها M, fol. 115b13 17 والخير : وقال الخير M. al-Ḥikam : الخير Ikhtiṣār | البدن : النفس Ikhtiṣār | النفس : البدن Ikhtiṣār

1 after منه break (∴) ind. and blank space of more than half a line M 6 بذاته : or read بذاتها ? | المهنة : المهية M 9 والآخر : والاحر M 10 آثَرُ : اير M 12 نصير : تصير M 14 والسعيد : 16 after بأسره blank space of three letters M 17 ثلاث : ثلث M

divine goods like permanence, eternity and perfection from Him.
(202) He says in the *Ethics*:[283] (a) It is well known that by every skill, every method, every action and every choice some good is desired and he who judges the good to be the thing which everybody desires does good.
(b) Just as naturally made things have a perfection which nature pursues 112b
by its essence,[284] and the intelligibles have a perfection which the intellect pursues by its essence, equally the artificially made things have a perfection which the human profession pursues by its essence. That perfection is the good which is pursued and is nobler than the other things which are wanted because of it and which cause it.
(c) The good is of two sorts: one of them is good by itself and the other is good on account of something else. What is sought for itself is nobler than what is sought for something else.
(d) We want ultimate happiness for itself, since it is our end and aim, whereas we want honours and virtues on account of happiness, because through these two we reach it. (e) Happiness is the sufficient thing through which man has everything to lead a praiseworthy life. Happiness is the ultimate of choices and the noblest of goods, since it is the perfection of the things done. (f) The happy [man] is he who performs his actions in accordance with the obligations of perfect virtue and is assisted by matters he receives from outside, not for an insignificant [period of]
time, but during his entire life. 113a
(g) The good is divided into three classes: the good in the body, the good in

283 Excerpted from *Ikhtiṣār al-Iskandarānīyīn*; cf. its fragments, ed. Dunlop, 260.1–3, 6–8 and 13–18; covering (a)-(d) and (g); cf. parallels in Aristotle, *al-Akhlāq*, ed. Akasoy and Fidora, 113.4–5, which has a closer parallel to (a) than the fragments of the *Ikhtiṣār*; and cf. Mubashshir, 210.3–8, who has closer parallels to (c) and (g) than the fragments of the *Ikhtiṣār*; and cf. *PR*, passage (208) to (f). Contrary to the previous case of fols. 23b and 26b in which the scribe of the Marsh MS obviously copied one and the same text twice, in the present case he seemed to have used different models here and in passage (208). Therefore the deviating readings are given in the apparatus of parallel passages this time.

284 It is also possible to understand the suffix *-hā* (here and in what follows) to refer to the naturally made things and thus translate: 'nature pursues with their essence(s)'.

وخير من خارج البدن . والذي هو أولى بمعنى الخير هو خير النفس ، ولا تظهر صورة هذا الخير إلا بالأفعال الصادرة عن الفضائل . والسعادة في اقتناء الفضيلة واستعمالها معاً .

(203) قال افلاطن في كتاب النواميس الخيرات الإلهيّة ما دبّرها الحكمة ، ثم اقتناء العفّة مع العقل ، ويحدث من هذين إذا تركّبا فضيلة ثالثة وهي [العدل] العدل ، والفضيلة الرابعة الشجاعة . والخيرات الإنسيّة الصحّة والجمال والشدّة والقوّة في جميع حركات البدن والرابعة اليسار ، ولست أعني باليسار الذي يكون بالأموال ، لكن حسن التدبير وجودة النظر والتفقّه .

113b (204) وقال الشيخ الخير المحض هو الأوّل الذي يفيض الخير على الأشياء ، فيلبسها الخير ، مثل ما تلبس الشمس الأجسام نوراً تشرق به . والخير الأوّل خير محض لا بإضافة إلى شيء آخر ، لأنّه ليس فوقه شيء آخر بل الأشياء كلّها تحته ومنه تنال الخير . وهو فاعل إلا أن فعله العقل والحياة والنفس وسائر الأشياء التي فيها حياة وعقل .

وكلّما تباعد الشيء من الهيولى كان الخير فيه أكثر ، فكلّما دنا من الهيولى وصار في حيّزها كان الخير فيه أقلّ .

فإن قال قائل إنّ الحياة ليست بخير لأنّها مملوءة تعباً ونصباً ، وليست بذات راحة ، قلنا إنّ هذه الحياة ليست الحياة المحضة الخالصة ، لكنّها الحياة الأرضيّة المستوية بالموت ، فأمّا الحياة الخالصة السماويّة فإنّها ذات راحة ولذّة غير فانية وليس فيها شيء من التعب والنصب .

114a فإن قال قائل فما الذي يفعل الآن الخير الأوّل ؟ قلنا إنّه يحفظ جواهر هذه

1 من ... البدن : في الأشياء الخارجة Ikhtiṣār | والذي : فالخير الذي M. al-Ḥikam | والذي ...
M. al-Ḥikam ما هو منها في النفس : هو² ... النفس | Ikhtiṣār وخير النفس أفضلها : النفس
التي : الصادرة ... الفضائل | Ikhtiṣār صورته : صورة ... الخير 2 Ikhtiṣār وهذا لا : ولا
: اقتناء الفضيلة 2–3 Ikhtiṣār فالسعادة : والسعادة | Ikhtiṣār الفضيلة : M. al-Ḥikam بالفضيلة
M. al-Ḥikam واستعمالهما : واستعمالها 3 Ikhtiṣār اقتنائها

3 after معاً break (∴) ind. and blank space of half a line M 4 دبّرها : دبره M 7 الذي : Ros2 : التي M 8 والتفقّه Ros2 : والتفقله with break (∴) ind. and blank space of half a line after it M 11 آخر¹ Ros, Bad : اخر M 13 after وعقل break (∴) ind. and blank space of ten letters M 14 فكلّما M, Ros : وكلما Bad 15 حيّزها Ros, Bad : خيرها M | after أقلّ break (∴) ind. M 16 مملوءة Ros, Bad : مملوة M 19 after والنصب break (∴) ind. and blank space of ten letters M

the soul and the good outside the body. The most adequate meaning of 'the good' is the good of the soul and the form of this good only manifests itself by actions proceeding from virtues. Happiness [lies] in the acquisition and the employment of virtue together.

(203) Plato says in the *Laws*:[285] What directs the divine goods is wisdom, then the acquisition of moderation together with intellect. When these two are composed, a third virtue occurs from them, namely justice. The fourth virtue is courage. The human goods are health, beauty and strength and power in all movements of the body, and the fourth [is] prosperity. By prosperity I do not mean the one which is due to possessions, but good direction, excellent reflection and insight.

(204) The Sage says:[286] the Pure Good is the First Who pours the good over 113b
[all] the things and then clothes them with good, just as the sun clothes the bodies with light which allows them to shine. The First Good is pure good, not [only good] in relation to something else.[287] For above Him there is nothing else; rather all things are below Him and attain the good from Him. He is acting, yet His action is intellect, life, soul and the other things in which there is life and intellect.

The more remote something is from matter the more good is in it, whereas the closer something is to matter and [the more] it comes to be in its realm, the less good is in it.

If someone said that life was not a good, because it was filled with hardship and exertion and was not in itself rest, we would say that this life was not the sheer pure life, but the earthly life which equals[288] death. As to the pure heavenly life it possesses imperishable rest and pleasure, and nothing of hardship or exertion exists in it.

If someone said: so what is it that the First Good does right now? We would
say that He preserves the substances and forms of these things and makes 114a

[285] See Plato's *Leges*, 631b-c; cf. ed. Rosenthal, p.396.

[286] Belonging to *GS*; see ed. Badawī, 194.1–21; and partial edn Rosenthal, pp.380–3.

[287] Rosenthal ('aš-Šayḫ', p.381) and Lewis ('Plotiniana', p.477) translate: '... and not (good) through contact with something else'.

[288] Lewis, apparently reading *mashūba*, translates ('Plotiniana', p.477): 'mixed with'.

الأشياء وصورها ويجعل الجواهر العاقلة تعقل والجواهر الحيّة تحيا ، وذلك أنّه يثبت في الأشياء العقليّة العقل وفي الأشياء الحيّة الحياة .

قال والشرّ إنما يكون أخيراً في الأشياء الأخيرة التي لم تنل من الخير الأوّل شيئاً البتّة ، وكما أنّه ليس فوق الخير الأوّل شيء آخر ، كذلك ليس تحت الأشياء الشرّيّة شيء آخر . ولا يظنّنّ ظانّ أنّ الشرّ الأخير ضدّ للخير الأوّل ، وذلك أنّه ليس بينهما وسط . فالخير الأوّل إذن هو الذي لا ضدّ له ، والخير إمّا ألا يكون موجوداً ، وإن كان فإنّه لا يكون له ضدّ البتّة ، ولا يمكن ألا يكون الخير موجوداً ، لأنّه هو علّة الأشياء .

(205) قال ارسطالس السعادة نيل فضائل النفس وفضائل البدن الذي لا قوام
114b للنفس في هذا العالم إلا به ، ونيل فضائل ما حول البدن ، إذ لا يستغني البدن في قوامه من القنية .

(206) وقال ارسطالس أقسام السعادة خمسة ، أحدها صحّة البدن ولطف الحواسّ ويكون ذلك من اعتدال المزاج . والثاني الثروة والأعوان وأشباهها حتّى يتسع لأن يضع المال في موضعه ويعمل به سائر الخيرات ويؤاسي منه أهل الخير خاصّة والمستحقّين عامّة ويعمل به كلّ ما يزيد في فضائله ويستحقّ الثناء والحمد عليه . والثالث أن تحسن أحدوثته في الناس وينتشر ذكره بين أهل الفضل فيكون ممدوحاً بينهم يكثرون الثناء عليه لما يتصرّف فيه من الإحسان والمعروف . والرابع أن يكون منجحاً في الأمور ، وذلك إذا استتمّ
115a كلّ ما روّى فيه وعزم عليه حتّى يصير إلى ما يأمله منه . والخامس أن يكون

12 وقال ارسطالس : فأمّا *Tahdhīb* | خمسة : خمسة أقسام فهي على مذهب هذا الحكيم *Tahdhīb* | after أحدها add. في *Tahdhīb* 13 after المزاج add. أعني أن يكون جيد السمع والبصر والشم والذوق واللمس *Tahdhīb* | after والثاني add. في *Tahdhīb* 16 والحمد : والمدح *Tahdhīb* | وينتشر : وينشر *Tahdhīb*

1 تعقل Bad: بعقل M 2 after الحياة blank space of five letters M 3 أخيراً M : خيراً Ros, Bad 4 آخر Ros, Bad : اخر M | كذلك Ros, Bad : لذلك M 5 آخر Ros, Bad : اخر M يظنّنّ Bad : يظتن M : يظنّ Ros 7 ألا يكون : ‹الا يكون› Ros 8 after الأشياء break (.·.) ind. and blank space of almost the entire line M 11 after القنية break (.·.) ind. and blank space of half a line M 12 after خمسة blank space of three letters M 13 after المزاج blank space of ten letters M | وأشباهها : واشيَا هُهَا M 16 after عليه blank space of five letters M 18 after والمعروف blank space of six letters M 19 after منه blank space of five letters M

the intelligible substances think and the living substances live. For He establishes the intellect in the intellectual things and life in the living things.
He says: evil is the last among the last things[289] which do not attain anything from the First Good at all. As there is no other thing above the First Good, in the same way there is no other thing below the evil things. No one should assume that the last evil is the opposite of the First Good, for there is no intermediate between the two. Therefore the First Good is what has no opposite. Either the good is not existent or, if it is, it will have no opposite at all. Yet, it is not possible that the good is not existent, because It is the cause of [all] things.
(205) Aristotle says: Happiness is the attainment of the excellences of the soul, the excellences of the body by which alone the soul has subsistence
in this world, and the attainment of excellences relating to the body, since 114b
the body cannot subsist without [their] acquisition.
(206) Aristotle says:[290] the parts of happiness are five. One of them is the health of the body and sensitiveness caused by a well-balanced [body] mixture. The second is wealth, attendants and similar things so that they suffice for spending money where it should be spent, producing other goods with it, using it to support good people especially and generally [all] who deserve it, and doing with it everything that increases one's virtues and entitles one to appreciation and praise. The third is having a good reputation and one's memory spread among excellent people, so that one is praised among those who bestow copious praise on [a person] for his charitable and beneficial actions. The fourth is being successful in [one's] affairs, i.e. after having completed what one has considered and decided
upon, so that the result is what one hoped for. The fifth is being faultless in 115a

[289] Rosenthal ('aš-Šayẖ', p.383) and Lewis ('Plotiniana', p.477), reading *khayr* instead of *ākhīr*, translate: 'Evil is (to be defined as the) good in the last things'. Cf. *Enn.* VI, 7, 23, 10–11: 'But if the evils come later, in things which do not participate in this Good in any way at all, and on the very last and lowest level'.
[290] Cf. source in Miskawayh, *Tahdhīb*, 79.7–80.12.

جيّد الرأي صحيح الفكر سليم الاعتقاد في دينه وفي غير دينه بريئاً من الخطأ والزلل جيّد المشورة . هي الآراء على مذهب هذا الحكيم .

فأمّا فيثاغورس وسقراط وافلاطن فإنّهم أجمعوا على أنّ الفضائل والسعادة كلّها في النفس وحدها . ولذلك لما قسموا السعادة ، جعلوها في قوى النفس ، وهي الحكمة والشجاعة والعفّة والعدالة ، فأجمعوا على أنّ هذه الفضائل هي كافية في السعادة ، ولا يحتاج معها إلى غيرها من فضائل البدن ولا ما هو خارج البدن . فإنّ الإنسان إذا حصّل تلك الفضائل لم يضرّه في سعادته إن يكون سقيماً ناقص الأعضاء مبتلى بجميع أمراض البدن أو يكون فقيراً خاملاً أو ساقط الجاه .

115b (207) وقال ارسطالس في موضع آخر أجناس السعادات ثلاثة ، سعادة في النفس ، وسعادة في البدن ، وسعادة من خارج البدن وفيما يطيف بالبدن . أمّا التي في النفس فهي العلوم والمعارف ونهايتها الحكم ، وهذه السعادة وهي أقصاها لأنّها تراد لذاتها لا شيء آخر . وأمّا التي في البدن فمثل الجمال واعتدال الأعضاء وصحّة المزاج ، وهذه تراد لنفسها وقد تراد لغيرها ، أعني لأن يتمّ بها أغراض أُخر من أفعال النفس وفضائلها . وأمّا التي من خارج البدن فمثل الأولاد النجباء والأصدقاء واليسار والنسب والكرامات .

1 الاعتقاد : الاعتقادات Tahdhīb 2 هي : في Tahdhīb | على ... الحكيم : فمن اجتمعت له هذه الأقسام كلّها فهو السعيد الكامل على مذهب هذا الرجل الفاضل ومن حصل له بعضها كان حظّه من السعادة بحسب ذلك Tahdhīb 3 فأمّا : وأما الحكماء الذين كانوا قبل هذا الرجل مثل Tahdhīb | after وافلاطن add. وأشباههم Tahdhīb 4 after جعلوها add. كلّها Tahdhīb | after النفس[2] add. التي ذكرناها في أوّل الكتاب Tahdhīb 5 فأجمعوا : وأجمعوا Tahdhīb | هذه om. Tahdhīb 7 فإنّ : وان Tahdhīb 8–9 أو[1] ... الجاه : اللهم إلا أن يلحق النفس منها مضرّة في خاصّ أفعالها مثل فساد العقل ورداءة الذهن وما أشبهها وأمّا الفقر والخمول وسقوط الجاه وسائر الأشياء الخارجة عنّا فليست عندهم بقادحة في السعادة البتّة Tahdhīb 10 وقال ... السعادات : وأرسطوطاليس رتّب أجناس السعادة فجعلها Tartīb al-saʿādāt 12 الحكم : الحكمة Tartīb al-saʿādāt | وهي : هي Tartīb al-saʿādāt 14 after واعتدال add. في Tartīb al-saʿādāt 16 واليسار : والإيسار Tartīb al-saʿādāt | والنسب : وشرف النسب Tartīb al-saʿādāt

1 بريئاً : برياً M 2 هي : في M | الآراء : الارا M | after الحكيم break (∴) ind. and blank space of more than half a line M 9 after الجاه break (∴) ind. and blank space of half a line M 10 ثلاثة : ثلثه M 11 after بالبدن break (∴) ind. and blank space of five letters M 13 آخر : اخر with blank space of five letter after it M 15 after وفضائلها break (∴) ind. and blank space of half a line M 16 والنسب : والنشب M | after والكرامات break (∴) ind. and blank space of ten letters M

one's opinion, correct in one's thought, sound in one's religious and other belief[s], free from mistake and error and faultless in advice. The[se] are the views according to the school of this sage.

Pythagoras, Socrates[291] and Plato have agreed on the virtues and happiness all being in the soul alone. Therefore when they have divided happiness, they have set it within the powers of the soul, namely wisdom, courage, moderation and justice, and have agreed on these virtues being sufficient for happiness and that one does not need, in addition to them, any of the other virtues of either the body or what is outside the body. For when man has attained those virtues it does not harm his happiness if he is ill, lacks any limbs, is afflicted by all illnesses of the body or is poor and unimportant or has lost status.

(207) Aristotle says somewhere else:[292] the genera of happiness are three: 115b
happiness in the soul, happiness in the body and happiness outside the body and in what surrounds the body. [The happiness] in the soul is sciences and knowledge, and its utmost degree is [all sorts of] wisdom. This happiness is the ultimate [happiness], for it is wanted for itself, not for something else. [Happiness] in the body is, for example, beauty, well-balanced body parts and a sound [body] mixture. These are wanted for themselves, but certainly [also] for something else. That is because through them other aims of the activities of the soul and its virtues become perfect. [Happiness] outside the body is, for example, outstanding offspring, friends, prosperity, [noble] lineage and honours.

[291] The Marsh MS's reading 'Socrates (Suqrāṭ)' confirms the emendation suggested by Walzer and Zurayk for some of the *Tahdhīb* MSS which read 'Hippocrates (Buqrāṭ)', see Zurayk, *Refinement*, p.202, n.12.

[292] Cf. source in Miskawayh, *Tartīb al-Saʿādāt*, 109.16–110.2.

(208) وقال السعيد الذي هو يفعل أفعاله على ما يوجبه الفضيلة الكاملة ، ويساعد عليه الأمور التي ترد عليها من خارج ، لا زماناً يسيراً ولكن في عمره 116a
بأسره .

(209) والفضيلة حالة تجعل الشيء الذي هي له فضيلة جيّداً وتجعل فعله كذلك ، مثاله أنّ فضيلة العين تجعل العين جيّدةً وفعلها جيّداً ، لأنّا بفضيلة العين ننظر جيّداً ، وكذلك فضيلة الفرس تجعل الفرس جواداً منقاداً لراكبه صبوراً ، وفضيلة الإنسان هي التي يصير بها جيّداً ويجعل بها فعله جيّداً .

(210) وقال أبو على مسكويه أجناس السعادات ثلاثة ، أحدها في النفس ، والثانية في البدن ، والثالثة فيما يطيف بالبدن من خارج .

فأمّا التي في النفس فهي العلوم الحقيقيّة والمعارف الصحيحة وجودة الرؤيّة والوقوف على الأمور الأبديّة ، ثم العمل بما يوجبه العلم والعدل . وهي السعادة القصوى والكمال الأخير التي بها يُوصل إلى معرفة البارئ جلّ ثناؤه 116b
بالصحّة والحقيقة ، ومنها ينال البقاء السرمديّ والنعيم الأبديّ .

وأمّا التي في البدن فهي صحّة المزاج وجودة التركيب وتناسُب الأعضاء مع سلامتها وجمالها ، وهي في المرتبة الثانية لأنّها إنما تراد لتلك الأولى ، وهي طريق إلى الترقّي إليها وسلّم للاطّلاع عليها ، وإن كانت تراد أيضاً لذاتها في بعض الأوقات .

وأمّا التي تطيف بالبدن من خارج فهي الجاه والمال والأهل والأصدقاء

1 وقال ... هو : والسعيد هو الذي M, fol. 112b11   2 ويساعد عليه : ويساعده M, fol. 112b12   عليها : عليه M, fol. 112b13   4 والفضيلة حالة : وليس ينبغي أن نقتصر على القول بأنّها حالة فقط بل نصف أيّ الحالات هي فنقول إنّ كلّ فضيلة K. al-Akhlāq   5 مثاله : مثل ذلك K. al-Akhlāq   5–6 بفضيلة ... جيّداً : فضلية العين الإبصار K. al-Akhlāq   6–7 جواداً ... صبوراً : جيّداً وحسناً للركوب ومعيناً راكبه على الحرب K. al-Akhlāq, but cf. Ullmann, *EN 2*, p. 139 who claims that this reconstruction does not correspond to the almost illegible MS reading   7 وفضيلة : فإذا كان هذا المعنى في كلّ شيء فعلى هذا تكون فضيلة K. al-Akhlāq | التي : حال K. al-Akhlāq | after بها[1] add. الإنسان K. al-Akhlāq | ويجعل ... جيّداً[2] : ويفعل بها فعلاً [...] K. al-Akhlāq   8 وقال ... مسكويه : فصل K. al-Ḥikma   12 جلّ ثناؤه : عزّ وجلّ K. al-Ḥikma

6 ننظر : ينظُر M   7 after صبوراً blank space of three letters M | after جيّداً[2] break (∴) ind. and blank space of three letters M   8 ثلاثة : ثلثه M   9 after خارج blank space of three letters M   10 الحقيقيّة : الحقيقة M   13 after الأبديّ break (∴) ind. and blank space of three letters M   15 لتلك : ليلك M   16 إليها : البهَا M   17 after الأوقات blank space of ten letters M   18 بالبدن : البدن M | الجاه : الحياة M

(208) He says:[293] the happy [man] is he who performs his actions in accordance with the obligations of perfect virtue and is assisted by external matters which profit [perfect virtue], not for an insignificant [period of] time, but during his entire life. 116a
(209) Excellence is a state[294] which makes the thing for which it is an excellence faultless as well as [this thing's] activity. An example of it is that the excellence of the eye makes the eye and its activity faultless, for through the excellence of the eye we see faultlessly. Similarly the excellence of the horse makes the horse swift, tractable for its rider and patient, and the excellence of man is that through which he becomes faultless and through which his activity is made faultless.
(210) Abū ʿAlī Miskawayh says:[295] the genera of happiness are three: one of them [lies] in the soul, the second in the body and the third in what surrounds the body from outside.
[Happiness] in the soul is true sciences, sound knowledge, excellence in deliberation and understanding of the eternal matters, then action in accordance with the obligations of knowledge and justice. [This] is the ultimate happiness and the final perfection by which knowledge of the 116b
Creator – Exalted be His praise – is reached in soundness and truth, and through which endless permanence and eternal bliss are attained.
[Happiness] in the body is the soundness of the [body] mixture, excellence of composition and good proportion of the body parts combined with their integrity and beauty. It is in the second degree [of importance], because it is desired for that first [happiness], is a way of ascending to it and a ladder to become acquainted with it, even though it is sometimes also wanted for itself.
[Happiness] which surrounds the body from outside is rank, wealth, family,

[293] This is a repetition of passage (202f).
[294] Cf. Aristotle, *al-Akhlāq*, 169.12–17.
[295] Excerpted from Miskawayh, alternative *Fawz;* cf. parallel in *K. al-Ḥikma,* MS E, fol.75a15–b19.

والأعوان . وهذه في المرتبة الثالثة لأنّها تراد للبدن والبدن يراد للنفس ، وكلّ ما أدّى إلى غرض صحيح فغرضه أشرف منه .

ومن أقرب ما يستدلّ به على أنّ هذه السعادة الثالثة تراد للبدن ، وأنّ الثانية تراد للأولى أنّ مَن جعل غايته القصوى الثروة والجاه أو سائر ما ذكرنا من [117a] الأمور الخارجة عنّا ، إذا ألم عضو من أعضاء بدنه رجع عن رأيه وترك غرضه الذي كان ظنّه غاية سعادته وجعل ذلك أجمع فداءً لسلامة ذلك العضو بل لعافية ساعة من عمره .

ومن أعجب الأمور وأبدعها في هذه السعادات أنّ الأشرف الأعلى منها وهي سعادة النفس لا تتمّ إلا بالسعي والاجتهاد ، وليس فيها شيء من الاتّفاق وما ينسب إلى البخت . فأمّا السعادة التي هي دونها في المرتبة الثانية فبعضها يحصل لنا بالسعي والاجتهاد وبعضها بالاتّفاق والبخت إلا أنّ حظّ الاجتهاد فيها أكثر من حظّ الاتّفاق . وأمّا السعادة التي دون هذه في المرتبة الثالثة فأكثرها بالبخت والاتّفاق . [117b]

(211) الحكمة على ما قسمها أهلها تنقسم بالقسمة الأولى إلى قسمين ، وهما الحكمة النظريّة والحكمة العمليّة .

(I) فأمّا النظريّة فإنّها تنقسم إلى ثلاثة أقسام ، (1) العلم الأعلى وهو العلم الإلهيّ ووجوده في العقل فقط (2) والعلم الأوسط وهو العلم الرياضيّ ووجوده في الوهم حسب (3) والعلم الأسفل وهو العلم الطبيعيّ ووجوده في الحسّ خارج الوهم والعقل . وينقسم كلّ واحد من هذه العلوم أقساماً كثيرة .

(1) أمّا العلم الأعلى فهو غاية ما يطلب ونهاية ما يرتقي إليه . وهو العلم الذي يقال فيه إنّه حرّ ، فأمّا العلوم الأخر فمستعبدة ، وذلك أنّ الحرّ هو

1 after لأنّها add. كلّها *K. al-Ḥikma* 3 هذه om. *K. al-Ḥikma* 6 ذلك أجمع : جميع ذلك *K. al-Ḥikma* 13 والاتّفاق om. *K. al-Ḥikma* 14 before الحكمة add. نكتة *K. al-Ḥikma*

2 after منه break (∴) ind. and blank space of six letters M 6 ذلك[1] : رحك M | فداءً : فِذآ M 7 after عمره break (∴) ind. M 9 الاتّفاق : الانفاق M 10 after البخت blank space of five letters M 11 بالاتّفاق : بالانفاق M | والبخت : والبخته M 12 الاتّفاق : بالبَحث والانفَاق with blank space of five letters after it M 13 بالبخت والاتّفاق : M | after والاتّفاق break (∴) ind. and blank space of more than half a line M 15 after العمليّة break (∴) ind. in M 16 ثلاثة : ثلثة M 19 after والعقل blank space of three letters M | after كثيرة break (∴) ind. and blank space of three letters M

friends and attendants. This is in the third degree [of importance], because it is wanted for the body and the body is wanted for the soul. Everything which leads to a sound end is less noble than its end.

What indicates immediately that this third happiness is wanted for the body and that the second [happiness] is wanted for the first one is that he 117a
who takes as his ultimate aim riches, rank or any other of the matters external to us we have mentioned, changes his mind as soon as one of his body parts is in pain. He abandons his end which he has assumed to be his utmost happiness and sacrifices all of that for the well-being of that body part to secure health even for an hour of his life.

What is most astonishing and exceptional with regard to these [three classes of] happiness is that the noblest and most elevated among them, namely the happiness of the soul, becomes perfect only by endeavour and effort. With regard to it, nothing is caused by luck or linked to chance. The happiness which is below it, in the second degree, partly accrues to us by means of endeavour and effort and partly by means of chance and luck, although its share of effort is greater than its share of chance. Most of the happiness which is below this, in the third degree, happens by luck and 117b
chance.

(211) According to the division by the people of wisdom,[296] wisdom is primarily divided into two parts: theoretical and practical wisdom.

(I) Theoretical [wisdom] is divided into three parts. (1) The most elevated science is metaphysics and it exists only in [the realm of] intellect. (2) The intermediate science is mathematics and it exists solely in [the realm of] imagination. (3) The lowest science is the natural science and it exists in [the realm of] sense perception outside imagination and intellect. Each one of these sciences is divided into many parts.

(1) As to the most elevated science, it is the utmost goal and the limit of one's ascent. It is the science which is said to be 'free', whereas the other sciences are ancillary. For the free [science] is one that does not need

[296] Excerpted from Miskawayh, alternative *Fawz;* cf. parallel in *K. al-Ḥikma,* MS E, fols76a1–78b15.

الذي لا يحتاج في قوامه وفي وجوده إلى شيء خارج عنه ، والعبد هو الذي لا بدّ له في قوامه ووجوده من شيء أو أشياء، وكأنّه موجود بغيره ولغيره ، 118a وبحسب حاجاته إلى كثرة الأشياء وقلّتها يكون رسوخه في العبوديّة ولزوم هذه السمة له . وجميع ما التمسناه وطلبناه إنما ارتَضَيْنا فيه وتهذّبنا به لنصل إلى هذا العلم وهو غير منقسم في ذاته .

(2) والعلم الرياضيّ ينقسم أربعة أقسام ، وهي علم العدد وعلم الهندسة وعلم النجوم وعلم الموسيقى . وإنما انقسمت بهذه الأقسام لأنّ الرياضة إنما تكون في أمور الأجسام الطبيعيّة المحسوسة التي هي أقرب الأشياء منّا ، فأوّل ما ننظر فيه من أمور الأجسام هو كمّيّتها وكيفيّتها ، أعني مساحتها وأشكالها لأنّ الأجسام أوّل ما يلحقها عند نظرنا هو العدد ، أعني تكرّر الواحد بعد الواحد على حاسّة البصر . ثم النظر فيما يصوّر به ، أعني ما به صار جسماً ، وهو الطول والعرض والعمق ، والناظر في القسم الأوّل هو الناظر في علم العدد ، 118b والناظر في القسم الثاني هو الناظر في علم الهندسة . وهما قسما الكمّيّة ، أعني المتّصل والمنفصل ، ولما كانت هذه الأبعاد الثلاثة ، أعني الطول والعرض والعمق منها ما يكون في الأمور السفليّة ⟨ومنها ما يكون في الامور العلويّة ، صار ما كان منها في السفليّة⟩ لأجل سكونها وحركاتها المستقيمة ذا خطوط وزوايا مستقيمة ، والأشكال المركّبة من ذلك هي أيضاً كذلك ، والناظر في هذا القسم مهندس إلا أنّ المهندس ينظر فيها من حيث هي ساكنة . وصار ما كان منها في العلويّة من أجل حركاتها المستديرة مستديراً فحدثت منها الأشكال المستديرة ، والناظر في هذا القسم منجّم إلا أنّ المنجّم ينظر فيها من حيث هي متحرّكة . ثم يحدث من إضافة الكمّ إلى الكيف علم آخر وهو 119a الموسيقى . فذلك أربعة أقسام ضروريّة وأولى هذه الأربعة بالتقديم علم العدد لأنّه بسيط غير مركّب ، والبسيط يتقدّم المركّب بالطبع .

1 وفي وجوده : ووجوده *K. al-Ḥikma*   2 ووجوده : وفي وجوده *K. al-Ḥikma*   4 ارتَضَيْنا : ارتضنا *K. al-Ḥikma*   8 فأوّل : وأوّل *K. al-Ḥikma* | ننظر : ينظر *K. al-Ḥikma*   11 ثم النظر : والنظر *K. al-Ḥikma*   22 الموسيقى : الموسيقيّ *K. al-Ḥikma*

1–2 إلى ... ووجوده add. in marg. M   2 ووجوده : وفي جوده *fī* corrected to *waw* M | بغيره : بعنده M   8 ننظر : ينظر M   13 القسم : قسم M   14 الثلاثة : الثلثه M   21 آخر : اخر M

anything outside itself for its foundation and existence, whereas the
ancillary is one that is inevitably dependent on something or some things
for its foundation and existence. It is as if it existed by virtue of and for 118a
something else. In accordance with its needs for many or a few things it is
fixed in its position of subservience and attached to this characteristic. We
only approve of and are improved by all we require and seek so that we
finally reach this science which is indivisible in itself.

(2) Mathematics is divided into four parts, namely arithmetic, geometry,
astrology and music. It is divided into these parts because mathematics
refers to matters of the natural sensible bodies, which are the things
closest to us. The first thing we examine in bodies is their quantity and
quality, i.e. their surface and shapes. For the first [thing] which follows the
bodies in our examination is number, i.e. the repetition of one [body] after
the [other] for the sense of vision. Next is the examination of what is
formed by [number], i.e. what becomes body by means of [number]. It is
length, width and depth. He who examines the first part examines 118b
arithmetic and he who examines the second part examines geometry.
These two are parts of quantity, i.e. the continuous and the discrete. When
these three dimensions, namely length, width and depth, are partly found
in the lower things ⟨and partly in the heavenly things⟩, they possesses
linear lines and angles ⟨in the lower things⟩[297] because of their rest and
linear movements. The shapes which are composed of these are also like
this. He who examines this part is a geometer, yet the geometer examines
them with regard to their being at rest. The [three dimensions] in the
heavenly [things] are circular because of their circular movements, and
from them the circular shapes emerge. The scholar of this part is an
astronomer, yet the astronomer examines them with regard to them being
moved. Next another science emerges from the addition of quantity to 119a
quality. This is music. So there must be four parts [of mathematics].
Arithmetic is the first of these four by precedence, for it is simple and not
composite, and the simple naturally precedes the composite.

[297] The additions are based on the parallel text of the *K. al-Ḥikma*.

(3) والعلم الطبيعيّ هو النظر في جملة الأجسام كلّها التي وجودها في الحسّ من السموات والأرض وما بينهما من النبات والحيوان وغيرهما ، ثم فيما يخصّ أجزآها العظام والخواصّ اللازمة لنوع نوعٍ منها ، وعن الأشياء التي قوامها بها ووجودها فيها وعن العلّة التي وجدت لأجلها فيها . والبحث الأوّل عن الأمور العامّة المشتركة لها وهي التي لا يخلو جسم طبيعيّ منها ، هي أربعة أشياء ما وجودها به وما وجودها فيها وما وجودها عنه وما وجودها لأجله . ولما كانت 119b الأجسام الصناعيّة قد يوجد أيضاً لها هذه الأربعة وهي فيها أبين وأظهر ، جعلناها مثالاً لما في الأجسام الطبيعيّة ليتصوّر ويتحقّق ، وأعني بالأجسام الصناعيّة ما كان مثل السرير والكرسي والثوب والخاتم وبالجُمل ما كان وجوده بالإرادة والرؤيّة . ونجعل مثال ما ذكرنا في الأجسام الطبيعيّة من الأجسام الصناعيّة السرير ، فإنّه إنما كان سريراً بصورته ، أعني هيئة التربيع ، وفي مادّته ، أعني الخشب ، وعن فاعله ، أعني النجّار، ولتمامه الذي من أجله عمل وهو الجلوس عليه . وتصوّر هذه الأشياء بعينها في الثوب والسيف وكلّ جسم صناعيّ لتجعلها مثالك في الأجسام الطبيعيّة ، فأمّا الأشياء التى قوامها 120a في الأجسام الصناعيّة ، كالنقوش في السرير والصقال في الثوب والبريق في السيف ، فإنّ لكلّ واحد منها أيضاً غرضاً وكمالاً لأجله وجد ، وذلك أنّ النقش أوجد في السرير ليحسن به منظره ، والحديد ليصير بها وثيقاً ، والصقال أوجد في الثوب ليتجمّل به وليزل عنه الغبار ، والبريق أوجد في السيف ليرهّب به العدو . ومن هذه الأمور الأربعة إثنان بهما يحصل كلّ موجود بالفعل وعلى الكمال وهما صورته ومادّته ، وأعني بالصورة هيئاته كالحدّة في السيف ، وبالمادّة الشيء الموضوع له كالحديد الحامل للحدّة . وإذ قد مثلنا ذلك في الأجسام الصناعيّة فتصوّرها في الأجسام الطبيعيّة ، فإنّ كلّ جسم طبيعيّ

5 هي : وهي *K. al-Ḥikma*   6 فيها : فيه *K. al-Ḥikma*   17 بها : به *K. al-Ḥikma*   21 وإذ : وإذا *K. al-Ḥikma*

3 أجزآها : اجْراؤها M   4 وجدت : وحدت M   5 يخلو : يخلوا M | after منها blank space of three letters M   6 after لأجله blank space of three letters M   10 ونجعل : ويجعل M   12 ولتمامه : واتمامه M   13 after عليه blank space of three letters M   15 كالنقوش : كالنفوس M   18 الثوب : السّرير written on the line M : الثوب written above the line M | ليتجمّل : ليتحمل M | وليزل : ولترك M   20 هيئاته : هياته M

(3) Natural science is the examination of the entirety of all the bodies, of plants, animals and other things which exist in the [realm of] sense perception [and] belong to the heavens, earth and what is between these two. Next it is the examination of what is characteristic of the parts [of these bodies] with regard to size and properties which are concomitant to each single species. [The discussion is further][298] of the things by which the subsistence of [these bodies] is [caused] and in which their existence is, and of the cause due to which they exist in them. The first discussion [is] of the general things which are common to [all natural bodies] and of which a natural body is never devoid. They are four things: whereby it exists, wherein it exists, whence it exists and wherefore it exists. Since these four are also found in artificial bodies, and are more evident and 119b
manifest in them, we take them as an example for what is in natural bodies in order to conceive and ascertain [them]. I mean by artificial bodies things like a bed, a chair, a cloth, a seal ring and on the whole what exists by way of will and deliberation. Among the artificial bodies we take the bed as an example of what we have mentioned in regard to the natural bodies. For it is truly a bed by its form, i.e. the rectangular shape, and in its matter, i.e. wood, and through its agent, i.e. the carpenter, and for the purpose because of which it is made, and that is to sit on. Imagine precisely these things in a cloth, a sword and every artificial body, so that you take them as your example with regard to the natural bodies. For each of the things the subsistence of which is in artificial bodies, like the 120a
engravings on the bed, the sheen of the cloth and the lustre in the sword, there is also an aim and a perfection because of which it exists. For the engraving is produced on the bed in order to decorate it and the iron structure in order to make it solid. The sheen is produced on the cloth in order to make it pretty and repellent of dust. The lustre is produced in the sword in order to intimidate the enemy. Among these four things there are two through which every existent reaches actuality and perfection. These two are its form and matter. I mean by form its shape, like the blade of the sword, and by matter its substratum, like the iron which bears the blade. Since we have already exemplified that with regard to the artificial bodies,

[298] The preposition *ʿan* probably corresponds to the following *baḥth*.

يشتمل على هذه المعاني كلّها ، وذلك أنّ كلّ واحد منها له صورة بها صار
120b هو ما هو ، ومادّة منها عمل وفيها وجدت تلك الصورة ، وفاعل عنه وجد
وهو صانعه وموجِده ، وكمال وتمام لأجله وجد وعمل . والفرق بين هذه
الأربعة الأشياء في الأجسام الصناعيّة والأجسام الطبائعيّة أنّها في الصناعيّة
محسوسة على الأمر الأكثر ، وهي غير محسوسة على الأمر الأكثر في الطبيعة
وإنما يصحّح وجودها بالبراهين والقياسات الصحيحة . وإنما قلتُ هي
محسوسة على الأمر الأكثر في الصناعيّة ، لأنّ الأدوية المركّبة وأشباهها لا
يظهر صورها وتماماتها للحسّ ، وإنما يحدث فيها قوى غير محسوسة تفعل في
الأبدان أفعالها التى من أجلها أوجدت ، كالترياق فإنّ هذا الجسم الصناعيّ
يشبه الأجسام الطبيعيّة من هذا الوجه ، وذلك أنّ للعين قوّة الإبصار طبيعيّة
لها وهي غير محسوسة
E77b ‹ومن أجلها وجدت العين ، وكذلك لكلّ جسم طبيعيّ مادّته وصورته وفاعله
وتمامه وتبيّن أحوال أغراضه التي بها قوامه .

والأجسام الطبيعيّة تنقسم بالقسمة الأولى إلي قسمين ، أحدهما البسيط
والآخر المركّب . فالبسيط منها ما كان وجوده لا عن جسم تقدّمه ولا قوامه
من جسمين فما فوقهما ، والمركّب منها ما كان وجوده عن أجسام أُخر
متقدّمة له وقوامه منها . فالأوّل مثل الأجرام العلويّة من الأفلاك والكواكب ،
E78a والثاني من الأجرام السفليّة مثل الحيوان | والنبات .

ثم انّ صاحب هذا العلم الذي رتبه وبرع فيه حتّى زاد على جميع من تقدّمه ،
أعني ارسطاطاليس قسمه إلى ثمانية أقسام .

(a) أحدها النظر في تلك الأشياء التي تشترك فيها الطبيعيّات كلّها بسيطها
ومركّبها ، أعني المبادئ الأول والأعراض التابعة لها . واستقصى ذلك غاية
الاستقصاء وتصفّح أقاويل القدماء فيها ، فأفسد ما أفسد منها ببراهين قويّة

12 لكلّ : كلّ *K. al-Ḥikma*

2 وجد : واحد M 3 after وعمل blank space of three letters M 4 الطبائعيّة : الطباعيه M 5 الطبيعة : الطبيعيّه M 7 وأشباهها : واشيَاهها M 10 يشبه : بشبه M 11 after محسوسة blank space of almost the entire line M 13 أغراضه : اعراضه E 21 أحدها rubricated in E | الأشياء : الاشيآء E 23 الاستقصاء : الاستقصَا E | القدماء : القدمَآءِ E

imagine [these four things] in the natural bodies. For every natural body contains all these features, as for every one of them there is a form by which it becomes what it is, and a matter out of which it is made and in which that form exists, and an agent through which it exists, namely its 120b
maker and producer, and a perfection and purpose because of which it exists and is made. The difference between these four things in the artificial bodies and in the natural bodies is that in artificial [bodies] they are more perceptible, whereas in natural [bodies] they are less perceptible and their existence is rather attested by sound proofs and demonstrations. I only said that they were more [yet not completely] perceptible in artificial [bodies], because the forms and purposes of compound drugs and their like are not evident to sense perception. There are imperceptible powers emerging in them, which perform in the bodies their activities on account of which they were produced, for example the theriac. Thus this artificial body resembles the natural bodies in this respect. For the power of vision is natural to the eye, yet not perceptible,

<[299][but] because of it the eye exists. In the same way every natural body has its matter, form, agent and purpose, and the conditions of its ends through which it subsists are evident.

Natural bodies are primarily divided into two parts, simple and composite. The simple [natural bodies] are those the existence of which is not from a body preceding them and the subsistence of which is not by two bodies. Thus they are above these two [possibilities]. The composite [natural bodies] are those the existence of which is through other bodies preceding them and the subsistence of which is by them. The first are, for example, the heavenly bodies of the spheres and the stars. The second belong to the lower bodies like animals and plants.

Then the master of this science, which he arranged [in a certain order] and in which he excelled so that he exceeded all who preceded him, i.e. Aristotle, divided [natural science] into eight parts.

(a) The first of them is the examination of those things which are common to all natural things, be they simple or composite, i.e. the first principles and the accidents subordinate to them. He enquired into that with the utmost intensity and examined the sayings of the Ancients about them. Then he demolished fallacious [doctrines] by powerful proofs and upheld

[299] The Marsh MS has a third gap here, starting from line 120b13. The folios 121 and 122 are missing completely and the account of the natural sciences is resumed on 123a1 in the middle of a sentence. The lost text can be supplied with the help of the *K. al-Ḥikma*, MS Esad Efendi, fols.77b13–78b3.

وصحّح الرأي الحقّ منها بمثل تلك البراهين .

(b) والثاني النظر في الأجسام البسيطة هل هي موجودة وما هي وكم هي وما حركتها وخواصّها وتوابعها ؟ وفَعل في هذا النظر مثل ما فَعله فيما تقدّم من إعطاء العلل والبراهين عليها ، ثم النظر في بسائط الأجسام المركّبة هل هي في جملة تلك البسائط التي تبيّن وجودها أم هي أجسام غيرها ، وإن كانت فيها فهل هي جميعها أو بعضها ، وإن كان بعضها فأيّما هي وما هي ؟ ثم النظر فيما تشترك فيه هذه البسائط كلّها ما كان منها أسطقسّات وما كان منها مركّباً عنها وتوابعها ؟ وأعطى البراهين عليها .

(c) والثالث النظر في الأجسام التي في الكون والفساد على العموم ، والنظر في كون الأسطقسّات وفسادها وأسبابها ومبادئها .

(d) والرابع النظر في مبادئ الانفعالات والأعراض التي تخصّ الأسطقسّات وحدها دون المركّبات .

(e) والخامس النظر في المركّبات عن الأسطقسّات وفيما كان متشابه الأجزاء
E78b ومختلف الأجزاء ، وفي أنّ المتشابه | الأجزاء منها ما هي أصول للأجسام المختلفة الأجزاء ، كاللحم والعظم ، ومنها ما ليست أصولاً لجسم طبيعيّ أصلاً ، كالملح والزاج والذهب والفضّة ، ثم النظر في جميع ما تشترك〉
123a فيه الأجسام المركّبة ، ثم النظر في جميع ما يشترك فيه المتشابهة الأجزاء والمختلفة الأجزاء .

(f) والسادس النظر فيما يشترك فيه الأجسام المركبة المتشابهة الأجزاء التي ليست أجزاء للمختلفة الأجزاء ، وهي الأجسام المعدنيّة والحجارة وأصنافها ، وما يخصّ كلّ نوع منها .

(g) والسابع النظر فيما يشترك فيه أنواع النبات وما يخصّ كلّ واحد منها ، وهو أحد جزئي النظر في المركّبة المختلفة الأجزاء .

17 في . . . ما : فيما *K. al-Ḥikma* 19 المركبة : المترّكبة *K. al-Ḥikma* 23 أحد : احدى *K. al-Ḥikma*

2 والثاني rubricated in E 4 إعطاء : اعطآءٍ E | بسائط : بسَآيط E 5 البسائط : البَسآيطِ E 7 البسائط : البسايط E 9 والثالث rubricated in E 11 والرابع rubricated in E 13 والخامس rubricated in E | الأجزاء[1] : الاجزآءِ E | الأجزاء[2] : الاجزَآ E 14 الأجزاء : الاجزآءِ E | الأجزاء : الاجزآ M 15 الأجزاء : الاجزآءِ E 19 التي : الذي M 20 أجزاء : اجزآ M 21 after منها blank space of three letters M 23 جزئي : جزي M

the true opinion by like proofs.
(b) The second is the examination of the simple bodies: whether they are existent, what they are, how many they are, what their motion is, their properties and their effects. He acted in regard to this examination as he had done with regard to what preceded [it] by presenting causes and proofs for them. Next is the examination of the simple [components] of the composite bodies: whether they are among the sum of those simple [bodies] the existence of which has [already] been explained or whether they are bodies other than those. And if they are among [the already described simple bodies], whether all of them are [components of composite bodies] or [only] some of them. And if [only] some of them, which they are and what they are. Next is the examination of what is common to all these simple [components]: which of them are elements and which of them are composed from them and [are] their effects. He presented proofs for that.
(c) The third is the examination of the bodies which fall under generation and corruption in general, and the examination of the generation and corruption of the elements, their causes and principles.
(d) The fourth is the examination of the principles of the passive [qualities] and the accidents which are characteristic of the elements alone and not [also] of the composites.
(e) The fifth is the examination of the composites of the elements and of what has similar parts, [i.e. is homoeomerous], and of what has different parts, [i.e. is heteromerous]. [It is also the examination] of the homoeomerous bodies, which are fundamental to the heteromerous bodies, like flesh and bone, and [the homoeomerous bodies], which are not at all fundamental to any natural body, like salt, vitriol, gold and silver. Next is the examination of all that[300]>
in which the composite bodies participate and thereupon the examination 123a
of all that is common to the homoeomerous and heteromerous bodies.
(f) The sixth is the examination of what is common to the composite homoeomerous bodies which are not parts for the heteromerous bodies. They are the mineral bodies, the stones and their [different] kinds. [The sixth is also the examination of] what is characteristic of each of their species.
(g) The seventh is the examination of what is common to all species of plants and what is characteristic to each of them. It is one of the two parts of the examination of the things composed of different parts.

300 Here the gap in the Marsh MS comes to an end.

(h) والثامن النظر في جميع ما يشترك فيه أنواع الحيوان وما يخصّ كلّ نوع
منها ، وأعطى أسبابها وعللها ، وهو الجزء الثاني من النظر في المركّبة المختلفة
الأجزاء .

(II) وأمّا الحكمة العمليّة فانقسامها إلى ثلاثة أقسام ، وهي في سياسات
ثلاث ، سياسة النفس بالأخلاق ليصير الإنسان فاضلاً ، وسياسة المنزل وما
123b يشتمل عليه بالتدبير ليصير منتظماً ، وسياسة المدن والعامّة بالسنن العادلة
لتجري على أفضل أحوالها .

(212) وقال الحكمة جزءان ، نظريّ وعمليّ . فبالنظريّ يمكن تحصيل الآراء
الصحيحة ، وبالعمليّ يمكن تحصيل الهيئة الفاضلة التي صدّر عنها الأفعال
الجميلة ، وبهذين الأمرين بعث الله الأنبياء عليهم السلم فدلّوا الناس عليهما
وكلّفوهم الأخذ بهما .

فأمروا الناس بالتوحيد ولزوم أحكام العقل وإقامة السياسات اللائقة بالأزمنة
والأوقات ، وحملوا الخواص على طريقة النظر والبحث وعوامّهم على طريق
الأدب والقهر ، فإنّ منزلة الأنبياء عليهم السلم من نفوس الناس منزلة الأطبّاء
من الأبدان ، فهم يعالجونهم معالجة المرضى ، وذاك أنّ كثيراً من المرضى يحتاج
124a أن يعالج بالكره وبالتهديد والضرب ليقبل ما ينفعه . ولا يلزم أن يفهم العلّة
التي من أجلها يناول المكروه ويمنع المحبوب لأنّ جدوى ذلك عليه قليلة ومعرفته
عن تحصيله بعيدة ، وكما أنّ كثيراً من المرضى إذا برئ على تدبير الطبيب

6 المدن : البدن K. al-Ḥikma 8 وقال الحكمة : نكتة الحكمة K. al-Ḥikma : والحكمة Fawz 9 صدّر : تصدر K. al-Ḥikma, Fawz 10 after الله add. تعالى K. al-Ḥikma | السلم : السلام إن جميعهم إنّما أمروا بالتوحيد Fawz 10–11 فدلّوا ... بهما om. Fawz 12 فأمروا ... بالتوحيد : Fawz 13 والأوقات : والأحوال Fawz | after الخواص add. من الناس Fawz 14 منزلة¹ om. Fawz | السلم : السلام منزلتهم Fawz 15 وذاك : وذلك Fawz | after يحتاج add. إلى Fawz 16 after بالكره add. منه K. al-Ḥikma, Fawz | وبالتهديد : أو بالتهديد K. al-Ḥikma | وبالتهديد : وربّما هدّد بالضرب بل ربما أوقع به : والضرب Fawz | ولا ... أن² : إذا لم يكن ذا منّة Fawz | العلّة : بها ما يشير به الطبيب ولذلك لا يشتغل معه بذكر السبب والعلّة Fawz 17–18 ومعرفته ... بعيدة : ولأن فهمه بعيد عن تحصيله Fawz

3 after الأجزاء break (∴) ind. and blank space of three letters M 4 ثلاثة : ثلثه M 5 ثلاث : ثلثٍ M 6 المدن : البدن M | بالسنن : بالسَّنتَنِ M 7 after أحوالها break (∴) ind. and blank space of three letters M 8 جزءان : جُزان with a faint *hamza* added above the *zāʾ* M | الآراء : الْارا M 9 الهيئة : الهية M

(h) The eighth is the examination of all that is common to all species of animals and of what is characteristic of each of their species. He presented their reasons and causes. It is the second part of the examination of the heteromerous composites.

(II) Practical wisdom is divided into three parts which belong to three sorts of governance: the governance of the soul by morality so that man becomes virtuous; the governance of the household and its contents by [household] management so that it becomes [well] ordered; and the 123b
governance of cities and the common people by just laws so that they are guided towards their best conditions.

(212) He says:[301] wisdom is of two parts, theoretical and practical. Through theoretical [wisdom] it is possible to attain sound opinions and through practical [wisdom] a virtuous form from which good actions proceed. With these two commands God has sent the prophets - upon them be peace - so that they show both to the people and charge them with executing both. They have commanded people to declare *tawḥīd*, to adhere to the judgements of intellect and to set up politics which suit the[ir] times and periods. They have set the elite on the way of examination and discussion, and the common people on the way of discipline and subjugation. For the prophets - upon them be peace - are with regard to the souls of people like the physicians with regard to the[ir] bodies. They treat them in the way sick people are treated, for many of the sick must be treated with something unpleasant, and with threats and beating, so that they receive 124a
what is beneficial to them. [The sick person] does not need to understand the reason on account of which something unpleasant has to be taken and something pleasant is forbidden, as that would be of little advantage [to him] and his knowledge is far from [being able to] grasp [that]. [This is] like the passion which prompts many a sick person when recovering as a result of the physician's regimen to interpret [this regimen] according to

[301] Excerpted from Miskawayh, alternative *Fawz;* cf. parallel in *K. al-Ḥikma,* MS E, fols.73b6–74a4; and *Fawz,* 93.17–94.1, 41.7–42.3.

يحمله الهوى على التأوّل لشهواته فيخرج لها طريقاً في مصالحه وإن كانت ضارّة له ، كذلك حال كثير من أهل النظر تدعوهم العادات ومساعدة النفس على أحكام الحسّ وصعوبة النظر بمجرّد العقل إلى تأوّل ما شرعه الرسول عليه السلم وأمر به الحكيم ، لا سيمّا إن انضاف إلى ذلك حبّ غلبة أو طلب رياسة .

(213) قال أبو الحسن العامري العلم فاتحة العمل والعمل تمام العلم ، والفاتحة بلا تمام ضائع والتمام بلا فاتحة محال .

والإنسان عبارة عن النفس المتّحدة بالجسم .

124b (214) والفضائل تنقسم قسمين ، جسمانيّة ونفسانيّة . وفضائل النفس أشرف من فضائل الجسم لمعان أربعة ، أحدها الدوام والاتّصال ، والثاني القوّة والازدياد ، والثالث الانتهاء والتمام ، والرابع المعونة والأرفاد . فأمّا الدوام والاتّصال فإنّ النفس متّى أقتنت فضيلة من فضائلها صارت دائمة الالتذاذ بها ، ولذّة البدن تزول وتستحيل . وأمّا القوّة والازدياد فلأنّ النفس كلّما استفادت فضيلة من فضائلها قويت على استجرار ما هو أفضل منها بل صارت عليه أقدر وعلى تحصيله أحرص وأصبر . وأمّا الانتهاء والتمام فإنّ النفس كلّما أفادت قنية فقد تمّ فعلها وانقضى سعيها ولم يبق لها شغل في
125a معناها ، والبدن لا ينتهي في مطلوبه إلى غاية بغية إلا عادت الحاجة مكرّرة إليه . وأمّا المعونة والأرفاد فلأنّ النفس في فضائلها تعيّن البدن وتكسبه الأزيان إمّا على جهة الصناعة والاكتساب وإمّا على سبيل النصرة بالطبع ، والبدن يقلّ رفده للنفس وتنزر معونته لها ، وإن وجد فليس على قصد منه فيحمد ،

1 لها : له *Fawz* 2 كذلك : وكذلك *Fawz* | تدعوهم : تحملهم *Fawz* 2–3 ومساعدة...على : واشتغال ما ذكرته من فطام النفس عن *Fawz* | شرعه : أمر به الحكيم وشرعه *Fawz* 3 إلى : على *Fawz* 4 السلم : السلام *Fawz* | وأمر ... الحكيم om. *Fawz* 5 طلب om. *K. al-Ḥikma*

2 ضارّة : صاره M 5 after رياسة break (∴) ind. and blank space of more than half a line M 8 below بالجسم add. in marg. as an explanatory gloss by a later hand اي جسم بسيط ; break (∴) ind. and blank space of more than half a line M 10 after والاتّصال blank space of three letters M 11 after والتمام blank space of three letters M 12–13 الالتذاذ بها : الالتذاذتها M 15 after وأصبر break (∴) ind. and blank space of half a line M 18 after إليه break (∴) ind. and blank space of half a line M | الأزيان : الازينُ M

their appetites and to explain following them as being in their interest, even if [their appetites] are harmful to them. Many a speculative thinker is urged in the same manner by habits, the support of the soul for the judgements of sense perception, and the difficulty of examining by means of pure intellect to interpret what the prophet - upon him be peace - has made law and what the sage has commanded to do, especially if love of predominance or pursuit of leadership is added to that.

(213) Abū l-Ḥasan al-ʿĀmirī says:[302] knowledge is the beginning of any action and action is the completion of knowledge. Beginning without completion is wasted and completion without beginning unthinkable.

Man consists of soul which is united with body.

(214) The virtues are divided into two parts: physical and psychic. The 124b
virtues of the soul are nobler than the virtues of the body due to four features. One of them is perpetuity and continuity, the second power and increase, the third end and completion, the fourth help and support. As for perpetuity and continuity, when soul has acquired one of its virtues it perpetually takes delight in it, whereas the pleasure of the body ceases and changes. As for power and increase, whenever the soul acquires one of its virtues, it becomes able to let its better [part] take the lead, and it even becomes stronger in [its better part] and more desirous of and patient in attaining it. As for end and completion, whenever the soul gains a property, its activity is already completed, its endeavour has come to an end and it no longer needs to be occupied with its feature. Yet the body, in
regard to its desire, does not reach the desired end unless the required 125a
object comes back to it repeatedly. As for help and support, the soul defines the body through its virtues and lets it obtain adornment either through skill and acquisition or by means of assistance according to [its] nature, whereas the support of the body for the soul is little and its help for it is negligibly small. Even if it exists, it is not pursued by [the body] so that it should be praised [for its help]; rather it is in the same way other

[302] This and all the remaining passages may be taken from al-ʿĀmirī's *al-Itmām li-faḍāʾil al-anām*, but as the work is not extant, it is impossible to be certain.

بل على مثال ما ينجد لها سائر الأجسام .

والفضائل النفسانيّة أربعة ، العلم والجود والشجاعة والسياسة . والثلاثة الأولى صادرة عن القوى الثلاث التي هي الشهوانيّة والغضبيّة والمنطقيّة . ففضيلة القوّة الناطقة العلم ، وفضيلة القوّة الشهوانيّة الجود ، وفضيلة القوّة الغضبيّة الشجاعة .

ومن كمل لاستصلاح هذه القوى وقوّم الأفعال الصادرة عنها سمّي عادلاً .
125b وإذا قدر على استصلاح نفسه قدر على استصلاح غيره فسمّي سائساً وسمّي فعله ذلك سياسة . فلن يستحقّ أحد الوصف بهذه الجلال على الإطلاق إلا إذا تتعدّاه فضيلته إلى جماعة يفيدهم فيصير سبباً لتخريجهم وعلّة لاستصلاحهم . فأمّا من علم ولم يعلّم غيره فلن يعدّ عالماً بل يسمّى مستبصراً ، كما أن من ذبّ عن نفسه وحرمته ولم يذبّ عى محلّته ومدينته لم يعدّ شجاعاً بل يسمّى ضابطاً ، ومن اقتصر بمعروفه على نفسه وعياله لم يعدّ جواداً بل يسمّى واسع النفقة .

(215) والعلم ينقسم قسمين ، أحدهما الملّيّ والوصول إليه من طريق الوحي وهو مقتبس من جهة الأنبياء ، والآخر الحكميّ وهو المستخرج من قضية العقل واقتباسه من جهة الحكماء . وكلّ نبيّ حكيم وليس كلّ حكيم نبيّاً .

126a والعلوم الملّيّة تنقسم ثلاثة أقسام ، أحدها المتناول بإتعاب الحسّ وهو صناعة الحديث ، والثاني المتناول بإتعاب العقل وهو صناعة الكلام ، والثالث المتناول بإتعابهما جميعاً وهو صناعة الفقه ، وعلم اللغة آلة من آلاتها ومعين على تحصيلها .

14 والعلم : قال ابو الحسن محمد العامري رحمة الله تعالى العلم *K. Ṣaḥāʾif* 18 المتناول[2] om. *K. Ṣaḥāʾif* 19 آلة om. *K. Ṣaḥāʾif*

1 ينجد : سخَّر M | after الأجسام break (.∴) ind. M 2 after والسياسة blank space of three letters M | والثلاثة : والثلثه له M 3 الثلاث : الثلث M | والغضبيّة : والعَضية M 5 after الشجاعة blank space of three letters M 6 after عادلاً blank space of five letters M 8 الجلال : الخلال M 9 تتعدّاه : تعدَّاه M 10 after لاستصلاحهم blank space of three letters M 13 after النفقة break (.∴) ind. and blank space of ten letters M 16 after نبيّاً blank space of three letters M 17 ثلاثة : ثلثه M | وهو : وهي M 18 after الحديث blank space of three letters M 18–19 صناعة ... الكلام add. in marg. M 19 آلاتها : الاتها M

bodies may be helpful to [the soul].

The psychic virtues are four: knowledge, generosity, courage and leadership. The first three proceed from the three faculties which are the appetitive, the irascible and the rational. The virtue of the rational faculty is knowledge, the virtue of the appetitive faculty generosity and the virtue of the irascible faculty courage.

He who is perfected because he seeks to act well with these faculties and who performs the actions proceeding from them properly is called just.
When he can improve himself, he can [also] improve someone else, and he 125b
is thus called governing and that action of his is called government. For nobody deserves to be described by this dignity absolutely, unless his virtue extends beyond himself to a group which he benefits so that he becomes a reason for their education and a cause for their improvement. He who knows, yet does not teach others, is not considered knowing, but called perceptive. In the same way he who defends himself and his wife, yet does not defend his quarter and his city, is not considered courageous, but called self-protective. And he who restricts his beneficence to himself and his family is not considered generous, but called lavish.

(215) Knowledge is divided into two parts:[303] one of them is religious [knowledge] which is reached via revelation and acquired from the prophets. The other is philosophical [knowledge] which is deduced from the assertion of intellect and acquired from the sages. Every prophet is a sage, but not every sage a prophet.

The religious sciences are divided into three parts: one of them is attainable through the effort of sense perception, namely the art of *ḥadīth*. The 126a
second is attainable through the effort of intellect, namely the art of *kalām*. The third is attainable through the effort of both of them together, namely the art of *fiqh*. Philology is one of their instruments and helps in attaining them.

[303] The passage (except for the last paragraph) is also quoted in ʿAlī al-Riyāḍī, *K. Ṣaḥāʾif al-laṭāʾif*, pp. 26–7.

والعلوم الحكميّة تنقسم ثلاثة أقسام ، أحدها العلم بالمحسوسات وهو صناعة الطبيعيّين ، والثاني العلم بالمعقولات وهو صناعة الإلهيّين ، والثالث العلم بما يتوسّط بينهما وهو صناعة الرياضيّين ، وعلم المنطق آلة من آلاتها .

ونسبة علم اللغة والمنطق إلى العلوم الأخر كنسبة العلوم أجمع إلى السعادة العظمى ، أعني أنّهما يطلبان لتلك كما أنّ تلك الأخر تطلب للسعادة . ولهذا قيل إنّه لا يتوصّل إلى ما يحتاج إليه من العلوم إلا بمعرفة ما لا يحتاج إليه ، فإذن صار ما لا يحتاج إليه محتاجاً إليه .

(216) ولن يتوصّل إلى شيء من المعلومات الحقيقيّة إلا من جهات ثلاث ،
إحداها المواد السماويّة ، والثانية الدلائل الاستنباطيّة ، والثالثة التجارب 126b
التكراريّة . فأمّا الموصول إليه من طرائق المواد السماويّة فكما اعتقد من حال البدء والإعادة وكيفيّة الثواب والعقاب وأبواب العبادات ومبادئ الصناعات ، فإنّها معان ينحبس العقل عن استدراكها إلا بتأييد علويّ . وأمّا الوصول إليه من طرائق الدليل الاستنباطيّ فكمعرفة الصانع جلّ جلاله ومعرفة حدوث أجرام العالم ومعرفة وجوب الجزاء ومعرفة جوهريّة النفس وغيرها من المعاني الواقعة في أصناف العلوم العقليّة . وأمّا الموصول إليه من تكرار التجارب فكعلم الفراسة وعبارة الرؤيا وعلم العيّافة والزجر والفأل وغيرها ممّا اشتهر في الأمم ، فكثر استعمالها من غير أن يعرف عللها أو يوقف على الأسباب
الموجبة لها . 127a

(217) الجود حمل النفس على بذل المال وإفاضة المعروف علماً بما يكتسب به من الثناء والثواب ، ففي طبع الإنسان محبّة الذخائر والضنّ بالنفائس غير أنّ النفس منقادة ، فمتى حُملت على الشيء بالإدمان رسخ فيها عروقه وصار كإحدى القوى الطبيعيّة . وحقيقة الجود إفاضة المعروف ، وكلّ عالم لم يبذل

2 العلم[2] om. *K. Ṣaḥāʾif*

1 ثلاثة : ثلثه M 3 آلة : آلة M | آلاتها : الأتها M 5 لتلك : ليلك M | للسعادة : السعادة M
7 after إليه[3] break (∴) ind. M 8 ثلاث : ثلث M 9 إحداها : احديها M | after الاستنباطيّة blank space of three letters M 11 البدء : البدْو M | وأبواب : ابواب M
12 ينحبس : ينحس M | after علويّ blank space of three letters M 14 الجزاء : الجزآ M
15 after العقليّة blank space of three letters M 18 after لها blank space of three letters M
22 after الطبيعيّة blank space of three letters M

The philosophical sciences are divided into three parts: one of them is the knowledge of the sensibles, namely the art of the natural scientists. The second is the knowledge of the intelligibles, namely the art of the metaphysicians. The third is the knowledge of what is intermediary between these two, namely the art of the mathematicians. Logic is one of their instruments.

The relation of philology and logic to the other sciences is like the relation of all the sciences to the greatest happiness, i.e. the two of them are sought for those [other sciences] just as those others are sought for happiness. Therefore it is said that the required science is only reached through non-required knowledge. So what has been non-required consequently becomes required.

(216) Any of the things that can be truly known is reached from three directions alone: One of them are [via] the heavenly matters; the second [via] deductive proofs; and the third [via] repeated experiences. What is 126b
reached via heavenly matters are, for example, the beliefs regarding the status of the first and second creation [provenance and destination], the quality of reward and punishment, the various sorts of acts of devotion and the principles of the crafts, for they are notions which the intellect is debarred from perceiving except with exalted support. What is reached via deductive proof is, for example, the cognition of the Creator – exalted His exaltedness –, of the emergence of the earthly bodies, of the necessity of recompense, of the substantiality of the soul and other notions which fall within the categories of intellectual knowledge. What is reached through repetition of experiences is like horsemanship, the interpretation of dreams, augury and ornithomancy and other things which are wellknown among the peoples. They are often employed without their causes being known or cognisance of the reasons necessitating them. 127a

(217) Generosity is the incitement of the soul to give possession freely and to pour forth benefit[s] in the knowledge of the appreciation and the reward acquired thereby. So it may lie in the human nature to love treasures and hold on to precious possessions, yet the soul is obedient and when it is incited to do something continually its roots are deeply rooted in [the soul] and it becomes like one of the [soul's] natural powers. The true essence of generosity is pouring forth benefit[s]. Every knower who

علمه ويخرّج به غيره فهو جواد ، وكلّ شجاع يبذل نفسه ويصون غيره فهو جواد ، بل الجود بالنفس أقصى غاية الجود ، والشجاعة صرامة القلب في الأهوال ورباطة الجأش في المخلوف . ولن يسمّى الرجل شجاعاً ما لم يمارس الحروب ، بل يقال إنه قوى القلب .

(218) السياسة صناعة بها تتهيّأ عمارة البلاد وحِمَايَة من فيها من العباد . وكلّ
127b صناعة فلا بدّ لها من مصنوع وكلّ مصنوع يتمّ كونه بخمسة معان هي كالعلل
لها والأسباب المخلصة لذاتها . أحدها المادّة الموضوعة وهي كالخشب للنجّار
والجسد المحتمل للصحّة والسقم ، ويقال له علّة عنصريّة . والثاني المعرفة بهيئة
ما يقصد إليه في الإيجاد وهي كتصوّر النجّار هيئة السرير وتصوّر الطبيب
صحّة البدن، ويقال لها علّة صوريّة . والثالث حركة يستعملها الصانع في
تأحيد الصورة بالمادّة كمباشرة الطبيب أمر المعالجة ومباشرة النجّار صناعة
النجر ، ويقال لها علّة فاعلة . والرابع غرض ينصبه الصانع في نفسه وهو المعنى
الذي لأجله يتعاطى إحداث المصنوع كالسكنى في اتّخاد البيت وبقاء النفس
في اجتلاب الصحّة ، ويقال لها علّة تماميّة . والخامس أداة يستعين بها الصانع
128a في تحصيل ما يباشره من عمله كالقدّوم للنجّار والأدوية للطبيب ، ويقال له
علّة آليّة . وإذا نقل المثال إلى صناعة السياسة وجدت المادّة فيها أمور الرعية ،
ووجدت الصورة اجتلاب المصلحة العامّة ، ووجدت الفاعلة ما يباشره الملك
من تثقيفهم وحمايتهم ، ووجدت الغرض دوام المصلحة وبقائها ، ووجدت
الآلة الترغيب والترهيب .

وفعل السائس بكلّيّة ينقسم قسمين ، أحدهما التعهّد وهو حفظ المستقيم من أمور رعاياه على استقامته وانتظامه ليطّرد على الأمن والهدوء ، ومثاله من صناعة الطبّ حفظ الصحّة على البدن ، والثاني الاستصلاح وهو ردّ الفساد

1 يبذل : يَذل M 4 إنه : ار M | after القلب break (∴) ind. and blank space of more than half a line M 6 كونه : لَونهُ M 8 بهيئة : بهَيه M 9 هيئة : هيةَ M 10 after صوريّة blank space of three letters M 11 تأحيد : تاجيد M 12 after فاعلة blank space of three letters M 13 وبقاء : وبقآ M 14 after تماميّة blank space of three letters M 16 الرعية : الرّغبة M 18 تثقيفهم : ينقيفهم M | الغرض : العرض M | وبقائها : وَبقاَهَا M 20 التعهّد : التعهيد M 21 أمور رعاياه : اموررعَاياة M 22 after البدن blank space of three letters M | الاستصلاح : للاستصلاح M

gives his knowledge freely and educates someone else through it is generous. Everyone courageous, who gives of himself freely and protects someone else, is generous. Yet, being generous of oneself is the ultimate end of generosity, and courage is being vigorous at heart in terrifying situations and retaining composure when being left behind. Man is not called courageous, unless he exerts himself in wars, rather he is said to be strong at heart.

(218) Leadership is a craft[304] that provides for the cultivation of a country and the defence of its people. It is inevitable that every craft is related to what is produced [by it], and the generation of every product is perfected 127b
by five features which are like causes for [the craft] and the reasons sincere to its essence. One of them is underlying matter, and it is like wood for the carpenter and the body bearing health and illness. It is called the material cause. The second is the cognition of the shape of what is intended in the production, and it is like the carpenter's conception of the shape of a bed and the physician's conception of the health of the body. It is called the formal cause. The third is the motion which the producer employs for uniting form with matter like the physician carrying out treatment and the carpenter carrying out carpentry. It is called the efficient cause. The fourth is the aim set up by the producer in his soul and the concept on account of which the product is brought forth, like habitability in the assembly of a house and permanence of the self by conveying health. It is called the final cause. The fifth is the tool which the producer uses for support in attaining what he carries out in his doing, like the adze for the carpenter and medication for the physician. It is called the 128a
instrumental cause. Now, applying the example to the craft of leadership, the matter here is the affairs of the subjects, the form conveying general welfare, the efficient [cause] the sovereign's practice of instructing and protecting [his subjects], the aim the perpetuity and permanence of welfare, and the instrument instilling desire and fear.

The entire activity of the leader is divided into two parts. One of them is care. It is the preservation of the subjects' affairs in a good state by virtue of his integrity and the organisation [of this state] so that there is continuous security and peace. An example of it from the craft of medicine is the preservation of the body's health. The second one is rectification. It is the

[304] For this passage al-ʿĀmirī has been inspired by a very similar account on leadership by Abū Zayd al-Balkhī preserved in al-Tawḥīdī's *al-Baṣāʾir* (see commentary).

إلى الصلاح والخلل إلى النظام ، ومثاله من صناعة الطبّ إعادة الصحّة إلى
البدن العليل .
(219) ويحتاج السائس الفاضل في تدبير أموره وتسديد أعماله إلى أربعة عشر
128b نفساً من أعوانه ، هو بمنزلة الأعضاء الطبيعيّة التي بها تظهر النفس أفعالها
وكاالآلات الصناعيّة التي بها يتمّ الصانع مصنوعه . وكلّهم ينقسم قسمين ،
أحدهما الصنف الذي يحتاج إليه الملك لحاجات نفسه وعوارض بدنه ، والثاني
الصنف الذي يحتاج إلى ارتباطهم لمصالح رعاياه وانتظام ممالكه .
فالسبعة المحتاج إليهم لحاجات النفس والبدن (1) الحكماء الذين يشحذون
عقله ويتفقون ذهنه ويفيدونه الفوائد الجليلة في أسباب دينه ودنياه ، (2)
والأطبّاء الذين يسوسون بدنه بحسب الطاقة ويتبعون صحّته بقدر الإمكان
فيدبرون له الأغذية والأدوية في خواصّ أوقاتها ، (3) والمنجّمون الذين يختارون
له الأيّام المسعودة لابتداءات الأعمال ويفيدونه بقدمة المعرفة بالكلّيّات من
129a الأمور الكائنة ، (4) والندماء الذين يسامرونه في خلواته ويحادثونه في أوقات
فراغه ليؤنسوا بذلك وحشته ويجددوا عليه نشاته ، (5) والملهون الذين يسمعونه
الألحان المنطقيّة والنغم المؤتلفة المشتملة على الحكم المحضة الجارية في قوى
نفسه وأخلاط جسده ، وإن يتحقّق كنه الفائدة بذلك إلا الخواصّ من أهل
الحكمة ، (6) والخدم المقيمون على رأسه المقربون له حاجات بدنه كالمطاعم
والمشارب والملابس والمراكب وغيرها ، فإنّ الملك لن يستغني عنهم في ترتيب
مملكته في أعين رعيته وتفخيم سلطانه في صدور أكفائه ، (7) والسابع
الحجاب الذين يرتبون لكلّ واحد من هذه الطبقات مراتبهم ويوصّلونهم إلى
خدمتهم في أوقاتهم ويفون بما إليهم حسب ما يلزمهم .
129b والسبعة المحتاج إليهم لمصالح رعاياه ، أوّلهم الوزراء فإنّ الملك الفاضل لا

2 after العليل break (∴) ind. and blank space of five letters M   3 وتسديد : وتَشديد M   4 الطبيعيّة : الطبيعة M   5 مصنوعه : مصنوعة M | وكاالآلات : وكاالآلَات M   6 after بدنه blank space of three letters M   9 after ودنياه break (∴) ind. and blank space of five letters M   11 والأدوية : الأدوية M | after أوقاتها blank space of three letters M   12 لابتداءات : لابتداات M   13 after الكائنة blank space of three letters M | خلواته : حلواتهِ M   14 نشاته : سَاوَتُه with blank space of three letters after it M   15 المنطقيّة M : or read المنظومة ? | والنغم : والنَعم M | الجارية : الحارية M   18 ترتيب : ترتبن M   19 رعيته : رعية M | after أكفائه blank space of three letters M   22 رعاياه : رعَاياهِم M

restoration of incorruptness from corruption and of order from disorder, and an example of it from the craft of medicine is the return of health to a sick body.

(219) In administering his affairs and acting well the virtuous leader needs fourteen persons supporting him. They are like the natural body parts by 128b which the soul manifests its actions, and like the instruments of the crafts through which the producer perfects his product. They are all divided into two parts: one of them is the class [of supporters] whom the king needs for the requirements of his body and soul. The second is the class [of supporters] whose commitment he needs for the welfare of his subjects and the organisation of his kingdom.

The seven needed for the requirements of [his] body and soul are: (1) the wise men who sharpen his intellect, conform to his understanding and benefit him greatly in matters concerning his religious and his worldly motives; (2) the physicians who have charge over his body in the best way possible and pursue its health as far as possible, so they administer food and medication for him at the right times; (3) the astrologers who choose for him the days which are favourable for beginning actions and benefit him because the knowledge of the universals [which they derive by means of astrology] precedes [the knowledge of] the generated [and corruptible] things; (4) the drinking companions who spend the night in pleasant conversation with him in his private quarters and talk to him in his leisure 129a time, in order to cheer him in his solitude by [their company] and revive his energy; (5) the entertainers who play for him melodious compositions and harmonious tunes, which include pure maxims which pervade the powers of his soul and the mixtures of his body; although only the most distinguished of the wise men confirm the full advantage of this [namely, music]; (6) the servants who wait on him and bring him his bodily requirements like food, drink, clothing, riding animals and the like. For the king is not able to dispense with them in arranging his kingdom in the view of his subject[s] and in magnifying [the standing] of his reign in front of his equals; (7) the seventh are the chamberlains who arrange for each one of these classes their ranks and get them to their tasks on time and to carry out their duty.

Of the seven which are required for the welfare of his subjects the first are the viziers. The good king is not able to dispense with a vizier of sound 129b

يستغني عن وزير حازم يرجع إلى إصابة رأي وجزالة تدبير ، والثاني القضاة المتولّون النظر في مظالم الخليقة وحملهم على حكم التصفية وإلزامهم سنن الدين وأحكام الشريعة ، والثالث قادة الجيوش الذين يتولّون حفظ الثغور وحماية الأطراف ، والرابع الكتبة الذين هم ألسنة الملوك وحافظو الحسبانات في الدخل والخرج وواضعو كلّ شيء من المعاني الكريمة مواضعها على الاحتياط ، والخامس حفظة الرعايا في المملكة عن عيب العتاة وصائنو حرمهم وأموالهم عن المردة والجناة ، والسادس الباحثون عن أخبار المملكة في القرب والبعد المتعرّفون أحوال العمّال والرعايا لينهوها ويجعلوا الملك من معرفتها كالشاهد لها ، والسابع المتردّدون في السفارة بينه وبين من يحتاج إلي مخاطبتهم من رعايا 130a وأهل مملكته أو من نظراء به وأكفائه .

(220) والواجب على الملك أن يعطي هو لأمن نفسه ستّة أشياء يستخلص بها عقائدهم في المناصحة والمحبّة ويحسم بها عنهم سبيل الشكاية والاستزادة ، ويأخذ منهم لها ستّة أشياء أخذاً لا يقاد واحد منهم على تقصير في شيء منها .

فأمّا الستّة التي يلزم بذلها لهم فأوّلها المتوسّعة في الأرزاق على قدر مراتب الأعمال ، فإنّه لا يكفي من لا يكفى ولا يغني من لا يغنى ، والثاني أن لا يستكدهم في الأعمال المنوطة بهم استكداد المستبعد لطوقهم المستنزف لغاية مجهودهم ، فإنّ الأنفس متى استكرهت ملّت ومتى استكدت كلّت ولا بقاء لها مع التعب الدائم والنصب المتّصل ، والثالث أن لا يؤاخذهم بكلّ تقصير 130b مؤاخذة من لا يقبل عثرةً ولا يقبل معذرة ، بل يتعقّبهم بالصفح والتجاوز والإغضاء والتجافي ما احتملته السياسة الفاضلة ، والرابع أن يترصّد أحوالهم في

1 وجزالة : وجَزَالَةَ M | after تدبير blank space of three letters M 2 النظر : للنّظر M | الكتبة : الكتَّه M | التصفية : اليصفه M 4 after الأطراف blank space of three letters M | وحافظو : وحَافِظوا M 5 وواضعو : وَواضعوا M | after الاحتياط blank space of three letters M 6 الرعايا : الرعاياه M | وصائنو : وصَاينوا M 7 والجناة : والحياة with blank space of three letters after it M 10 وأكفائه : والفايه with blank space of three letters after it M 12 والاستزادة : والاستراده M 13 أخذاً : احداً M | واحد : وَاحِداً M 16 after يغنى blank space of three letters M 17 المنوطة : المبوطة M | المستبعد : المستنعد M | لطوقهم : لطوفهم M 19 after المتّصل break (∴) ind. and blank space of three letters M 21 والتجافي : والتَّحافِي M | after الفاضلة blank space of three letters M

management who resorts to finding the right opinion and excellent administration. The second are the judges who are entrusted with the examination of injustice done to men, inflict on them judgements to rectify [wrongs], and compel them to [obey] the traditions of religion and the ordinances of law. The third are the leaders of the armies who are entrusted with the preservation of the frontiers and the protection of the borders. The fourth are the secretaries who are the tongues of the kings, the keepers of the accounts of income and expenditure, and those who implement every noble concept with due caution. The fifth are the protectors of the subjects in the kingdom from the vice of the wrongdoers, and the guardians of their women and their possessions against rebels and criminals. The sixth are the informants [who gather] information of what happens in the kingdom near and far, are knowledgeable about the situations of the governors and the subjects in order to communicate them
and make the king aware of them as if he were witnessing them. The 130a
seventh are the emissaries moving to and fro between him and those of the subjects and inhabitants of his kingdom or his peers and equals who need to be addressed.

(220) It is the king's duty to grant, for his own security, six things [to his subjects] by which he secures their loyalty with sincerity and love and prevents them [taking] the path of complaint and excessive requests. In exchange for them he takes from [his subjects] six things in a way which lets none of them go short in any of [these six].

The first of the six [things] which it is imperative upon him to bestow on them is to ensure that they have sufficient means of subsistence corresponding to the[ir] professional positions. For he who is not given sufficient means will not meet what is required of him, and he who is in need will not be of any use. The second is not to wear them out in the works entrusted to them [by expecting] what is unattainable for them even if they exhaust their power in greatest effort. For when the people are forced, they become weary and when they are worn out, they become weak and there is no survival for them as long as there is perpetual hardship and
continuous exertion. The third is not to punish them for every shortcom- 130b
ing as if he were someone who accepts neither slips nor excuses, but rather to discipline them by pardoning them, passing over without punishing and by overlooking and turning away from [their shortcomings] as far as good leadership tolerates. The fourth is to observe their attitudes

آثار الطاعة وينعم الفحص عن دلائل المناصحة ، ومتى وقف من أحدهم على حالة مرضية أذاقه حلاوة سعيه وثمرة كدحه ليقوّى عزمه على الازدياد من الخدمة ويظهر لاقرانه أنّ سعي المجتهد لا يضيع ونصيحة المشفق لا تبور ، فالخامس أن يرعى حرماتهم على مرّ الأيّام ويحفظ وسائلهم على تصرّم الأعوام ويقضي حقّ مشهدهم في مغيبهم وعملهم في عطلتهم ومحياهم في مماتهم وأنفسهم في اعقابهم ، فإنّ من ضيع صاحبه في وقت الغيبة عنه لم يجده عند
الحاجة إليه ، ومن حافظ على حرمات المتّصلين به دائماً فقد بسط بذلك 131a
آمالهم وقوى به رجآهم ، والسادس أن لا يقبل في أحد منهم مكيدة ساعٍ يروم استفساد حاله ولا إغراء واشٍ يقصد به حط رتبته ، بل يستبرئ الأمر على وجهه وينعم التنقير على كنهه ويعلم أنّ من شأن الدهاة الحسدة الإفساد بأغمض حيلة .

وأمّا التي يعتاضونها ويؤاخذون أصحابهم بها فأوّلها الإخلاص ، فإنّ المدخول في العقيدة ربما قدح في الملك وذلك ذنب لا يغفر وجرم لا يصفح ، والثاني الوفاء وهو أن يتصرّف معه في حالات السرّاء والضرّاء وأطوار الشدّة والرخاء معتقداً أنّه من أيسر حقّ إنعام الملك عليه وأهون أذمّته واختصاصه أيّاه ، وقد
قال لله ما كان لأهل المدينة ومن حولهم من الأعراب ⟨ما⟩ إليه ، والثالث 131b
إمحاض النصح وهو أن يمشي أعماله على أفضل سبلها وأجمل طرقها ويجتهد في تربيب أسباب سلطانه وتوطيد قواعد ملكه وإصلاح أمور رعيّته غير مفرق بين ما يتولاه ويتولاه غيره في صدق العناية عارفاً بأنّ الملك سائس الكلّ والكلّ شامل للجزء والجزء في ضمن الكلّ ، ولا يسعي ترك العناية مما هو

3 after تبور blank space of three letters M 5 مغيبهم : مَعيبهم M 8 آمالهم : امَالَهم M رجآهم : رجَاهُم M 10 الدهاة : دهَاه M 12 التي : الذي M | after بها blank space of three letters M 14 السرّاء : السرّآ M 15 after أيّاه blank space of three letters M 16 إليه : الأيه M : or read الآية ? 19 ويتولاه : يَتولّى M

for signs of obedience and to make an effort to inquire closely into the proofs of [their] sincerity; and when he meets in one of them with a satisfying attitude, he allows him to taste the sweetness of his effort and the fruit of his labour so that he will strengthen his determination to give yet more service and demonstrate to his peers that neither the effort of the diligent is wasted nor the advice of the concerned futile. The fifth is to respect their honour in the course of the days, guarantee their means in the passing of the years, respect their tombs when they are gone, their [accomplished] work when they have stopped [working], their dwelling places when they are dead and their selves in their offspring. For someone who neglects his associate when absent will not find him [by his side] when he needs him, whereas he who has always protected the honour of
those associated with him will thereby have enlarged their hopes and 131a
strengthened their expectations. The sixth is not to be taken in either by tricks of a slanderer who wishes to destroy [the king's] prosperity or by the allurements of a traitor who thereby pursues undermining his standing, but to make a full investigation upon the issue in the correct way, to make an effort to examine its core and to know that it is the way of envious and cunning people to use concealment and trickery to destroy.

As to [the things] they [i.e the kings] receive in exchange and by which they gain mastery over their associates, the first of them is sincere loyalty. For someone of false loyalty may offend against the king and that is an offence not forgiven and a crime not pardoned. The second is sincerity, i.e. that [the subject] stands by [his king] in situations of prosperity and distress and at times of trouble and ease believing that his entitlement to the king's favour is less significant and his claims and demands for it are less important. Already God has said: 'The people of Medina and their
neighbouring desert Arabs should not' and so on.[305] The third is honesty in 131b
counsel, i.e. that [the subject's] deeds proceed according to the most virtuous of paths and most beautiful of ways, and are made in an effort to foster the fundamentals of [the king's] reign, the consolidation of the foundations of his kingship and the improvement of the affairs of his subjects without distinguishing between one's own tasks and those of others in providential care, as he knows that the king leads the whole, that

[305] This is from *Surat al-Tawba* (9:120): 'The people of Medina and their neighbouring desert Arabs should not have held back from following God's Messenger, nor should they have cared about themselves more than him: if ever they suffer any thirst, weariness, or hunger in God's cause, take any step that angers the disbelievers, or gain any advantage over an enemy, a good deed is recorded in their favour on account of it. God never wastes the reward of those who do good' (tr. Haleem).

قطعة منه ، والرابع الإجلال وهو أن لا يطرح مهابته عن قلبه في خفض صوته وضمّ أطرافه واستعمال حسن الأدب في قيامه وقعوده والإقبال على حديثه بذهنه وسمعه ، فإنّ الملوك ربّما استعادوا المخاطبين حديثهم امتحاناً لهم ، والخامس كتمان السرّ فإنّه ركن من أركان التدبير ، والسادس ترك الدالّة والسحب في الطلبة ، بل يصبر إذا منع ويشكر إذا أعطى ولا يزيده صاحبه ادّناءً وتقريباً إلا زاده إجلالاً وتوقيراً . 132a

(221) قد تقدّم القول بأنّ الفضائل التي يحصرها الإنسان من جهة نفسه راجعة إلى القوى الثلاث ، أعني قوّة الشهوة وقوّة الغضب وقوّة التمييز ، وقوّة الشهوة بإزاء النفس المغتذية التي بها ينمو الإنسان ، وقوّة الغضبيّة بإزاء النفس الحسّاسة التي بها يدفع المضارّ ، وقوّة التمييز هي النفس الناطقة التي بها يعلم ويعقل . ولكلّ واحدة من هذه القوى الثلاث خاصّيّة فضيلة تتفرّد بها عن غيرها ، وفضيلة القوة الشهوانيّة العفّة وهي حالة متوسّطة بين طرفي الشرة وخمود الشهوة ، وفضيلة القوّة الغضبيّة الشجاعة وهي حالة متوسّطة بين طرفي التهوّر والجبن ، وفضيلة القوّة التمييزيّة الحكمة وهي حالة متوسّطة بين طرفي المكر والغباوة . ثم يتولّد من مجموع هذه الفضائل على الاعتدال فضيلة رابعة ، 132b أعني العدل وهي حالة متوسّطة بين طرفي التسلّط والاسخذاء . فقد نزل العدل منزلة الصورة الموجودة في النفس عند اعتدال تلك القوى الثلاث ، ثم يتفرّع من هذه القوى الثلاث فضائل أخر ترجع في القسمة إليها إمّا كالأنواع لها أو كالمركّب منها .

ومثال ذلك أنّ القوّة الشهوانيّة متى ضبطت عن الاشتغال بما هو خارج عن مقدار الحاجة سمّي ذلك قناعة ، ومتى سلست لبذل ما يحوزه الإنسان من الأملاك يحسن المواساة على جهة يقتني به الحمد والأخذ سمّي ذلك جوداً ، وهما جميعاً متّصلان بالعفّة .

وبمثله متى ضبطت القوّة الغضبيّة ضبطاً لا يستفزّها ألم المكروه سمّي ذلك صبراً

1 after منه blank space of three letters M 3 حديثهم : حد بينهم M | after لهم blank space of three letters M 4 after التدبير blank space of three letters M 5 after يصبر add. above the line M 6 ادّناءً : ادّنآ M | after وتوقيراً break (∴) ind. and blank space of three letters M 8 الثلاث : الثلث M | التمييز : التَميز M 11 الثلاث : الثلث M 15 after والغباوة blank space of three letters M 17 الثلاث : الثلث M 18 الثلاث : الثلث M

the whole contains the part and the part is within the whole. He does not attempt to stop caring for what he is part of. The fourth is reverence, i.e. not to diminish the awe in his heart, by speaking in hushed tones, bowing, displaying the finest manners while sitting and standing and attending to [the king's] speech with his mind and hearing. For the kings may ask the ones addressed to repeat their speech, testing them. The fifth is secrecy, for it is one of the essentials of government. The sixth is to avoid boldness and to step back from making demands. Rather [the subject] is patient when [his demand is] refused and grateful when it is granted. His master does not increase his closeness and nearness, unless he increases the 132a
honour and respect [he shows his king].

(221) The doctrine has already been presented before that the virtues which man comprises on the part of his soul derive from three faculties, i.e. the faculty of appetite, the faculty of anger and the faculty of discrimination. The faculty of appetite corresponds to the nourishing soul by which man grows. The faculty of anger corresponds to the sensible soul by which one avoids harm. The faculty of discrimination is the rational soul by which one knows and thinks. Each one of these three faculties has a specific virtue by which it is distinguished from the others. The virtue of the appetitive faculty is moderation, which is the intermediary condition between the two extremes of greed and the extinction of passion[s]. The virtue of the irascible faculty is courage, which is the intermediary condition between the two extremes of recklessness and cowardice. The virtue of the discriminating faculty is wisdom, which is the intermediary condition between the extremes of deception and ignorance. From the wellbalanced sum of these virtues a fourth virtue is then brought forth, i.e. 132b
justice. It is the intermediary condition between the two extremes of tyranny and servility. So justice is certainly like the form that exists in the soul when those three faculties are well-balanced. From these three faculties, then, other virtues branch out which derive from them by division, either as kinds of them or as composites of them.

For example, when the appetitive faculty is restrained from excess, that is called self-restraint. When it is compliant to give freely of man's possessions to such an extent that he can be charitable in a way by which praise and acceptance are acquired, that is called generosity. These two [i.e. self-restraint and generosity] together are joined to moderation.

In the same manner when the irascible faculty is restrained from being provoked by some inconvenience, that is called patience, magnanimity or

أو سعة الصدر أو رحب الذراع ، ومتى ضبطت ضبطاً يستمسك به عن 133a المبادرة إلى قضاء وطرّ الغضب ممن يجني عليه سمّي حلماً أو عفواً أو كرماً ، وهما يتعلّقان بالشجاعة .

وبمثله متى ضبطت النفس التمييزيّة ضبطاً لا تقدم على العمل في الحوادث الواقعة إلا بما هو أقرب إلى السلامة وأبعد من الضرر واستعقاب الندامة سمّي حزماً ، ومتى هذّبت تهذيباً أسرعت الهجوم على حقائق ما أدّته الحواس إلى نفسها سمّي ذلك فطنةً أو فهماً ، ومتى لاحظت ما ينتج ارتكاب القبائح من سوء الأحدوثة فأحسنت الارتداع عن تعاطيها وانخزلت عن الإقدام عليها سمّي ذلك حياء ، ومتى منعتها عن الرفع والاستطالة سمّي ذلك تواضعاً .

(222) والفضائل الجسمانيّة تنقسم قسمين ، أحدهما ما يتّصل بالمرفق ، نحو قوّة البدن وسلامته عن العاهات ، والثاني يتّصل بالزينة ، نحو الصورة الحسنة 133b ونزاهتها عن الدمامة . والرجال إلى فضائل الزينة أحوج منهنّ إلى فضائل المرفق . والمعاني التي يتعلّق بها الفضائل الجسمانيّة تكون من وجهين ، أحدهما النسب ، والآخر التربة ، وكما أنّ الوالدين ينسلان الولد على شكليهما في الدمامة والملاحة كذا المكان الذي هو موضع الكون ، وكثيراً ما يوجد وطن الرجل ومسكنه شيئاً لطول عمره وصحّة بدنه وعلّة لشدّة أشدّه وجمال صورته وداعيةً لطهارة طباعه وسجاحة أخلاقه وذريعةً إلى تهذيب سيره ورضى أنحائه . واتّفاق الموطن الجيّد يكون باتّفاق الأسباب ، الأسباب العلويّة والسفليّة جميعاً ، أمّا العلويّة فبأن لا تكون مفرطةً في ناحيتي الشمال أو الجنوب فيستولي عليها فرط الحرارة أو البرودة ، وأمّا السفليّة فأن تكون بريّة بحريّة سهليّة جبليّةً .

7 نفسها : نفسه M 8 سوء : سُؤ M | تعاطيها : تعاطبها M 9 حياء : حيَآ M | after تواضعاً break (∴) ind. and blank space of half a line M 12 ونزاهتها : ويزانها M 14 والآخر : والاخر M 18 واتّفاق : وانفاق M 21 after جبليّةً break (∴) ind. and blank space of three letters M

generousness. When it is restrained from reacting to the acts and outbursts of someone pitiable, it is called clemency, forgiveness or kindness. The two [i.e. patience and clemency] are attached to courage. 133a

In the same manner, when the discriminating soul is restrained from proceeding [from thought] to action under normal circumstances unless it is in the interest of wellbeing, avoiding harm and regrettable consequences, it is called prudence. When it is refined in a way in which it immediately grasps the truth of what the senses convey to it, that is called intelligence or understanding. When it observes the results of committing shameful deeds of slander, so that it acts well to prevent them from being undertaken and restrains from venturing upon them boldly, that is called shame. When it prevents itself from presumption and arrogance, that is called modesty.

(222) The physical virtues are divided into two parts. One of them concerns the benefits of the physique, for example bodily strength and freedom from physical ailments. The second concerns beauty, for example
a lovely form and freedom from ugliness. Men are more in need of the 133b
virtues of beauty than [women] are in need of the virtues of benefits of the physique. The features on which the physical virtues depend are twofold: one [set] of them is lineage and the other environment. Just as parents procreate offspring who reflect their own appearances in respect to ugliness and beauty, in the same way the birth-place [is reflected in the offspring]. Often the homeland of a man and his dwelling place have a bearing on his longevity, the soundness of his body, the factor[s] behind his physical strength and the beauty of his form. It also induces purity in his nature and gentleness in his character, and is a means towards the refinement of his conduct and the graciousness of his manners. To be given a fortunate living-place is caused by chance, which includes the upper and the lower causes together. As for the upper causes, they should not be extreme with regard to north or south, so that an excess of heat or coldness would prevail over them. As for the lower causes, they are land, sea, plain and mountainous.[306]

[306] The catchword on fol.133b, which most probably reads *qāla Abū Zayd*, is a clear indication that the quotation in the name of al-ʿĀmirī is preserved in full.

Commentary

(1) 5a1 (end of unidentified passage)[1]

Edition: ed. Cottrell with French tr. in 'L'Anonyme', pp. 435–6.

Source: unidentified; Ps.-Platonic?

As the beginning of the Marsh manuscript is lost, the text of our *Philosophy Reader* (*PR*) starts, mid-sentence, at the end of a discussion about how to talk about God appropriately. The topic is further elaborated in the following passages, which all sound Ps.-Platonic, but are introduced simply by 'He says'. This seems to allow for the assumption that the incompletely preserved first quotation must also derive from some (Ps.-)Platonic source.

(2) 5a2–13 He says in his *Letter to Dionysius*

Edition: ed. Cottrell with French tr. in 'L'Anonyme', pp. 435–6.

Source: unidentified; Arabic version of a commentary on the Ps.-Platonic *Letter to Dionysius*?

Parallel (partial): *R. fī l-Radd*, Istanbul, Üniversite Kütüphanesi, MS 1458, fol. 106a15–18;[2] ed. Türker in 'Traité', 61.6–12; ed. Cottrell with French tr. in 'L'Anonyme', pp. 435–6.[3]

This is one of the rare instances in which the compiler of the *PR* provides his readers with the – although corrupted – title of the source he quotes. However, the provision of this title is not particularly helpful, even if Cottrell convincingly argues that the addressee of the letter must be

[1] The text of the *PR* starts on what is presently numbered as fol. 5a, but whether the current foliation originates from a time when the manuscript was still complete and thus indicates how many folios of text are missing at the beginning must remain in doubt (see pp. 7–8 above).

[2] The manuscript has two different foliation numbers in the upper-left-hand corner of the recto page. I give the foliation in Hindu-Arabic numerals and ink, the corresponding European foliation in pencil is 104a15–18.

[3] Cottrell's edition ('L'Anonyme', pp. 431–6) with a French translation is only a partial edition, but includes the parallel to the *PR*. Badawī's edition of the *R. fī l-Radd* (*Aflāṭūn*, pp. 337–9) is based on an Iranian MS which does not include the passage parallel to the *PR*. This passage is only found in the Istanbul MS indicated above, p. 47, n. 132.

Dionysius and the reference thus to the Ps.-Platonic *Second Letter*.[4] The Greek version, *Epistolae* 310b-315a, has no parallel to the topics of our passage. Proclus' discussion of the letter in his *Theologia Platonica* II, chapters 8 and 9, touches, at least, on the inconceivability and inexpressibility of the One and stresses that the soul must leave all multiplicity aside in order to join It.[5] The similarities are not strong enough to assume that this Proclean passage is a likely source for the quotation in the *PR*, but they suggest that our compiler draws on a commentary of the letter rather than the letter itself.[6]

So far, no Arabic version of or commentary on a *Letter to Dionysius* has been discovered, but that it must have existed is suggested by another excerpt ascribed to it in the Ps.-Platonic *R. fī l-Radd*.[7] The entire excerpt is, as Cottrell shows, parallel to the quotation in the *PR*, yet the latter is slightly longer.

(3) 5b1–5 He says

Source: unidentified; Ps.-Platonic?

As this passage is introduced by '(and) he says', the reader is led to believe that the same authority as in the previous passage is quoted again. The material may thus be Platonic or rather Ps.-Platonic. It is presented in the form of an exhortation or advice based on personal experience: engaging in divine matters and turning away from the bodily world ensures divine providence.

(4) 5b5–8 He says

Source: unidentified; Ps.- or Neoplatonic?

Parallel: Anonymous, *R. fī Ārā*ʾ, ed. Badawī in *Aflāṭūn*, 318.16–19.

[4] 'L'Anonyme', p. 423, n. 29.

[5] See Proclus Diadochus, *Théologie*, ed. Saffrey and Westerink with French tr., II, pp. 51–7, esp. 55–6.

[6] Plotinus also mentions the *Second Letter* of Plato several times (for the passages, cf. *ibid.*, pp. XLIII–XLIX), but his references have nothing in common with the passage in the *PR*. The problem of resorting to other things when thinking or talking about the One is, however, addressed in the *Enneades* (e.g. V 3, 13 and V 5, 10).

[7] The unique MS, which contains the parallel to the *PR* (cf. n. 3 above), however, refers to a letter to *W-s-ā-s-w-s*. This reading can easily be explained as a misspelling of Dionysius, *Diyānisiyūs* in Arabic. The *Fihrist* attests to 'extant letters by Plato' (246.20: *wa-lahū rasāʾil mawjūda*) which is repeated by Ibn al-Qifṭī (18.10).

The statement also occurs in the *R. fī Ārāʾ al-ḥukamāʾ al-yūnāniyīn*, where it is explicitly ascribed to Plato. As the explanation of the meaning of the three heavens appears in both texts, although in different places, it is obvious that it is not the compiler of the *PR* commenting on his excerpt.

The four heavens may stand for the realms of nature, soul, intellect and the One of the Neoplatonic tradition, according to which Intellect keeps nature and soul from approaching the One directly. The explanation of the three heavens presented at the end of the passage is somewhat puzzling as one would expect the first heaven to correspond to the natural sciences and the second to psychology and/or mathematics.

(5) 5b8–6a1 Hermes says

Source: unidentified.

Parallel: Stobaeus, *Anthologium* II, 1, 26.1–9; Ibn Durayd, *K. al-Mujtanā*, ed. Rosenthal with Greek text of Stobaeus and tr. in 'Sayings', pp. 183 and 54; ed. al-ʿAwwād, 194.1–6; ed. al-Dālī, 144.5–10: no. 488.

The quotation in the name of Hermes ultimately derives from a Greek source, as the parallel to Stobaeus' *Anthologium* indicated by Rosenthal clearly shows. Its Arabic version appears also in the *K. al-Mujtanā* by the Arab philologist and lexicographer Ibn Durayd (837–933), who may or may not be the source for our compiler.[8] Ibn Durayd's version is closer to the Greek, but has already suffered from some corruption.[9] It is, in any case, the only parallel between the *K. al-Mujtanā* and the *PR*.

The passage takes up the same topic as passages (1) and (2), namely the indescribability of God.

[8] On which, see also Van Bladel, *Arabic Hermes*, pp. 198–9, who argues that the transmission of this particular Hermes passage from Greek into Arabic may have happened either through the translation of a Greek gnomologium as Rosenthal has already suggested or by the translation of various Christian authors who used Stobaeus. The appearance of the passage in the *PR* somehow contradicts Van Bladel's verdict (p. 199) that 'Moreover, it does not seem to have been repeated in later florilegia by other Muslim authors; this particular Hermetic citation is ultimately a rare one in Arabic.'

[9] See Rosenthal, 'Sayings', p. 54, n. 1: 'This [i.e. the reading: 'For it is impossible for a comprehensible body to be described as something incomprehensible'], of course, is non-sense. The Greek text suggests the reading: *an yūṣafa* ⟨*bi-*⟩*jismin mudrakin mā*, "for an incomprehensible body to be described as something comprehensible". But the Arabic text may possibly have read originally: *an yaṣifa* (*sic* Bodl. ms.) *jismun mudrakun mā*, "for a comprehensible body to describe something incomprehensible".' This textual corruption may also have existed in the source text of the *PR*, be it the *K. al-Mujtanā* or not, and explain the shortening of the quotation in the *PR*.

(6) 6a1–10 Pythagoras says

Source: some commentary on the *Dicta aurea*?

A possible source for this Pythagorean quotation may be some version of the *Dicta aurea*. Elsewhere the *PR* seems to make use of the standard version of the *Dicta aurea* (passage 102) and of the no longer extant Proclean commentary on them (passage 16). The standard version and the extant commentary on the *Dicta aurea* attributed to Iamblichus do not contain our passage, but do have material on which a commentator could have commented in part by saying what is quoted in the *PR*.[10] The commentary ascribed to Proclus as preserved in Ibn al-Ṭayyib's *Istithmār* does not have a striking parallel to our passage. Yet, it does use the terms 'divine' and 'natural number' (*ʿadad ilāhī/ṭabīʿī*), but within a list of four categories of numbers: the divine, intellectual, psychic and natural numbers. Consequently, Ibn al-Ṭayyib (78.10–11, 80.3) states that the existents begin with the divine One and end with the natural four.

The *PR*'s explicit ascription of goodness to the divine numbers and of the complete lack of goodness to the natural numbers, is absent in Ibn al-Ṭayyib and any other text I have looked at. It is also surprising that the number One is said to stand for the intellect and not for the One as the First Principle, as it does, for example, in Ibn al-Ṭayyib (78.6–7) and the *Ikhwān al-Ṣafāʾ*.[11]

(7) 6a10–b3 Aristotle says in his book called *Metaphysics*

Source: unidentified; some commentary on Aristotle, *Met. Lambda*, 7?

The passage may stem from some commentary on Book *Lambda* of Aristotle's *Metaphysica*,[12] in which case the underlying Aristotelian text is

[10] The material in question is an invocation of the tetractys, the Arabic rendering of which is faithful to the Greek original. It occurs in the *Dicta aurea* as preserved in Miskawayh's *Ḥikma al-khālida* (227.17), the *Muntakhab Ṣiwān al-ḥikma* (ed. Dunlop, line 532), and even twice – in a shorter and a longer version – in the lemmata of Iamblichus' commentary (pp. 88–90), but not in al-Anṣārī's *Ādāb al-falāsifa*:
إی والذي وهب لأنفسنا الينبوع ذا الأربع من الطبيعة التي لا تفتر !
The interpretation of the tetractys is manifold (see, e.g., Delatte, 'Tétractys') and what is preserved in the *PR* may well be a late Neoplatonic interpretation.

[11] In Ibn al-Ṭayyib the first numbers represent the Gods. The *Ikhwān al-Ṣafāʾ*, I, p. 53, further list the active universal intellect, the universal soul and first matter as two, three and four.

[12] On the Arabic transmission of the *Metaphysica* in general, see Bertolacci, 'Translations'. There are no parallels to our passage in the anonymous paraphrase of Book *Lambda*, chs. 6–10 (on which see *ibid.*, pp. 256–7) or in the fragments of Alexander's comment-

probably 1073a3–13.[13] There the existence of 'a substance which is eternal and unmovable and separate from sensible things' (tr. Ross) is stated. This substance is further said not to have any magnitude, which corresponds to the claim of God's incorporeality in the passage of the *PR*. There, impassivity[14] and inalterability are deduced from the substance's immovability.

Non-Aristotelian, however, are the attribution of wisdom (taking *al-ʿilm wa-l-ḥikma* to express one single concept), generosity and omnipotence to God and the exclusion of any anthropomorphism from Him. As to the three divine attributes 'wisdom', 'generosity' and 'omnipotence' their usage is at least attested in some Ps.-Aristotelian text, namely the *R. al-Siyāsa al-ʿāmmīya*.[15]

ary preserved in Ibn Rushd, or in the extant parts of Themistius' paraphrase which is quoted in passage (17) below, or in Theophrastus' paraphrase, or in the fragments preserved of Nicolaus Damascenus' *De philosophia Aristotelis*, Books II and III, which discuss the *Metaphysica*, or in the paraphrase by Thābit ibn Qurra, or in the one by ʿAbd al-Laṭīf al-Baghdādī.

[13] Cf. the Arabic text in the lemmata of Ibn Rushd's *Long Commentary* (Averroès, III, 1625.10–1626.4):
قال ارسطاطاليس فأمّا إنّ جوهر ما أزليّ وغير متحرّك مفارق للمحسوسات فبيّن مما قيل وقد أوضح أنّه لا يمكن أن يكون لهذا الجوهر عظم البتّة بل هو لا جزء له ولا قسمة لأنّه يحرّك زماناً لا نهاية له وليس شيء متناه له قوّة لا متناهية فإذا كلّ عظم إمّا لا متناه وإمّا | متناه فلهذه العلّة لا يكون في العظم المتناهي ولا يكون لا متناهي لأن ليس عظم لا متناه البتّة وأيضاً بيّن أنّه لا ينفعل ولا يتغيّر فإنّ جميع الحركات الأخر بعد المكانيّة فهذه الأشياء بيّنة أنّها على هذه الحال .

[14] Following the suggested emendation 'affections' (*al-infiʿālāt*) instead of the manuscript reading 'translocations' (*al-intiqālāt*).

[15] The passage in question is found in the *R. al-Siyāsa* and partly quoted in passage (19) below. Curiously enough, the sentence which describes the creator as omnipotent, wise and generous (ed. Maróth, 50.19–20) is omitted by our compiler. A possible reason may be the similarity to the alleged *Metaphysics* quotation. Yet, in the course of the *PR* repetition is ubiquitous.
The three divine attributes (*jūd, qudra, ḥikma*) are commonplace in Arabic philosophy. They occur, e.g., in an Arabic version of Late Alexandrian *Prolegomena* preserved by Ibn al-Ṭayyib in the context of the fifth definition of philosophy, i.e. likeness to God (ed. Gyekye, 22.15, and tr. Dunlop, 'Existence', p. 89). They are further used in an invocation contained in the *Praise [of God] by the Philosophers* (*Tamjīd al-falāsifa*) which is preserved in Istanbul, MS Köprülü 1608, fols. 32a16–33a18. There (fol. 32b8) we read:
سبحانك يا تامّ القدرة والجود والحكمة .
The attributes can also be found in ʿAbd al-Laṭīf's *K. fī l-ʿIlm mā baʿd al-ṭabīʿa*, ch. 22 (ed. Badawī in *Aflūṭīn*, 217.10–13). This chapter is supposedly a summary of the *ThA*, but in the extant version of the *ThA* there is no corresponding passage. In Miskawayh's *Fawz* (40.6–9) it is claimed that 'none of the Ancients … rejected any of [the Creator's] attributes which He deserves [to be attributed] by men according to their ability [to do so], namely generosity, omnipotence and wisdom'. See also Rowson, *Muslim Philosopher*, pp. 217–18 and 226. The three attributes also figure prominently in al-Isfizārī's *K. fī Masāʾil al-umūr al-ilāhīya*, pp. 236–7, 247, 249.

(8) 6b3–7a9 He [Aristotle] says

Source: unidentified; some commentary on Aristotle, *Met. Lambda*, 6–10 and *Phys.*?

It is not clear whether this passage derives from the same source as the previous and the following passage which are both explicitly marked as quotations from Aristotle's *Metaphysics*, but are, in fact, free re-workings of the text. Be that as it may, the present section covers a few selected issues starting with time (*Met.* 1071b8) and closing with the order in the universe (*Met.* 1075a-b). Aristotle (*Met.* 1071b6–9) pictures motion and time as what can neither come into being nor cease to be, expounding 'for there could not be a before and an after if time did not exist' (tr. Ross).[16] The definition of time as 'number' and 'measure' of motion goes back to the *Physica* (219b1, 220b32).[17] The motion which time measures is continuous, eternal and circular, as the circular motion is the only continuous one (*Met.* 1071b9–11).[18] Aristotle claims that 'the first mover produces' 'motion in a circle' (*Met.* 1072b9–10, tr. Ross).[19] Our passage apparently understands this motion to be the rotation of the outermost sphere around itself, as the sphere is said to have no starting-point as all its parts are equally bent towards each other. This argument seems to be a transfer of the Aristotelian argument of the circular motion not having definite points (*Phys.* 265a29–b3) to the body which moves in such a circular motion.

Then the passage takes up the problem of the spatial motions of the

[16] Cf. the corresponding lemma of Aristotle's *Metaphysica* in Ibn Rushd's *Long Commentary* (Averroès, *Tafsīr*, III, 1556.4–7):
إلا انّه غير ممكن أن تكون الحركة حدثت حدوثاً أو تفسد فساداً وذلك أنّه قد كان دائماً ولا الزمان أيضاً إذ لا يمكن معنى الأكثر تقدماً والأكثر تأخراً إذا لم يكن الزمان .

[17] The text of the only extant Arabic translation of the *Physica* reads *al-zamān huwa ʿadad al-ḥaraka* (*al-Ṭabīʿa*, ed. Badawī, 420.3) and *miqdār al-ḥaraka wa-l-taḥarruk* (448.4).

[18] Cf. the corresponding lemma of Aristotle's *Metaphysica* in Ibn Rushd's *Long Commentary* (Averroès, *Tafsīr*, III, 1556.7–10):
والحركة إذا على هذا الوجه متّصلة كما أنّ الزمان أيضاً وذلك أنّه إمّا أن يكون واحداً بعينه وإمّا تاثيراً ما وانفعالاً للحركة وليس توجد حركة متّصلة إلا التي في المكان ومن هذه المستديرة .

[19] Cf. the corresponding *lemma* of Aristotle's *Metaphysica* in Ibn Rushd's *Long Commentary* (Averroès, *Tafsīr*, III, 1608.3–4):
والنقلة هي أوّل التغيّرات وهي التي على الاستدارة وهذا هو محرّك هذه ...
This Arabic version of the Aristotelian text does not explain the peculiar expression in the Marsh manuscript, namely 'the motion ... is shaped in the shape of rotation (*tushakallu bi-shakli l-istidārati*) by the Creator'. Interestingly, the *R. al-Siyāsa* (ed. Maróth, 51.5–10) uses a similar expression in a comparable context:
فجرم السماء أكرم جرم الأجرام لا حارّ ولا بارد ولا رطب ولا يابس ولا ثقيل ولا خفيف مركّب من الهيولى والصورة النقلية حرّكه بحركة دائمة متّصلة وهي أكرم الحركات المستديرة وشكّله بأكرم الأشكال وبأوّل الأشكال وهو الشكل العالي جعل الجزء الأعلى أسرعها حركة وبحميّة الحركة تدور الأفلاك جميعها .

planets which differ from the motion of the universe (*Met.* 1073a23–b1), i.e. the motion of the outermost sphere, and which would each require a separate unmoved mover. The solution presented in the *PR* offers no explanation for the different planetary motions, but follows the teleological argument for God's existence: the obviously existent order in the universe proves that there can only be one single cause for that order and thus only one first principle and unmoved mover. The argument may partly be inspired by *Metaphysica Lambda*, 10, but does not have any noteworthy parallels to the Aristotelian text.

(9) 7a9–13 [Aristotle] says in the *Metaphysics*

Source: unidentified; some summary of Aristotle's philosophy?

Parallel: *Mukhtār min Kalām al-ḥukamāʾ*, ed. and tr. Gutas in *Literature*, 164.7–12.

Cf.: beginning of passage (13).

The passage may be taken from the same *Metaphysica* Source as the two previous passages, but it also occurs in the *Mukhtār min Kalām al-ḥukamāʾ*. The latter's editor Gutas suggests that the passage 'is most likely drawn from a summary account of Aristotle's philosophy'.[20] A slightly different version of the same quotation, but probably deriving from a different source is presented in the first half of passage (13) which, in its turn, has a word-by-word parallel to the *K. al-Ḥaraka*.

The substance of the unmoved mover is said to be above the substance of heaven and described by features that can all be found, at least germinally, in *Metaphysica Lambda*, 7. The whole presentation, however, has a very distinctive Neoplatonic ring, especially the last sentence which mentions this substance's power spreading in the world. Passage (13) further down elaborates this thought and discusses the receptivity which things in the world have for this power.

(10) 7b1–3 He [Aristotle] says

Source: Aristotle, *Physica* 198a35–198b2; *vetus* translation entitled *K. Samʿ al-Kiyān*?

Parallel: *K. al-Ḥaraka*, appendix 2.

[20] Gutas, *Literature*, p. 387. He also draws attention to similar statements on God being above heaven and having unending activity in the *Maqāla fī l-Mudkhal ilā ʿilm al-akhlāq*, attributed to Nicolaus, see ed. Badawī in *al-Akhlāq*, 398.7–13 and 428.2–3; and Lyons, 'Treatise', pp. 54–5. On the *M. fī l-Mudkhal*, see Zonta, 'Nicolas (Pseudo)', pp. 680–3.

Cf.: Isḥāq ibn Ḥunayn's translation *al-Samāʿ al-ṭabīʿī* as *al-Ṭabīʿa*, ed. Badawī, 140.1–5.

The quotation must be taken from an Arabic translation, possibly with some commentary of the Aristotelian *Physica* 198a35–198b2, although it is not the preserved translation by Isḥāq ibn Ḥunayn. As the *K. al-Ḥaraka* refers to the *Physica* by the name of *K. Samʿ al-Kiyān*, it seems reasonable to assume that the translation used is the so-called *vetus* translation.[21] The context in which the statement appears in *Physica* II, 7 is the discussion of the causes which may answer the question why something comes into being, i.e. the efficient causes. These causes which move and change[22] something are either natural, e.g. nature, or not natural, as they are themselves unmoved, like the first mover or the form of something. The context into which our compiler transposes the Aristotelian quotation reduces it to a further statement on the first mover.

In the *K. al-Ḥaraka* the context of the quotation becomes much clearer, as it appears imbedded in a summary of Book II of the *Physica* where Aristotle is said to 'describe and classify the causes and to distinguish the efficient cause from the other causes. Then he describes how many causes are necessary to be connected to the being of the whole world and what "cause" means and how it is necessary to distinguish some of them from others.' The subsequently inserted quotation from the *Physica* is slightly longer than the one given in the *PR*, i.e. *Phys.* 198a35–198b4, and is followed by a short explanation that Aristotle's Unmoved Mover is 'the principle of the whole world's being, the mover of the spheres and the organizer (*nāẓim*) of their order (*sharḥ*)'.[23]

(11) 7b3–6 He [Aristotle] says

Source: unidentified; some commentary on Aristotle?

Parallel: *K. al-Ḥaraka*, appendix 2.

Whereas the previous passage has stuck to the genuine Aristotelian wording, the present statement may, at best, be described as inspired by the Stagirite. *Physica* VIII, 6, 258b10, for example, states that 'there must always be motion without intermission' (tr. Hardie and Gaye). The notion that circular motion is effortless may go back to the statement that it can-

[21] On which, see Peters, *Aristoteles Arabus*, p. 32.
[22] Aristotle understands change as one kind of motion.
[23] For the Arabic text, see appendix 2.

not have a contrary motion in *De caelo* I, 3, 270a18–20. In *Phys.* III, 5 and *DC* I, 5–6, Aristotle shows that no body can be infinite. The *K. al-Ḥaraka* provides its readers with references to these two Aristotelian works.

(12) 7b7–10 He [Aristotle] says

Source: unidentified; some commentary on Aristotle?

Parallel: *K. al-Ḥaraka*, appendix 2.

This passage summarises the main findings of *Physica* VIII, 10, namely 'that it is impossible for a finite magnitude to have an infinite force, and also that it is impossible for a thing to be moved by a finite magnitude during an infinite time', and 'that it is impossible for the finite to cause motion during an infinite time' (tr. Hardie and Gaye).[24] The second conclusion is missing in the parallel passage in the *K. al-Ḥaraka* where it may have fallen out due to *homoeoteleuton*. In the context of the *PR* the importance of this passage is that it implicitly demonstrates the First Mover as being of no finite magnitude, but having infinite power.

(13) 7b10–8b2 He [Aristotle] says

Source: unidentified; some commentary on Aristotle?

Parallel: *K. al-Ḥaraka*, appendix 2.

Cf.: passage (9).

The first part of the passage which characterises the most elevated substance, i.e. God, repeats passage (9) with some changes and without ascribing it to Aristotle's *Metaphysics*. This suggests that the two texts may actually be taken from different sources.

The entire passage (13) occurs also in the *K. al-Ḥaraka* where it ends with a reference to Aristotle's *Theology*. The second part of the passage is indeed reminiscent of *Neoplatonica* such as the Arabic *Plotiniana* which are partly preserved in the *ThA* (ed. Badawī, 108.5–17). The power spread by the most elevated substance is said to be received by everything according to its receptiveness. The different degrees of receptiveness and the order

[24] Cf. *Phys.* 267b22–4 in its Arabic version (*al-Ṭabīʿa*, 934.7–9):
وبيّنا الآن أنّ المتناهي لا يمكن أن يكون له قوّةٌ بلا نهاية ؛ وأنّه لا يمكن أن يتحرّك شيءٌ غير متناه زماناً بلا نهاية .

And *Phys.* 266a23–4 in its Arabic version (925.9–10):
وليس يمكن إذن أن يتحرّك شيءٌ أصلاً عن متناهٍ حركةً بلا نهاية .

of the world in general cannot come about incidentally, but must have an efficient cause. This argument for the existence of God has already been presented in passage (8).

(14) 8b3–9a1 Proclus says in his book known as the *Minor Elements*

Edition: ed. Badawī in *al-Aflāṭūnīya*, 257.3–11.

Source: Proclus Procleius' commentary on the *Dicta aurea*?

This is another one of the few occasions, occurring mainly at the beginning of the *PR*, on which the compiler provides his readers with the exact title of his source. Unfortunately, in this case the title *Usṭūkhūsīs al-ṣughrā* (*Minor Elements*) is not as telling as one might expect at first, since it is not clear to which Proclean writing it may refer.[25] The Greek tradition does not know a work by Proclus with that title. The *Fihrist* lists it, but does not give any further information which could help to identify it. A tentative suggestion is, however, possible if we assume that the next two passages introduced by a consecutive '(and) he says (*wa-qāla*)' derive from these *Usṭūkhūsīs al-ṣughrā* as well. For the second of these passages has a striking parallel in Ibn al-Ṭayyib's *Istithmār li-Maqālat Fīthāghurūs al-maʿrūfa bi-l-dhahabīya*. So Ibn al-Ṭayyib and the compiler of the *PR* may draw on the same source, i.e. an Arabic version of the commentary on the *Dicta aurea* attributed to Proclus.[26]

As to the content, most of the topics raised in the passage also occur in Book II of the Greek text of Proclus' *Theologia Platonica* (pp. 41–2, 52):[27] since the first principle is unknowable, we try to know it through what proceeds from it; in this way we ascribe to it the name of the Good and the One; and

[25] Dodds (*Proclus, Elements*, p. xxix, n. 3) and Rosenthal ('Knowledge', p. 402, n.1) suggest the *Elementatio theologia*. Rosenthal notices some similarities between our passages and Proclus' *Theologia Platonica*, but, because of the mention of the *Minor Elements*, was more inclined 'to seek the origin of the citation in "*Elements of Theology*", propositions XII and XIII'. Endreß (*Proclus Arabus*, p. 23), however, sees only faint resemblances between these propositions and the *PR*, but indicates some similarities to the *R. fī l-ʿIlm al-ilāhī*. Badawī, in his edition of the passages (*al-Aflāṭūnīya*, p. 257, n. 1) only remarks that the *Usṭūkhūsīs al-ṣughrā* must refer to the *Elementatio physica* as these are shorter than the *Elementatio theologica*, but that the passages cannot be found there.

[26] As Westerink ('Proclus commentateur', pp. 74–8) shows, it is possible that the commentary is not a work by Proclus Diadochus, but by Proclus Procleius (see above, p. 38).

[27] On the question whether an Arabic translation of this Proclean work, or at least a translation of Book II, existed, see Jolivet, 'Dossier'; and Endreß, 'Theology', p. 561.

Socrates teaches that the sun in its sovereignty over the world resembles the Good.

(15) 9a1–9 He says

Edition: ed. Badawī in *al-Aflāṭūnīya*, 257.12–258.2.

Source: Proclus Procleius' commentary on the *Dicta aurea*?

This passage resembles the previous one with regard to its possible source as well as to its content.

The idea that we call God by the best name we can think of is found as well in the *R. fī l-ʿIlm al-ilāhī* which contains Plotinian material.[28] The realisation that every appellation must still fall short of His majesty is also expressed in the *R. al-Siyāsa al-ʿāmmīya*, which is attributed to Aristotle and quoted in the passage after the next two.[29]

(16) 9a9–9b10 He says

Edition: ed. Badawī in *al-Aflāṭūnīya*, 258.3–12.

Source: Proclus Procleius' commentary on the *Dicta aurea.*

Cf.: Ibn al-Ṭayyib, *Proclus' Commentary*, ed. Linley, 8.2–10.

This passage has a striking similarity to Ibn al-Ṭayyib's *Istithmār*. Whereas the wording is completely different, the content is extremely close; next to every sentence in the *PR* has a corresponding one in Ibn al-Ṭayyib. This suggests that both texts derive from a common source, which, according to the title of Ibn al-Ṭayyib's treatise, can only be a commentary on the Pythagorean *Dicta aurea* attributed to Proclus. This is partly confirmed by the *PR* since the latest authority quoted by name, in passage (14), is Proclus from whose *Usṭūkhūsīs al-ṣughrā* the excerpt is said to be taken. So if we assume that the compiler of the *PR* stuck not only to one authority for his last three quotations, but also to one and the same source text, he would

[28] Cf. the following passage in *R. fī l-ʿIlm* (175.17–18):
لكنّا لا نجد له شبيهاً في الأشياء كلّها ولا أسماء من أسمائها ويليق بشيء من الأشياء إلا به وحده ، ولذلك نسميه أفضل الأسماء التي يمكنّا أن نسميه بها .

[29] See the passage in *R. al-Siyāsa al-ʿāmmīya* (ed. Maróth 50.20):
تقصر الألفاظ عن عظمته ويضيق المنطق عن بلوغ كنهه واحد ...
Interestingly, this very statement is omitted in passage (19), although the compiler of the *PR* quotes the context in which it occurs (50.18–20 and 50.20–51.2). He may have omitted it simply because of its similarity to the present passage. On the other hand, he normally does not shy away from repetition.

have known the commentary on the *Dicta aurea* attributed to Proclus as *Usṭūkhūsīs al-ṣughrā*. The fact that Ibn al-Nadīm lists, for Proclus, not only the *Usṭūkhūsīs al-ṣughrā*, but also a commentary on the *Dicta aurea*, neither of which he has necessarily seen, however, weakens this assumption. As for the commentary, the *Fihrist* (252.16–8) only attests a Syriac version and we can not, therefore, rule out the possibility that Ibn al-Ṭayyib summarised the *Istithmār* on the basis of a Syriac model. In that case his Arabic *Istithmār* could then have been reworked preserving Proclus' name and finally quoted by the compiler of the *PR*. Thus there would be no common source, but Ibn al-Ṭayyib would be the ultimate source of the *PR*. The probability of this scenario is, however, slim, as it would presuppose a rapid course of events: since Ibn al-Ṭayyib (d. 1043) was a later contemporary of Miskawayh, his *Istithmār* would probably not have been composed before the eleventh century, in which the *PR* was assumedly compiled as well. Furthermore, the *Istithmār* itself is unlikely to be our compiler's immediate source, for this would require an extensive reworking of his source which is uncharacteristic for him. Thus we would have to allow for a third party rewriting the *Istithmār* in the eleventh century and then this revision being quoted by the *PR*.

The parallel text to passage (16) of the *PR* is found at the beginning of Ibn al-Ṭayyib's *Istithmār* and is presented as the first of the 'Golden Exhortations'. Thus it may well be a commentary on the first verse as presented in the standard version of Pythagoras' *Dicta aurea*.[30] The first belief one should hold about God, namely that He is good, the giver of the goods and not the cause of evil, takes up the divine attributes of Plato, *Res Publica* II, esp. 379a-e.[31]

(17) 9b10–10a1 Themistius says

Source: Themistius' paraphrase on Aristotle's *Metaphysica*, Book *Lambda*, chs 7 and 9; cf. abridged Arabic version of *Sharḥ li-ḥarf al-lām*, ed. Badawī in *Arisṭū*, 18.4–5, 21.10–12; partial tr. Pines in 'Themistius' Commentary', pp. 182 and 185; French tr. Brague in *Thémistius*, pp. 94 and 113; and Arabic-Hebrew translation, ed. Landauer with Lat. tr. in *CAG* V, 5, 33.24–8.

[30] It reads, in *Ādāb al-falāsifa* (116.4–6), *al-Ḥikma al-khālida* (225.13–14), the *Muntakhab* (ed. Dunlop, lines 500–2) and slightly enhanced, in one of the two versions of the sentence in the lemmata of Iamblichus' commentary (pp. 40–2):
أوّل ما أوصيك به . بعد تقوى الله . تبجيل الذين لا يحل بهم الموت : من الله وأوليائه وإكرامهم
بما توجبه الشريعة ، وتوقى اليمين .

[31] See Westerink, 'Proclus commentateur', p. 63.

Cf.: Ps.-ʿĀmirī, *as-Saʿādah*, facsimile edn Minovi, 182.10–17.

The passage is taken from Themistius' paraphrase of Book *Lambda*, of which neither the Greek text nor a complete Arabic version is known to be extant. However, most of the quotation has word-to-word parallels in two different chapters of the extant part of an abridgement of Themistius' *Sharḥ li-ḥarf al-lām*. The first passage is also taken up in Ps.-ʿĀmirī's *K. al-Saʿāda*, which, interestingly, quotes two slightly different versions of which only the first (182.10–14) is parallel to the abridgement of the *Sharḥ* (18.4–7). The *PR* is, however, closer to the second version (182.14–17) not preserved in the abridgement.[32]

With regard to the second parallel passage in ch. 9 of the abridged *Sharḥ*, the *PR* has an additional sentence which is also attested by the Hebrew translation of the complete Arabic version.[33] The reason for this omission in the abridgement may be that it is more or less a repetition of a statement in ch. 7[34] and this may be the very reason why the compiler of the *PR* combined these two passages – omitting the statement in ch. 7, but taking up its repetition in ch. 9.

The quotation which the compiler of the *PR* selects is devoid of any indication of Aristotle's and Themistius' identification of the intellect with the First Cause. Further it does not refer to Themistius' peculiar doctrine[35] according to which the intellect, as First Cause, not only knows itself – as in Aristotle – but also knows all intelligibles, because they are in it. This interpretation, which is influenced by Plotinian and Stoic views, merges Plotinus' second principle, the intellect, with his First Principle, the One.

The *PR*'s Themistius quotation is thus one of the few examples in which

[32] The Themistius quotation in the *K. al-Saʿāda* (182.10–17) reads:
وقال في حرف اللام أخرجناه من تفسير ثامسطيوس الناموس هو الله . قال وإنّه السبب لنظام الأشياء الموجودة ولترتيبها . قال وإنّه ناموس حيّ وحياته أفضل حياة وهي حياة دائمة .
وفي حرف اللام الله قدرة وناموس وسبب لنظام العالم وترتيبه وإنّه حقّ وأنّه عقل وإنّه الخير على الحقيقة . قال وهو المبدأ والكمال ...

[33] The *PR* even suggests an emendation of the Hebrew translation which reads 'The statement that God is strong …', but should probably read 'The statement that God sees …' as in the *PR*. The Arabic *yarā* (يرى) must have been misread as *qawīy* (قوي). See also the statement in the abridgement of ch. 7 (18. 5), quoted in the next footnote.

[34] The text of the abridgement (18.4–6) reads:
وذلك أنّ الله ناموس وسبب نظام الأشياء الموجودة وترتيبها . وهو ناموس حيّ ، كما لو أمكن أن يكون الناموس متنفّساً يرى ذاته ويعقل ذاته .
And its end is parallel to the sentence of the *PR* which is only attested in Hebrew:
والقول أيضاً بأنّه يرى ويعقل ذاته قول صواب .

[35] See Pines, 'Themistius' Commentary', pp. 177–90; also Thémistius, *Paraphrase*, tr. Brague, pp. 37–8.

it seems safe to assume that the presented excerpt was carefully selected and censored by the compiler.

(18) 10a2–5 Aristotle says in the *Letter on Gouverning in General*

Source: Ps.-Aristotle, *R. al-Siyāsa al-ʿāmmīya*, ed. Maróth in *Correspondence*, 25.1–3, 48.12–16; ed. Grignaschi in '*Siyâsatu-l-ʿâmmiyya*', 97.9–10, 137.8–11.

Once again the compiler of the *PR* indicates to his readers the author and exact title of his source, namely (Ps.-)Aristotle's *Letter on Gouverning in General* to Alexander. This letter, which in itself consists of various parts (33 chapters in Grignaschi's edition), forms the eighth section of the larger collection of 16 *Letters of Aristotle to Alexander* which were combined as an epistolary novel. Our quotation is taken from *R. al-Siyāsa*, ch. 17, which is presented as a letter Alexander wrote, following Aristotle's advice, to the Persian king Darius. Whereas the material quoted in the *PR* appears in the letter as Alexander's description of his people to Darius, our compiler rephrases it as maxims Aristotle commends to Alexander, namely to be truthful and to seek God through intellect alone.

(19) 10a5–b2: Then he says

Source: Ps.-Aristotle, *R. al-Siyāsa al-ʿāmmīya*, ed. Maróth in *Correspondence*, 49.19–50.7, 50.15–51.2;[36] ed. Grignaschi in '*Siyâsatu-l-ʿâmmiyya*', 140.5–141.1, 141.11–142.7; partial edn Grignaschi with French tr. in 'Rasāʾil', pp. 78–80.

Parallel (partial): Mubashshir, *Mukhtār al-Ḥikam*, 185.2–7.

The passage is taken from the same source as the previous one, Ps.-Aristotle's *Letter on Gouverning in General*, but from ch. 19 the beginning of which purports to be the speech that an emissary of Alexander addresses to the Persians. The beginning of the quotation also occurs in Mubashshir in the section on Aristotle.

Grignaschi interprets the first half of ch. 19 as a *logos hermétique* which has been incorporated into the *R. al-Siyāsa*.[37] An explicit ascription to Hermes, however, only occurs on the second page of the chapter leaving it unclear whether the material before derives from a hermetic source as

[36] In two instances Grignaschi's reading agrees with the reading of the *PR* against Maróth who gives Grignaschi's variants in his footnotes.

[37] See his 'Rasāʾil', p. 49. Maróth approves only partly of Grignaschi's hermetic hypothesis in his *Correspondence*, pp. 58–66.

well.[38] Interestingly, the compiler of the *PR* quotes from both sections, but as he skips material about one page long in between them, the attribution to Hermes is lost. To his readers it is thus Aristotle who is speaking to Alexander throughout, and neither Alexander's emissary addressing the Persians nor the emissary quoting Hermes.

The material the compiler of the *PR* uses from the first section, not explicitly ascribed to Hermes in the *R. al-Siyāsa*, describes God's creation to be *ex nihilo* and to bring forth the principles and, through them, different 'natures' (*ṭabāʾiʿ*). Grignaschi suggests that 'nature' here means 'element' corresponding to the Greek στοιχεῖον.[39] Another possibility may be to read 'nature' as 'substance' or 'being', corresponding to οὐσία.[40] As to the principles, the *PR* quotes neither their number, namely three, nor their identity, i.e. form, matter and privation, as given in the *R. al-Siyāsa* which is thus perfectly in line with Aristotle.[41]

As to the section explicitly ascribed to Hermes, the compiler of the *PR* draws on it for yet another description of the exaltedness and transcendence of God.[42]

[38] The attribution to Hermes reads (ed. Mároth, 50.9–10):

وقد أحسن هرمس المقدّم حيث يقول : ...

Mubashshir's quotation only covers part of the first page of ch. 19, i.e. the material not attributed to Hermes.

[39] 'Rasāʾil', p. 78, n. 2, where he does not present any other example for such a meaning of *ṭabīʿa*.

[40] There is at least one example of such a translation, namely in the Arabic version of Aristotle's *De caelo* 268a3; see Endreß, *Proclus Arabus*, p. 90.

[41] Grignaschi ('Rasāʾil', p. 78, n. 4) claims that these three principles must correspond to the Proclean triad πέρας, ἄπειρον and μικτόν. His claim seems to be founded on his conviction that this section of the text is Hermetic and is, in part, based on the *R. al-Siyāsa* identifying matter with 'potentiality' (*quwwa*), form with 'actuality' (*fiʿl*) and privation with 'turning matter into actuality' (*inqilāb al-hayūlā ilā mā huwa bi-l-fiʿl*). Grignaschi further understands *inqilāb* as equivalent of the Greek ἐπιστροφή and therefore suggests an emendation of the third principle *ʿadam* to *qadm*. A less fanciful interpretation is to read the three principles as Aristotelian – and this is what the *R. al-Siyāsa* claims to be first and foremost – namely matter (*hayūlā*), form (*ṣūra*) and privation (*ʿadam*) as in *Met.* 1069b32–4 (Averroès, *Tafsīr*, III, 1452.3–5):

فالعلل ثلث والمبادي ثلثة اثنان منها الضدّان واحد هذين شيء هو القول والصورة وشيء هو العدم وشيء ثالث هو المادّة

and 1070b17–19 (*ibid.*, III, 1517.7–8):

فليس لها لاكن تساوي المقايسة بمنزلة أن قال إنسان إنّ المبادي ثلثة أي الصورة والعدم والهيولى

as well as, e.g., in al-Isfizārī's *Masāʾil* (239.23–240.1):

فهذه الثلاثة إذاً هي أوائل الطبيعيّات الهيولي والصورة والعدم . أمّا الهيولي فهو الموضوع الأوّل لكون الأجسام الموجودة فيما هي بالقوّة . وأمّا الصورة فهي التي عند الاتحاد بها صارت الهيولى موجودة بالفعل ...

[42] The topic has already been addressed in passages (2), (5), (14) and (15). As for the terminology used, see Grignaschi, 'Siyâsatu-l-ʿâmiyya', pp. 84–7.

(20) 10b3–7 The second sect are the people of the porch

Source: unidentified; some doxography arranged according to topics?

In this passage the compiler of the *PR* displays a remarkable degree of unwillingness to tamper with his source, as he presents the Stoic view on God's oneness (*tawḥīd*) and corporeality as a second opinion on the matter and contrasts it to the view of a not specified 'them', apparently of a first group not quoted by him. His source is possibly a doxographical text, maybe inserted into a more comprehensive philosophical treatise, arranged according to topics. Our quotation would then derive from the section on *tawḥīd*.

The oneness of God has already been stressed in several previous passages, although the Koranic term *tawḥīd* has never been used. God's non-corporeality has been expressed through citing Aristotle in passage (7). This might be a reason why our compiler does not feel obliged to change his source text, as he perceives the Greek philosophers so far quoted as a 'first' group which is united by agreeing on views of God in contrast with the Stoic views mentioned here.

The Stoic doctrine is presented accurately. This is one of the few occasions on which the compiler of the *PR* quotes a view apparently not in line with his own and thus takes the role of a doxographer. It is only possible to speculate about his intention in doing so. He may, for example, want to stress the importance of *tawḥīd* by the fact that, although the Stoics believe in God's corporeality, they do not deny His oneness.

(21) 10b8–10 Parmenides says

Source: unidentified; some doxography?

This short statement in the name of Parmenides has a distinct Neoplatonic ring. A similar idea, namely that the First Cause is in everything in the same way, whereas not everything is in the First Cause in one and the same way, is found in the Plotinian passage (33) below, in the two versions of the *Liber de causis* (*Ldc* ch. 23, *Ldc II* ch. 12) and, interestingly, also in the *R. fī l-Radd* (ed. Türker in 'Traité', pp. 58–60), which has a parallel to the *PR*, passage (2).

(22) 10b10–11a1 He says

Source: unidentified; some doxography?

This statement is probably still meant to be understood as uttered by Parmenides, yet the scribe of the Marsh manuscript, who most probably follows the compiler of the *PR*, leaves an obvious gap, by which he separates it from the previous one and by which he might want to indicate the turning to another passage in the same source. The topic, however, is the same as in the previous passage, namely the Ultimate Cause and the possessive pronoun (*-hā*) used in the present passage can only be understood as referring to that Ultimate Cause as well.

The indescribability of the First Cause is also addressed in the *Liber de causis* (*Ldc* ch. 5, *Ldc II* ch. 23) and the *R. fī l-Radd* (ed. Türker in 'Traité', pp. 60–1), which presents an argument and terminology quite similar to our passage: a description is said to depend on speech (*nuṭq*), intellect (*ʿaql*), thinking (*fikra*), imagination (*wahm*) and the senses (*ḥawāss*), which are all said to be caused by the First Cause.

Imagination (*wahm*) in our passage should probably be understood as referring to the soul's faculty – or its acts – and thus, in good Neoplatonic tradition, the intellect is implicitly presented to be the cause of the soul.

(23) 11a1–5 Pythagoras says

Source: unidentified; some doxography?

Cf.: Plotinus, *Enn.* V 1, 9, 1–3 on Anaxagoras.[43]

This short passage in the name of Pythagoras may derive from the same doxographical source as the following passage and perhaps as the previous passages as well. A possible model for the passage could be seen in Plotinus, *Enn.* V 1, 9 which quotes Anaxagoras with a similar view to the one found in our passage. This Anaxagoras quotation is preceded by a mention of Parmenides as in the *PR*, although different in content, and followed by a passage in the name of Heraclitus which is quite similar to the following passage in the *PR*. However, the hypothesis that Plotinus is the source of the *PR* here would require an emendation of 'Pythagoras' to 'Anaxagoras' which, although not completely improbable, cannot be supported by any evidence in the *PR*.

As to the content of our passage, two interpretations are possible. The intellect may either be considered as the immanent principle of this world

[43] Cf. tr. Armstrong: 'And Anaxagoras also, when he says that *Intellect is pure and unmixed*, posits that *the first* principle is *simple* and that the *One* is *separate*, but he neglects to give an accurate account because of his antiquity.' I put all similarities to the passage in the *PR* in italics.

above which a transcendent First Principle exists, or the intellect itself may be understood as this First Principle.

(24) 11a5–10 Heraclitus says

Source: unidentified; some doxography?

Cf.: Plotinus, *Enn.* V 1, 9, 3–5 on Heraclitus.[44]

The passage in the name of Heraclitus sounds similar to Plotinus' doxographical account in *Enn.* V 1, 9. It seems to fit with the knowledge we have of Heraclitus' teaching, but may well be phrased by some Neoplatonic writer.

(25) 11a10–15a3 Plato says in one of his discourses

Source: unidentified.

This long account of the creation of the universe is presented as one of Plato's 'discourses' which may or may not refer to his dialogues. The creation myth in the *Timaeus*, for example, of which our passage is slightly reminiscent, is also introduced as a 'discourse' (λόγος). Another '*khuṭba*' of Plato can be found in the *R. al-Siyāsa al-ʿāmmīya* (ed. Maróth, pp. 40–2), but that one is clearly phrased as a speech. However, our passage also assumes the character of a speech in its last third. After the creation of the universe and man's position in it has been expounded, man is addressed with a series of exhortations.

The goodness of the Creator and His desire for everything to be good is already stated in Plato's *Timaeus* 29e-30a. The souls which are mentioned in our passage at first may refer to the souls of the planets which correspond to Aristotle's celestial movers. As the soul is then spoken of in the singular, the reference is possibly to the world-soul. Then the intellect is introduced, so that the account conforms to the Neoplatonic hierarchy of being, i.e. the One, intellect, soul and nature.

The clothing metaphor, popular in the early stages of the Arabic translation movement, also occurs, for example, in Galen's *Jawāmiʿ* of the *Tim-*

[44] Cf. tr. Armstrong: '*Heraclitus* also knows that the *One is eternal and intelligible*: for bodies are *always* coming into being and *flowing away*.' I put all similarities to the passage in the *PR* in italics.

aeus where God is said to have created the rational soul immortal and clothed it (*kasā*) with a mortal body,[45] and in passage (100) below.

Surprising, however, is the claim that the soul perceived the material world as dark and therefore decided to descend into it. This reminds one of the view which Plotinus characterises as Gnostic and to which he addresses the following objection (*Enn.* II 9, 12, 39–41): 'For the soul which declined saw, they [the Gnostics] say, and illuminated the darkness already in existence. Where, then, did the darkness come from?' (tr. Armstrong).[46] On the other hand, Plotinus himself uses the expression 'soul sees this darkness' when talking about the soul coming into the universe and the body (*Enn.* IV 3, 9).

The 'word (*kalima*)', most probably translating λόγος, is reminiscent of the Stoic 'spermatic *logos*' which has been taken up by Plotinus and incorporated into his own philosophical system.[47] In Plotinus, who is much more likely than any Stoic material to be the ultimate source of the present passage, the *logoi* are in the soul as the ideas are in the intellect and by them soul forms and animates matter. Plotinus, *Enn.* V 1, 3, 7–8 also calls the expressed *logos* a kind of *logos* in the soul, which, according to Armstrong, recalls the Stoic distinction between '*logos* in mind' and 'expressed *logos*' (λόγος ἐνδιάθετος/προφορικός).[48] This may be the concept behind the soul's manifestations being 'expressed by spiritual speech' in our passage.

The interpretation of the term *sakīna* and its plural form *sakīnāt* poses some difficulties. *Sakīna* occurs in the Koran,[49] where it expresses the calmness that God bestows upon or sends down into the hearts of the believers. It is also used in some Graeco-Arabic translations, for example in two excerpts from Plato's *Leges* preserved in al-Bīrūnī's *Taḥqīq mā li-l-Hind*, to render divine entities.[50] In the *Dicta aurea*, *sakīna*, translating the Greek

[45] See *Galeni Compendium*, 22.17.

[46] A further indication that this passage is most probably not Plotinian is its use of the term *ḥilya* (on which, see below, p. 369, n. 77). For the Plotinian account of the bringing forth of the sensible world by the desiring soul, and its Arabic version in the *ThA*, see D'Ancona, 'Porphyry', pp. 70–6.

[47] See Witt, 'Plotinian Logos'.

[48] Plotinus, *Enn.* V, p. 19, n. 3.

[49] See *Sura* 48, 4 and 18:
هو الذي أنزل السكينة في قلوب المؤمنين ليزدادوا أيماناً مع أيمانهم ولله جنود السموات والأرض وكان الله عليماً حكيماً .
لقد رضي الله عن المؤمنين إذ يبايعونك تحت الشجرة فعلم ما في قلوبهم فأنزل السكينة عليهم وأثابهم فتحاً قريباً .

[50] In the first case, a rendering of *Leges* 717b2–4, *al-sakīnāt* most certainly translates the Greek *daemon*, see *Alberuni's India*, ed. Sachau, 59.17–19 (disregarding his emendation

daemon (ed. Diehl and Young, line 62), occurs in an invocation of the Lord Who is able to deliver men from many evils by showing them the *sakīna* (*daemon*) He has put in them.[51] Iamblichus' commentary on these verses identifies the *sakīna* with intellect,[52] and it seems most probable that these very verses have also triggered the following passage in Ibn al-Ṭayyib's *Proclus' Commentary*: 'God has knowledge of each demon (*kull sakīna*), and of where it is to be found, just as a father knows his own offspring. The name "demon" is applied to all intellectual souls; those which are on the earth belong to the mortal species, whilst those that are above the earth are of the heavenly species' (tr. Linley, p. 95).[53] The fact that *sakīna* as equivalent for the Greek *daemon* could survive is probably due to the connotation of the term in the Koran.

The reoccurrence of the term *sakīna* at the end of the present passage in a quotation attributed to Homer may be seen as an indication for the unity of this lengthy passage and for its being taken from one single source.

of the transliteration of *heroes* to *la-yuraʾʾisu*. The transliteration is attested in Ibn al-Ṭayyib's *Proclus' Commentary*, e.g. 18.10, 20.6). The passage reads:
وهذا افلاطون يقول في المقالة الرابعة من كتاب النواميس واجب على من أعطى الكرامات التامّة أن ينصبَ بسرّ الآلهة والسكنيات وايراوس اصناما خاصّة للآلهة الابويّة .
In the second case, the Greek text (653c9–d5) does not contain the term *daemon* and *al-sakīnāt* renders the *Muses*. The Arabic reads (51.15–18):
وقال الاثينيّ في المقالة الثانية من هذا الكتاب [النواميس] لما رحم آلالهةُ جنس البشر من أجل أنّه مطبوع على التعب هيّؤا لهم أعياداً للآلهة وللسكينات ولافوللن مدبّر السكينات ولديونوسيس مانح البشر الخمرة دواءً لهم من عفوصة الشيخوخة ليعودوا فِتْيانا بالذهول عن الكآبة وانتقال خُلْق النفس من الشدّة إلى السلامة .

51 The Arabic version of the *Dicta aurea* reads in the *Ādāb al-falāsifa* (118.14–5), in the *Muntakhab* (ed. Dunlop, lines 542–3), in *al-Ḥikma al-khālida* (228.10–11) and in the lemma of Iamblichus' commentary (106.7–10):
يأيّها الربّ [أو : الآب] الواهب للحياة ! حقاً أقول إنّك القادر على أن تدفع عنهم بلايا كثيرة إن أُظهرت لهم السكينة التي جعلتها فيهم .
This is reminiscent of Plato's *Timaeus* 90a3–4 and 90c5–6, where God is said to give each of us a *daemon*.
At the beginning of the *Dicta aurea*, the 'daemons' are rendered as 'angels' (*malāʾika*) or 'earthly inhabitants' (*ʿummār al-ʿarḍ*) in the lemmata of Iamblichus commentary, yet in the corresponding commentary the trace of the *daemons* is still preserved, at least in the latter case, when the 'earthly inhabitants' are described as follows (42.3–5):
قال المفسّر : أما عمار الأرض فيعني بهم الأنفس التي قد فارقت الأبدان وصارت إلى الموقف لتحاسب هنالك ؛

52 The passage in question (106.12–16) reads:
قال المفسّر : ... وأمّا قوله "السكينة" فإنّه إمّا أن يكون يعني به العقل الذي نصبه الله في واحد واحد من الناس .

53 Further occurrences of *sakīna*, *sakīnāt* or *sakāʾin* in Ibn al-Ṭayyib's *Proclus' Commentary* are found on 18.2–3 and 7, 20.3–4, 92.10 and 94.8–9.

(26) 15a4–16b9 Aristotle says in one of his discourses

Source: unidentified; same source as for the previous passage?

The passage has a distinct Neoplatonic ring which casts some doubt on the attribution to Aristotle. This reminds one, however, of the *ThA* and its Plotinian source. There is furthermore a striking resemblance to the previously quoted passage. Both passages claim to be drawn from a 'discourse' (*khuṭba*), one either by Plato or by Aristotle, use the same very unusual terminology (for example *dh-w-t* II. and V.; *w-ḥ-d* V.) and concepts (esp. the *logos*, matter perceived as dark and low by the soul) and seem to give, in part, an account of the creation of the universe and the position of man and his soul in it.

(27) 16b9–12 Among the indications for their belief in resurrection and the hereafter is what Socrates says in the book attributed to Phaedo

Source: Plato, *Phaedo* 63b6–8, 63c2–4 in an unidentified Arabic version.

Cf.: al-Bīrūnī, *Taḥqīq mā li-l-Hind,* ed. Sachau in *Alberuni's India*, 32.21–33.1; and ed. Badawī in *Aflāṭūn*, 124.18–20.

This *Phaedo* passage is correctly quoted as a statement by Socrates. However, the compiler of the *PR* puts it into a larger context when he claims it is evidence for 'their' belief in resurrection and the hereafter. He is doubtlessly referring to the Greek philosophers here.[54]

Whereas the question which is posed to Socrates has no exact parallel in the *Phaedo*,[55] his answer is, apart from some omissions, a faithful rendering of 63b6–8 and 63c2–4.[56]

The first part of the same Platonic passage also occurs among the *Phaedo* quotations in al-Bīrūnī:

وقال في المجامع : لو لم أرنى صائراً أوّلاً إلى آلهة حكماء سادة أخيار ، ثمّ من بعد

[54] An account of Empedocles', Pythagoras', Socrates', Plato's and Aristotle's views on the topic is, e.g., preserved in al-ʿĀmirī's *K. al-Amad*, ch. 4.

[55] The context in the *Phaedo* 62d-63b is not whether one should be concerned about leaving one's fellow men upon death, but whether one should be concerned about leaving this world when one has been appointed to one's position therein by the wise gods.

[56] The Platonic text reads: ... εἰ μὲν μὴ ᾤμην ἥξειν πρῶτον μὲν παρὰ θεοὺς ἄλλους σοφούς τε καὶ ἀγαθούς, ἔπειτα καὶ παρ' ἀνθρώπους τετελευτηκότας ἀμείνους τῶν ἐνθάδε, ... ὅτι μέντοι παρὰ θεοὺς δεσπότας πάνυ ἀγαθοὺς ἥξειν, εὖ ἴστε ὅτι εἴπερ τι ἄλλο τῶν τοιούτων διισχυρισαίμην ἂν καὶ τοῦτο.

إلى ناس ماتوا خير ممن هاهنا [ممن هاهنا Sachau : مما هنا Bad] ، لكان تركى الحزن على الموت ظلماً .

Further he says about the gathering-places of the souls after death: 'If I did not think that I am going first to gods who are wise, ruling, and good, then afterwards to men, deceased ones, better than those here, I should be wrong not to be in sorrow about death.' (tr. Sachau, I, p. 65)
The quotation in the *PR* and al-Bīrūnī's text share sufficient similarities to suggest an ultimate common source of which both may quote different versions.[57]

(28) 16b12–17a5 He also says in this book

Source: Plato, *Phaedo* 69c3–7 in an unidentified Arabic version.

The source of this quotation and of the previous one is almost certainly the same and must probably be some doxographic account of the beliefs of the Greek philosophers. The compiler of the *PR* indicates the ultimate source correctly as the *Phaedo* to which he refers as 'this book', i.e. the book just mentioned before. The rendering of 69c3–7 is, once again, close to the Greek,[58] but has, this time, no parallel in al-Bīrūnī's extracts.

(29) 17a6–b10 Aristotle says

Source: unidentified.

This allegedly Aristotelian passage is another piece of clearly Neoplatonic doctrine according to which there are four worlds: the world of the Divine, intellect, soul and nature, each of which depends on the next higher one.[59]

[57] Sachau suggests that al-Bīrūnī's *Phaedo* source was the commentary on the dialogue by Proclus, the existence of which is attested by the *Fihrist*, and Rosenthal assumes that Sachau is right (Bürgel, 'Quotation', p. 284). The transmission of Platonic dialogues into Arabic in general, as well as of the *Phaedo* in particular, still remains a puzzle. There is some evidence for an Arabic tradition of the *Phaedo* attested by quotations in al-Ruhāwī and a Persian paraphrase translated from the Arabic which must share a common source with al-Ruhāwī's material. However, this material has no similarities to the quotations in al-Bīrūnī (*ibid.*, pp. 285–90). For the existence of a narrative of the death of Socrates attested by Ibn al-Qifṭī and Mubashshir which does not seem related to either of the two other traditions, see Rowson, *Muslim Philosopher*, pp. 35–7.

[58] The Platonic text reads: καὶ κινδυνεύουσι καὶ οἱ τὰς τελετὰς ἡμῖν οὗτοι καταστήσαντες οὐ φαῦλοί τινες εἶναι, ἀλλὰ τῷ ὄντι πάλαι αἰνίττεσθαι ὅτι ὃς ἂν ἀμύητος καὶ ἀτέλεστος εἰς Ἅιδου ἀφίκηται ἐν βορβόρῳ κείσεται, ὁ δὲ κεκαθαρμένος τε καὶ τετελεσμένος ἐκεῖσε ἀφικόμενος μετὰ θεῶν οἰκήσει.

[59] The notion recalls the description of the *Timaeus* as an account of the order of the

The soul is given an intermediary position between the intellectual and the natural things and it is said to help perfect the latter. The passage ends with an allegedly Platonic quotation which, in fact, sounds Stoic as it claims that the life of the world is a psychic spirit or pneuma.[60]

(30) 17b10–18a4 The Greek Sage says

Edition: ed. and tr. Rosenthal in 'Aš-Šayḫ', pp. 474–5: no. A1; ed. Badawī in *Aflūṭīn*, 184.3–9: no. 1; tr. Lewis in 'Plotiniana', p. 275.

Cf.: *Enn.* V 1, 7, 1–17 and V 2, 1, 7–13.

The first passage in the name of the 'Greek Sage' (*GS*), whose name has become synonymous with part of the Arabic paraphrase of Plotinus' *Enneades*, sets off with a statement on the immobility of the first agent which sounds rather Aristotelian and recalls Aristotle's unmoved mover. The ensuing discussion of the generation of the intellect as image (εἰκών, *mithāl*) of the One is, as Rosenthal has already partly indicated, inspired by *Enn.* V 1, 7, 1 and 5–6 and V 2, 1, 9–11. According to the Plotinian concept expressed in the third passage, the overflow of the One causes something else and 'Its halt and turning towards the One constitutes being [ὄν rendered as 'substance (*jawhar*)' in the Arabic], its gaze upon the One, Intellect' (tr. Armstrong).[61]

(31) 18a4–18b1 He says

Edition: ed. Rosenthal in 'Aš-Šayḫ', p. 474: no. A2.

Parallel: *ThA*, ed. Badawī in *Aflūṭīn*, 114.7–14; and ed. Dieterici in *Uthūlūjiyā* 111.15–112.4; tr. Lewis in 'Plotiniana', pp. 273–5; and German tr. Dieterici in *Theologie*, pp. 112–13.

three intelligible worlds, the world of Divinity, intellect and soul, in Ibn Abī Uṣaybiʿa, 53.17–18:

كتاب طيماوس الروحانيّ في ترتيب العوالم الثلاثة العقليّة التي هي عالم الربوبيّة وعالم العقل وعالم النفس .

A similar reference to the *Timaeus* occurs in the *R. fī Ārā'* (309.1–12):

كلامه في العوالم العالية ، يعني عالم النفس وعالم العقل وعالم الربوبية ... كلامه في العقل ....
هذه الفصول انتزعتها من كتاب أفلاطون المعروف بـ"طيماوس" في هذه المعاني ...

[60] A similar idea is expressed in the *Rasā'il Ikhwān al-Ṣafā'* II, 133.8:

واعلم أنّ النفس الكلّيّة هي روح العالم .

[61] If our passage were, indeed, in part a paraphrase of *Enn.* V 1, 7, 5–6 it would provide some evidence for the Greek reading 'by its [i.e. intellect's] return to it [i.e. the One] it sees: and this seeing is intellect' (tr. Armstrong) instead of the alternative interpretation 'by its [i.e. the One's] return to itself it sees: ...', on which, see Armstrong in his tr. Plotinus, *Enn.* V, pp. 34–5, n. 1, and Rosenthal, 'Aš-Šayḫ', p. 475, n. 2.

Cf.: *Enn.* V 1, 6, 6 and 12–19.

This passage belongs not only to the *GS*, but has also an exact parallel in the *ThA*. Rosenthal presents the passage as a proof that the *GS* and the *ThA* share a common source.[62] For the *ThA* renders the complete text of *Enn.* V 1, 6, 1–22, whereas the *GS* omits V 1, 6, 19–22 which is an excursus within the Plotinian text and continues with V 1, 6, 25[63] after that excursus in the next passage, (32), which is not preserved in the *ThA*.

The passage takes up the question posed in *Enn.* V 1, 6, 4–7 which, as Plotinus puts it in the Greek, was 'repeatedly discussed also by the ancient philosophers, how from the One, if it is such as we say it is, anything else, whether a multiplicity or a dyad or a number, came into existence' (tr. Armstrong). Plotinus commends readers to imitate the worshipper in a temple who takes the statues standing outside[64] to indicate what is in the sanctuary. As all the things in the world are moving towards an end, the One Itself unmoved must be that end.

(32) 18b1–13 The First Agent must be at rest

Edition: ed. and tr. Rosenthal in 'Aš-Šayẖ', pp. 476–7: no. A3; ed. Badawī in *Aflūṭīn*, 184.10–185.2: no. 2; tr. Lewis in 'Plotiniana', p. 275.

Cf.: *Enn.* V 1, 6, 25–37; also passage (44).

In the model of the *PR* this passage may have constituted a unit with the previous passage, as our compiler gives no indication – usually, but not always he leaves a clear gap – of the beginning of a new section. This would fit well with Rosenthal's argument that the Plotinian material in *Enn.* V 1, 6, 19–22 'disrupts the sequence of Plotinus' thought'.[65]

Although the Arabic version seems to be close to the Greek text, it introduces two new concepts: the mediation of the intellect which results in a

[62] 'Aš-Šayẖ', p. 467.

[63] Rosenthal (*ibid.*, p. 467, n. 1) mentions the possibility that the rendering of V 1, 6, 22–5 may have dropped out in the Marsh manuscript through *per homoeoteleuton*.

[64] The Greek τὰ ἔξω ἤδη ἀγάλματα ἑστῶτα (in V 1, 6, 14) explains the use of the Arabic term *ṣanam* 'idol, image' as translation of ἄγαλμα (see Ullmann, *WGAÜ* Suppl. 1, p. 53) and supports the reading *muthbata* which Lewis (*Re-examination*, p. 172) suggests (but for which he does not quote MS authority, as the difference from the earlier editors is only in pointing; see *ibid.*, p. xxiii).

[65] 'Aš-Šayẖ', p. 467.

creation *mediante intelligentia*[66] or a mediated creation, and the explicit distinction between substances and accidents.

(33) 18b13–19b8 He says

Edition: ed. and tr. Rosenthal in 'aš-Šayḫ', pp. 478–81: no. A4; ed. Badawī in *Aflūṭīn*, 185.3–19: no. 3; tr. Lewis in 'Plotiniana', p. 281.

Cf.: *Enn.* V 1, 7, 18–26.

This lengthy passage is related to the previous one by the concept of creation, especially the creation *mediante intelligentia*. There are, however, ambiguous statements on the topic, and as it is introduced in the Arabic version alone, we have no means of checking the text against the Greek. The intellect is explicitly said to be only the formal cause, whereas the One is the final cause of everything. However, in what follows the One is presented as 'the cause of the thing's being[67] and its form without mediation' and 'the cause of the soul's being and of the form of things by mediation of the intellect'. If we interpret the first part of the sentence to refer, as I suggest in the translation, to the 'first created' thing alone and thus understand the One as causing intellect's being and form,[68] we are still left with the problem that not only the forms of all things, but also the being of the soul is said to be caused by the One only via the intellect.[69] The latter disagrees with the claim that the One is the final cause and thus the sole cause of the being of all things. The ensuing account may be read as implying that the One creates the being of everything first, but then uses intellect to define and form them.

As to the One, this passage is a good example of 'predication by causality', which is, as Adamson shows, one way in which the Arabic Plotinus attempts to apply language to the One and, to a lesser degree, to intellect,

[66] On which see D'Ancona, 'Doctrine', pp. 73–95; her 'Pseudo-Theology', pp. 101–9; and her 'Profilo', pp. 122–3; and Adamson, *Arabic Plotinus*, pp. 141–2.

[67] I translate the Arabic term *huwīya* as 'being', whereas Rosenthal and Lewis use 'identity'. Adamson (*Arabic Plotinus*, pp. 125–6) points out that in the Arabic *Plotiniana*, *huwīya* is, with one exception, employed in the same meaning as *annīya*.

[68] If we understand 'the thing' to refer to everything, we are facing an even more obvious contradiction. Then the One would be said to be the cause of the being and form of everything without mediation, but also the cause of the soul's being and everything's form via mediation.

[69] A way to avoid this problem would be to read 'by mediation of the intellect and the soul. All things are formed in the intellect ...' as I point out in the notes to the translation. However, I do not think this reading is particularly probable, especially as there is no evidence that any entity other than the intellect was ever thought to be a possible mediator between the Creator and His creation.

although these two metaphysical levels are completely above the discursivity of the soul.[70]

(34) 19b8–20a6 It is said

Edition: ed. and tr. Rosenthal in 'aš-Šayḫ', pp. 480–83: no. A5; ed. Badawī in *Aflūṭīn*, 186.1–9: no. 4; tr. Lewis in 'Plotiniana', p. 367.

Cf.: *Enn.* V 6, 4, 14–22.

In the Marsh copy of the *PR* this passage is in no way separated from the previous one, so only the fact that it paraphrases a different *Ennead* sets it apart. The rendering is very close to the Greek original, the only addition in the Arabic being the reference in the last sentence to the Neoplatonic concept that each thing receives the outflow of the higher principles according to its own potential (see, for example, *Ldc*, chs. 19 and 23; *Ldc II*, chs. 12 and 26).[71]

(35) 20a7–b6 He says

Edition: ed. and tr. Rosenthal in 'Aš-Šayḫ', pp. 482–3: no. A6; ed. Badawī in *Aflūṭīn*, 186.10–13: no. 5; tr. Lewis in 'Plotiniana', p. 474.

Cf.: *Enn.* V 4, 1, 5–8?

This short statement bears hardly any resemblance to the Greek text of the *Enneades*, but faintly recalls passage (33) where it is said that everything is in the Creator, but He is in things only as cause. The statement that He is not separate from all things seems to contradict passage (23) in the name of Pythagoras at first, but it is then modified in a way to preserve His complete transcendence.

[70] Adamson (*Arabic Plotinus*, p. 86) even detects 'a technical phrase' used in the Arabic Plotinus 'to mark predication by causality, which is to say that something possesses a predicate "in the manner of a cause" (*bi-nawʿi ʿillatin*)'. He further mentions (*ibid.*, p. 215, n. 1): 'This would correspond in Greek to *kat' aitian*, which does not appear in Plotinus but can be found in Proclus and the Pseudo-Dionysius', as has already been discussed by D'Ancona, 'Profilo', pp. 128–9, who uses the term *per modum causae*. As the second method of talking about the One, Adamson indicates 'predication by eminence'. See *Arabic Plotinus*, pp. 85–7, 117–19 and 131.

[71] Adamson (*Arabic Plotinus*, pp. 167–170) points out the similarity of the language used in this passage and the language used by *Kalām*-writers.

(36) 20a10–b6 He says

Edition: ed. and tr. Rosenthal in 'Aš-Šayḫ', pp. 482–5: no. A6; ed. Badawī in *Aflūṭīn*, 186.14–187.3: no. 6; tr. Lewis in 'Plotiniana', p. 474.

Cf.: *Enn.* VI 9, 6, 7–16 and 39–54.

The first part of the passage is a close rendering of the Greek text of the *Enneades*. The Plotinian elaboration of the One being without need is then omitted, before the rest of ch. 6 of *Enn.* VI 9 is paraphrased, faithfully to the sense, but not to the wording, of the original. The final definition of the One as knowledge, which is contradictory to the former statement that It is prior to knowledge, occurs also in the Plotinian Source and may be connected to the exceptional description of the One as intellective in *Enn.* V 4, 2, and probably meant as a reference to the concept of divine providence. For, as Adamson puts it, 'if the First does not know, there can be no providence'.[72]

(37) 20b7–13 Someone may say

Edition: ed. and tr. Rosenthal in 'Aš-Šayḫ', pp. 484–5: no. A7; ed. Badawī in *Aflūṭīn*, 187.4–10: no. 6; tr. Lewis in 'Plotiniana', pp. 474–5.

It is not clear whether this and the following passage belong to the Arabic *Plotiniana*. They do not correspond to anything in the *Enneades*. Furthermore, the compiler of the *PR* introduces the passage after the next with the words 'The Greek Sage says' which may or may not indicate that the passages before do not belong to the Sage.[73]

As for its contents, this present passage is clearly linked to the following one which presents the answer to the question raised here, namely whether the intellect acquired its forms from the One all at once or one

[72] Adamson, *Arabic Plotinus*, p. 105; see also pp. 153–5. See also D'Ancona, 'Knowledge', pp. 435–8, where she states that the translator (of our passage) was probably aware of his mixing Aristotelian and Neoplatonic traditions and that Arabic readers did not perceive Aristotelian and Plotinian metaphysics as incompatible, which probably explains why they adopted the Plotinian rather than the Proclean model. For in the former, the One and the (Aristotelian) intellect are closer to each other, whereas the latter puts many layers between the One and the intellect.

[73] Rosenthal includes passage (37) in his edition and translation of the 'Greek Sage', but omits the following (38), for which he is rebuked by Badawī (*Aflūṭīn*, p. 186, n. 1 and n. *), who interprets both passages as comments on the previous Plotinian passage (36) and reasonably argues that Rosenthal should have either omitted both or edited them together.

after the other as this is the way in which intellect passes the forms on to what is below.

(38) 21a1–12 We say

Edition: ed. Badawī in *Aflūṭīn*, 187.11–18: no. 6; tr. Lewis in 'Plotiniana', p. 475.

This section elaborates on the problem posed in the previous passage and expounds how the One acts all at once and how His action renders the intellect good all at once, but how the latter is nevertheless not able to know the former's virtues other than one after the other.

The idea that the First has or is all virtues is not Plotinian, but it is, as Adamson shows, a distinctive feature of the Arabic *Plotiniana* which is presented at length in the *ThA*.[74] The notion that the intellect receives the virtues from the One is also present in the *Ldc*, ch. 18.[75]

(39) 21a13–b9 The Greek Sage says

Edition: ed. and tr. Rosenthal in 'Aš-Šayẖ', pp. 486–7: no. A8; ed. Badawī in *Aflūṭīn*, 188.1–10: no. 7; tr. Lewis in 'Plotiniana', p. 475.

Cf.: *Enn.* V 1, 7, 40–3; VI 7, 17, 40–2.

The concept of form and *logos* presented in passage (39) is clearly Plotinian. However, the parallels to the *Enneades* suggested by Rosenthal (V 1, 7) and Lewis (VI 7, 17) contain only a few sentences which may be detected in the Arabic. The passage is rather a concise summary of Plotinus'

[74] *Arabic Plotinus*, pp. 69–75. Whereas Plotinus follows Aristotle in denying that the intellect or what is above may have virtue (*Enn.* I 2, 3), in ch. 9 of the *ThA* (ed. Badawī, 130.3–10) we read the following (tr. Adamson, *Arabic Plotinus*, pp. 73–4): 'All the virtues are in the intellect eternally ... they are acquired, since the intellect acquires them from the first cause. The virtues are only in the intellect eternally because the intellect does not subside in its vision of the first cause ... For they come to be in it from the first cause with no intermediary ... As for the first cause, the excellences are in it in the way of a cause (*bi-nawʿi ʿillatin*). Not that it is a receptacle for the virtues, but rather its whole is a being (*anniyya*) that is all the virtues, even though the virtues flow from it without its being divided'. The motivation for the inclusion of this non-Plotinian concept may, as Adamson explains, be 'to place ethics within a religious context. By making God the highest principle of the virtues, he [i.e. the Adaptor of the Arabic Plotinus] provides a philosophical argument for the theological view that the source of human goodness is ultimately divine' (p. 74).
The question whether human and divine virtues or excellences differ is discussed in the *Maqāla fī l-Mudkhal ilā ʿilm al-akhlāq*, ed. Badawī in *al-Akhlāq*, pp. 420–6.

[75] See *Ldc*, ch. 18, where the first intellects are said to have abundant reception of the virtues, whereas the second intellects only receive them through the first intellects.

doctrine on form and *logos* which seems to be, at least partly, influenced by the Stoic notion of the *logos spermatikos*.[76]

Whereas the One is not limited in any way, the intellect is clearly limited by not being the One. Therefore it can, somehow, be said to have form.[77] If one transferred this doctrine to the level of the soul, which is not done in our passage, the soul, less perfect than the intellect, would be said not to be form, but word (*logos*). So form is what exists in the intellect, but becomes a *logos* when it is passed on to the lower level and instantiated in the soul.[78]

(40) 21b10–22b9 The First must not be multiple

Edition: ed. and tr. Rosenthal in 'Aš-Šayḫ', pp. 488–92: no. A9; ed. Badawī in *Aflūṭīn*, 188.11–189.12: no. 9; tr. Lewis in 'Plotiniana', pp. 475–6.

Cf.: *Enn.* VI 7, 17, 42–3 and 18, 2–5.[79]

The passage seems to take up the end of ch. 17 of *Enn.* VI 7 and the beginning of ch. 18, and may thus form one unit with the previous passage, especially as there is no indication in the manuscript of the *PR* for any kind of break in the text.[80]

The terminology clearly derives from *Enn.* VI 7, 18, 3, where everything coming from the Good is said to have a 'trace' (ἴχνος) or an 'impression' (τύπος) from It.[81] The example of fire is also spelt out in Plotinus. The listing of the four different impressions and their motions, however, sounds

[76] For discussion of the passage, see also Adamson, *Arabic Plotinus*, pp. 113–14.

[77] The Arabic hendiadys *ḥilya wa-ṣūra* is used and *ḥilya* is thus obviously understood as equivalent to 'form'. This use also occurs in the *Ldc*, end of ch. 8 (Badawī's edition has *kullīya* instead, but the Latin translation of the Arabic which reads 'helyatun' indicates the necessary emendation), and *Ldc II*, end of ch. 22. This understanding is in contrast with the use of the term in passage (25) above, where it apparently expresses 'adornment' or something similar. This is an indication of the different source texts of these two passages, i.e. passage (25) is unlikely to derive from the Arabic *Plotiniana*.

[78] For the concept of the 'intellectual soul-word', which is the order of the sensible world according to the *ThA* (ed. Badawī, 127.14–15) and based on *Enn.* IV 7, 3, 28–31, which criticised the Stoic doctrine of pneuma, see D'Ancona, 'Porphyry', pp. 66–8.

[79] Rosenthal ('Aš-Šayḫ', p. 489, n. 1) also gives *Enn.* V 3, 6, 14–15 as a possible parallel for the beginning of our section, but the relevance seems insignificant to me.

[80] D'Ancona quotes part of this passage in her discussion of the concept of creation in the Arabic Plotinus ('Doctrine', pp. 81–2).

[81] The notion 'impression' (*athar*) and the idea that the intellect is the 'first impression of the Creator' is also found in the *Ikhwān al-Ṣafāʾ*, I, 278.9, and 'action' (*fiʿl*) in general is defined (*ibid.*, III, 386.2) as an 'impression by someone producing [an impression]' (*athar min muʾaththir*).

like a later systematisation of the Plotinian doctrine and may well derive from some Neoplatonic commentary tradition.

(41) 22b10–25a8 Plato says

Edition: ed. D'Ancona with Italian tr. in *Plotino*, pp. 239–41; tr. Lewis in 'Plotiniana', pp. 235–7.

Cf.: *Enn.* IV 8, 3, 6–30.

This lengthy passage is a faithful, although slightly elaborated, rendering of *Enn.* IV 8, 3, yet attributed to Plato whose doctrines are clearly expounded, in *Enn.* IV 8, but whose name is not mentioned in the Greek text. Therefore Zimmermann suggests that the entirety of *Enn.* IV 8 was titled 'The soul's descent into the body according to Plato' in the Greek model on which the Arabic Plotinus is based. D'Ancona, however, assumes that the compiler of the *PR* had an Arabic source in which our passage was preceded by a non-Plotinian doxography ascribed to Plato (preserved at the end of *ThA*, ch. 1), from which he derived his ascription.[82] As to the content of the passage, the reason why creation as such exists is expounded as well as how intellect and soul are one and many at the same time.[83] The explanation of why the soul also thinks is more elaborate than in its Greek model and introduces the technical Arabic term of the 'acquired intellect'. This term recurs in passages (45–6) below, in the rendering of Alexander's *De intellectu* and a commenting passage by al-Fārābī.

The only major doctrinal addition to the Greek Plotinus in the present passage is the discussion of different kinds of necessity. Although this account seems, at first, contradictory as it stands, it may be interpreted consistently as stating that natural necessity brings forth the sensible

[82] See Zimmermann ('Origins', pp. 143–9, esp. pp. 146 and 148) and D'Ancona ('Greek Sage', pp. 170–6). However, in his modern English translation from the Greek, Armstrong supplies Plato's name at exactly the same position where it occurs in the Arabic. Could it thus simply be that any learned translator might feel obliged to provide Plato's name there? As for the compiler of the *PR*, he certainly found the Platonic attribution in his Plotinian source and was perfectly aware that he excerpted not from Plato directly, but from someone else, i.e. the Greek Sage, quoting Plato. He proceeded similarly when he excerpted Miskawayh quoting Aristotle in passages (201), (206) and (207). The compiler of the *PR* obviously did not mind that his readers would be under the impression of reading Plato's or Aristotle's words directly, presumably because he considered their words to be faithfully transmitted and knowledge of who transmitted them to be dispensable.

[83] D'Ancona (*Plotino*, p. 163) explains that the divine intellect is one by nature, but many, because it encompasses the multiplicity of the intelligibles.

world, wilful necessity its cause, i.e. the soul, and necessity above nature and will, i.e. the necessity of the First Cause, brings forth intellect.[84]

(42) 25a9–27a11 He says

Edition: ed. D'Ancona with Italian tr. in *Plotino*, pp. 242–4; tr. Lewis in 'Plotiniana', pp. 237–9.

Cf.: *Enn.* IV 8, 4, 1–42.

This passage is the immediate continuation of the previous passage and renders ch. 4 of *Enn.* IV 8 more or less completely and faithfully. As most of the similes used derive from Plato (*Rep.* 532b-c, 515c; *Phaedrus* 246c; *Phaedo* 62b, 109a-110a; *Timaeus* 41c-d),[85] it comes as no surprise to find the passage indirectly ascribed to Plato, i.e. introduced by 'He says' which, in the context of the *PR*, has to be understood to refer to the last authority quoted, namely Plato.

(43) 27a11–28b9 So if you hear him [i.e. Plato] say

Edition: ed. D'Ancona with Italian tr. in *Plotino*, pp. 244–6; tr. Lewis in 'Plotiniana', pp. 239–41.

Cf.: *Enn.* IV 8, 5, 1–14.

The passage renders the first part of *Enn.* IV 8, 5, and is thus consecutive to the previous passages. The aim is to present a method for combining the philosophers' differing views on the descent of the soul, especially with regard to the question of whether this descent happens voluntarily or by necessity. On this topic our passage also has material additional to the Greek Plotinus and offers slightly different explanations.

Adamson suggests that this passage and passage (41) should be read together and in the light of *Enn.* VI 8, which is 'On Free Will and the Will of the One', yet no Arabic adaptation of this text has surfaced so far.[86]

[84] The exposition of what is meant by 'necessity' is positioned after the description of the emergence of intellect, but may, in fact, refer to the aforementioned necessity of the existence of the cause of the sensible world, i.e. the soul.
On necessity, see D'Ancona, *Plotino*, pp. 326–7; see also passage (43) below and Adamson's discussion of both *GS*-passages together (*Arabic Plotinus*, pp. 147–9).

[85] Cf. D'Ancona, *Plotino*, pp. 168–73.

[86] See his *Arabic Plotinus*, pp. 147–9; p. 148: 'Taking these two texts together, we have the following view. As in *Enn.* VI.8, the First or pure cause operates above both will and necessity. This is because the First is the cause of both volition and necessity ... Below God are the intellectual causes, such as soul, which performs its act both freely *and*

As to the statements by Empedocles and Heraclitus, Plotinus only refers to them shortly at this point, as their more elaborate discussion is found in *Enn.* IV 8, 1, which is taken up in the *ThA* (ed. Badawī, 23.1–11)[87] and to which the author of the Arabic Plotinus may have turned when rendering the present passage.

(44) 28b10–29b1 The Greek Sage says

Edition: ed. Badawī in *Aflūṭīn*, 189.13–17: no. 9; tr. Rosenthal in 'Aš-Šayḫ', p. 477, n. 2; tr. Lewis in 'Plotiniana', p. 275.

Cf.: *Enn.* V 1, 6, 30–7; end of passage (32).

The reason why this passage, which repeats, with slight variations, the end of passage (32), is taken up again here is not apparent. One also wonders whether the compiler of the *PR* is responsible for this repetition or whether he has already found two versions based on the same Plotinian passage in his model.

(45) 29a1–6 Alexander says

Source: *Maqālat al-Iskandar al-Afrūdīsī fī l-ʿAql*, ed. Badawī in *Commentaires*, 31.15–34.2; ed. Finnegan in 'Texte arabe du ΠΕΡΙ ΝΟΥ', 181.3–184.10.[88]

The passage summarises the beginning of the Arabic translation of Alexander of Aphrodisias' *De intellectu*. Oddly, this passage's use of the expression 'acquired intellect' (*ʿaql mustafād*) seems much closer to the underlying Greek text than the expression in the Arabic Alexander.[89]

necessarily. If these causes were purely free, they would act only some of the time (...). If they were purely necessary, they would be simple links in a chain of causality, like natural causes and effects'; and p. 149: the One has 'a freedom so constant that it resembles necessity. This resemblance with necessity explains why the effects of God are "necessarily" good and must "necessarily" occur, that is, because there is no change or variation in this willing.' Adamson (*Al-Kindī*, p. 206) also detects an influence of this passage on al-Kindī, who understands God as acting 'using "volition and choice," but also "necessarily". The reason for this, apparently, is that God's wisdom ensures that He will choose what is best. Al-Kindī says that "unconstrained" choice would lead to worse results.'
On the present passage, see also D'Ancona, *Plotino*, pp. 335–7, and, on Plotinus on the voluntary/necessary descent of the soul, *ibid.*, pp. 175–8; on its Greek model, Leroux, 'Freedom', pp. 296–7.

87 See De Smet, *Empedocles Arabus*, pp. 144–5.

88 For manuscripts overlooked in these editions, see Geoffroy, 'Tradition', p. 196, n. 20.

89 See Finnegan, 'Texte arabe du ΠΕΡΙ ΝΟΥ', pp. 171–8; Davidson, *Alfarabi*, pp. 11–12; Geoffroy, 'Tradition', pp. 213–21.

Three different levels of intellect are distinguished: the lowest level is the 'material intellect' (*al-ʿaql al-hayūlānī*; νοῦς ὑλικός), which may also be called 'potential intellect' as it has, according to Aristotle, no nature on its own, but the capacity to become all intelligibles by thinking about them (*DA* 429a10–29 and 430a10–17). On the next level this material and potential intellect becomes actualised by actually thinking about intelligibles and thereby acquires a disposition or habitus for thought. Thus it is called 'intellect *in habitu*' (νοῦς ἕξιν ἔχων, νοῦς ἐν ἕξει) or 'acquired intellect' (*ʿaql mustafād*; νοῦς ἐπίκτητος).[90] As everything potential needs something already actual to make it actual, there must also be an 'active intellect' (*al-ʿaql al-faʿʿāl,* νοῦς ποιητικός) which brings about disposition within the potential intellect. The simile of light which makes potential colours actually visible is Aristotelian (*DA* 430a14–17).[91]

The passage in the *PR* thus provides evidence that the term 'acquired intellect' was used to refer to the second stage of intellect, the material human intellect which has already acquired a disposition or habitus for thinking.[92] In the Arabic version of Alexander this is far less clear as the second stage is not referred to by any specific name at first,[93] and the term

[90] The term 'acquired intellect' (νοῦς ἐπίκτητος) does not occur in Aristotle, but in Alexander's *De anima*, 82.1, which may be the ultimate source for the Arabic expression, especially given the fact that Isḥāq ibn Ḥunayn translated Alexander's *De anima* (unfortunately only preserved in a Hebrew translation) as well as his *De intellectu* (see Finnegan, 'Texte arabe du ΠΕΡΙ ΝΟΥ', pp. 171–8; Davidson, *Alfarabi*, p. 11; and Geoffroy, 'Tradition', p. 217). Aristotle further distinguishes two states within this level of intellect, namely the intellect of someone who has the ability or disposition to think but does not exercise it at the moment, i.e. someone who is potentially thinking, and of someone who is actually thinking (cf. *DA* 429b5–9 and 417a21–b1).

[91] Similar, especially in part of the phrasing, is the quotation in the name of Alexander, but implicitly ascribed to his commentary on the *De anima*, in al-Maqdisī, *K. al-Badʾ wa-l-taʾrīkh*, I, 24.5–11, on which, see also Peters, *Aristoteles Arabus*, p. 43, and Gätje, *Psychologie*, pp. 72–3:
وقال [ارسطاطاليس] في كتاب النفس بخلاف هذا وقسّمه [العقل] إلى ثلاثة أقسام إلى العقل الهيولانيّ والعقل الفعّال والعقل المستفاد وفسّره الاسكندر [الاسكندر MS : لاسكندر ed. Huart] فقال إنّ العقل الهيولانيّ هو ما يوجد في شخص الإنسان من إمكان التهيّؤ لتأثير العقل الفعّال وإنّ العقل المستفاد هو المصوَّر والعقل الهيولانيّ بمنزلة العنصر وإنّ العقل الفعّال هو المخرج للعقل المستفاد على الوجوه بالفعل .

[92] The *PR* can not, of course, prove that this understanding was also that of the Arabic translator, but only that there existed a philosophical tradition which interpreted the translation in this way. However, I think it is perfectly plausible that the translator also held this view.

[93] The implicit reference in the Arabic version of Alexander is only preserved in Finnegan's edition ('Texte arabe du ΠΕΡΙ ΝΟΥ', 184.3–4) which describes the second type of intellect as follows:
وهذا العقل هو العقل الهيولانيّ ومن بعد أن صارت له ملكة واستفاد أن يعقل وأن يفعل
translating (Alexander, *De intellectu*, ed. Sharples, 44.5–6) καὶ ἔστιν οὗτος ὁ ὑλικὸς ἕξιν ἤδη καὶ τὸ νοεῖν τε καὶ ἐνεργεῖν προσειληφώς. Badawī (*Commentaires*, 33.19) reads:

'acquired intellect', oddly, occurs first to render the Greek expression 'from without (θύραθεν)',[94] by which Alexander characterises 'the productive [intellect], not being a part or power of our soul, but coming to be in us from outside, whenever we think of it' (tr. Sharples).[95] In the rest of the Arabic *De intellectu* the Greek term '[intellect] from without' is rendered as 'acquired [intellect]' or '[intellect] acquired from without' when its relation to the human intellect is expressed, but simply as '[intellect] from without' (*min khārij*) where the reference is solely to the divine intellect.[96] These renderings make perfect sense as the translator of *De intellectu*, Isḥāq ibn Ḥunayn, surely expected the text to refer to the three levels of intellect introduced at the beginning of the treatise and may therefore have been well inclined to read the second one, the intellect *in habitu*, into the Greek expression '[intellect] from without' and call it 'acquired intellect' (*ʿaql mustafād*).[97]

istiʿdād for *istifād* and *wa-an yataqabbala* (for προσειληφώς) instead of *wa-an yafʿala*. However, Marwan Rashed has informed me (personal communication) that MS Taskhent 2385 on which Badawī's edition is based reads *istifād* as well (399b6). *Istifād* thus renders προσειληφώς.

[94] The term 'from without' (θύραθεν) is first used by Aristotle in *De generatione animalium* 736b28 to express that intellect alone enters [bodies] from without and is alone divine. The Arabic translation ascribed to Yaḥyā ibn al-Biṭrīq, but maybe done by Usṭāt (see Brugmann and Drossaart Lulofs, p. 10), reads *min khārij* (*ibid.*, 63.22–3):
ويتبع قولنا إن نقول العقل فقط يدخل من خارج وإنّه وحده إلاهيّ ، ومن أجل أنّ فعاله لا يشارك فعالاً جسدانيّاً بنوع من الأنواع .

[95] The Arabic version reads (Badawī, 34.21–2 and, with slight variations, Finnegan, 186.5–6):
... العقل المستفاد الفاعل وليس هو جزءاً ولا قوّة للنفس منّا ، ولكنّه يحدث فينا من خارج إذا كمل عقلنا به
(or the last part, according to Finnegan's edition and Badawī, p. 34, n. 2: إذ نحن عقلنا به).

[96] See Davidson's assessment, *Alfarabi*, p. 12: 'the Arabic translator of Alexander saw that the expression "intellect from without" denotes the active intellect – or other incorporeal forms – insofar as it somehow enters the human intellect; and for unknown reasons he rendered the expression as *acquired intellect*. The Arabic does not misrepresent the intent of the original, although, by coining the new name, it does draw additional attention to, and permits a misunderstanding of, the aspect of the active intellect – or of another incorporeal form – that enters man.' This understanding of Alexander may well be influenced by Themistius' *De anima*-paraphrase which was also translated by Isḥāq and which understands the active intellect as joining with the human potential intellect and thus effecting the intellect *in habitu* (see *ibid.*, p. 27, and also p. 10, n. 22, and p. 40).
For a detailed discussion of all the occurrences of *al-ʿaql al-mustafād* and related terms in the Arabic *De intellectu* see Finnegan, 'Texte arabe du ΠΕΡΙ ΝΟΥ', p. 172; and Geoffroy, 'Tradition', pp. 214–16.

[97] The intellect *in habitu* occurs only twice in Alexander's *De intellectu*; first in the presentation of the three levels of intellect as ὁ .. νοῶν καὶ ἕξιν ἔχων (44.1), translated as [العقل] الذي قد صار يعقل وله ملكة أن يعقل (ed. Badawī, 33.15), and then as ὁ ἐν ἕξει νοῦς (46.6), translated as والعقل الذي بالملكة (ed. Badawī, 35.10). The term ἕξις (44.6,

(46) 29a7–13 Al-Fārābī says

Source: unidentified; possibly al-Fārābī's commentary on either Alexander's *De anima* or his *De intellectu.*

This passage displays none of the characteristic features of the Farabian doctrine on the intellect, namely the fourfold distinction of intellect, the acquired intellect as the highest level human intellect can possibly reach which grants immortality, and the active intellect as the tenth intelligence which exercises providence over the sublunar sphere.[98]

The description of human intellect as matter may be compared to the *R. fī l-ʿAql* (13.4–5, 14.1). However, the passage in its entirety reads like a commentary on the previous Alexander quotation and shows striking similarities to Alexander's *De anima*, where matter is said to be receptive of what its agent generates, the relation matter-agent is applied to intellect, and the habitus the active intellect causes in the material intellect is explained as form and perfection.[99] Thus the passage may derive from al-Fārābī's commentary on Alexander's *De anima*, which is attested in the

44.10, 44.13, 50.3, 52.1), translated as *malaka*, occurs five times (ed. Badawī, 33.19, 33.22, 34.1, 39.10, 41.6–7). Towards the end of the text, the material intellect is not mentioned in the Greek anymore, so Isḥāq translates ὁ δυνάμει νοῦς 'potential intellect' (52.5) as *al-ʿaql al-hayūlānī* 'material intellect' (ed. Badawī, 41.9), most probably in order to provide consistency with the three levels of intellect presented at the beginning.

[98] The Farabian doctrine on the intellect is especially expounded in the *R. fī l-ʿAql*, the *K. al-Siyāsa al-madanīya* and the *Madīna al-fāḍila*. Cf., e.g., Davidson, *Alfarabi*, pp. 44–73. As Geoffroy ('Tradition', pp. 229–30) shows, the *ThA* has some importance for al-Fārābī's understanding of the term *ʿaql mustafād*. The term occurs there twice and, interestingly enough, in a passage which immediately follows a passage taken up in the *PR* as passage (51). The Plotinian material in the *PR* in the name of the Greek Sage does not use the term as such, but explains that 'the intellect is acquired in [the soul]' (*al-ʿaql fīhā mustafād*) – see passage (41), 106.9. Might this be the reason for the inclusion of the two passages on the intellect in our compilation? Alternatively, the reason for the inclusion of the Farabian passage might be to present his doctrine on the intellect as if it were in line with the doctrine of the Kindian tradition from which the *PR* most likely stems.

[99] See the following passages of Alexander's *De anima* (88.19–24), to which Davidson refers (in *Alfarabi*, p. 20): τὸ δέ τι ποιητικὸν τοῦ ἐν τῇ ὕλῃ γίνεσθαι ταῦτα ὧν ἐστι δεκτική ... ἀναγκαῖον δοκεῖ καὶ ἐπὶ τοῦ νοῦ ταύτας εἶναι τὰς διαφορᾶς. καὶ ἐπεί ἐστιν ὑλικός τις νοῦς, εἶναί τινα δεῖ καὶ ποιητικὸν νοῦν, ὅς αἴτιος τῆς ἕξεως τῆς τοῦ ὑλικοῦ νοῦ γίνεται. And 85.10–11: καὶ τοιοῦτος μὲν ὁ ὑλικὸς νοῦς, ὁ δὲ ὡς ἕξις λεγόμενος εἶδός ἐστι [καὶ δύναμις] καὶ τελειότης τούτου ...
The Arabic translation of Alexander's *De anima* seems lost. The surviving Arabic-Hebrew translation names Isḥāq b. Ḥunayn as the Arabic translator (Gätje, *Psychologie*, p. 69).

bio-bibliographical literature, or his commentary on Alexander's *De intellectu*, if it ever existed.[100]

(47) 29a13–b12 One of them says

Source: Miskawayh, alternative *Fawz*.

Parallel: *K. al-Ḥikma*, MS Esad Efendi 1933, fol. 56b2–13.

Cf.: Miskawayh, *Fawz* I, 9, ed. ʿUḍayma, 55.2–56.3, with French tr. Arnaldez, pp. 28–9; tr. Sweetman, pp. 113–14.

The passage is set apart from the previous al-Fārābī quotation by a stroke of the pen, which must be an indication that the scribe forgot to leave an intended gap. It is then introduced by the phrase 'One of them says', and has a word-for-word parallel in a philosophical compilation similar to the *PR*, i.e. the *K. al-Ḥikma*, where it occurs without any indication of author or source. However, the two texts are very close to the wording of Miskawayh's *Fawz* I, 9, but considerably shorter. As there are a considerable number of similar passages, namely sixteen, it seems reasonable to suggest that there existed a second recension of the *Fawz*, which is the source for the *PR* and the *K. al-Ḥikma*.

The content of the passage fits well with the material of the Arabic *Plotiniana* presented before, as it presents Plotinus' threefold hierarchy of beings, i.e. the One, intellect and soul, merged with the genuine Arabic addition of mediated creation.[101] Nature is not mentioned in this passage, although it will be in the next, but is here replaced by the sphere. This merging of the role of nature with the sphere is not unusual in works by philosophers of the medieval Islamic world,[102] and may eventually go back to the ambivalence with which Plotinus himself treated nature.

[100] Ibn Abī Uṣabiya (II, 138.28–9) mentions the following:
شرح مقالة الاسكندر الافروديسي في النفس على جهة التعليق .

[101] On the three- or fourfold Plotinian hierarchy of being, i.e. One, intellect, soul (and nature) as it is used in the Arabic Plotinus, see D'Ancona, 'Porphyry', pp. 63–4.

[102] In al-Kindī's *Kalām fī l-Nafs* (*Rasāʾil*, ed. Abū Rīda, I, 281.14–15) we read that the soul's activities in the generated bodies only become manifest by the mediation of the sphere. Al-ʿĀmirī replaces nature by 'the sphere of the spheres' in the hierarchy of being in his *Fuṣūl fī l-Maʿālim al-ilāhīya* which is identical to the one presented in our passage (ed. Khalīfāt, *Rasāʾil*, p. 364; ed. Wakelnig, *Feder*, pp. 84–6, with German tr. and commentary, pp. 162–7). And Isaac Israeli, expounding the different meanings of the 'equivocal term' 'nature', claims that it also stands 'for the sphere and the spherical power which is appointed by the Creator, may He be exalted, in nature for the sake of influencing coming-to-be and passing-away, increase and decrease, movement and rest' (tr. Altmann and Stern, *Isaac Israeli*, pp. 50–1).

(48) 29b12–30a5 The bodies of the existents

Source: Miskawayh, alternative *Fawz.*

Parallel: *K. al-Ḥikma*, MS Esad Efendi 1933, fols. 56b13–57a1.

Cf.: Miskawayh, *Fawz* III, 1, ed. ʿUḍayma, 111.10–112.5, with French tr. Arnaldez, pp. 69–70; tr. Sweetman, pp. 158–9.

In the *PR* this passage is not separated from the previous one, but the parallel in the *K. al-Ḥikma* starts with a few rubricated words and the corresponding section in Miskawayh's *Fawz* derives from a different book and chapter (III, 1) from the previous one. As to its contents, the passage smoothly continues where passage (47) stops, namely with a discussion of the corporeal universe which is said to be a unified one. The ninth sphere is the outermost, starless sphere which comprises the seven spheres of the planets and the sphere of the fixed stars or the zodiac,[103] and it is the last corporeal existent.

The entire concept recalls the Platonic account of the creation of the universe in the *Timaeus* 30b-34c, where the *topos* of the cosmos as one single living being is stressed and the role of the four elements and the sphere is discussed.

(49) 30a5–32a7 All existents are divided

Source: Miskawayh, alternative *Fawz.*

Parallel: *K. al-Ḥikma*, MS Esad Efendi 1933, fols. 57a1–58a8.

Cf.: Miskawayh, *Fawz* II, 10, ed. ʿUḍayma, 100.1–101.13, 102.4–15, with French tr. Arnaldez, pp. 60–2; tr. Sweetman, pp. 149–52.

Except for one minor omission, this passage, as well as its parallel in the *K. al-Ḥikma*, corresponds to the beginning of ch. 10 of Miskawayh's *Fawz*, Book II. The division of the universe in the previous passage is taken up, but rephrased as a distinction between the corporeal and the spiritual. While describing the spherical, onion-like[104] structure of the corporeal

[103] This is the Ptolemaic addition to the Aristotelian system of the spheres and was taken over widely in the Islamic world – e.g. among the philosophers, by al-Fārābī, Ibn Sīnā, al-Ghazzālī (see Walzer in al-Fārābī, *Perfect State*, p. 364) and al-ʿĀmirī. The sphere of the zodiac was often identified with the divine pedestal (*kursī*) and the ninth sphere with the divine throne (*ʿarsh*) (see Heinen, *Islamic Cosmology*, pp. 77–81).

[104] Plato states in the *Timaeus* 33b, that the most fitting shape for the cosmos is the sphere. The simile of the onion is not used by Miskawayh, but by the *Ikhwān al-Ṣafāʾ* (e.g. in I, 115.9; and II, 28.5–7, which reads):

part of the universe, Miskawayh, in passing, denies the existence of a void. He turns the structure of the universe into an 'argument from design', as Galen does with the shape of the human body in some passages below. Divine providence is thus apparent in the eccentricity due to which the centre of the sun is not in line with the centre of the earth, but to its south. Miskawayh uses this Ptolemaic model to explain that the sun draws the water to the south where it collects in the form of an ocean, leaving the northern region of the earth devoid of water, i.e. the earthly sphere which is below the sphere of water becomes uncovered, and thus inhabitable. This process functions through the sun heating up the water which then evaporates, so that more water is pulled in to replace what has evaporated.[105]

The idea that each sphere is like sediment to the next higher one, is reminiscent of the simile used by the *Ikhwān al-Ṣafāʾ*, who liken the lower sphere to earth and its corresponding higher sphere to heaven.[106]

In the present passage the hierarchy of the spiritual beings is given as God, intellect, soul and nature, not sphere as in passage (47). Nature is defined as divine power which permeates the entire corporeal world. In analogy to the corporeal world, the spiritual beings are also thought to encompass each other, yet in a purely spiritual way. Miskawayh compares their mutual encompassing with the joining of light and thus implicitly states that light is not a body. This question is discussed below, esp. passage (180) and, unsurprisingly, the opinion emerging as the true one is in line with Miskawayh.

(50) 32a7–33a6 He says

Translation: Lewis in 'Plotiniana', p. 263.

Cf.: *Enn.* V 1, 2, 10–25.

The passage is part of the Arabic *Plotiniana* and renders Plotinus' *Enneades* V 1, 2, 10–25 with some elaboration. It describes how the corporeal world is ensouled. The link to the previous passage by Miskawayh is probably the

اعلم يا أخي أنّ هذه الأكر محيطات بعضها ببعض كإحاطة طبقات البصل ، مماسّ سطح الحاوي يسطح المحويّ ، وليس بينهما فراغ ولا خلاء إلا فصل مشترك وهميّ .

105 A similar concept is found in al-Bīrūnī's *Taḥdīd nihāyāt al-amākin*, pp. 56–60, esp. 56.20–57.4, 58.7–12, 60.11–14.

106 The passage of the *Ikhwān* reads (II, 26.17–19):
واعلم يا أخي أنّ كلّ واحد من هذه السبعة المقدّم ذكرها سماء لما تحته وأرض لما فوقه ، ففلك القمر سماء الأرض التي نحن عليها وأرض لفلك عطارد ...

simile of light which both texts use to illustrate the function of the spiritual beings.

(51) 33a7–b3 For although the soul is in the condition of nobleness and refinement

Translation: Lewis in 'Plotiniana', p. 267.

Parallel (partial): *ThA*, ed. Badawī in *Aflūṭīn*, 108.17–109.6; ed. Dieterici in *Uthūlūjiyā*, 105.5–13; German tr. Dieterici in *Theologie*, pp. 106–7.

Cf.: *Enn.* V 1, 3, 1 and 6–12.

Between this passage and the one before some Plotinian material, namely *Enn.* V 1, 2, 26–50, is omitted. The present passage refers back to the previous topic by inserting the description of the soul as 'producing good impressions in bodies'. This may be an attempt to gloss over the omitted material. Interestingly, the passage also occurs in the *ThA*, though with a different beginning (ed. Badawī, 108.5–17) which is not parallel to Plotinus.[107] However, the passage following in the *ThA* (109.7–110.4) is parallel to *Enn.* V 1, 3, 12–23, whereas the *PR* continues with a rendering of IV 3, 18 which is particularly puzzling as the rendering of V 1, 3, 12–23 in the *ThA* calls the intellect in the soul 'an acquired intellect' (*ʿaql mustafād*). One can only speculate why the compiler of the *PR* did not take this opportunity to link his excerpts of the Arabic Plotinus with the previously quoted excerpt by Alexander in passage (45): might the passage have been missing in his model?

(52) 33b4–34b13 When the soul comes to be in bodies

Translation: Lewis in 'Plotiniana', pp. 37–9.

Cf.: *Enn.* IV 3, 18, 2–19.

The passage renders *Enn.* IV 3, 18 closely, but omits the simile of the eye: in this world we can look into someone's eyes and, although he remains silent, are able to know many things. In the other world it works similarly as there everything is as pure as an eye.

[107] Zimmermann discusses this particular 'jarring joint', as he calls the places where the extant version of the *ThA* glosses over textual loss or disorder in its source text, in 'Origins', pp. 152–5 and esp. pp. 164–5. For a discussion of the description of the relationship between soul and intellect as image and model, see D'Ancona, 'Porphyry', pp. 68–9.

The text also presents additional material, which is not in Plotinus, namely a discussion of the souls in the heavenly bodies. They are said to need no speech, which is probably against Plotinus' own views as he ends his chapter 18 of *Enn.* IV 3 stating that 'there is nothing absurd in spirits and souls in the air using voices' (tr. Armstrong).[108]

A discussion of 'deliberation and thought (*rawīya wa-fikra*)' which is similar to the present passage is preserved in the *ThA* (ed. Badawī, 66.12–67.15) drawn from *Enn.* VI 7, 1. There the main focus is on the First Creator and His providence which cannot be said to be by 'deliberation and thought', as He acts through His being alone.[109]

(53) 34b13–35a4 When the soul is in the body

Translation: Lewis in 'Plotiniana', p. 49.

Cf.: *Enn.* IV 3, 22, 7–10? *Muntakhab Ṣiwān al-ḥikma* (ed. Dunlop, lines 878–9; ed. Badawī, 151.3–4); *Mukhtār min Kalām al-ḥukamā*ʾ (ed. Gutas in *Literature*, 200.4–5).

The concept that the body is in the soul and not vice versa is probably inspired by *Enn.* IV 3, 22. Further aspects of this are discussed in the next passage and passages (76–77) below.

An almost word-for-word parallel to the second part of the passage is found in the section on Aristotle in the *Muntakhab* and, probably citing it from there, in the *Mukhtār min Kalām al-ḥukamā*ʾ.[110] One may wonder whether the attribution to Aristotle in both of these texts has anything to do with the *ThA* which, in its present shape, does not contain the quotation, but could have done in an earlier more comprehensive recension, for example at the end of the current second chapter.

(54) 35a4–36a7 Each of the soul's powers

Translation: Lewis in 'Plotiniana', p. 43.

[108] In an epistle for which the ascription to al-Kindī is not certain, as it is not mentioned in the bio-bibliographical literature (Veccia Vaglieri and Celentano, 'Trois épîtres', pp. 526–30), Ptolemy, Heraclitus and Nicomachus are reported to have held a similar view, namely that the planets have speech (*mutakallima*) (French tr., *ibid.*, p. 537, and see Plate II, b, lines 1–3 for the Arabic).

[109] For a discussion of this passage see Adamson, *Arabic Plotinus*, pp. 151–2.

[110] The passage in the *Mukhtār min Kalām al-ḥukamā*ʾ reads:
وقال النفس ليست في البدن بل البدن في النفس لأنّها أوسع منه وأبسط .
'The soul is not in the body but rather the body is in the soul because the soul is wider and more spacious than the body' (tr. Gutas, p. 201).

Parallel: *ThA*, ed. Badawī in *Aflūṭīn*, 41.7–16, 42.4–10, 42.14–18; ed. Dieterici in *Uthūlūjiyā*, 27.18–28.10, 28.19–29.7, 29.12–17; German tr. Dieterici in *Theologie*, pp. 28–30.

Cf.: *Enn.* IV 3, 20, 10–19.

The passage is not set apart from the previous passage, but has, in contrast to passage (53), a clear model in the Greek version of the *Enneades* and a parallel to the *ThA*. The *ThA*, in turn, contains versions of IV 3, 19 and the beginning of IV 3, 20, which distinguish between divisible and indivisible powers or faculties of the soul and which are not taken up in the *PR*. It is unclear whether this omission is a conscious decision on the part of our compiler or whether he had a source text in which passage (53) was used to introduce passage (54) in a different way from its context in the Arabic Plotinus as attested in the *ThA*. The compiler also skips two sections which the *ThA* has; in the first instance he indicates this by a gap, in the second instance there is no indication at all.

The first half of the passage on the soul preparing the body parts in specific shapes to manifest its different activities is not based on Plotinus, but is additional material from the Arabic *Plotiniana*. The second half, on why the soul cannot be in a place, follows the *Enneades* quite closely, except for the back-reference to the first half and the discussion of a certain place for each power of the soul.

(55) 36a7–37a3 He says

Translation: Lewis in 'Plotiniana', p. 47.

Cf.: *Enn.* IV 3, 21, 5–17.

The passage is a close rendering of most of *Enn.* IV 3, 21, which refers to Aristotle's discussion of the problem 'whether the soul may not be the actuality of its body in the sense in which the sailor is the actuality of the ship' (*DA* 413a8–9, tr. Smith). The Arabic stresses the difference between accidental and natural presence much more strongly than the Plotinian original. There, it is only said that the presence of the navigator as a passenger would be 'by accident (κατὰ συμβεβηκός)', but how his presence as a navigator should be defined is not spelt out. The Arabic version further adds separability as a characteristic which the soul has in the body in contrast to a skill in its instrument. The notion that the soul needs an instrument makes a smooth connection to the next passage which renders part of *Enn.* IV 3, 23, so the question arises whether the Plotinian material in-between was ever taken up in the Arabic version.

(56) 37a3–13 The beginning of [the soul's] activity

Translation: Lewis in 'Plotiniana', pp. 47[111] and 49.

Cf.: *Enn.* IV 3, 23, 20–7.

The somewhat enigmatic first sentence of the passage becomes clearer when we look at the context in which it appears in the *Enneades*. In *Enn.* IV 3, 23 the whole body is said to be the organ or instrument of the soul's perception, especially touch. As touch takes place in the nerves and nerves start in the brain, 'they [i.e. most probably the physicians – EW] established the principle of perception and of impulse and in general of the whole living being in the brain, assuming that obviously that which was going to use the organs would be there where their beginnings were – but it would be better to say that the beginning of the actualisation of the potency [of perception] is there' (tr. Armstrong).

The localisation of the different parts or faculties of the soul in the body is presented in the physiological section below, esp. passages (116–18), in more detail. Plotinus, however, refers clearly to the Platonic doctrine of the rational soul being located in the brain and not to the Aristotelian cardiocentric model.

(57) 37a13–b11 If someone says

Translation: Lewis in 'Plotiniana', pp. 65–7.

Parallel: *ThA*, ed. Badawī in *Aflūṭīn*, 32.3–9; ed. Dieterici in *Uthūlūjiyā*, 18.4–11; German tr. Dieterici in *Theologie*, p. 18.

Cf.: *Enn.* IV 4, 1, 31–6.

This passage is not set apart from the previous one, but renders a different *Ennead* and is paralleled in the *ThA* which preserves much more Plotinian material from *Enn.* IV 4, 1–4, than does the *PR*. The passage is difficult to interpret in the Greek,[112] and the Arabic version presents one attempt to

[111] Although the compiler of the *PR* indicates the transition from the rendering of *Enn.* IV 3, 21 to 23 quite nicely with a gap, in Lewis' translation the first sentence of passage (56) still appears as the last sentence of passage (55) on p. 47.

[112] See Armstrong's English translation which reads: 'But when the soul looks first at one [intelligible object], and then possesses the whole multiplicity of them, how does it possess one first and another next? The one power is one in such a way that it becomes many in something else, and does not comprehend all things by one act of intelligence. For its acts are individual, but always together in a power which remains unchanged, but is divided in other things. For that intelligible object is able in virtue

make sense of it. In the light of the Greek one may read the question as referring not to the soul's knowledge of the simple and multiple things, but to the intelligible object which is one and many at the same time. The soul's powers would mean different acts of the soul's intellection of the intelligible. However, at the end of the passage the Arabic version decisively talks about the corporeal things which receive the soul's activity and thereby turn it into a number of activities distinguishable from each other. In the context of the *PR* this seems to describe how each of the soul's activities manifests itself in a different body part.

(58) 37b12–38b1 He says

Translation: Lewis in 'Plotiniana', p. 255.

Cf.: *Enn.* IV 9, 3, 6–9?

This passage offers yet another explanation, in addition to passages (25) and (43), of why the soul enters this world, namely its long travel with the heavenly body, which somehow fails to provide any satisfactory reason for its descent, unless we are to assume that it got tired of wandering. This is in any case as un-Plotinian as the ensuing account of the soul's presence in this world, namely that it attaches itself only to one man through whom all others may take part in the soul. The concept seems to imply the idea of a chosen human being, most probably a prophet and thus indicates a religious, monotheistic, Christian or Islamic origin. The simile of voice or sound is mismatched to illustrate the presence of the soul, which is said to be only in one person, but may be meant to exemplify the doctrine or creed which this chosen person spreads among his fellow men. A simile of voice also occurs in the *Enneades*, yet there it illustrates the functioning of spells and magic which is not dependent on the actual sound reaching the recipient, but on the unity of all souls. The definition of hearing in actuality is similar to the one given below, in passage (149).

(59) 38b2–10 He says

Translation: Lewis in 'Plotiniana', p. 255.

Cf.: *Enn.* IV 9, 4, 2–3.

This passage and the following three all discuss the problem of the rela-

of its not being one to perceive in itself the nature of the many which did not previously exist.'

tionship between the one All-Soul and the many individual souls. This is certainly an important issue for Plotinus, yet the single passages are, at best, inspired by the Greek text of *Enn.* IV 9, 1–4.

The present passage may draw on *Enn.* IV 9, 4 which tackles the question of how the many souls are one and suggests that the many come from the one. The notion of a creation of the souls is added in the Arabic version as is the discussion of divisibility, which is in line with general Neoplatonic doctrine (see, for example *Proclus Arabus*, prop. 80).

(60) 38b11–39a1 He says

Translation: Lewis in 'Plotiniana', p. 255.

Cf.: *Enn.* IV 9, 4, 19–20.

This passage is linked to the previous one by discussing how many souls proceed from the one soul and compares the process with the Plotinian simile of many pieces of wax which take the impression of one single seal-ring.

(61) 39a1–5 He says

Translation: Lewis in 'Plotiniana', p. 255.

Cf.: *Enn.* IV 9, 3, 1–4.

These examples, demonstrating how our souls must somehow be one, because we would otherwise not feel compassion and sympathy for each other or fall in love at first sight, are all found in Plotinus, to whose account our passage does not add much.

(62) 39a6–b8 Someone may say

Translation: Lewis in 'Plotiniana', p. 255.

Cf.: *Enn.* IV 9, 1, 15–18; IV 9, 2, 3–8.

Plotinus presents a possible objection to the notion that the individual souls are one in *Enn.* IV 9, 1, which he resolves in the following chapter, and the present text is modelled after these Plotinian passages. The main addition is the simile of sunlight, which is a commonplace in Neoplatonism.

(63) 39b9–40a1 He says

Translation: Lewis in 'Plotiniana', p. 89.

Cf.: *Enn.* IV 4, 11, 5–11.

Having discussed the relation between the All-Soul and the individual souls, our compiler turns to the relationship between the metaphysical level of the soul and the next lower level, nature, which he covers in passage (63) and what follows. The present text is a heavily revised version of part of ch. 11 of *Enn.* IV 4, in which Plotinus does not mention the soul at all and the relationship he focuses on is the one between the one nature and the individual natures.

(64) 40a2–41b5 If someone asks

Translation: Lewis in 'Plotiniana', pp. 89–91.

Cf.: *Enn.* IV 4, 13, 1 – 14, 8.

This is a lengthy and reasonably faithful version of ch. 13 of *Enn.* IV 4 and the beginning of ch. 14. The Arabic version renders the Plotinian material in a more simplified and straightforward manner. So in the introductory question on the difference between the soul and nature, the term 'soul' (*nafs*) is a specification of the imprecise Greek notion φρόνησις, which, in Plotinus, most likely refers to the intellectual activity of both intellect and soul. This specification is kept in what follows when nature is said to be 'the image of the soul' (ἴνδαλμα ... φρονήσεως), but the definition of nature as the lowest part of the soul is omitted.

(65) 41b6–42b1 We want to know

Translation: Lewis in 'Plotiniana', p. 97.

Cf.: *Enn.* IV 4, 18, 4–15.

The passage is another reasonably faithful rendering of a part of *Enn.* IV 4, namely the first half of ch. 18. The Plotinian thought is presented in a more formalised way: the concept of mediation is stressed and the abstract principle of the cause's presence in its remotest effect via intermediate effects is presented. This presentation may make use of some Proclean source – as we read, for example, in the *Ldc II*, ch. 4 (based on Proclus, *ET* prop. 81) – that the cause is only in its effect via an intermediary which is itself an impression or effect of that cause.[113] The idea that the rational soul only experiences pleasure and pain through its body is also reminiscent of *Proclus Arabus* (prop. 80), where incorporeal things are said to be acted upon and divided only because of the bodies in which they are or to which they are joined.

(66) 42b2–43a3 He says

Translation: Lewis in 'Plotiniana', p. 245.

Parallel: *ThA*, ed. Badawī in *Aflūṭīn*, 86.15–87.6; ed. Dieterici in *Uthūlūjiyā*, 78.17–79.10; German tr. Dieterici in *Theologie*, p. 80.

Cf.: *Enn.* IV 8, 6, 18–20.

This passage is paralleled in the *ThA* which contains a complete and highly elaborated version of *Enn.* IV 8, 6. The compiler of the *PR* quotes mainly the additional material of the Arabic version which gives yet another description of the genesis of the universe, which is quite close to the account at the beginning of the Ps.-Platonic passage (25).

(67) 43a4–6 He says

Translation: Lewis in 'Plotiniana', p. 245.

The short text has no obvious parallel in the *Enneades*. As to content, it is clearly related to the previous passage, but in contrast to the former, it states that 'matter', and not 'nature',[114] is the last of the intellectual things. However, in the material quoted in the *PR*, these three notions, i.e. nature, matter and generation, are obviously closely related to each other.

(68) 43a6–43b11 Similarly Plato said

Translation: Lewis in 'Plotiniana', p. 281.

Cf.: *Enn.* V 1, 8, 1–10.

This is a reasonably faithful Arabic version of *Enn.* V 1, 8, 1–10, in which

[113] The passage of the *Ldc II* (323.2–4) reads:
... وذلك أنّه إن كانت العلّة مباينة لمعلولها ، أعني أنّها ليست بمعلولة فيه ، فإنّه لا يمكن أن يكون فيه إلا بمتوسّط يجمع بين العلّة وبين معلولها وينبغي للمبسوط أن يكون المتوسّط أثراً من آثار العلّة ...

[114] 'Generation' according to the varying *ThA* reading. Zimmermann takes this variation in reading as an indication that the text of the *GS* is the older one here, but also states that in another instance the case seems to be the opposite ('Origins', pp. 163–4).

Plotinus quotes and interprets part of the *Second Letter* attributed to Plato. Our passage preserves the attribution to Plato, but introduces a more detailed interpretation of the Ps.-Platonic passage identifying the secondary things with matter and the heavenly bodies, and the third things with this corporeal world. Intellect and soul are thus understood as being primary things which represents the general tendency of the Arabic Plotinus to compress intellect and soul into one single level of 'the intelligible world'.[115] Yet, this is somehow at odds with the end of the passage where the three levels with which the text operates are intellect, soul, and matter together with the material things. The additional material of the Arabic version also refers to the doctrine of mediated creation.

(69) 43b12–44a6 He says

Translation: Lewis in 'Plotiniana', p. 281.

This passage is clearly set apart – by starting on a new line – from the previous one, although it is similar in content, as it also presents the interpretation of some doctrine ascribed to Plato. Neither the doctrine nor the interpretation sounds particularly Plotinian, so it seems questionable whether this text is still part of the Arabic *Plotiniana*. They rather recall the beginning of passage (25) in the name of Plato, where the soul perceives matter as dark and is willing to change this situation by acting upon it.

(70) 44a7–b4 Aristotle says in the *Physica*

Source: Aristotle, *Physica* 192b20–3 with some commentary.

Cf.: Isḥāq ibn Ḥunayn's translation *al-Samāʿ al-ṭabīʿī* as *al-Ṭabīʿa*, ed. Badawī, 79.7–9.

The text starts with a faithfully translated sentence from Aristotle's *Physica*, 192b20–3,[116] which is followed by a commentary section introduced by 'By that he means'. As in passage (10) above, it is not clear which version of the *Physica* our compiler uses, but in the present context some translation with commentary seems the most likely candidate. In any case the translation is not the one by Isḥāq b. Ḥunayn. The commentary stresses a differ-

[115] See Adamson, *Arabic Plotinus*, pp. 78 and 140.

[116] The Greek reads: ὡς οὔσης τῆς φύσεως ἀρχῆς τινὸς καὶ αἰτίας τοῦ κινεῖσθαι καὶ ἠρεμεῖν ἐν ᾧ ὑπάρχει πρώτως καθ' αὑτὸ καὶ μὴ κατὰ συμβεβηκός – 'that nature is a principle or cause of being moved and of being at rest in that to which it belongs primarily, in virtue of itself and not accidentally' (tr. Hardie and Gaye).

ent point from Aristotle's in the context of this passage, as he distinguishes between the elements, plants and animals, which are constituted by nature, and artificial products, which only have a natural impulse to move or to rest insofar as they are created of the natural materials, i.e. accidentally. The commentary which the compiler of the *PR* provides, however, focuses on nature in contrast to the soul, by which this passage is nicely joined to the previous Plotinian passages.

(71) 44b4–5 He says

Source: unidentified; section on nature in some doxography, or the same source as in passage (70)?

Cf.: Aristotle, *Physica* 250b14–15.

The entire passage sounds purely Aristotelian and may thus derive from the same source as the previous one. The description of nature as life recalls *Physica* at the beginning of Book VIII where Aristotle discusses whether nature is 'a sort of life as it were to all naturally constituted things' (tr. Hardie and Gaye).[117]

(72) 44b6–7 Hippocrates says

Source: unidentified; section on nature in some doxography?

In the name of Hippocrates nature is again defined as the principle of motion, i.e. the mover. The second part of the statement recalls the Neoplatonic concept of an All-Nature which encompasses the single individual natures.

(73) 44b7–9 Galen says

Source: unidentified; section on nature in some doxography?

Cf.: Abū l-Ḥasan al-Ṭabarī, *al-Muʿālajāt al-Buqrāṭīya*, MS Marsh 158, fol. 4b14–15; facsimile edn Sezgin, 6.16–17.

The definition of nature given in the name of Galen stresses its function as an active and intellectual power innate in everything. Part of this definition resembles the statement Abū l-Ḥasan al-Ṭabarī presents in his *Muʿāla-*

[117] See the Arabic version (*al-Ṭabīʿa*, 801.13–4):
... كأنّها [أي الطبيعة] حياة ما لجميع ما قوامه بالطبيعة .

jāt al-Buqrāṭīya as Galenic, namely that the term ‘nature’ refers to the directive power of the living beings and the other bodies.[118]

(74) 44b9–45a1 One of them says

Source: unidentified; section on nature in some doxography?

Cf.: al-Majūsī, *Kāmil* I, 1.3, 11.27–8.

‘Nature’ is said to be homonymous and the examples given for its possible meanings are essence, form and matter, which are different from Aristotle’s list in *Metaphysica* V, 4. The homonymy of ‘nature’ is also stated, for example, by al-Majūsī, who presents different examples.[119]

(75) 45a2–45b9 He says

Translation: Lewis in ‘Plotiniana’, p. 285.

Cf.: *Enn.* V 1, 10, 21–30.

Lewis includes this, as well as the following two passages, in his translation of the Arabic *Plotiniana*, although it can, at best, be said to be inspired by *Enn.* V 1, 10, 21–30. The Greek text is an interpretation of the Platonic statement that the demiurge wrapped the soul around the universe. Plotinus takes this to indicate the part of the soul which remains in the intelligible realm. A reference to this may be found in our passage which states that part of the soul’s form remains unmoved and within the intellectual world, and the Plotinian interpretation may have, furthermore, inspired passage (76). Plotinus also mentions the claim that the soul is in the head which may have found an echo in the following passage in the *PR*, and he explains that the exhortation to separate ourselves from the bodies is not meant spatially.

(76) 45b10–46a9 He says

Translation: Lewis in ‘Plotiniana’, p. 285.

Cf.: *Enn.* V 1, 10, 23–4?

[118] The passage in *al-Muʿālajāt al-Buqrāṭīya* reads:
قال جالينوس ومتى قلنا طبع أو طبيعة فإنّا نريد به القوّة المدبّرة لأبدان الحيوان والمدبّرة لسائر الأجسام .

[119] Al-Majūsī’s passage (*Kāmil* I, 11.27–8) reads:
وربما استعملنا قسمة الاسم المشترك إلى معان مختلفة كقولنا اسم الطبيعة ونحن نريد بذلك إمّا القوّة المدبّرة للبدن وإمّا ماهية البدن وإمّا المزاج ...

As mentioned just above, this account of the soul's dwelling place as being in the head may be inspired by *Enn.* V 1, 10, 23–4, but it also recalls passage (56) above, which draws on *Enn.* IV 3, 23, 20–7. The Platonic localisation of the three parts of the soul and the resulting tripartite physiological system with its centres in the brain, the heart and the liver will be expounded in the words of Galen below, passage (120). The idea that the soul needs the body to act in this world, and the way of describing the relationship between cause and effect are commonplaces in Neoplatonism, which have already occurred in the *PR*, e.g., passages (54–57) and (65).

(77) 46a9–b1 Plato has already said

Translation: Lewis in 'Plotiniana', p. 285.

Parallel (partial): *K. al-Ḥikma*, MS Esad Efendi 1933, fol. 79b17–80a3.

Cf.: *Enn.* V 1, 10, 21–22; *Enn.* IV 3, 20, 10–19 in passage (54), and *ThA*, ed. Badawī in *Aflūṭīn*, 42.4–12; ed. Dieterici in *Uthūlūjiyā*, 28.19–29.10?

This text may be inspired by the aforementioned Plotinian interpretation of Plato's statement that the soul is wrapped around the universe in *Enn.* V 1, 10, 21–2. Another possible source of inspiration may have been *Enn.* IV 3, 22, 8–9, where Plato is said to be right 'not [to] put the soul in the body when he is speaking of the universe, but the body in the soul' (tr. Armstrong). However, this Plotinian wording is much better represented in passage (53). The present text further recalls the previous discussion of the relationship between soul and place in passage (54), drawn from *Enn.* IV 3, 20, 10–19.[120]

Interestingly the *K. al-Ḥikma* has a word-for-word parallel for most of the passage, but puts it in a different context. Instead of being about Plato stating that the soul is inside and outside this world, it is about intellect being the first thing created and the soul being posterior to it.[121]

[120] The passage in the *ThA* is not completely taken up in the Marsh manuscript, passage (53). Omitted is, e.g., a statement which is very close in wording to our present passage (ed. Badawī, 42.13–14):
والنفس تحيط بالمكان ، والمكان لا يحيط بها لأنّها علّة له ، والمعلول لا يحيط بالعلّة ، بل العلّة تحيط بالمعلول .

[121] The passage in the *K. al-Ḥikma* reads as follows (the parallels to the *PR* passage are set in italics): 'Another issue. The intellect is the noblest and most perfect of all things, because it is the first action performed by the agent (*awwal fiʿl faʿalahū l-fāʿil*). In it there is more of the light of the agent than in all the other things. It has the strongest likeness to the First Agent among the entire creation. The soul is posterior (*baʿd*) to it. It is *spiritual and not a body. It enters and leaves bodies, because bodies do not encompass it,*

(78) 46b2–47b9 Aristotle says

Source: *DA*-paraphrase, ed. Arnzen with German tr., 215.5–217.8; 219.1–4, 7–16; MS Aya Sofya 4156, fols. 67a4–b15; 68a11–b1, 68b4–69a3.[122]

Cf.: passage (107).

This lengthy excerpt in the name of Aristotle derives from the anonymous *De anima*-paraphrase, which is a Late Antique Alexandrian commentary translated and reworked in the Kindī circle. The passage deals with the Aristotelian concept of entelechy, its different kinds and, most importantly, the question of in which way the soul can be said to be the entelechy, i.e. the actuality or perfection of the body. The *DA*-paraphrase lists four different kinds, of which only the first and the last ones are perfectly Aristotelian (see *DA* 417a22–b1 and 412b23–413a9), whereas the other two are heavily influenced by the Arabic *Plotiniana*, as Arnzen shows.[123] The compiler of the *PR* presents only the first, second and fourth kind, namely actual or potential perfection, separable or inseparable perfection,[124] and perfection localised in the whole body or in a certain body part alone. He omits the third kind of perfection, which is either something being the perfection of something else (like heat in fire) or something causing perfection in something else (like the builder in a house or the soul in the body).[125] The discussion of the fourth kind is crucial, as, for the first time, different parts of the soul are distinguished and the soul which is the separable perfection of the entire body is said to be the rational soul alone.

(79) 47b10–48a6 He says

Source: *DA*-paraphrase, ed. Arnzen with German tr., 213.10–215.4; MS Aya Sofya 4156, fols. 66b6–67a3.

but it is [the soul] which encompasses bodies. For it is their cause, and the cause encompasses its effect[s] without any of them escaping it.'

122 Arnzen's edition and the Aya Sofya represent two slightly different recensions of the *DA*-paraphrase. The *PR* is more often in accordance with the former, but at times also with the latter or with neither of the two.

123 For this and the following, see Arnzen, *Aristoteles' DA*, pp. 377–82.

124 This section of the *DA*-paraphrase (217.3–6) draws on *Enn.* V 4, 2, 27–33 (cf. Arnzen, *Aristoteles' DA*, p. 117, and the testimonia given in his edition, p. 217). It is, in turn, taken up in ʿAlī ibn Rabban's *Firdaws* (ed. al-Ṣiddīqī, 68.15–19; ed. Jundī, 54.1–4); on which, see Arnzen, *Aristoteles' DA*, pp. 71–5.

125 This is the kind of perfection which the *ThA*, ch. 3, uses to transform Plotinus' criticism of the Aristotelian concept of entelechy (*Enn.* IV 7, 8^5) into a criticism directed only at the wrong interpretation of Aristotle, using the *DA*-paraphrase, as Zimmermann ('Origins', pp. 124–5) and D'Ancona ('Profilo', p. 117) show.

In the *DA*-paraphrase this passage precedes the previous one directly and introduces the term 'perfection' (*tamām* or *kamāl*), which renders the Greek ἐντελέχεια. It explains why Aristotle would rather use this term than 'form' when describing the relationship between soul and body, which, in fact, resembles the relationship between form and matter.

(80) 48a6–48b9 He says

Source: *DA*-paraphrase, ed. Arnzen with German tr., 203.9–205.5; MS Aya Sofya 4156, fols. 62b10–63b3.

In the now following passages the compiler of the *PR* presents several arguments for the incorporeality of the soul taken from the *DA*-paraphrase, most of them from the reworking of the doxographical section in Aristotle, *DA* 403b20–408a34. The argument of the present passage is, as Arnzen explains,[126] not taken from Aristotle or his commentator John Philoponus, on whose commentary the paraphrase draws heavily, but rather from Plotinus and thus most probably from the Arabic *Plotiniana*. However, the extant Arabic Plotinian material does not contain the possible source text, i.e. *Enn.* IV 3, 2–4. There the relationship between the All-Soul and the individual souls is discussed, which is, in our passage, transferred to the relationship between the soul and its different powers or faculties. Plotinus also uses the example of eyes and ears and states that the different perceptions only occur due to the different organs involved, not due to different parts of the soul, as the soul is one, even in its different functions (*Enn.* IV 3, 3, 12–25). He also discusses the case of numerical units or geometrical figures which, although incorporeal and homoeomerous, will still change in quantity and become more or less by addition or division (*Enn.* IV 3, 2, 24–9). The second argument is based on *DA* 408b18–25, but reinterprets Aristotle who explicitly admits that old age may affect the soul, but only via its vehicle, the body.[127]

(81) 48b10–49a12 He says

Source: *DA*-paraphrase, ed. Arnzen with German tr., 207.18–209.13; MS Aya Sofya 4156, fols. 64b8–65a14.

As Arnzen points out, this passage of the *DA*-paraphrase is quite faithful to

[126] For this and the following, see his *Aristoteles' DA*, pp. 367–9, also pp. 116–17 and p. 202, n. 29.

[127] This reinterpretation follows the tradition of Plato and Philoponus (see *ibid.*, p. 369).

Aristotle's *DA* 405b13–14 and 410b7–15.[128] In the context of the *PR* the arguments show, in combination with the previous and the following section, that the soul has no bodily aspect.

(82) 49a12–49b9 He says

Source: *DA*-paraphrase, ed. Arnzen with German tr., 315.8–18; MS Aya Sofya 4156, fols. 95b15–96a14.

Whereas the excerpts from the *DA*-paraphrase have so far been taken from the second book, this passage derives from the third book and a discussion of the relationship between the rational soul or intellect and the body.[129] In the context of the *PR*, the text supplies yet another argument for the incorporeality of the soul and its complete difference and separability from the bodies. The soul is said to be affected not by the objects of sense perception and knowledge as such, but by the movement they cause within the organ of sense perception and knowledge. As Arnzen points out, the concept of affection, i.e. the act of affecting something, being a movement is not Aristotelian and was therefore avoided by his commentators Philoponus and Stephanus, although it was taken up by Sophonias, who must have had access to some Greek model of the *DA*-paraphrase.[130]

(83) 49b9–50a2 He refutes someone who says

Source: *DA*-paraphrase, ed. with German tr. Arnzen, 209.14, 211.5–10, (211.12–15);[131] MS Aya Sofya 4156, fols. 65a14–15, 65b11–66a3, 66a7–10.

The compiler of the *PR* excerpts these passages, which demonstrate that the soul exists through neither the harmony nor the mixture of the body, from the end of the first book of the *DA*-paraphrase, which, in turn, is based on Aristotle, *DA* 407b27–408a18.[132] In both cases, he omits the first

[128] See *Aristoteles' DA*, pp. 371–2. The Aristotelian passages read '… all those who define the soul by its power of knowing make it either an element or constructed out of elements' and 'In general, why has not everything a soul, since everything either is an element, or is formed out of one or several or all of the elements? … What is that which unifies the elements?' (tr. Smith).

[129] See Arnzen, *Aristoteles' DA*, pp. 433–5 and 119.

[130] See *Aristoteles' DA*, p. 105.

[131] Here the Arabic text of the *DA*-paraphrase is only extant in the Aya Sofya version. In his edition, Arnzen supplies his own retranslation from the Persian witness.

[132] See Arnzen, *Aristoteles' DA*, pp. 372–3.

part of the argument in which it is argued that, as the harmony and the mixture of each body part is different, a multitude of souls would result.

(84) 50a6–51a7 Abū ʿAlī Miskawayh says

Source: Miskawayh, alternative *Fawz*.

Parallel: *K. al-Ḥikma*, MS Esad Efendi 1933, fols. 62a11–b14.

Cf.: Miskawayh, *Fawz* II, 1, ed. ʿUḍayma, 61.3, 62.12–63.8, 64.4–14, with French tr. Arnaldez, pp. 33–5; tr. Sweetman, pp. 118–20; **(for the first part):** Miskawayh, *Tahdhīb*, ed. Zurayk, 3.9–5.9; tr. Zurayk in *Refinement*, pp. 5–6; French tr. Arkoun in *Traité*, pp. 7–9; and *R. fī l-Nafs*, ed. Arkoun in 'Deux épîtres', pp. 47–8; ed. Badawī in *Dirāsāt*, pp. 89–90.

This passage, which is attributed to Miskawayh and most probably taken from the alternative *Fawz*, starts off with the statement that the soul is a simple substance which also introduces passage (79) taken from the *DA*-paraphrase. The two introductory sentences differ in the term for 'simple' since the *DA*-paraphrase uses the form *mabsūṭ* which is replaced by *basīṭ* in later texts,[133] as, for example, in our Miskawayh passage. Keeping these two different terms is another example of the reluctance of the compiler of the *PR* to tamper with his sources.

The first proof for the incorporeality of the soul, namely its reception of a multitude of forms, is also presented at the beginning of Miskawayh's *Tahdhīb al-akhlāq* and with less striking similarity in his *R. fī l-Nafs wa-l-ʿaql*. In the former, the discussion is much more elaborate and the process during which the soul receives these many forms, namely by conceiving sensibles and intelligibles, is explicitly referred to. The argument is probably inspired by Aristotle's approval of calling the soul 'the place of forms',[134] but Miskawayh's source may actually be Plotinus, *Enn.* IV 7, 6, where the argument as such is presented.[135]

[133] See Endreß, *Proclus Arabus*, pp. 112–15. Most of the early translations which Endreß discusses use both terms, whereas the *DA*-paraphrase has only *mabsūṭ*.

[134] See *DA* 429a27–8, which is taken up twice in the *DA*-paraphrase (ed. Arnzen, 253.16 and 307.11).

[135] In the extant fragments of the Arabic Plotinus no rendering of *Enn.* IV 7, 6 is attested. The crucial section of the Greek reads in Armstrong's translation as follows: 'Again, since the object being perceived is a body, perception could not occur in any other way than that in which seal-impressions are imprinted in wax from seal-rings, whether the sense-objects are imprinted on blood or on air. And if this happens as it does in fluid bodies, which is probable, the impression will be obliterated as if it was on water, and there will be no memory. But if the impressions persist, either it will not be possible for others to be imprinted because the first will prevent them, so that there will be no

The idea that the soul uses the body like a tool has already occurred in several passages above, namely in passages (54–7) and (76),[136] and the assumption that the soul does not constrain bodily instruments in space while using them recalls Miskawayh's description of the joining of spiritual existents in passage (49).

(85) 51a8–12 [The proof] that [the soul] is not an accident

Source: Miskawayh, alternative *Fawz*?

Parallel: Miskawayh, *Tahdhīb*, ed. Zurayk, 5.10–14; tr. Zurayk in *Refinement*, pp. 6–7; French tr. Arkoun in *Traité*, p. 9.

Cf.: Miskawayh, *R. fī l-Nafs*, ed. Arkoun in 'Deux épîtres', 48.8–11; ed. Badawī in *Dirāsāt*, 90.25–91.3.

Whereas the end of passage (79) only states that the soul is not an accident, this passage provides the corresponding argument. The argument certainly derives from Miskawayh as the word-for-word parallel in his *Tahdhīb* and the looser parallel in his *R. fī l-Nafs* clearly show. Whether it is also part of the alternative recension of his *Fawz* cannot be said with certainty, as it is missing from the standard *Fawz* and the *K. al-Ḥikma*, which is the only other witness to the alternative recension beside the *PR*. It would be the most probable assumption,[137] but it is, of course, also possible to assume that the compiler of the *PR* extracted the argument from the *Tahdhīb*, which presents it as the conclusion of a section which is of very similar content to the previous passage in the *PR*. It is particularly possible since the *Tahdhīb* was clearly a source text for the compiler of the *PR*, who included several excerpts from this work in what follows.

(86) 51a12–51b6 The proof that [the soul] is simple

Source: Miskawayh, alternative *Fawz*.

Parallel: *K. al-Ḥikma*, MS Esad Efendi 1933, fols. 62b14–63a1.

Cf.: Miskawayh, *Fawz* II, 4, ed. ʿUḍayma, 74.5–6, 74.4–5, 74.1–3, 73.8–10, with

other sense-impressions, or if others are made, those former impressions will be destroyed: so that there will be no possibility of remembering. But if it is possible to remember and to perceive one set of things after another without the previous ones hindering, it is impossible for the soul to be a body.'

[136] One may also think of Aristotle, *DA* 407b26 and 415b18–19.

[137] For which I have argued in Wakelnig, 'New Version', pp. 93–4.

French tr. Arnaldez, pp. 41–2; tr. Sweetman, pp. 127–8; Miskawayh, *R. fī l-Nafs*, ed. Arkoun in 'Deux épîtres', 52.20–53.3; ed. Badawī in *Dirāsāt*, 95.4–9.

This section must have been part of the alternative recension of the *Fawz* as is indicated by the parallel with the *K. al-Ḥikma* and the existence of most of the textual material in the standard *Fawz*, although in re-arranged order. The order of the *PR* is, however, parallel to the order in Miskawayh's *R. fī l-Nafs* where part of the argument, i.e. that intellect has no form as air has no colour, appears as well. As to contents, the passage rephrases the argument presented in passage (84) above. The example of air having to be colourless in order to be receptive of colours occurs in Aristotle, *DA* 418b26–7.[138]

(87) 51b7–52a6 [The proof] that [the soul] is self-moving

Source: Miskawayh, alternative *Fawz*.

Parallel: *K. al-Ḥikma*, MS Esad Efendi 1933, fol. 63a1–13.

Cf.: Miskawayh, *Fawz* II, 7, ed. ʿUḍayma, 84.16–85.1, 84.11–16, 85.15–86.4, with French tr. Arnaldez, pp. 48–9; tr. Sweetman, pp. 136–7.

This Platonic proof for the soul being self-moving is also taken from Miskawayh, who, in the standard version of the *Fawz*, explicitly refers his readers to Plato's *Leges*.[139] The two opposing directions of the motion of the soul sound Plotinian,[140] but are probably presented here in Miskawayh's own phrasing.

(88) 52a6–10 Plato says

Source: Miskawayh, alternative *Fawz*.

Parallel: *K. al-Ḥikma*, MS Esad Efendi 1933, fol. 63a13–17.

Cf.: Miskawayh, *Fawz* II, 8, ed. ʿUḍayma, 89.11–90.1, with French tr. Arnaldez, p. 52; tr. Sweetman, p. 140; *R. fī Ārāʾ*, ed. Badawī in *Aflāṭūn*, 325.9–20, on which,

[138] See Sweetman, *Fawz*, p. 127, who, however, does not translate 'air' but 'eye': 'In detail, the eye is appointed for the reception of all colours and therefore is the negator of every colour.'

[139] Sweetman (*Fawz*, p. 136, n. 2) suggests more convincingly that this reference must be to 895c-896c, whereas Arnaldez (*Salut*, p. 110, n. 43) assumes it is to 966e. If the reference is only to Plato as in the *PR* and not explicitly to the *Leges* one may also think of *Phaedrus* 245c-d, which is also taken up by Plotinus in *Enn.* IV 7, 9. However, this Plotinian section is not rendered in the extant Arabic *Plotiniana*.

[140] See Sweetman (*Fawz*, p. 137, n. 2), who gives as examples *Enn.* V 1, 3; IV 3, 9 and IV 3, 17, and refers to the second chapter of the *ThA*.

see passage (89) below. **(Partly):** *Tahdhīb*, ed. Zurayk, 211.23–212.8; tr. Zurayk in *Refinement*, p. 187; French tr. Arkoun in *Traité*, p. 316; *R. fī l-Nafs*, ed. Arkoun in 'Deux épîtres', 55.7–10; ed. Badawī in *Dirāsāt*, 96.25–97.3; ed. Arkoun in 'Textes inédits de Miskawayh', 195.10–12; French tr. Cuvelier in 'Les textes inédits', p. 226.

The context in which this passage occurs in the standard version of the *Fawz* is to present the upwards motion of the soul as its happiness, while its downward motion brings it misery. This is then linked to the alleged Platonic saying that 'philosophy is the habituation to voluntary death (*al-falsafa hiya l-tadarrub bi-l-mawt al-irādī*)', which probably goes back to the fourth definition of philosophy in the Alexandrian *Prolegomena*. In the Arabic version of the *Prolegomena* preserved in Ibn al-Ṭayyib, philosophy is defined as 'concern with death (*muʿānāt al-mawt*)' and the definition is rightly ascribed to *Phaedo* 64a.[141]

The ensuing discussion of the two lives is most probably also inspired by the *Prolegomena*,[142] although Miskawayh combines it with the previous passage on the two motions of the soul. In his *Tahdhīb*, in the chapter 'On the Fear of Death', which became known independently as *R. fī l-Khawf min al-mawt*, he reports the philosophers' identification of voluntary life with 'what man seeks in this world of such things as food, drinks, and desires', and continues that 'natural life was to them the eternal existence of the soul in everlasting bliss through one's acquisition of the true sciences and his purification from ignorance' (tr. Zurayk, p. 187).

141 Ibn al-Ṭayyib's passage (*Commentary on Porphyry's* Eisagoge, ed. Gyekye, 21.8–9) reads:
فأمّا الحدّ الرابع فهو القائل إنّ الفلسفة معاناة الموت . وهذا الحدّ حدّها به فلاطن في كتابه الموسوم بفادن .
According to an earlier Arabic witness to the survival of the Alexandrian *Prolegomena* tradition, namely Qusṭā ibn Lūqā, the definition is given as *al-ihtimām bi-l-mawt* (in a fragment possibly deriving from his introduction to logic, see Daiber, 'Qosṭā', 106.15 and p. 118). The definition was still used in later times (see Badawī, *Aflāṭūn*, 169.6–9).

142 See the passage in Ibn al-Ṭayyib (*Commentary on Porphyry's* Eisagoge, ed. Gyekye, 21.15–22.5):
ونقول إنّ الموت والحياة عند الفلاسفة على ضربين : موت طبيعيّ وموت إراديّ وحياة طبيعيّة وحياة إراديّة . فالموت الطبيعيّ هو مفارقة الصورة للمادّة ، أعني النفس للجسم ، والموت الإراديّ هو إماتة الإنسان شهواته وتغليبه قوّته العقليّة ويصيرها كالملك في بدنه حاكم على جميع القوى الجسميّة وتعديل أفعالها . والحياة الطبيعيّة هي مقارنة الصورة للمادّة ، أعني النفس للجسم ، والحياة الإراديّة هي التمتّع في الشهوات وترك الالتفات إلى ما توجبه سنّة العقل . ففلاطن إنما عنى بقوله إنّ الفلسفة معاناة الموت ، أي معاناة الموت الإراديّ ، لا الموت الطبيعيّ .
وبالجملة فلكلّ واحد من الناس أحوال أربع : إن يكون موجوداً وألا يكون موجوداً وإن يكون موجوداً وجوداً حسناً وألا يكون موجوداً وجوداً حسناً . فإن يكون الإنسان موجوداً إنما هو بالحياة الطبيعيّة . وإن لا يكون موجوداً إنما هو بالموت الطبيعيّ . وإن يكون موجوداً حسناً إنما هو بالموت الإراديّ . وألا يكون موجوداً حسناً إنما هو بالحياة الإراديّة . فلذلك لما لم يفهم اومبرطس هذه القسمة صعد إلى أسّ جبل شاهق وجهر بصوته وقال ايتها الشمس إنّ اومبرطس قد سمع بنفسه لكيما يرث الحياة الدائمة ، وتخلص من الآلام الجسدانيّة وتردى .

The maxim to die voluntarily so as to live by nature is quite widespread in Arabic;[143] Miskawayh also quotes it in his *Tahdhīb* immediately after the discussion of the philosophers' understanding of the two lives, and in his *R. fī l-Nafs*. In this latter work the maxim is strangely reversed and reads 'Die naturally so you will live voluntarily', which may well be a scribal error, as the following commentary is in line with what Miskawayh says elsewhere.[144] An interpretation of the first half of the maxim is further found in an excerpt from his doctrines (*Faṣl ākhir min kalāmihī*) which has been edited by Arkoun.[145]

The entire present passage must also be regarded in close connection with the next one.

(89) 52a10–12 And they [i.e. the philosophers] says

Source: Miskawayh, alternative *Fawz*.

Parallel: *K. al-Ḥikma*, MS Esad Efendi 1933, fol. 63a17–19.

Cf.: *Tahdhīb*, ed. Zurayk, 211.23–212.4; tr. Zurayk in *Refinement*, p. 187; French tr. Arkoun in *Traité*, p. 316; Anonymous, *R. fī Ārāʾ*, ed. Badawī in *Aflāṭūn*, 325.9–20; *Muntakhab Ṣiwān al-ḥikma*, ed. Dunlop, lines 1275–6 and 3513–14; ed. Badawī, 185.5–6 and 353.1–3.

The exposition of the two deaths is not preserved in the standard version of the *Fawz*, but fits well with the previous passage, so that its inclusion in the alternative recension is not surprising. It also occurs, although phrased slightly differently, in the *Tahdhīb*. There Miskawayh ascribes it explicitly to the philosophers, which explains the use of the plural form of the verb (*qālū*) in our passage.[146]

In his *Sententia* 9, Porphyry states that death is twofold and this is prob-

143 See Rosenthal, 'Knowledge', p. 409; and Gutas, *Literature*, pp. 309–11. Cf. also the *Mukhtār min Kalām al-ḥukamāʾ*, where the statement is ascribed to Socrates (Gutas, *Literature*, pp. 102–3).

144 In the *R. fī l-Nafs* (ed. Arkoun, 'Deux épîtres', 55.7–10) he writes:
وكذلك ما يُحكى عن افلاطون . فإنّه كان يوصّي تلامذته بأن يقول : "مُت بالطبيعة تحيَ بالإرادة" وقد فسّر هذا بتفسير كبير يعود إلى ما ذكرناه وهو إنّه يريد أن يُميت الإنسان شهوات بدنه الذي هو لا محالة ميّت ، ليُحيي نفسه السرمديّة الوجود والبقاء .

145 The passage reads (ed. Arkoun, 'Textes inédits', 195.10–12):
ولهذا نجد الفلاسفة يقولون "مت بالإرادة" لأن الموت الإراديّ هو المدرّب في هجر الحسّيّات والملاذّ الجسمانيّة وأطراح الشهوات والتصرّف مع العقل والعقليّات .

146 See Zurayk's translation (p. 187): 'For this reason, the philosophers affirmed that death is of two kinds: the voluntary and the natural, and that life also is of the same two kinds. By voluntary death, they meant the suppression of desires and their abandonment, and by natural death the separation of the soul from the body.'

ably the origin for the same notion in the Alexandrian *Prolegomena*.[147] In Arabic literature the statement is ascribed to authorities as diverse as Hermes, Socrates and Plato.[148]

(90) 52a13–52b7 Aristotle says

Source: *DA*-paraphrase, ed. Arnzen with German tr., 219.17–221.5; MS Aya Sofya 4156, fol. 69a3–12.

The argument for the soul's separability from its body is extracted from the *DA*-paraphrase which, in turn, is based on Proclus' *Elementatio theologica*, prop. 16, and the application of this proposition to the soul in props. 186–90.[149] This section of the *DA*-paraphrase is also used in the *Firdaws*.[150]

In the context of the *PR* this argument links the discussion of the different kinds of life and death with the following proofs for the immortality of the soul. It further provides the step from the incorporeality of the soul,

[147] See Gutas, *Literature*, p. 310. For the Arabic text of the *Prolegomena*, see above, n. 142.

[148] The *Muntakhab* ascribes it to Hermes (ed. Dunlop, lines 1275–6):
وقال : إنّ الموت موتان : موت إراديّ ، وموت طبيعيّ . فمن أمات نفسه موتاً إراديّاً فإنّ موته الطبيعيّ له حياة .
However, it also presents a shortened maxim, 'Die voluntarily', together with a definition of voluntary death in the entry on Miskawayh which is not identical to any of the parallels in Miskawayh's existent works.
In a collection of sayings entitled *Fuṣūl falsafīya majmūʿa min nawādir kathīra* preserved in the Istanbul MS, Köprülü 1608, fol. 44b1–2, it is presented as Socratic:
وقال الموت موتان موت طبيعيّ وموت إراديّ فمن امات نفسه موتاً إراديّاً كان موته الطبيعيّ له حياة .
In the *R. fī Ārāʾ al-ḥukamāʾ* a discussion of the twofold death and life is linked to Plato and his follower Xenocrates. The text is probably a later reworking of the topic as its definition of natural life is the opposite of the one given in the Greek tradition (ed. Badawī in *Aflāṭūn*, 325.10–20):
فقوله إنّ الفلسفة اختيار الموت الإراديّ على الحياة الطبيعيّة وفسّر هذا الحدّ كسنقراطيس تلميذه فقال : الموت نوعان : إراديّ وطبيعيّ . فالموت الإراديّ إماتة الشهوات الرديئة التي تنتجها النفس الحيوانيّة من اللذّات الدهريّة والأسباب الجسدانيّة ، مثل الغضب والانتقام . ومعنى الحياة الطبيعيّة ملازمة الأشياء السبعيّة من المأكول والمشروب والمنكوح ، وترك الأشياء التي تلذّ بها النفس الناطقة من العلوم الفلسفيّة والأمور العقليّة . فالموت الإراديّ ، على ما بيّنّا ، مضادّ للحياة الطبيعيّة . والموت الطبيعيّ مضادّ للحياة الإراديّة . وموت الطبيعة مفارقة النفس الجسد . وموت الفلسفة ملازمة النفس الشهوات . وحياة الطبيعة ملازمة كلّ ما يصلح به الجسد . وحياة الفلسفة ملازمة كلّ ما يفرح به العقل .

[149] See Arnzen, *Aristoteles' DA*, pp. 382–3. Arnzen further discusses how an earlier, no longer extant version of *PA* must have been the source of the *DA*-paraphrase, which, in turn, must have been used by the editor of the preserved recension of *PA*; see also Endreß, *Proclus Arabus*, pp. 196–200.

[150] See *Firdaws* (ed. al-Ṣiddīqī, 62.10–16; ed. al-Jundī, 50.4–8); and Arnzen, *Aristoteles' DA*, p. 73.

argued for especially in passages (80–5), to its persistence after its separation from the body at death.

(91) 52b7–54a5 Abū ʿAlī [Miskawayh] says

Source: Miskawayh, alternative *Fawz*.

Parallel: *K. al-Ḥikma*, MS Esad Efendi 1933, fols. 63a19–64a14.

Cf.: Miskawayh, *Fawz* II, 6, ed. ʿUḍayma, 81.10–82.13, 83.2–84.3, with French tr. Arnaldez, pp. 47–8; tr. Sweetman, pp. 133–5.

The proofs for the immortality of the soul are taken from Miskawayh who, in turn, rephrases some Arabic version of Proclus' monograph on Plato's three proofs for the immortality of the soul.[151] The Platonic sources are *Phaedo* 105b5–107a1 (soul gives life), *Res Publica* X, 608c9–611a3 (soul is not corrupted by its own badness), and *Phaedrus* 245c5–246a2 and *Leges* X, 893b1–898c5 (soul is self-moved).[152] Sweetman, Westerink and Hasnawi point out that Miskawayh's phrasing of the second proof is far more Plotinian than Proclean, as it is Plotinus who believes in a sinless soul (*Enn.* I 1, 12) and regards matter as evil, two doctrines which were rejected by Proclus.[153] The Platonic proof for the self-motion of the soul has already been taken up in passage (87), which also derives from Miskawayh.

(92) 54a6–54b8 Aristotle says

Source: *DA*-paraphrase, ed. Arnzen with German tr., 311.8–313.5; MS Aya Sofya 4156, fol. 95a4–15.

[151] The Greek original is lost, but, by comparing fragments of it preserved in Priscianus with Miskawayh's text, Westerink ('Proofs') established the Proclean work as their common source. Hasnawi ('Textes') discussed Miskawayh's text as well as two further Arabic texts which are independent witnesses of an Arabic version of Proclus on Plato's three proofs for the immortality of the soul. The texts are *Ḥujaj Aflāṭūn ʿalā baqāʾ al-nafs* (ed. Badawī in *Arisṭū*, pp. 73–4) and *Thamara laṭīfa min maqāyīs Aflāṭun fī anna al-nafs lā tafsud* (ed. Badawī in *Aflāṭūn*, pp. 331–2). The existence of an Arabic translation of the Proclean work is attested in the *Fihrist* (252.15–16), which gives the title as *K. Sharḥ qawl Flāṭūn inna al-nafs ghayr māʾita, thalāth maqālāt*.
A further text ascribed to Miskawayh (ed. Arkoun in 'Textes inédits', 195.16–21) argues for the immortality of the soul on the grounds that it is a substance and, as such, has no opposite. Proofs for the soul being a substance are said to be its giving life, its self-movement and its self-knowing.

[152] See Westerink, 'Proofs', p. 296.

[153] See esp. *ibid.*, pp. 305–6; also Sweetman, *Fawz*, p. 134, n. 1; and Hasnawi, 'Textes', pp. 403–4.

The compiler of the *PR* excerpts this description of how the (rational) soul knows the intelligibles and sensibles from the *DA*-paraphrase. The section in the paraphrase elaborates on Aristotle's words in *DA* 429b10–22, making use of Philoponus' commentary as well as of Proclean thought.[154] ʿAlī ibn Rabban also draws on this passage in his *Firdaws*.[155]

(93) 54b8–55a6 He says

Source: *DA*-paraphrase, ed. Arnzen with German tr., 317.15–319.9; MS Aya Sofya 4156, fol. 97a1–b1.

Whereas the body has often been referred to as the soul's instrument, this passage, drawn from the *DA*-paraphrase, demonstrates that the body cannot be said to be the tool of the intellect. The material used in this section is mainly Plotinian and has little to do with the Aristotelian *DA*.[156]

(94) 55a7–56b8 Abū ʿAlī Miskawayh says

Source: Miskawayh, alternative *Fawz*.

Parallel: *K. al-Ḥikma*, MS Esad Efendi 1933, fols. 64a14–65a13.

Cf.: Miskawayh, *Fawz* II, 2, ed. ʿUḍayma, 65.4–67.3, with French tr. Arnaldez, pp. 35–7; tr. Sweetman, pp. 120–2.

This section is an almost exact reproduction of the standard version of the *Fawz* II, 2, although the immediate source of the *PR* is most probably, once again, the alternative recension attested independently also by the *K. al-Ḥikma*. The first part of the section resembles the discussion of the way in which the soul knows the intelligibles and sensibles in passage (92) above, excerpted from the *DA*-paraphrase. Whereas there the objects of thought in the soul are divided into two categories, the simple intelligibles and the composite sensibles, Miskawayh's division introduces a third category, namely mathematical concepts which are abstracted from their material instantiations but still do not reach the rank of intelligibles. Therefore Plato has already situated them 'in between' (μεταξύ) the sensible and the

154 See Arnzen, *Aristoteles' DA*, pp. 430–1.

155 *Firdaws*, ed. al-Ṣiddīqī, 69.17–22; and ed. al-Jundī, 54.22–5; see Arnzen, *Aristoteles' DA*, p. 73.

156 See Arnzen, *Aristoteles' DA*, pp. 435–6.

intelligible realm, mathematics being thus the intermediate science between physics and metaphysics.[157]

Miskawayh may, in fact, draw on some version of Aristotle's *DA*, as his account of resolving and composing objects of thought is paralleled in 430a26–b6.[158] However, this aspect is not taken up in any of the two extant recensions of the *DA*-paraphrase. The Aristotelian concept that each sense corresponds to one of the four elements is taken up and expounded in passages (157) and (159).[159]

(95) 56b8–59b8 The soul does not have parts

Source: Miskawayh, alternative *Fawz*.

Parallel: *K. al-Ḥikma*, MS Esad Efendi 1933, fols. 65a13–66a13, 60b4–61a15, 67a3–9.[160]

Cf.: Miskawayh, *Fawz* II, 3, ed. ʿUḍayma, 67.9–71.12, with French tr. Arnaldez, pp. 37–40; tr. Sweetman, pp. 122–5.

The passage, most probably taken from the alternative *Fawz*, corresponds closely to book II, ch. 3 of the standard version. The first issue raised is that the soul cannot obtain its perceptibles through different parts resembling body parts, as it has only one power which judges all its perceptibles. This is probably inspired by Aristotle, *DA* 426b25–9.[161] Thereupon the possible, but erroneous, view that the composite perceptibles are perceived by composite organs, whereas the simple perceptibles are perceived by simple ones, is expounded. The refutation is presented in the name of Aristotle and repeats several points which are discussed in passage (92) drawn from the *DA*-paraphrase. The soul perceives everything with one power, otherwise it could not correct the errors of sense perception. The misperception of the sun as small is already mentioned by Aristotle (*DA* 428b3–4),[162] and

157 Arnaldez (*Salut*, p. 106, n. 9) suggests that Miskawayh's criticism of people believing the mathematical objects to be intelligibles is directed against the Pythagoreans.

158 See Sweetman, *Fawz*, p. 121, n. 1.

159 Aristotle discusses the topic at length in the second chapter of his *De sensu*.

160 The three different places in which this continuous passage occurs in the *K. al-Ḥikma* are due to a misplacement of some folios in the MS (see Wakelnig, 'New Version', p. 96, n. 35).

161 See Arnaldez, *Salut*, p. 107, n. 13.

162 The *topos* of the erring senses is also addressed in the Arabic version of the Alexandrian *Prolegomena* as preserved in Ibn al-Ṭayyib's commentary on Porphyry's *Eisagoge*, in the first definition of philosophy as 'knowledge of the truths of all existent things'. The sun is said to be 'perceived as a foot in size, because of the distance, though the

the *DA*-paraphrase (313.10–13) adds the rudder which is perceived as crooked when under water and the images seen in a mirror as further examples. Miskawayh himself presents the same, as well as additional, examples of misperceptions in his *Tahdhīb* (pp. 7–9) and the *R. fī l-Nafs* (ed. Arkoun in 'Deux épîtres', pp. 14–15). That the existence of the sense organs is a *sine qua non* for sense perception has already been explained in passage (92) above and exemplified by the same example of the man blind since birth. The same holds true for the description of how intelligibles are perceived by self-reversion, but sensibles by extension towards the sense organs. Therefore it seems plausible that Miskawayh himself draws here on the *DA*-paraphrase. The motion of the soul (termed *jawalān* 'revolving') occurs already in passage (87).

(96) 59b8–61b6 The Greek Sage says

Translation: Lewis in 'Plotiniana', pp. 201–3.

Parallel: *ThA*, ed. Badawī in *Aflūṭīn*, 46.7–11, 46.14–48.8, 49.11–14; ed. Dieterici in *Uthūlūjiyā*, 33.17–34.2, 34.6–35.17, 37.5–9; German tr. Dieterici in *Theologie*, pp. 34–7.

Cf.: *Enn.* IV 7, 8^1, 23–8 and 32–4, IV 7, 8^2, 1–10 and 20–2.

This section belongs to the *Sayings of the Greek Sage* as the compiler of the *PR* explicitly indicates and is parallel to most of the rendering of *Enn.* IV 7, 8^1 and 8^2 in *ThA*. The text presents yet more arguments for the incorporability of the soul and its powers. Firstly, it is argued that the strongest powers are often not linked to the largest bodies and that power therefore does not depend on body. Secondly, it is shown that soul is not bodily as blood, breath or the body's mixture and therefore not a material, but an efficient cause for the body. Thirdly, it is demonstrated that in the mixture of bodies the essences of the components mixed together are not preserved, and the soul could not remain soul if it were mixed with the body in the way bodies are mixed with each other. Fourthly, it is argued that the soul does not increase the body's size by joining it. The last argument is presented in a shorter version than in Plotinus and the *ThA*, as it is

mathematicians have shown that it is 166 times and a fraction greater than the earth' (tr. Dunlop, 'Existence', p. 87). The entire passage reads (ed. Gyekye, 19.14–19):
وذلك أنّ المعرفة تقال على معان كثيرة : على المعرفة الحسّيّة بمنزلة جميع الأشياء التي تدركها الحواسّ الخمس وهذه المعرفة قد يقع فيها الزلل فإن الحسّ قد يعرض له أن يدرك الأشياء على غير ما هي عليه بنسبة إلى المحسوس غير ملائمة . فإنّه يدرك من الشمس أنّ مساحتها كالقدم لبعد المدى ، وقد بيّن التعليمون أنها مثل الأرض مائة وستّة وستّون مرّة وكسر . ويدرك الأشياء البعيدة سودا .

reduced to the impossibility of complete permeation of one body by another. The aspect of infinity, i.e. that bodily permeation would amount to an infinite division of the body which is impossible, is omitted.

(97) 61b7–62a10 Plato says

Edition: ed. and tr. Zimmermann, 'Origins', pp. 220–1.

Cf.: *Enn.* IV 8, 2, 24–38.

As Zimmermann shows, this passage may also belong to the *GS* and render the second chapter of *Enn.* IV 8.[163] For this entire *Ennead* he further supposes a heading referring to Plato which would explain the attribution in the *PR*.[164] The passage looks, in fact, like a reworking of the Plotinian text in light of the second of Plato's proofs of the immortality of the soul presented in the reworking of Miskawayh in passage (91).

(98) 62a11–12 He says

Source: unidentified.

These two short statements, clearly perceived as a unity by the compiler of the *PR*, still discuss the relationship between soul and body.

(99) 62a13–62b2 He says

Source: unidentified.

This passage sums up the main findings about the soul which are that it does not move spatially, is incorruptible and indivisible, and defines, in a condensed and therefore difficult phrase, the soul's relationship to nature, which is exemplified in the following section. Whether this following section actually forms a unity with the present passage is not clear.[165]

(100) 62b3–13 And he says

Source: unidentified.

Nature is likened to a potter who forms his creation, but has to hand it

[163] See the comparison between the Greek and the Arabic in his 'Origins', pp. 219–20.
[164] See *ibid.*
[165] See note to the translation of the passage.

over to fire which, in turn, completes what he has created. Fire is compared to the soul. The passage makes use of the clothing metaphor already employed in passage (25).

(101) 63a1–5 He says

Source: unidentified.

Cf.: *DA*-paraphrase, ed. Arnzen with German tr., 309.10–11, 309.15–17.

This is yet another argument for the immortality of the soul, which may, in fact, be inspired by Aristotle's *DA* 429a29–b5, where we read: 'in case of thought thinking about an object that is highly thinkable renders it more and not less able afterwards to think of objects that are less thinkable' (tr. Smith), or the corresponding passage in the *DA*-paraphrase. Contrary to thinking, a strong sense perception impairs the perceiving sense organ, which demonstrates that sensation is dependent on the body, but thought, and thus the soul – or more precisely the rational soul – are separable. The idea also occurs below in the context of the discussion of sense perception, in passages (155) and (160).

(102) 63a6–8 Pythagoras says

Source: Ps.-Pythagoras, *Dicta aurea*, ed. Diehl and Young in *Theognis*, lines 63, 68–71;[166] *Waṣāyā Fūthāghūrus* as lemmata in Iamblichus' commentary in *Pythagorica*, ed. Daiber with German tr., 108.7–9, 116.14–15, 118.8–11; in Miskawayh, *al-Ḥikma al-khālida*, 228.11–12, 228.17–19; *Muntakhab Ṣiwan al-ḥikma*, ed. Dunlop, lines 543–4 and 547–9; and ed. Badawī, 123.10–11 and 123.16–18; al-Anṣārī, *Ādāb al-falāsifa*, 118.15–16 and 118.19–21.

Cf.: *R. Ikhwān al-Ṣafāʾ*, I, 138.14–16 and, deriving from there but slightly different, *Muntakhab Ṣiwan al-ḥikma*, ed. Dunlop, lines 3673–5; and ed. Badawī, 363.21–3.

The passage combines three different sections of Pythagoras' *Dicta aurea*. The first one encourages man to be brave, since, by his divine element, he takes part in the divine genus and has thus the ability to know everything. In the commentary of Iamblichus, the 'divine genus' in man is explained by the presence of intellect, discrimination and free will. Second is the imperative to employ one's discrimination in obeying some aforemen-

[166] The translation of the Greek by Thom (*Verses*, p. 99) reads: 'But take courage, for mortals have a divine origin, ... and consider each thing by putting the excellent faculty of judgment in control as charioteer. Then, if you leave the body behind and go to the free *aither*, you will be immortal, an undying god, no longer mortal.'

tioned rules concerning the consumption of food. This is directly followed by the description of the condition of the soul after having left the body in the third part.

The version closest to the *PR* is in the lemmata in the commentary of Iamblichus, whereas the Proclean commentary most probably used above, passages (14–16), does not have any corresponding texts if we can trust the evidence of Ibn al-Ṭayyib's summary of the work. The description of the status man may reach after having separated from his body is also taken up in the *Rasāʾil Ikhwān al-Ṣafāʾ*, in the fourth epistle on astronomy in a section on the soul's longing for the heavenly world, which also quotes the famous Plotinian account of his leaving his body in the version of the *ThA* (ed. Badawī in *Aflūṭīn*, 22.1–5).[167]

(103) 63a9–11 Zeno says

Source: unidentified.

Cf.: Mubashshir, *Mukhtār al-Ḥikam*, 43.17–19; German tr. Rosenthal, 'Arabische Nachrichten', p. 59.

The saying ascribed to Zeno mentions the education of the body as the reason why the soul leaves its lofty dwelling place and becomes embodied. The first part of the passage which describes the soul also occurs in a saying ascribed to Zeno by Mubashshir, which focuses on the opposition of life and death and thus uses the statement to portray the soul's life.[168]

(104) 63a12–63b1 Plato says

Source: unidentified.

Parallel: Anonymous, *R. fī Ārāʾ*, ed. Badawī in *Aflāṭūn*, 319.4–6.

Cf.: Mubashshir, *Mukhtār al-Ḥikam*, 140.16–17.

This saying attributed to Plato is also preserved in the *R. fī Ārāʾ al-ḥukamāʾ al-yūnānīyīn* and its first part is similar to a Platonic statement in Mubashshir. The *PR* and the *R. fī Ārāʾ al-ḥukamāʾ* describe the opposition between

[167] Thee passage in *R. Ikhwān al-Ṣafāʾ* (I, 137.1 and 138.14–16) reads:
فصل في تجرّد النفس واشتياقها إلى عالم الأفلاك : ... وقال فيثاغورس في الوصية الذهبيّة : إذا فعلتَ ما قلتُ لك يا ديوجانس ، وفارقتَ هذا البدن حتى تصير نحلاً في الجوّ ، فتكون حينئذ سائحاً غير عائد إلى الإنسانيّة ولا قابل للموت .

[168] Mubashshir's passage reads:
وقال : إن حياة النفس عالية رفيعة بعيدة من الموت . ولن تقع النفس في يد الموت إلا أن تدعوه ذاتها وتُبْدِي له كامنها وتمكنه من عيبها فيعلم بذلك كيف يتشبّث بها .

the rational and the brute soul, whereas Mubashshir's version only implies a contrast between the soul and things harmful to it.[169] By mentioning death, which cannot befall the soul, this passage is linked to the previous ones.

(105) 63b2–3 He [Plato] says

Source: unidentified.

Parallel: Anonymous, *R. fī Ārāʾ*, ed. Badawī in *Aflāṭūn*, 319.7–8, and cf. also 319.8–13.

This short statement occurs in the *R. fī Ārāʾ al-ḥukamāʾ al-yūnānīyīn*, where it is still attributed to Plato and enhanced with a lengthy exposition which reads:[170]

'That indicates that when the soul is unimpaired by passions, employs intellect alone in order not to be yielded to misdeed and mistake, perfects the performance of the intellectual virtues, and then departs from the body, it returns to its first world which is the source of joy and happiness and [where it dwells] with the spiritual beings. When it grows dark through [its] embroilment with passions and employs bodily appetites, and then departs from the body, it is returned to this kind of earthly body which is the source of distress and affliction. In this there is one example of setting the rewards for good and bad deeds.'

The compiler of the *PR* may not have felt tempted to include this exposition as it implies reincarnation, but the main topic of the opposition between intellect and the sensible world and the soul's intermediate position is a recurring theme in the *PR*, esp. in the *GS* passages above.

[169] Mubashshir's passage reads:
وقال إنّ حياة النفس وقوامها بأعمالها المحصنة لها من الآفات حتى لا يدنو منها شيء يمسها فيكون ذلك قتلاً للنفس .

The 'killing' (*qatl*) probably refers to something similar to the 'killing of the appetites of the growing soul' in passage (87) above, i.e. if the soul no longer has any longing for the world, if it has killed this longing, no harm can befall it any more.

[170] The Arabic reads (319.8–13):
وهذا يشير إلى (أنّ) النفس إذا سلمت من الهوى واستعملت العقل مجرداً حتّى لا تقارف [تقارف Wak : يقارف Bad] ذنباً ولا زلّة ، وتستكمل فعل الفضائل العقليّة ثم فارقت الجسد . عادت إلى عالمها الأوّل ، معدن السرور والفرح ، مع الروحانيّين . وإذا أظلمت بمشابكة الهوى واستعمال الشهوات الجسدانيّة ، ثم فارقت الجسد ، ردّت إلى مثل ذلك الجسد الأرضيّ ، معدن الهم والحزن . . وفي هذا ضربٌ من الأقرار بالمجازاة بأفعال الخير والشرّ ...

(106) 63b3–6 One of them says

Source: unidentified.

The distinction between soul and spirit, or pneuma, which is addressed in this short passage, is dealt with at greater length in the words of Qusṭā ibn Lūqā in passage (145) below. The spirit has already been alluded to in passage (96) from the *GS* by the term 'innate breath'. There it is shown that the soul is not a bodily aspect like blood, innate breath or the body's mixture, and it is this difference that is the main focus of the present passage.

(107) 63b7–11 Aristotle says

Source: unidentified.

Cf.: *DA*-paraphrase, ed. Arnzen with German tr., 219.5–6, 4–5, 3–4; MS Aya Sofya 4156, fol. 68b2–3, 1–2, 68a15–b1; passage (78).

The introductory part of this passage seems badly matched with the second part which looks like a rearrangement of examples presented in the *DA*-paraphrase to answer the question why the soul alone is not called a living being, if it is the body's perfection. The *DA*-paraphrase, in turn, is based on *DA* 412b10–413a3, and there the passage occurs in the context of the discussion of the soul as entelechy or perfection of the body. It is excerpted in the *PR* in passage (78) above, which is, in fact, partially parallel to the present passage. However, a comparison between the two passages makes it seem unlikely that the present passage is also quoted directly from the *DA*-paraphrase, as the terminology used is different (*badan* vs *jism*). Some reworking of the *DA*-paraphrase, therefore, is a more likely candidate for the source of the present passage.

The concept that the compound of body and soul alone is called a living being is already expressed in the Platonic *Phaedrus* 246c.

(108) 63b12–64a4 He says

Source: Nemesius, *De natura hominis* 3, ed. Morani, 40.8, 40.10, 40.20–41.2; tr. Sharples and van der Eijk in *Nemesius On the Nature*, pp. 81–2; *Fī Ṭabīʿa*, ed. Haji-Athanasiou in 'Le traité de Némésius', 80.2–5, 80.18–81.4.

This passage is a merging of different sentences from the Arabic version of Nemesius' *De natura hominis*. It demonstrates, once again, that the soul is not a body and therefore does not mix with the body in a bodily way, but affects it like the sun affects the air.

(109) 64a5–65a4 The powers of the soul are divided into three parts

Source: Miskawayh, *Tahdhīb*, ed. Zurayk, 15.10–12, 16.2–3, 16.4–5, 15.12–14, 16.5, 16.3, 15.14–16, 16.4, 15.16–17, 16.6–19; tr. Zurayk in *Refinement*, pp. 14–16; French tr. Arkoun in *Traité*, pp. 23–5.

This account of the powers of the soul and their corresponding virtues is paralleled in Miskawayh's *Tahdhīb*, but the text is arranged in a completely different order. As the compiler of the *PR* is rather unlikely to have re-arranged the passage himself and Miskawayh was in the habit of reusing his own texts, one may wonder whether this excerpt is actually taken from a recension of the *Tahdhīb* different from the one we know today. The passage starts off with the Platonic tripartition of the soul into a rational, irascible and appetitive part and their corresponding seats in the body, i.e. brain, heart and liver. This division reoccurs in passage (116) where the activities ascribed to each part differ slightly, although the main ones, i.e. thought, anger and appetite, remain the same. Following the view expressed in the *Res Publica* IV, and repeated at the beginning of the Ps.-Aristotelian *De virtutibus et vitiis*, each part is linked to a certain virtue and vice. Miskawayh gives no reason why he, in fact, relates two virtues to each part. The idea behind it could be a progression from the first mentioned virtue to the second one which would be considered to be the nobler one, as is the case of knowledge and wisdom.

The concept that all three virtues together constitute a fourth one, namely justice, is, of course, Platonic and results in the common doctrine of the four cardinal virtues.

A possible model for Miskawayh's account may be the Ps.-Platonic *Maqāla fī Ithbāt faḍāʾil al-nafs*, which is his source for at least one passage in the *Tahdhīb*.[171]

(110) 65a4–12 Each one of these virtues

Source: Miskawayh, *Tahdhīb*, ed. Zurayk, 25.5, 26.9–13, 27.11–12, 27.21–2, 28.2–3, 28.6; tr. Zurayk in *Refinement*, pp. 22–4; French tr. Arkoun in *Traité*, pp. 39–41.

Like the previous passage, this one is taken from Miskawayh's *Tahdhīb* either in the known or some unknown recension. It presents the Aristotelian idea (see *EN* II, 6, esp. 1106b36–1107a8) according to which each

171 See Ps.-Plato, 'Pseudoplatonischer Text', ed. Daiber, pp. 28–9, 35, 37, 39–40; and for the *Tahdhīb* passage, see ed. Zurayk, 80.1–7 and 80.13–17.

virtue is a mean between two vices which either exceed or fall short of their corresponding virtue.

(111) 65a13–66a2 One of these three powers

Source: Miskawayh, *Tahdhīb*, ed. Zurayk, 51.18–19, 51.21, 51.19–20, 52.1–2, 51.20–1, 52.2–3, 52.5–9, 52.13–22; tr. Zurayk in *Refinement*, pp. 46–7; French tr. Arkoun in *Traité*, pp. 84–5.

This section is, once again, paralleled in Miskawayh's *Tahdhīb*, where the arrangement of the text is different. The view that two parts of the soul, i.e. the rational and the irascible, are better than the third, appetitive part, is held by Plato, whose positive assessment of the θυμός, i.e. the irascible which may side with the rational against the irrational appetitive, is apparent in *Res Publica* IV, 439d-440e.

A version of the simile of the hunter, his dog and his riding animal is found in the Arabic Galen, as Walzer shows.[172] Since Miskawayh does not ascribe it to him, but simply to the Ancients, Walzer suggests that he derived it from a different source.

(112) 66a2–8 The powers of the soul follow the mixture of the body

Source: Qusṭā ibn Lūqā, *R. fī l-Farq*, ed. Troupeau, lines 280–6; ed. Gabrieli with Italian tr. in 'Risālah', 653.11–654.7; ed. Cheikho in 'Risāla', 108.13–20; ed. Ülken in *Ibn Sina*, 94.2–8; and tr. Livingston in 'Qusṭā', p. 73.

The passage is taken from Qusṭā ibn Lūqā's *R. fī l-Farq bayn al-rūḥ wa-l-nafs*, but the first sentence recalls the title of Galen's treatise *K. fī anna Quwā l-nafs tābiʿa li-mizāj al-badan*, and Qusṭā is certainly indebted to the Galenic tradition.[173] Galen's 'tripartite physiological system', as Nutton puts it, 'linked body and soul together, explaining how and why physical changes in the body could alter one's mental balance and behaviour, and vice versa'.[174] In his *De methodo medendi*, Galen states that the powers or fac-

[172] Walzer, 'Diatribe', p. 171, n. 2. In the Galenic text as it survives in Arabic, namely in a summary of *De moribus*, ch. 3 (ed. Kraus in 'K. al-Akhlāq', 42.10–19; ed. Badawī in *Dirāsāt*, 203.10–18), the riding animal is only referred to as some kind of strong beast and the exposition of the simile is different.

[173] At the beginning of the *R. fī l-Farq*, Qusṭā mentions a number of Galenic writings the doctrines of which he summarises in his treatise, but the *K. fī anna Quwā l-nafs* is not among them.

[174] *Ancient Medicine*, p. 234.

ulties (*quwā*) of the soul depend on the spirit or pneuma (*rūḥ*) and the temperament or mixture (*mizāj*) of the solid body parts.[175] The pneuma is air which is refined within the body until it becomes psychic pneuma, or spirit, which is the finest bodily element and 'the organ or instrument of the rational soul', as our compiler quotes Galen saying in his *De propriis placitis* further down, passage (146). The mixture of the four elements within the body, which is also called temperament, is highly dependent on the food consumed. Using the concept of pneuma and temperament, Galen thus finds a perfectly reasonable way to explain how the soul is influenced by the body and why the quality of inhaled air and ingested food matters.

A passage similar to the present one occurs in Qusṭā's *K. fī ʿAlal ikhtilāf al-nās fī akhlāqihim*, in which it is, however, the mixture of brain alone which is said to determine the condition of the pneuma and of its activity.[176]

The position of the passage within the *PR* is slightly unfortunate as many concepts used here will only be expounded in what follows. However, it provides a link between the previous purely philosophical discussion of the soul and the following section on physiology.

(113) 66a9–66b2 Aristotle says

Source: Ps.-Aristotle, *Physiognomonica* 808b11–27; *K. fī l-Firāsa*, ed. Ghersetti with Italian tr., 29.8–12, 30.4–8.

Parallel: Abū Saʿīd ibn Bakhtīshūʿ, *R. fī l-Ṭibb*, ed. Klein-Franke, 35.1–5, 35.10–13.

The source of this extract is the Ps.-Aristotelian *Physiognomonica* on the sympathetic affection of soul and body on each other, in its Arabic translation by Ḥunayn ibn Isḥāq. The same Pseudo-Aristotelian passage is also quoted by the tenth- and eleventh-century physician Ibn Bakhtīshūʿ in his

175 See Temkin, 'Pneumatology', p. 158. The Greek text in *De methodo medendi* reads (X, 837.17–838.3): ἐπεὶ δὲ ἡ οὐσία τῶν διοικουσῶν ἡμᾶς δυνάμεων ἔν τε τῷ πνεύματι καὶ τῇ τῶν στερεῶν σωμάτων ἐστὶ κράσει, ταῦθ' ἡμῖν φυλακτέον ἐστὶ παρόντα καὶ ἀνασωστέον διαφθειρόμενα.

176 See 'Livre des caractères', ed. Sbath, 120.18–121.5, where the passage occurs after locating the different psychic activities in the three ventricles of the brain:
فالروح التي في هذه التجويفات قد تختلف في الناس على اختلاف أمزجة أدمغتهم : فإذا كان الدماغ معتدل المزاج ، كانت الروح معتدلة ، وكانت أفعالها معتدلة مستوية ؛ وإذا كان مزاج الدماغ مائلاً عن الاعتدال إلى البرد ، كانت الروح مشاكلة أيضاً لذلك ، وكانت أفعالها مائلة إلى الثقل والإبطاء ؛ وإذا كان مزاج الدماغ مائلاً إلى الحرارة ، كانت الروح أيضاً أزيد حرارة ، وكانت أفعالها سريعة أكثر من المقدار المعتدل يشوبها طيش وتهور . فهذه علة اختلاف الناس في جودة التمييز ورداءته وسرعته وإبطائه بالجملة .
The entire treatise shares a number of parallel passages with the *R. fī l-Farq*, cf. esp. pp. 119–25.

R. fī l-Ṭibb. In contrast to the previous passage from Qusṭā's *R. fī l-Farq*, this one stresses the impact of psychic conditions upon the body. It also draws attention to the simultaneity with which a certain psychic condition affects the body and vice versa.[177]

(114) 66b3–4 Plato says

Source: unidentified; some reworking of the *Timaeus*?

Cf.: Plato, *Tim.* 86b and 88a on diseases of the soul caused by the body and vice versa.

This passage does not add much information to what has been said in the two previous passages (112–13). The reason for including it may be to link the already presented doctrine to Plato as well.

(115) 66b5–68a4 Someone says: the natural matters

Source: some reworking of Ḥunayn ibn Isḥāq, *Masāʾil*, ed. Abū Rayyān, ʿArab and Mūsā, 2.4–5.8, 252.3–253.2, 9.5–11.8, 12.1–3, 255.11–256.8, 13.7–15.5.

The ultimate source of this lengthy passage is most probably Ḥunayn ibn Isḥāq's *Masāʾil fī l-Ṭibb*.[178] However, the compiler of the *PR* seems to have

[177] This aspect is dealt with at the end of the passage. It also occurs in the material of the Ps.-Aristotelian Arabic text which the compiler of the *PR* omits (29.12–30.4), probably due to its similarity with the passage (114) in the name of Plato, presented next.

[178] The work also exists in tabular form entitled *Masāʾil ʿalā Ṭarīq al-taqsīm wa-l-tashjīr*, which contains only part of the material quoted in the *PR*. Yet the passage on the powers of the soul is, for example, similar to the *Masāʾil fī l-Ṭibb* and the *PR*, see Berlin, MS 6258 (= Sprenger 1885), fol. 14a. The text starts with the heading under which there are three distinct sections, the last of which is further divided into three columns. The other copy, Bodleian, MS Huntington 602, fol. 61b, however, shortens the passage and lists only the serving powers (*al-quwā al-makhdūma*) and the locomotive faculty (*al-quww⟨a⟩ al-muḥarrika fī l-makān jinsuhā wāḥid*). The passage reads in Berlin, MS 6258:

أقسام القوى ثلثة
⟨1⟩ طبيعيّة مخدومة وهي المولدّة والمربّية والغاذية والمولّدة تحيل النطفة وتغيّرها وتعمل منها متشابهة ثم آليّة ثم جملته البدن ويخدمها المغيّرة والمصورة والمربّية تمدد الأعضاء طولاً وعرضاً وعمقاً وتنقلها من ⟨الصغر⟩ إلى الكبر والغادية تشبه الغاذي بالمغتذي ويخلف عليه مكان ما نقص منه ويخدمها الجاذبة والماسكة والهاضمة والدافعة .
⟨2⟩ حيوانيّة شأنها بسيط القلب لإدخال الهواء وقبضة لإخراج البخار .
⟨3⟩ نفسانيّة وتنقسم إلى :
⟨3أ⟩ الحسّيّة وتتفرّع إلى خمسة أقسام : البصر السمع الشمّ الذوق اللمس .
⟨3ب⟩ والحركة وهي تجري من الدماغ والنخاع في العصب إلى العضـ⟨ـل⟩ المحرّك للأعضاء المتحرّكة بالحركة الإراديّة وعدد على عدد الأعضاء المتحرّكة .
⟨3ج⟩ والسياسة وتنقسم إلى التخيّل والفكر والذكر .

drawn on a version different from the one edited, or on a summary of the treatise, and must have copied from a corrupted model as the many unintelligible omissions show.

In his *Masāʾil*, Ḥunayn sets out to canonise Galenic doctrines which are often only implied or hinted at by Galen himself, as for example the 'natural matters', i.e. aspects or attributes of man's physical constitution. Ḥunayn lists seven of them: elements, temperaments or mixtures, humours, body parts, powers or faculties of the soul, their activities, and spirits or pneuma. The *PR* has only the first six of these aspects, but does mention the spirits at the end of the passage. Thus the omission does not seem to be a conscious decision on the part of the compiler, especially as he farther down includes quite a few passages dealing with spirit(s) or pneuma, but must rather be due to the defective model from which he copied his excerpts.

The elements fire, air, water and earth are presented as the remotest matter or material from which the entire body is gradually composed. How they mix with each other defines the temperament of the body. The compiler of the *PR* does not quote Ḥunayn's division of nine different temperaments (3.10–4.6), but only stresses that they provide the material from which the living beings are formed. He (or his source) further deviates from the *Masāʾil* when describing the four humours – yellow bile, blood, phlegm and black bile – as material for the animals which have blood. The resulting body parts are divided twice by Ḥunayn, according to their genesis and according to their function within the body. Both divisions are preserved in the *PR*, although the first into simple, i.e. homoeomerous, and composite body parts lacks any introductory statement and is thus not very obvious. As to function, four kinds of body parts can be distinguished: (1) four principal organs, namely brain, heart, liver[179] and testicles, from which (2) different vessels arise, i.e. the nerves, arteries, veins and spermatic ducts. There are further (3) body parts which have one specific function to fulfil in the body, for example bones which provide the body's structure, and (4) organs in which a specific power operates. The listing of four principal organs contradicts the more traditional view of three principal organs, namely brain, heart and liver which is presented in the following passages of the *PR*.

In the discussion of the powers of the soul the *PR* follows the *Masāʾil*

[179] The omission of the liver in the *PR* is further evidence for the bad copy the compiler must have had of this particular source, as it is highly unlikely that he would have deliberately omitted the mention of just the liver.

closer than anywhere else in this lengthy passage. Thus it is possible to add in the edition of the *PR* a few lines which the compiler omits, apparently unaware of any corruption in his source.[180] The three main powers or faculties of the soul are traditionally said to be natural, vital and psychic, and can thus be brought in line with the Platonic parts of the soul presented in the next passage in the *PR*. The natural powers originate in the liver, the vital ones in the heart and the psychic ones in the brain, which Ḥunayn explicitly points out in his *Masāʾil*, in a passage not taken up in the *PR*.[181]

The last two natural matters, i.e. the activities of the soul and the spirits are only mentioned in passing in the *PR*. Ḥunayn, however, distinguishes two kinds of activities, simple and composite ones. The former are executed by only one power of the soul, the latter by more than one. As for the spirits, he lists the natural, vital and psychic ones.[182] The natural one originates in the liver, circulates in the veins and serves the natural powers of the soul. The vital spirit is brought forth by the heart, spreads via the arteries and assists the vital powers. The psychic spirit develops in the brain, operates in the nerves and serves the psychic powers.

(116) 68a5–9 Plato says

Source: unidentified.

Cf.: Plato, *Timaeus* 43a-45b, 70a-73d; and Aristotle, *DA* 414a31–2; *Multaqaṭāt Aflāṭūn*, ed. Badawī in *Aflāṭūn*, 253.7–10.[183]

This passage is a remarkable, but not entirely smooth, merger of Platonic and Aristotelian doctrine. In the *Timaeus* Plato locates each part of the

[180] If the compiler suspects a gap in his model, he usually indicates that by leaving space blank.

[181] For the passage, see *Masāʾil*, pp. 15–16. The description of the directive faculty is a good example of how close the *Masāʾil* often are to Galen's own words in their Arabic translation in the *K. fī Manāfiʿ* VIII (National Library of Medicine, MS A 30.1, fol. 123b9–10):
الشيء الذي به يكون التخيّال والحفظ والذكر والفكر وهي القوّة المدبّرة من قوى النفس .

[182] For the activities, see *Masāʾil*, pp. 16–17, and for the spirits, pp. 17–18.
Galen himself seems to have been in doubt about the possible existence of a natural spirit or pneuma (see Temkin, 'Pneumatology', p. 160; Galen, *On the Usefulness*, tr. May, pp. 48–9). The passages taken up in the following (by Galen, Nemesius, Qusṭā) only speak of the vital and the psychic pneuma, so this particular omission may have been an active decision by our compiler.

[183] The passage in *Multaqaṭāt Aflāṭūn* reads:
وقال : أنواع النفس ثلاثة في ثلاثة أماكن : أحدها إلهيّ ، وبه يكون الفكر والرويّة ، وهو حيّ ، وموضعه ومسكنه : الدماغ ؛ والثاني به يغضب المرء ، ومسكنه القلب ؛ والثالث به شوق المرء إلى اللذّات التي ينالها البدن ومسكنه الكبد .

tripartite soul in a specific organ of the body: the part which renders man rational (44b7: ἔμφρων) in the head or, later on, more specifically in the brain (73d1), the part which partakes of courage and spirit/anger (70a2–3: τὸ μετέχον οὖν τῆς ψυχῆς ἀνδρείας καὶ θυμοῦ) in the heart, and the desiring or appetitive part (70d7: ἐπιθυμητικός) in the liver (71a-b).[184] This localisation is taken up in the quotation in the *PR*, but the terminology for the soul is Aristotelian, taken from *DA* II, 2–3. There, the powers of (self)-nutrition, sensation and thinking are distinguished and nutrition is linked to growth.[185] This combination of Platonic and Aristotelian doctrine results in attributing sensation to the rational soul which resides in the brain and not to the sensible soul which has come to replace the Platonic irascible soul. The passage has a definite Galenic ring to it, as Galen links sensation and movement to the brain, as for example in passage (120), and states in his *De placitis Hippocratis et Platonis* VI, 3, 7 that the desiring power of the soul is called nutritive, vegetative or reproductive by Aristotle. The positioning of the excerpt within the *PR* seems quite carefully chosen as it introduces a long section on human physiology just as the Platonic passages referred to introduce the physiological account in the *Timaeus*.

(117) 68a10–11 The brain is the source of sensation

Source: unidentified.

This sentence of unknown provenance provides the transition between the previous and the following quotation and is clearly singled out in the *PR* as it begins on a new line and is followed by a gap. It is in line with the Platonic doctrine of the *Timaeus* presented in the previous passage and expounds the role of the brain in sensation, movement, imagination and thinking.

(118) 68a11–68b3 The rational soul dwells in the brain

Source: Galen, *UP* IX, 4, ed. Helmreich, II, 12.23–13.10; ed. Kühn, III, 700.12–16; tr. May in *Galen On the Usefulness*, p. 432; *K. fī Manāfiʿ al-aʿḍāʾ* IX, appendix 1, excerpt 1.

[184] The expressions used in the *PR*, *wa-minhu al-ghaḍab wa-l-shajāʿa* and *wa-minhu al-shahwa*, are particularly close to the Greek τὸ μετέχον οὖν τῆς ψυχῆς ἀνδρείας καὶ θυμοῦ and ἐπιθυμητικός.

[185] Cf. also the introductory part of the *DA*-paraphrase (ed. Arnzen, 185.2–3), which lists *al-nafs al-nāmiya*, *al-ḥissīya* and *al-nāṭiqa*.

This is a short passage taken verbatim, but heavily rearranged (see appendix 1, excerpt 1), from the Arabic version of Galen's *De usu partium* Book 9. The link to the previous discussion is the localisation of the rational soul in the brain, but the main topic is the psychic spirit, which is also contained in the brain and which is produced from the vital spirit through refinement.[186]

(119) 68b3–11 The brain is cold and moist

Source: some reworking of Ḥunayn ibn Isḥāq, *Masāʾil*, ed. Abū Rayyān, ʿArab and Mūsā, 258.1–259.9.

The passage most probably derives from the same source as passage (115), i.e. some reworking of Ḥunayn's *Masāʾil*.[187] Therefore it seems safe to assume that the compiler of the *PR* relied on the same defective copy of his source text, which would explain why his extract breaks off before presenting the second reason for the brain being moist.

The condition of the brain is fitted to its activities such as thought, memory and sense perception, which are all motions and as such produce heat. This heat is compensated for by the brain's coldness and moisture. Further, coldness provides stability for the processes of thought and memory, whereas moisture guarantees the softness of the brain. This latter argument is missing in the *PR*, but preserved in Ḥunayn. The Galenic tradition considers softness and the formability that comes with it as prerequisites for sense perception, which is the reason why sensory nerves must be soft, whereas motor nerves are hard.[188]

In his *Firdaws*, ʿAlī b. Rabban quotes Aristotle[189] claiming that the reason

[186] See commentary on passage (112).

[187] There is also a remarkably similar passage in the *Masāʾil ʿalā Ṭarīq al-taqsīm*, Berlin, MS 6258 (= Sprenger 1885), fol. 14a, divided into an introductory section and below it into two columns (and similarly, though less exactly divided and with some omissions in the text, Bodleian, MS Huntington 602, fol. 61b):
والدماغ هو الأصل والمبدأ والذي ينبعث منه الحسّ والحركة الإرادية وجعل بارداً رطباً بالطبع والعلّة التي من اجله〈ا〉 صار الدماغ بارداً رطباً بالطبع مع كونه مبدأ وأصلاً ينبعث منه الحسّ والحركة الإراديّة وهذين يتمّ أمرها بالحرارة كان ذلك لسببين :
〈1〉 لأنّ الحركات فيه ومنه والتي فيه حركة التخيّل والتفكّر والذكر والتي من حركة الحسّ والحركة الإراديّة والحركة يتبعها الإسخان فجعل بارداً لتقاوم هذه الحركات كلّها ولا يلتهب ويحترق .
〈2〉 ولأنّ الدماغ هو أصل ومبدأ للفكر ويحتاج تثبّت والحرارة شأنها سرعة الحركة والنقل والبرودة شأنها الثبات فلهذا جعل بارداً ليثبت بالذكر لسبب برودة وجعل رطباً حتى لا يجفّ بسبب كثرة الحركات فالحركة يتبعها الجفاف فجعل رطباً ليحمل الحركات ولا يجفّ بإفراط وكما يكون سهل التغيّر وسريع الحركة لما يتطبع فيه ولينبت منه أعصاب ليّنة تصلح للحسّ .

[188] See Savage-Smith, 'Galen's Account', p. 80.

why the brain is cold and moist are its sensations and movements. For were it warm, it would be inflamed by them and dry out.

(120) 68b11–69a8 Galen says

Source: Galen, *UP* I, 16, ed. Helmreich I, 32.23–33.10; ed. Kühn, III, 45.10–46.4; tr. May in *Galen On the Usefulness*, p. 89; *K. fī Manāfiᶜ* I, Bibliothèque Nationale de France, MS Arabe 2853, fol. 12a23–12b8; John Rylands Library, MS Arabic 809, fols. 11b24–12a7; National Library of Medicine, MS A 30.1, fols. 11b24–12a7.

This Galenic passage, drawn from Book I of *K. fī Manāfiᶜ al-aᶜḍāʾ*, summarises the three physiological systems operating in the human body. As each of them originates in one of the three Platonic principal organs the section sits well with passage (116) above. In determining the appropriate vessels for each system, Galen turns to Herophilus and Erasistratus for the nervous system and to Praxagoras and his distinction between veins and arteries.[190] It is probably due to Galen's frequent lack of consistency in expressing his doctrines that he talks of the psychic, vital and natural powers or faculties (δύναμις) as being conveyed in these vessels and not the corresponding spirit or pneuma. For it must be the pneuma, as the material substance of these powers, which physically moves within the body.[191] The heart, and more exactly its left ventricle, the pneumatic cavity,[192] is specifically linked to respiration and thus is responsible for the regulation of the innate heat and the nutrition of the psychic pneuma, whereas the liver is mainly in charge of changing food into blood and thus making it available to the entire body. These two basic functions of the living body, 'digestion' of air and of nutriment, are discussed in the following lengthy section implicitly ascribed to Galen, but actually compiled from Galen's *De usu partium* and Nemesius of Emesa's *De natura hominis*.

(121) 69a9–69b10 He says

Source: Nemesius, *De natura hominis* 24, ed. Morani, 85.8–21; tr. Sharples and van der Eijk in *Nemesius On the Nature*, pp. 151–2; *Fī Ṭabīᶜa*, ed. Haji-Athanasiou in 'Le traité de Némésius', 160.3–17.

[189] The reference is actually to the author of an aforementioned Aristotelian book (*ṣāḥib dhālika l-kitāb*), see ᶜAlī ibn Rabban, *Firdaws*, ed. al-Ṣiddīqī, 80.17; ed. al-Jundī, 61.23.

[190] See Sharples and van der Eijk in *Nemesius On the Nature*, p. 151, n. 766.

[191] On the Galenic use of pneuma and its different kinds, see Temkin, 'Pneumatology'.

[192] On which, see *ibid.*, p. 156.

The way this passage is presented in the *PR* implies that it is still Galenic as the introductory 'he says' would be understood as referring back to the last authority quoted and that was Galen. However, although the content is definitely Galenic, the author is actually Nemesius of Emesa. The first part of the section is repetitive in that it discusses the three physiological systems of the human body anew, but at the same time differs in ascribing the blood flow merely to the veins and the spirit or pneuma to the arteries. What, at first, seems like a return to Erasistratus' denial of the existence of blood in the arteries, is modified in what follows, according to Galen's anatomical demonstration to the contrary.[193] Thus it is claimed there is exchange between the three systems and it is said that the arteries and nerves are nourished by the veins, whereas the veins receive natural heat and vital pneuma from the arteries.[194]

Galen's definition of the pulse as 'a movement contained within the coats of the arteries'[195] is perfectly well captured by the Arabic term *ʿurūq ḍawārib*, i.e. 'pulsating blood-vessels', for the arteries, and 'non-pulsating' (*ghayr ḍawārib*) for the veins. During the expansion of the arteries blood from the veins flows into them and is then, by the heat of the heart, turned into a vaporous substance which provides the material for the production of the psychic pneuma. Through the reverse movement of contraction, the arteries are cleansed of residual vapours – a process which is compared to the exhalation by which the heart is emptied of smoky vapours.

(122) 69b10–70a5 The starting point for voluntary movement

Source: Nemesius, *De natura hominis* 27, ed. Morani, 88.3–13; tr. Sharples and van der Eijk in *Nemesius On the Nature*, pp. 158–9; *Fī Ṭabīʿa*, ed. Haji-Athanasiou in 'Le traité de Némésius', 166.2–11.

After the presentation of the three physiological systems, the compiler – once again – sets out to describe the main function of each system in detail. The passage on the brain, which is the present one, is rather short, whereas respiration and digestion are dealt with at great length in a number of subsequent passages. The discussion of the brain and the instruments of motion, for example nerves and tendons, is taken from Nemesius'

[193] On the discussion of this issue, see Sharples and van der Eijk in *Nemesius On the Nature*, p. 151, n. 766; Nutton, *Ancient Medicine*, p. 234.

[194] On the anastomoses of arteries and veins in Galen, see Temkin, 'Pneumatology', p. 157.

[195] See Nutton, *Ancient Medicine*, p. 232.

chapter on voluntary movement. This exposition, again, draws heavily on Galen.

(123) 70a6–12 The living being needs respiration

Source: Galen, *UP* VI, 1, ed. Helmreich, I, 301.3–12; ed. Kühn, III, 412, 7–16; tr. May in *Galen On the Usefulness*, p. 280; *K. fī Manāfiʿ* VI, Bibliothèque Nationale de France, MS Arabe 2853, fol. 95a1–95a5; John Rylands Library, MS Arabic 809, fols. 96a23–b1; National Library of Medicine, MS A 30.1, fol. 65a2–6 and parallel fol. 76a17–22.

The basic description of respiration is taken from a passage in Galen's *K. fī Manāfiʿ al-aʿḍāʾ* VI, which explains respiration's impact on the heart, namely cooling it. Similarly to the previously mentioned double motion of the arteries, the heart undergoes expansion (*diastole*) and contraction (*systole*). During the first it attracts air, whereas it rids itself of waste products during the second.

(124) 70a12–71a2 If the living being lacks respiration

Source: Nemesius, *De natura hominis* 28, ed. Morani, 90.6–11, 90.15–25; tr. Sharples and van der Eijk in *Nemesius On the Nature*, pp. 162–4; *Fī Ṭabīʿa*, ed. Haji-Athanasiou in 'Le traité de Némésius', 169.17–170.6, 170.13–171.5.

The passage is taken from Nemesius' chapter on respiration and partly repeats what has just been stated in Galen's own words. Additionally, the lack of respiration is shown to result in death and it is therefore obvious that breathing must continue during sleep.

In line with the Aristotelian tradition,[196] three causes for respiration are presented. Its usefulness, i.e. the final cause, is the preservation and regulation of innate heat, and the provision of material for the production of the psychic pneuma. Its efficient cause is the power causing respiration, namely the psychic power. The term 'psychic power' must be understood in a much broader sense here than above in passage (115) – namely, to mean a power simply belonging to the soul. According to Ḥunayn's distinction of natural, vital and psychic powers, seen above, the power of respiration would range among the vital powers. As for the instruments or organs of respiration, i.e. its material cause, they are the thorax, the lungs,

[196] See Sharples and van der Eijk in *Nemesius On the Nature*, p. 163, n. 832.

the windpipe or the rough artery which corresponds to the trachea. They are all moved by the psychic power through the muscles.

(125) 71a3–13 The heart does not draw air

Source: Galen, *UP* VI, 2 and VI, 10, ed. Helmreich, I, 301.21–6, 303.3–10, 327.10–17; ed. Kühn, III, 413.6–11, 415.2–9, 448.13–449.2; tr. May in *Galen On the Usefulness*, pp. 280–1, 298; *K. fī Manāfi*ᶜ VI, Bibliothèque Nationale de France, MS Arabe 2853, fols. 95a9–12, 95b3–6, 102a20–23; John Rylands Library, MS Arabic 809, fols. 95b5–8, 23–5, 103b13–17; National Library of Medicine, MS A 30.1, fols. 65a10–13, 65b2–4, and, due to repetition in the MS, also at 76b1–4, 18–20, 82b18–21.

This passage on the interaction of the heart, lungs and thorax is taken from Galen. Nemesius does not have anything similar and the compiler of the *PR* obviously deemed it necessary to go into greater detail here. The incoming air passes through the larynx, and does not then directly reach the heart, but fills the lungs which operate as storage for the air. The idea that the lungs also function as padding for the heart is already expressed in Plato's *Timaeus* 70d3 and is referred to by Galen, but our compiler skips this reference as he does many others. The thorax, in turn, protects the lungs and heart, and its double motion of contraction and expansion moves the lungs which have no movement on their own. In the case of the lungs, contraction brings about the emergence of breath and voice, whereas expansion causes the attraction of air.

(126) 71a13–71b9 The cartilaginous part of the windpipe

Source: Nemesius, *De natura hominis* 28, ed. Morani, 90.25–91.3, 91.12–17; tr. Sharples and van der Eijk in *Nemesius On the Nature*, pp. 164–5; *Fī Ṭabīᶜa*, ed. Haji-Athanasiou in 'Le traité de Némésius', 171.5–8, 171.16–172.4.

The windpipe or rough artery consists of two structurally different parts, which act as an organ of the voice and an organ of respiration.[197] The air moves along the windpipe and through the larynx, the throat, mouth and nose. Towards the brain, the air is filtered through the ethmoid bone[198] in

[197] Galen discusses the topic in his *UP* VII, 4–6, which also includes the source of the following passage in the *PR*. See especially *UP*, VII, 6 (ed. Helmreich I, p. 384; ed. Kühn, III, p. 526): 'I have said that the cartilage of the rough artery is an instrument of the voice; that the membranous bands are instruments of respiration; that what both structures combine to form, namely the [rough] artery, is a part serving both respiration and the voice at the same time' (tr. May, *Galen On the Usefulness*, p. 341).

[198] Galen objects to the description of the bone as ethmoid (colander- or sieve-like) and

order to protect the brain from harm. This air which reaches the brain through the nose functions as the major nourishment for the psychic pneuma.[199]

(127) 71b9–72a11 The windpipe adapts the air

Source: Galen, *UP* VII, 5 and VII, 13, ed. Helmreich, I, 382.1–6, 411.19–412.12; ed. Kühn, III, 525.16–526.3, 566.3–567.5; tr. May in *Galen On the Usefulness*, pp. 340, 360–1; *K. fī Manāfiʿ* VII, Bibliothèque Nationale de France, MS Arabe 2853, fols. 118b18–20, 127a17–127b5; John Rylands Library, MS Arabic 809, fols. 120a10–12, 127b15–128a2; National Library of Medicine, MS A 30.1, fols. 102a13–5, 110a4–15.

This passage, taken from Galen's *K. fī Manāfiʿ al-aʿḍāʾ* VII, describes the production of voice, which is not dealt with in any detail by Nemesius. However, in his very basic account of the topic in ch. 14 of *De natura hominis*, Nemesius reproduces the Galenic simile and compares the function of tongue and uvula to that of a plectrum and the function of the palate of the mouth to that of a sound-board.

The same simile is also used to demonstrate why the glottis must be moist, adhesive and greasy. Galen gives examples for unnatural conditions in which the glottis is nevertheless dried out, namely high fever or blazing heat, but these are not taken up in the *PR*.

(128) 72a11–13 Animals possessing lungs

Source: Galen, *UP* VIII, 1, ed. Helmreich, I, 443.9–18; ed. Kühn, III, 611.9–18; tr. May in *Galen On the Usefulness*, p. 385; *K. fī Manāfiʿ* VIII, Bibliothèque Nationale de France, MS Arabe 2853, fol. 136b2–6; John Rylands Library, MS Arabic 809, fol. 136b19–23; National Library of Medicine, MS A 30.1, fol. 118a12–16.

The excerpt takes up a passage at the beginning of Book VIII of *De usu partium*, yet a quite similar passage occurs before, in Book VI. The passage in Book VIII stresses that animals without neck and thus without larynx and throat have no voice,[200] but this specification is lost in the Arabic version. The Arabic states that animals lacking lungs must necessarily also lack

prefers the comparison to a sponge the perforation of which is intricate and not straight (see Sharples and van der Eijk in *Nemesius On the Nature*, p. 165, n. 850).

199 See Temkin, 'Pneumatology', p. 156.

200 A similar statement is found in Aristotle's *DA* 421a4–6, where fish are said to be voiceless because they have no windpipe.

voice as they cannot store air by means of which voice may be produced, and do not channel air through the larynx where the voice emerges.

(129) 72b1–73a13 The first of the instruments of the nutritive faculty

Source: Nemesius, *De natura hominis* 23, ed. Morani, 83.17–84.18; tr. Sharples and van der Eijk in *Nemesius On the Nature*, pp. 146–50; *Fī Ṭabīʿa*, ed. Haji-Athanasiou in 'Le traité de Némésius', 157.3–158.4, 158.10–159.3.

The passage is taken from Nemesius, who, in turn, draws heavily on Galen[201] to describe the way the food is taken from the mouth, via the gullet or oesophagus,[202] to the stomach and the liver which concocts it into blood. Already Hippocrates has likened the function the stomach performs in animals to the one the soil performs for plants.[203] Blood coming from the liver is further refined by the spleen, gall bladder and kidneys, and then distributed via the veins to the whole body.

(130) 73b1–8 The spleen ... and the kidneys

Source: Nemesius, *De natura hominis* 28, ed. Morani, 92.8–16; tr. Sharples and van der Eijk in *Nemesius On the Nature*, pp. 166–7; *Fī Ṭabīʿa*, ed. Haji-Athanasiou in 'Le traité de Némésius', 173.5–8, 173.9–174.3.

Although spleen and kidneys have been mentioned in the previous passage, a further, more detailed description of their functions is extracted from the same source, i.e. Nemesius. The spleen's help for the stomach is reported, but its role in clearing the liver is omitted by the compiler of the *PR*. The kidneys are pictured as clearing blood and inciting sexual appetite. These two functions are obviously linked, as the sharp element which the liver draws from the blood is transported in the veins from the liver to the testicles where it causes a sensation which is compared to the itching of

[201] See especially *UP* IV, 1–2, ed. Helmreich, I, pp. 195–8; ed. Kühn, III, pp. 266–270; tr. May, pp. 204–6.

[202] Cf. *UP* IV, 1 (ed. Helmreich, I, 195.7–13; ed. Kühn, III, 266.7–267.5): 'The first and largest of the main thoroughfares extends from the mouth to the stomach [i.e. γαστήρ], as to a central storehouse established in the middle of the animal for the benefit of all the parts. Οἰσοφάγος [the oesophagus] is the special name for this avenue, which is commonly called στόμαχος; for στόμαχος is the general term for any narrow passage or isthmus, so to speak, leading to a cavity' (tr. May, p. 204).

[203] Cf. *De humoribus* 11, 1–2 (ed. Littré, V, 490.17) and the Arabic text in *Fī l-Akhlāṭ* (ed. Mattock, 23.5–6):

بمنزلة ما الأرض للنبات كذلك البطن للحيوان ، يغذو ويسخّن ويبرّد .

something under the skin and the resulting urge to scratch. The compiler of the *PR* abbreviates his source particularly at this point. The subsequent sentence from Nemesius[204] would surely have helped to make the statement clearer.

The role of the kidneys in arousing sexual appetite is similarly described by Galen in *UP* XIV, 9,[205] where the serous humour which the liver removes from the blood is said to be heated, especially when acrid humours collect under the skin, and then stimulate and provoke their own evacuation. Galen also provides the simile of itching and scratching, and an explicit explanation for the sensation of the pleasure which is felt, namely that warm pneuma is produced as well and consequently exhaled in the process.

(131) 73b9–74a5 The intestines

Source: Galen, *UP* V, 3, ed. Helmreich I, 253.9–26; ed. Kühn III, 345.11–346.9; tr. May in *Galen On the Usefulness*, pp. 247–8; *K. fī Manāfiᶜ al-aᶜḍāʾ* V, appendix 1, excerpt 2.

This list of the intestines is taken from Galen's *K. fī Manāfiᶜ al-aᶜḍāʾ* V, but also includes some information Galen provides before listing the names of the six intestines and thus rearranges the Galenic original (cf. appendix 1, excerpt 2). The Greek terms are all translated into Arabic expressions except for the colon, which is still called *qūlūn* in modern Arabic.

(132) 74a6–9 Hippocrates says

Source: ᶜAlī ibn Rabban al-Ṭabarī, *Firdaws*, ed. al-Ṣiddīqī, 34.12–14; ed. al-Jundī, 31.25–6; German tr. Siggel in *Propädeutische Kapitel*, p. 399.

Parallel: *K. al-Ḥikma*, MS Esad Efendi 1933, fol. 101b12–14.

Cf.: Hippocrates, *De genitura / De natura pueri* 6, ed. Littré VII, 478.6–7; *K. al-Ajinna*, ed. Lyons and Mattock, 39.9–16.

204 The sentence reads (174.3–6):
وبمقدار زيادة ليّن لحم الأنثيين على ليّن الجلد بذلك المقدار يزيد ما يعرض لهما من اللذع بالشيء الحادّ على ما يعرض منه〈ـه〉 في الجلد حتّى يحرّكا الشهوة لنقص المنى حركة يخرج منها الإنسان عن عقله .
This can be translated as follows: 'In as much as the flesh of the testicles is softer than the skin, so much more are the testicles affected by the burning of something sharp than the skin so that they set the desire to decrease the sperm in motion and by doing that man departs from his mind.'

205 Ed. Helmreich, II, 313.21–315.4; ed. Kühn, IV,180.10–182.7, tr. May, *Galen On the Usefulness*, pp. 640–1.

The source for this and the following passages is most probably ʿAlī ibn Rabban's *Firdaws*, which contains a good deal of rephrased physiological material by Hippocrates, Galen, Plato and Aristotle. The present statement about the determination of the embryo's gender is based on Hippocrates' *De genitura / De natura pueri*, which is preserved in Arabic under the title *K. al-Ajinna* (*On Embryos*) in 'a mixture of translation, paraphrase and comment'.[206] ʿAlī ibn Rabban shortens the Hippocratic passage which deals with male and female sperm individually, without any substantial loss of information.

The *PR*, as well as the *K. al-Ḥikma*, skips the following Hippocratic quotation in the *Firdaws*, which is still based on the same section of *K. al-Ajinna* and which claims that women may have only female offspring with one partner, but male offspring when changing their partner, and vice versa. The fact that both the *PR* and the *K. al-Ḥikma* omit this passage seems to indicate that they have a common source which is probably not the entire *Firdaws*, but rather a selection from it.

(133) 74a10–75a2 Aristotle says

Source: ʿAlī ibn Rabban al-Ṭabarī, *Firdaws*, ed. al-Ṣiddīqī, 34.17–35.1; ed. al-Jundī, 31.28–32.7; German tr. Siggel in *Propädeutische Kapitel*, pp. 399–400.

Parallel: *K. al-Ḥikma*, MS Esad Efendi 1933, fols. 101b14–102a1.

Cf.: Aristotle, *De generatione animalium* 763b20–766b26; *Fī Kawn al-ḥayawān*, ed. Brugman and Drossaart Lulofs, pp. 134–42.

ʿAlī ibn Rabban's source for the Aristotelian material he presents dealing with the determination of a child's gender and the abundance or lack of offspring seems to be mainly *De generatione animalium* IV. The material is taken up completely in the *PR*, passages (133–5) and the *K. al-Ḥikma.*

In *GA* 763b20–766b26, Aristotle discusses three different explanations for whether the embryo develops as male or female, which are ascribed to Empedocles (heat or cold of the uterus), Democritus (prevalence of which parent's sperm) and an unspecified group of people (embryo's being in the right or the left of the uterus). In contrast to ʿAlī ibn Rabban's presentation, Aristotle does not opt for any of these explanations exclusively. Yet he claims that the right side of the body is hotter than the left and that the male is hotter than the female. He also stresses the difference between the male and the female sexual organs without providing any detail. ʿAlī ibn

[206] See Lyons and Mattock in their edn of *K. al-Ajinna*, p. ii.

Rabban's description of the sexual organs (protruding versus hollow) may actually derive from Aristotle's *De historia animalium* 493b2–4. The difference between male and female with regard to the cessation of the sexual powers and to voice are mentioned, for example, in *HA* 585b2–5 and *GA* 786b16–22.

(134) 75a2–11 He says

Source: ʿAlī ibn Rabban al-Ṭabarī, *Firdaws*, ed. al-Ṣiddīqī, 35.1–9; ed. al-Jundī, 32.7–13; German tr. Siggel in *Propädeutische Kapitel*, p. 400.[207]

Parallel: *K. al-Ḥikma*, MS Esad Efendi 1933, fol. 102a2–9.

Cf.: Aristotle, *GA* 766b26–767a35; *Fī Kawn al-ḥayawān*, ed. Brugman and Drossaart Lulofs, pp. 142–3.

In this passage ʿAlī ibn Rabban clearly summarises *GA* IV, 2, where all aspects of his presentation are present. The explanation Aristotle provides for the effect of the south wind on procreation is slightly different. He claims that it makes the bodies more liquid, so they produce more residues, which are harder to concoct, and thus the sperm remains more liquid. He also gives more detail of the shepherds' observation – namely, not only do the winds matter, but also whether the animals copulate facing north or south.[208]

[207] The same *Firdaws* passage is also quoted by Saʿīd b. Hibat Allāh (d. 1101) in his *Khalq al-Insān* (20.14–20), yet without any reference to ʿAlī b. Rabban:
وأرسطوطاليس يقول : إن من علل الذكر والأنثى هبوب الرياح لأنّ الجنوب ترخي الأبدان وتذيب الزرع وتخرجه رقيقاً مائياً غير نضج ، والشمال تصلب الأبدان وتمنع الحرارة من الانتشار ويخرج المني وقد نضج ، والرعاة يعرفون ذلك من فعل الرياح في نتاج أغنامهم . ولذلك صار المشايخ والغلمان يولدون الإناث على الأكثر ، والشباب يولدون الذكور على الأكثر لقوّة حرارة الشباب وضعف حرارة المشايخ والغلمان . وبالجملة فإن الأمر كلّه ... [editor's omission due to textual loss in his model] بالاعتدال ، فإنّ الحرارة الشديدة تحرق الزرع والحرارة الضعيفة تعجز عن إنضاجه .

[208] In his *Tadbīr al-ḥabālā wa-l-aṭfāl wa-l-ṣibyān* (88.12–13), the tenth-century author Aḥmad ibn Muḥammad ibn Yaḥyā al-Baladī (on whom, see Ullmann, *Medizin*, pp. 146–7) also quotes Aristotle's opinion that the south wind causes female offspring, the north wind male offspring. Interestingly, he quotes Aristotle in the words of Nicolaus (to which the edited *S-y-q-ā-l-ā-w-s* must certainly be emended) from his summary of *De philosophia Aristotelis* (*wa-qāla Nīqālaws fī kitābihī fī khtisārihī li-falsafat Arisṭūṭālīs*). As the *Firdaws* slightly deviates from the original Aristotelian passage, Nicolaus may well be its source (see above, p. 21, n. 39).

(135) 75a11–75b6 He says

Source: ʿAlī ibn Rabban al-Ṭabarī, *Firdaws*, ed. al-Ṣiddīqī, 35.9–15; ed. al-Jundī, 32.13–18; German tr. Siggel in *Propädeutische Kapitel*, p. 400.

Parallel: *K. al-Ḥikma*, MS Esad Efendi 1933, fol. 102a9–15.

Cf.: Aristotle, *PA* 651b13–17; *Fī Aʿḍāʾ al-ḥayawān*, ed. Kruk, 35.21–36.2; *GA* 727a32–4, 771a17–b14; *Fī Kawn al-ḥayawān*, ed. Brugman and Drossaart Lulofs, 38.7–8; 154.1–155.6.

For this section, ʿAlī ibn Rabban seems to merge Aristotle's claim that fat creatures produce less sperm (*GA* 727a32–4, also *PA* 651b13–17) with his discussion as to why some animals produce many young, but others only one (*GA* 771a17–b14). He does not, however, present Aristotle's explanation for the first claim, namely that fat and sperm are two different forms of concocted blood. So if that blood is taken up as fat, there is not much of it left to produce sperm.

Unlike the shepherds of the previous passage, the farmers appearing here have no parallel in the Aristotelian sources.

(136) 75b7–75b10 Hippocrates says

Source: ʿAlī ibn Rabban al-Ṭabarī, *Firdaws*, ed. al-Ṣiddīqī, 35.16–19; ed. al-Jundī, 32.19–21; German tr. Siggel in *Propädeutische Kapitel*, p. 400.

Parallel: *K. al-Ḥikma*, MS Esad Efendi 1933, fol. 102a15–18.

Cf.: Hippocrates, *Aphorismi* II, 44, ed. Littré IV, 482.11–12; *K. al-Fuṣūl*, MSS Fatih 3565, fol. 8b13–14; Aya Sofya 3724, fol. 4b12; Raghib Pasha 1482, fol. 164a7.[209]

The Marsh manuscript, as well as the *K. al-Ḥikma*, is still drawing on the *Firdaws* for this and the following passages attributed to Hippocrates. ʿAlī ibn Rabban most probably elaborates on one of Hippocrates' *Aphorismi* here. The topic is linked to the previous Aristotelian excerpt on fat people. However, the explanation ʿAlī ibn Rabban presents for the short life expectancy of fat people must derive from another source.

[209] I have not been able to get hold of Tytler's 1832 edition of the *K. al-Fuṣūl* and the aphorism is not contained in al-Yaʿqūbī's selection of aphorisms in his *Taʾrīkh*. The Arabic reads with slight variations in three Istanbul MSS:
من كان بدنه سميناً غليظاً جدّاً بالطبع فالموت إليه أسرع منه الى القضيف .

(137) 75b10–76b3 He says

Source: ʿAlī ibn Rabban al-Ṭabarī, *Firdaws*, ed. al-Ṣiddīqī, 35.21–36.13; ed. al-Jundī, 32.23–33.7; German tr. Siggel in *Propädeutische Kapitel*, p. 401.[210]

Parallel: *K. al-Ḥikma*, MS Esad Efendi 1933, fol. 102a18–b15.

Cf.: Aristotle, *GA* 765a3–6, 771b15–18, 768b15–36; *Fī Kawn al-ḥayawān*, ed. Brugman and Drossaart Lulofs, 137.13–15, 155.9–10, 147–8; Hippocrates, *De genitura / De natura pueri* 9–10, ed. Littré, VII, 482.3–484.15; *K. al-Ajinna*, ed. Lyons and Mattock, 42.7–12, 43.5–44.3, 45.2–5. ʿArīb ibn Saʿīd, *K. Khalq al-janīn*, ed. Abdelkader and Jahier, 24.25–8, 26.26, 32.10–11, 31.28–32.2, without parallel to the last section.

The *PR* and the K. *al-Ḥikma* skip a short statement in their assumed source, the *Firdaws*,[211] which states that Hippocrates and Aristotle agreed in likening the development of man to that of trees and animals. Therefore they also do not pick up the opposition expressed by ʿAlī ibn Rabban, who explicitly claims that the following passage is by someone other than Hippocrates and Aristotle. Thus in the *PR* and the *K. al-Ḥikma*, the excerpt becomes implicitly attributed to Hippocrates. As pointed out above, Aristotle himself (*GA* 765a3–6) does not ascribe the doctrine that the male comes from the right, the female from the left, to any particular group. However, this assumption was quite widespread among Hippocratic writers and the strict schematism of the present passage seems to indicate a later period. The same statement appears, for example, in the *K. Khalq al-janīn* of the tenth-century author ʿArīb ibn Saʿīd, who presents it anonymously and omits the case of female gender.[212]

The fact that some animals are able to conceive many offspring from a single copulation is addressed by Aristotle in a passage in *GA* 771b15–18,[213] which follows the one taken up above, passage (135). In *GA* IV, 3, Aristotle also addresses different kinds of monstrosities, i.e. nature's departure from what an embryo should look like; the first degree of this is actually

[210] The second half of this *Firdaws* passage is also taken up, although in rephrased form, in Saʿīd b. Hibat Allāh's *Khalq al-Insān*, ch. 31 (124.11–20).

[211] If the *PR*'s source had been a shortened version of the *Firdaws*, the omission could have already happened there.

[212] On ʿArīb ibn Saʿīd, see Ullmann, *Medizin*, pp. 139–40. The passage in his *K. Khalq al-janīn* (24.25–8) reads:
وقد قيل أيضاً إن من المنى ما إذا نزل من يمين بدن الرجل إلى يمين رحم المرأة كان الولد ذكراً وإن جرى من يسار الرجل إلى يمين الرحم كان الولد أنثى قويّة مذكرة وإن جرى من يمين الرجل إلى يسار المرأة كان الولد ليناً مؤنثاً .

[213] The Arabic version of Aristotle (ed. Brugmann, Drossaart Lulofs, 155. 10) uses the term *sifād* instead of *jimāʿ* in the *PR*.

the female embryo, and the second the child who does not look like its parents. The irregularity of certain body parts may be caused by too great a quantity of semen which cannot be concocted and formed properly, as Aristotle explains by comparing the situation to athletes eating too much food (768b15–36). This may have been the trigger for ʿAlī ibn Rabban to mention oversized and deficient body parts.[214]

For the second half of the passage discussed at present, ʿAlī ibn Rabban seems to have turned back to his Hippocratic source, the *K. al-Ajinna*, but without indicating it. There, when addressing the reasons for children being born weak and small, Hippocrates first mentions the shortage of nutriment in the womb and then the constriction of the womb, which he illustrates by a small bottle in which a cucumber may not grow to its normal size. Hippocrates also mentions that an embryo may be deformed if the womb is too narrow, as can be observed in plants growing between stones which become twisted and in some places thin, in others thick.

ʿArīb ibn Saʿīd in his *K. Khalq al-janīn* also explains the weakness and smallness of an embryo's body by the constriction of the womb or the shortage of food, for which he is most probably relying on the *Firdaws*, as his wording is very similar. He also illustrates his statement by rephrasing ʿAlī ibn Rabban's example of the citron in a bottle.

(138) 76b4–77a8 He says

Source: ʿAlī ibn Rabban al-Ṭabarī, *Firdaws*, ed. al-Ṣiddīqī, 45.6–20; ed. al-Jundī, 39.2–13; German tr. Siggel in *Propädeutische Kapitel*, p. 412.[215]

Parallel: *K. al-Ḥikma*, MS Esad Efendi 1933, fols. 102b15–103a10.

The *Firdaws*' account of the roundness of the head which is taken up by the *PR* and the *K. al-Ḥikma* does not contain any reference to possible sources. However, ʿAlī ibn Rabban may have merged three distinct texts dealing with the formation of the embryo's head within the womb, the physiognomic indications of a head's shape and the head's spherical form

[214] *K. Khalq al-janīn* (32.10–11) has a passage which may well draw on the *Firdaws*, yet is closer to the Aristotelian wording as it reads 'the excess and lack of body parts of the embryo':

وأمّا زيادة أعضاء الجنين ونقصانها فيكون بما يجري إلى تلك الأعضاء من كثرة مادّة المنى أو قلّته والله أعلم .

May this be the original reading of the *Firdaws* and thus strengthen the assumption of ʿAlī ibn Rabban's use of Aristotle?

[215] Also this *Firdaws* passage is found in the *Khalq al-Insān* by Saʿīd b. Hibat Allāh, ch. 22 (81.2–9), yet again without reference to ʿAlī ibn Rabban.

in imitation of the heavenly sphere. The last discussion is probably inspired by the Platonic *Timaeus* 44d.

(139) 77a9–12 Hippocrates says

Source: some reworking of Hippocrates, *Aphorismi* V, 42, ed. Littré, IV, 546.4–5; *K. al-Fuṣūl*, MSS Fatih 3565, fol. 23a15–b1; Aya Sofya 3724, fol. 12b14–15; Raghib Pasha 1482, fol. 175b19–21.[216]

Parallel: al-Majūsī, *Kāmil* I, 3.34, 121.18–20.

Cf.: al-Baladī, *Tadbīr al-ḥabālā*, ed. Qāsim Muḥammad, 120.7–23.

The Hippocratic indications for the embryo's gender seem to have been quite popular in ancient times as well as in the Islamic world. The present passage is an elaboration of *Aphorismi* V, 42, which claims that a good colour in a woman with child is a sign for her child to be male, a bad colour a sign for it to be female. This is even stated by Aristotle in his *HA* 584a12–15, and in its Arabic version where we also find, in addition to the Greek, the mention of slow walking as an indication of women with female child.[217] The indicative colour is also mentioned by ʿAlī ibn Rabban in his *Firdaws* (ed. al-Ṣiddīqī, 37.15–18; ed. al-Jundī, 33.29–34.2), and explained as caused by the heat of the male child or the cold of the female. As for the size of a woman's breasts, Hippocrates alludes to its indicative value only indirectly, stating in *Aphorismi* V, 38,[218] that the loss of fullness in the right breast in a woman with twins shows she will lose the male child. Al-Majūsī

[216] The *K. al-Fuṣūl* reads as follows:
إذا كانت المراة حبلى بذكر كان لونها حسناً وإذا كانت حبلى بأنثى كان لونها حائلاً .

[217] The passage in *Ṭibāʿ al-Ḥayawān* (470.1–4) reads:
وإذا كان المحمول ذكراً ، تكون الأنثى الحامل أحسن حالاً ولونها أجود ، إلى أن تلد . وإذا كان الحمل أنثى ، يعرض للحامل خلاف ذلك ، أعني أنها تكون رديئة اللون ، سيئة الحال ، بطيئة الحركة ، وتعرض لساقيها أورامٌ وقروح .

Al-Baladī (121.1–6) also lists the complexion of the pregnant woman among the indications which Aristotle has given for the gender of the child (other indications are an easy deliverance in the case of a male child and swollen feet in the case of a female). He quotes Aristotle, once again (see above, p. 425, n. 208) in the words of Nicolaus from the eight chapter of his book *De philosophia Aristotelis*.

[218] The *K. al-Fuṣūl* reads (in the three Istanbul MSS, Fatih 3565, fol. 23a5–9; Aya Sofya 3724, fol. 12b7–10; Raghib Pasha 1482, fol. 176a3–6) as follows:
إذا كانت المراة حاملاً فضمر أحد ثدييها وكان حملها توأماً فإنها تسقط أحد طفليها فإن كان الضامر هو الثدي الأيمن أسقطت الذكر وإن كان الضامر هو الأيسر أسقطت الأنثى .

In Galen's *K. fī Manāfiʿ* the aphorism is quoted as well (Bibliothèque Nationale de France, MS Arabe 2853, fol. 246b321–3):
وكذلك قال بقراط في الموضع الذي قال فيه إنّ المراة الحامل إن تضمر ودقّ أحد ثديهما وهي حامل توأم فإنّها سقط إحداهما وان كان الثدي الذي دقّ وتضمر التدي اليمين اسقطت الذكر منهما وإن كان الثدي الأيسر فالأنثى وهذا القول موافق لقوله حيث نقول .

produces exactly the same list of indications as the *PR* in his *K. Kāmil al-Ṣināʿa al-ṭibbīya*, although without an explicit ascription to Hippocrates.[219] The two texts may thus have a common source. Al-Baladī presents colour as a criterion and even mentions his exact source for this information, i.e. Hippocrates in the fifth book of his *Aphorismi*, the *Fuṣūl*. ʿArīb ibn Saʿīd has a different list which starts off with exactly the same Hippocratic aphorisms and provides an explanation for the appearance of these signs.[220]

(140) 77a13–77b1 He says

Source: Galen, *UP* XIV, 4, ed. Helmreich II, 293.2–4; ed. Kühn IV, 153.11–13; tr. May in *Galen On the Usefulness*, p. 626; *K. fī Manāfiʿ* XIV, Bibliothèque Nationale de France, MS Arabe 2853, fols. 246b21–247a1; John Rylands Library, MS Arabic 809, fol. 249a13–17; Escorial, MS Arab. 850, fol. 91b22–3; National Library of Medicine, MS A 30.1, fol. 205a21–4.

Cf.: Hippocrates, *Aphorismi* V, 48, ed. Littré, IV, 550.1–2; *K. al-Fuṣūl*, MSS Fatih 3565, fols. 23b14–24a2; Aya Sofya 3724, fol. 13a2–3; Raghib Pasha 1482, fol. 175b21–2.[221]

In the *PR* this Hippocratic aphorism is phrased exactly as it is in the Arabic version of Galen's *De usu partium*, which suggests that this was our com-

219 His last reference (120.24–5) is, however, to Hippocrates' *De octimestri partu*, to which he refers by the title *K. fī al-Janīn al-mawlūd li-thamāniya ashhur*.

220 Al-Baladī's passage in *Tadbīr al-ḥabālā* (120.7–23) reads:
قال ابقراط في الخامسة من كتاب الفصول إذا كانت المرأة حبلى بذكر كان لونها حسناً وإن كانت حبلى بأنثى كان لونها حائلاً وقال في فصل آخر إذا كانت المرأة حاملاً فضمر أحد ثدييها وكان حملها توأماً فإنها تسقط أحد طفليها فإن كان الضامر هو الثدي الأيمن أسقطت الذكر وإن كان الضامر هو الثدي الأيسر سقطت الأنثى . وقال في فصل آخر من كان من الأطفال ذكراً فأحرى أن يكون تولده في الجانب الأيمن وما كان أنثى ففي الجانب الأيسر فدل ابقراط بقوله هذا أنّ من كان لونها من الحبالى حسناً صافياً فإنّ حملها بذكر ومن كان لونها حائلاً كدراً غير حسن فإنّ حملها بأنثى فإنّ من كان حملها في الجانب الأيسر كان حملها بأنثى ودليل الحمل في الجانب الأيمن والأيسر أن يكون الثدي في ذلك الشقّ كما قال ابقراط أعظم ولا سيما جملته إن تكون العروق التي في ذلك الشقّ أيضاً أشدّ امتلاء من عروق الشقّ الآخر ولا سيما العروق التي تحت اللسان . ومن علامات الحمل بالذكور فإنّ الحبالى بهن أسهل حركات وأخفّ نهضات وإنهّن نشيطات وأصحّ الدلائل وأقواها وأسلمها من الغلط في ذلك أنّ بعض العروق في الحبالى بالذكور متوتراً ممتلئاً وارتكاض الجنين وحركته ومن العلامات الدالّة على الحمل بالأناث فإنها تكون بخلاف العلامات في الذكور في جميع الأحوال .
ʿArīb ibn Saʿīd, *K. Khalq al-janīn* (24.19–24) has:
وإن كان لون الحامل حسناً وماء وجهها صافياً وحركتها خفيفة ونفسها نشيطة ذكية دلّ على أن المولود ذكر . وإن كان اللون متغيّراً والبشرة غير نقية والكلف كثيراً والحركة بطيئة والحواسّ بليدة دلّ على الأنثى لأنّ الذكر أقوى حرّاً والحرارة تحسن اللون وترقّ الدم وتذكي الحواسّ وتعين على خفّة الحركات ، والأنثى أميل إلى البرد والبرودة تفتح اللون وتجمد الدم وتخضر لونه وتفسد البشرة منه .

221 In the three manuscripts the text reads as follows:
ما كان من الاطفال ذكرا فاحرى ان يكون يولدهم في الجانب الايمن وما كان انثى من الجانب الايسر .

piler's source. The aphorism is also quoted, although phrased slightly differently, by al-Majūsī and al-Baladī.[222]

(141) 77b2–78b8 Hippocrates says

Source: unidentified; some reworking of Hippocratic and Galenic accounts of embryonic development?

Parallel: al-Majūsī, *Kāmil* I, 3.34, 117.19–21 (partially repeated in 118.21–2, which is quoted in the apparatus for readings which are closer to the *PR*), 117.33–118.1, 118.4, 6–7, 10–12, 15–17, 18–19, 23–7, 119.9–11, 12–16, 19–20, 23–4, 26, 30, 120.14–16, 121.10–11.

Cf.: Hippocrates, *De genitura / De natura pueri* 12, ed. Littré, VII, 488.10–21; *K. al-Ajinna*, ed. Lyons and Mattock, pp. 45–8; Hippocrates, *De alimento* 42, ed. Littré, IX, 112.12–114.3; *K. al-Ghidhā*ʾ, ed. Mattock, 17.6–12; Hippocrates, *Aphorismi* V, 48 as above.

The ultimate source of the first part of the passage is probably Hippocrates' account in *De genitura / De natura pueri*, which talks about the mingling of the male and female sperm in the womb, its thickening because of the womb's heat, the pneuma which finds a path through the sperm and the formation of a membrane which is likened to the outermost layer of baked bread.[223] The day count may be based on Hippocrates' *De alimento*, which presents four different calculations which do not distinguish between male and female embryos. The second of them is taken over in our passage for the female embryo, the last one for the male.[224] The end of

[222] See end of the next passage (141), where the *PR* repeats the statement in a different version, and for al-Baladī, above p. 430, n. 220.

[223] The Hippocratic passage reads in the Arabic version (*K. al-Ajinna*, 48.6–10):
فنقول إنّه من بعد ما اختلط مع الرطوبة فلذلك يصير منه حجاب ولانّه أيضاً محبوس ممسك مثل ما يكون على الخبز إذا ما نضج فانّه يكون عليه خارجاً شيء رقيق شبيه بالحجاب وذلك لأنّه يسخن وتكثر فيه الحرارة وينتفخ منها ويرتفع وما يرتفع منه هو الحجاب .
And Lyons' translation (p. 9): 'We say that after it [i.e. the sperm – EW] is mixed with moisture, there is formed from it a membrane, and because it is also imprisoned and held, like what is on bread when it is baked, there is upon it, outside, a thin thing like a membrane. This is because it is heated, and heat becomes abundant in it, and it swells and rises because of it and that part of it which rises is the membrane.'

[224] The Arabic version in *K. al-Ghidhā*ʾ (17.6–12) reads:
انّ الجنين تتمّ خلقته في خمسة وثلاثين شمساً ويتحرّك في سبعين ويتمّ في مائتي وعشرة وقال آخرون انّ خلقة الجنين تتمّ في خمسة وأربعين يوماً ويتحرّك في تسعين ويخرج في مائتي وسبعين وقال آخرون انّ خلقته تتمّ في خمسين ويتحرّك في مائة ويتمّ في ثلاثمائة وقال آخرون انّ صورة الجنين تتمّ في أربعين ويتحرّك في ثمانين ويخرج في مائتين واربعين .
The same passage is quoted with some variations in Galen's *Maqāla fī l-Mawlūd li-sabʿa ashhar*, 335.3–9. The idea that the female embryo needs longer to develop than the

the passage seems to be inspired, once again, by *Aphorismi* V, 48. The combination, reworking and elaboration of these Hippocratic sources are attested by the *PR* as well as by al-Majūsī. The textual material preserved in al-Majūsī is introduced with the words 'Hippocrates and Galen agree that', which may suggest that we are actually dealing with a Galenic reworking of Hippocrates.

However, the introductory statement that the male sperm functions as matter and agent whereas the female sperm functions only as matter seems to imply the Aristotelian theory of the male sperm acting as form and the female as matter, which is refuted by Galen.[225] Yet other aspects of the passage have a distinct Galenic ring.[226]

The passage in the *K. Kāmil al-Ṣināʿa al-ṭibbīya* is substantially longer than the parallel account in the *PR*, but this is probably due to the tendency of our compiler of the *PR* to shorten his source texts by omitting sentences.

(142) 78b8–83b11 The wise men say

Source: *Kitāb/Rasāʾil Ikhwān al-Ṣafāʾ*, II, 421.2–17, 422.8–11, 434.19–20, 422.11–16, 436.6–7, 422.16–423.11, 423.22–425.4, 425.15, 439.24–440.2, 425.15–20, 441.7–10, 425.20–426.22, 451.20–452.1, 427.1, 446.2–8, 446.14–447.10, 447.15–18, 447.22–448.4, 448.7–10, 449.3–11, 448.11–19; German tr. Dieterici in *Anthropologie*, pp. 67–96.

This long section is taken from the 25th Epistle, entitled *Masqaṭ al-nuṭfa* (*On*

male depends on the assumption that the male is hotter, and is therefore widespread. See, e.g., Yūḥannā ibn Māsawayh's *Maqāla fī l-Janīn wa-kawnihī fī l-raḥim* (122.13–124.3).

225 See *De semine* I, 5, 8–17 (ed. and tr. De Lacy, pp. 80–3).

226 Cf. *De semine* II, 4, 19: the female semen 'being moister and colder, whereas that of the male is thicker and hotter' (tr. De Lacy, p. 177).
Cf. *De semine* I, 6, 4 – 7, 7: 'And presently this membrane was completely hard and continuous, surrounding the whole fetus on all sides, pierced only by those passages through which there was movement of the materials. For since there was at no time a respite in the pull, the perforation of the membrane always remained open; it could not grow together because nothing else either can grow together, one part with another, when constantly moving bodies stand in between. And not only does the perforation of the membrane fail to grow shut; it grows even wider the whole time, in proportion to the quantity of the matter that flows through it' (tr. De Lacy, pp. 85–7).
Cf. *De semine* I, 4, 34–8: 'This is Hippocrates' testimony on the point that the semen remains inside and has a membrane ... and in the company of warm bodies it easily forms a membrane, like those on the outside that bakers make by gently pouring moist dough on a hot flat pan. ... the membrane of the semen ... is separated off from the uterus, like the cake from the bronze pan' (tr. De Lacy, p. 79).
For the formation of the three principal organs and what follows, cf., e.g., *De foetuum formatione* III, 8 (ed. and tr. Nickel, pp. 66–7); III, 21 (pp. 72–3); and VI, 28 (pp. 102–3).

insemination), of the encyclopaedic *Kitāb* or *Rasāʾil Ikhwān al-Ṣafāʾ*. Whether the compiler of the *PR* had access to the entire encyclopaedia or whether he relied on a separate copy of this particular treatise cannot be known for certain. The lack of any further quotations in the *PR*, which can only have been taken from the *Ikhwān al-Ṣafāʾ*, seems to suggest the latter.[227]

The *Ikhwān al-Ṣafāʾ* claim that the period which elapses from the moment of insemination to the birth of the child is normally 240 days, i.e. 8 months. In the first 4 months, the structure of the embryo's body is established and the powers of the vegetative soul start acting in it. In the second set of 4 months, the embryo's body is completed and prepared for the activities of the animal soul which become apparent in it. The four elements play an important role in this development as their natures influence the embryo, starting with the nature of the element assigned to the sign of the zodiac where the sun was at the time of insemination and then changing to another nature each month until the embryo has received the influence of each nature twice.[228] After birth it takes 4 further years, in which the body is perfected, before the rational soul enters it.[229] As the months of the embryo's time in the womb are under the influence of different planets, so are the years of a person's life. These two sequences of heavenly influences are also found in ʿArīb ibn Saʿīd's *K. Khalq al-Janīn* (pp. 38–9, 85–7), where they are presented in the name of the astrologers, and the year count for the planets' influences deviates slightly.

The day count of 240 days for the embryo's development agrees with that of the previous passage, and the seven vegetative powers or faculties are in accordance with Hunayn ibn Isḥāq's presentation quoted above, passage (115). In his *Masāʾil fī l-Ṭibb* Hunayn lists four powers which serve the nutritive faculty, namely the attractive, retentive, digestive and repellent, and which the compiler omitted in his quotation. Ḥunayn's 'augmentative (*murabbiya*) power' equals the *Ikhwān*'s 'growing (*nāmiya*) one', and his 'reproductive (*muwallida*) power' their 'formative (*muṣawwira*) one'.

[227] However, the attribution of the passage to 'the wise men (*al-hukamāʾ*)' may have something to do with the ascription of the next, 26th Epistle to the wise men in the title, which reads *Fī l-Qawl al-Ḥukamāʾ inna l-insān ʿālam ṣaghīr*.
There is one further parallel to the *Ikhwān al-Ṣafāʾ* in the *PR*, in passage (167), which contains various sayings on vision and hearing. In the *Rasāʾil Ikhwān al-Ṣafāʾ* these and other sayings are quoted under the heading *Fī Nawādir al-falāsifa fī l-mūsīqā*. The source of the *PR* may, therefore, be the original collection of sayings or their quotation in al-Kindī. See commentary on passage (167).

[228] Since Antiquity the twelve zodiac signs have been divided into four triplicities, i.e. groups of three signs, each group linked to one of the four elements.

[229] See *Rasāʾil*, II, pp. 428–9.

For short explanations of the astrological terms used in this passage al-Qabīṣī's *K. al-Mudkhal ilā ṣināʿat aḥkām al-nujūm* seems appropriate as it was composed in the tenth/eleventh century as a summary of the common astrological tradition.

(143) 83b12–84b4 He says

Source: Galen, *UP* XIV, 2 and 4, ed. Helmreich II, 285.9–286.12, 291.1–292.6; ed. Kühn IV, 143.7–144.19, 151.2–152.11; tr. May in *Galen On the Usefulness*, pp. 621–2, 625; *K. fī Manāfiʿ al-aʿḍāʾ* XIV, appendix 1, excerpt 3.

The excerpt combines two sections taken from Galen, *UP* XIV, which address the topic of the continuity and immortality of nature's work, which becomes God's creation in the Arabic version. The simile of the good founder of a city who tries to ensure the longevity of his foundation is omitted by the compiler of the *PR*. Procreation, which is nature's or God's way of preserving creation not individually, but by species, is shown to be the best possible design. This is one of the many 'arguments from design' which Galen presents to convince his readers of the existence of providence and a wise Creator. The Galenic text is phrased to refute Sophistic objections to the claim of a divine plan, but this phrasing is not kept by the compiler of the *PR*.

(144) 84b5–85a8 If the body parts always remained in the same condition

Source: Galen, *UP* XVI, 14, ed. Helmreich II, 433.2–434.13; ed. Kühn IV, 339.13–341.16; tr. May in *Galen On the Usefulness*, pp. 721–2; *K. fī Manāfiʿ al-aʿḍāʾ* XVI, ed. Savage-Smith in *Galen On Nerves*, 103.3–14, 104.4–6, and tr., pp. 189–91.

Destruction, which has been shown to affect individuals within a species in the previous passage, is now dealt with on an individual level and thus presented as affecting the body of each single individual. The doctrine fits well with passage (119), where Ḥunayn claims the coldness of the brain is necessary to counterbalance the heat produced by the brain's multiple motions.

(145) 85a9–86a13 Qusṭā ibn Lūqā says

Source: Qusṭā ibn Lūqā, *R. fī l-Farq*, ed. Troupeau, lines 146–8, 54–6, 148–50, 141–3, 99–100, 102–14, 144–5; ed. Gabrieli with Italian tr. in 'Risālah', 641.3–5,

631.10–632.2, 641.6–8, 640.9–641.1, 636.6–8, 637.1–638.4, 641.2–3; ed. Cheikho in 'Risāla', 102.26–103.2, 99.16–17, 103.2–4, 102.21–2, 101.6–7, 101.10–23, 102.23–5; ed. Ülken in *Ibn Sīnā*, 88.22–4, 85.7–9, 88.24–6, 88.17–19, 86.25–87.1, 87.3–15, 88.20–2; and tr. Livingston in 'Qusṭā', pp. 65–8.

Cf.: *Jawāmiᶜ al-Maqāla al-thālitha min kitāb Jālīnūs wa-hiya fī tashriḥ al-ᶜaṣab*, ed. al-Dubayan in *Galen: "Über die Anatomie"*, 121.4–123.1.

The passage is put together from different sections of Qusṭā's treatise *R. fī l-Farq*. The presentation of the two spirits or two kinds of pneuma and their functions fits well with the Galenic statements in (118) and (120), and Nemesius' in (121). For the description of the seven pairs of nerves Qusṭā follows Galen closely.[230] With regard to the first, fourth, fifth, sixth and seventh pair of nerves the *PR* faithfully quotes Qusṭā, who followed Galen in listing what corresponds to the optic nerve, the auditory nerve, the nerve extending to the palate, a complex of the glossopharyngeal, accessory nerve and vagus, and the hypoglossal nerve.[231] As to the second and third pair of nerves, the *PR* differs from Qusṭā. In the *PR* Qusṭā's second pair, i.e. the oculomotor nerve, is replaced by his third pair, i.e. the sensory nerves in the tongue. The third pair of nerves which the compiler of the *PR* lists is the olfactory nerves. Galen does not regard the olfactory extensions as proper nerves and Qusṭā seems to be following him since he does not mention olfactory nerves.

Within the *PR* in its entirety this detailed presentation of the nerves serves as the physiological basis for the following discussion of the senses and sense perception. Within the present passage this description of the nerves is imbedded in a discussion of the ventricular localisation of the psychological faculties within the brain. The doctrine is implied in Galenic writings,[232] and explicitly expounded in Nemesius. However, the scheme presented by Qusṭā is slightly different from the latter's. In the front ventricles, where Nemesius places imagination, Qusṭā puts sensation. Both locate thought in the middle ventricle. As to the rear ventricle, Qusṭā follows Nemesius in placing memory there, but he also adds motion.

230 See esp. the *Jawāmiᶜ* in the translation of Ḥunayn as indicated above. For an account of the Galenic nerves and their correspondence to modern terminology, see Savage-Smith, 'Galen's Account', pp. 81–8, 173–88, whom I closely follow here.

231 A short omission occurs with regard to the fourth pair extending to the palate. The compiler of the *PR* chose to omit Qusṭā's addition that this pair causes the sensation of taste (*ṭaᶜm*), probably deeming it unnecessary as taste (*dhawq*) has already been described as resulting from the pair of nerves joined to the tongue.

232 See Rocca, *Galen*, pp. 245–7. Interestingly, further down, in passage (184), the *PR* quotes a text on love in the name of Galen in which the ventricular localisation is explicitly stated.

(146) 86b1–87a3 Galen says

Source: Galen, *De propriis placitis*, ed. Boudon-Millet and Pietrobelli with French tr., 179.30–4, 185.35–8, 186.3–8, 181.28–32, 186.1–3.

Cf.: Arabic-Latin translation in Nutton, *Galen*, 80.10–13, 102.14–17, 102.21–104.2, 88.9–15, 102.19–21.

The passage seems to be taken from a no longer extant Arabic translation of Galen's *De propriis placitis*[233] and deals once again with the three principal organs of the animal body. The information on spirit and soul which this quotation provides in addition to the previous one is that whereas the spirit is located in the ventricles of the brain, the soul resides in the mass of the brain. As to the heart, it produces heat in animal bodies which is much warmer than the heat in plants. The liver is said to be like the stem of a tree, which fits well with passage (129). There the same simile is used for veins described as growing out of the liver, which are compared to the roots branching out from a stem.

The knowledge of plants implicitly conveyed in this passage is a prerequisite for the following discussion of why plants do not have sensation.

(147) 87a4–8 Aristotle says

Source: *DA*-paraphrase, ed. Arnzen with German tr., 293.5–9; MS Aya Sofya 4156, fol. 89a1–8.

This and the following passages give an account of sense perception in general, and of each sense in particular, following the text of the *DA*-paraphrase, but bringing together statements dispersed throughout Book II.

In this first section the general principle of the function of the senses is described, which occurs in the *DA*-paraphrase twice, first when discussing the potentiality of the senses based on Aristotle's *DA* II, 5,[234] and second as a summary of the discussion of sense perception at the end of Book II of the paraphrase which is taken up in the *PR*.

[233] Ḥunayn seems to attest the existence of at least two Arabic translations (see above, p. 27). See Wakelnig, 'Fragments', where I discuss the entire passage in great detail and in comparison to the Greek and Latin versions.

[234] *Aristoteles' DA*, ed. Arnzen, p. 251, and see his commentary (*ibid.*, pp. 394–5) on the passage which draws on *DA* 417a21–418b6 and 426a15–19.

(148) 87a9–11 He says

Source: *DA*-paraphrase, ed. Arnzen with German tr., 265.11–13; MS Aya Sofya 4156, fol. 78a6–10.

The compiler of the *PR* starts his discussion of the individual senses with vision and thus follows the Aristotelian order. This expresses the high rank and value Aristotle attributes to vision among the senses, stated as well in passage (166).[235]

The quoted text of the *DA*-paraphrase is a summary of Philoponus' commentary on Aristotle's *DA* 419a11–20, and the stress on the process of vision happening without time is directed against Empedocles and maybe Stoic views as well.[236]

(149) 87a11–87b1 Voice is the stroking of air

Source: *DA*-paraphrase, ed. Arnzen with German tr., 269.16–271.1, 271.4–6; MS Aya Sofya 4156, fols. 81b8–10, 81b14–82a2.

In contrast to his discussion of vision, the compiler of the *PR* begins his account of hearing with a description of what can be perceived through hearing, i.e. voice. Interestingly, he does not give the more general definition of sound first, as does the *DA*-paraphrase, but only mentions voice which is 'sound with a meaning' and 'must be accompanied by an act of imagination' (*DA* 420b32–3, tr. Smith). Another point worth mentioning is that the *PR* agrees with the text of the Aya Sofya version of the *DA*-paraphrase on the rather vague reference to the 'instruments of voice', whereas the version edited by Arnzen specifies the windpipe and the larynx.[237]

235 Passage (166) discusses vision and hearing together, and passage (167), which does not derive from an Aristotelian source, stresses the priority of hearing to vision. In passage (160), containing material from Miskawayh's alternative *Fawz* which may or may not have been meant to fill the gap left blank in the *PR*, vision and hearing are ranked as the highest senses.

236 See Arnzen, *Aristoteles' DA*, p. 404.

237 If one accepts Sebti's assumption, that the version edited by Arnzen represents a later stage of the transmission of the *DA*-paraphrase, its specification 'windpipe and larynx' would have to be explained either by a later insertion (although the Philoponus commentary, which Arnzen indicates as a source for the passage, provides a number of anatomical details which may well be the origin of the mention of windpipe and larynx in the Arabic) or by being the original reading which has dropped out of the Aya Sofya version, which seems, indeed, not to be a very accurate copy. The Arabic translation of Aristotle's *DA* is not particularly specific either, cf. *Fī l-Nafs* (51.16–21):
وإنما يكون الصوت من ضارب ذي نفس مع توهّم . وذلك أن الصوت قرع 〈له〉 دليل على

(150) 87b1–5 The stroke comes to hearing

Source: *DA*-paraphrase, ed. Arnzen with German tr., 267.1–4; MS Aya Sofya 4156, fol. 78a13–b3.

Cf.: passage (169).

This passage occurs in the *DA*-paraphrase before the previous one and proves that hearing needs air to perceive sound. The *DA*-paraphrase also mentions that the sense of smell needs air, but this is not taken up by the compiler of the *PR* at this point. Strangely enough he quotes it later on when again discussing hearing based on some other source text in passage (169).

(151) 87b6–9 He says

Source: *DA*-paraphrase, ed. Arnzen with German tr., 273.15–275.2; MS Aya Sofya 4156, fol. 83a6–11.

Regarding the sense of smell, the compiler of the *PR* only excerpts the application of the general principle expounded in passage (147) to smell from his source text, i.e. the *DA*-paraphrase.

(152) 87b10–88a4 He says

Source: *DA*-paraphrase, ed. Arnzen with German tr., 275.4–12; MS Aya Sofya 4156, fol. 83a13–b6.

The difference of taste from the senses mentioned so far, i.e. vision, hearing and smell, is stated by Aristotle in *DA* II, 10, and in the *DA*-paraphrase on which the *PR* draws. The *DA*-paraphrase, following Philoponus, further likens taste to touch,[238] which is not taken up by the compiler of the *PR*. He thus avoids a contradiction with passage (154), where touch is shown to be mediated by air as well. He also omits the mention of the liquid which is introduced in the *DA* and its paraphrase as a quasi-medium: 'But nothing excites a perception of flavour without the help of liquid' (*DA* 422a17–18, tr. Smith), and 'The flavours belong to the liquids and the liquid is something touched and touchable' (my tr. based on ed. Arnzen, 275.15–6).

The example of the visible object which when placed directly on the

شيء ، ... لكنّه هو صدمة هواء التنسم هواءَ الوريد وجرم الوريد ... وبهذا يستبين لم كانت السمكة لا صوت لها ، لأنّه ليس لها حنجرة .

[238] See Arnzen, *Aristoteles' DA*, p. 408.

surface of the eye is imperceptible occurs in Aristotle, *DA* 419a12–13 and 423b21.

(153) 88a4–6 Next he describes it saying

Source: *DA*-paraphrase, ed. Arnzen with German tr., 277.9–11; MS Aya Sofya 4156, fol. 79b11–13.[239]

This is, once again, the application of the general principle of passage (147) to taste, quoting the *DA*-paraphrase which is based on Aristotle, *DA* 422b14–16.

(154) 88a8–13 He says

Source: *DA*-paraphrase, ed. Arnzen with German tr., 281.12–17;[240] MS Aya Sofya 4156, fol. 84b6–13.

This passage, demonstrating that touch is perceived through a medium of either air or water, is taken from the *DA*-paraphrase, which, in turn, draws heavily on Philoponus' commentary on Aristotle.[241]

(155) 88a13–b11 He says

Source: *DA*-paraphrase, ed. Arnzen with German tr., 285.16–287.4; 345.16–347.1; MS Aya Sofya 4156, fol. 86a15–b10; without parallel to the second part.

Here the compiler of the *PR* combines two distant sections of the *DA*-paraphrase, linking them by the phrase 'on account of that (*wa-min ajli dhālika*)'. His reason for combining them seems to be that in the *DA*-paraphrase the first passage, corresponding to *DA* 424a28–30 and taken up in the *PR*, is followed by a discussion of why plants lack the sense of touch. Our compiler replaces this discussion with the more general account, which occurs later on in the *DA*-paraphrase and corresponds to *DA* 434a-b, of why plants do not have sense perception at all.[242] This replacement is

[239] Here the Aya Sofya manuscript suffers from some disorder: up to 78b4 the parallel to Arnzen's edition is up to 267.5, then it suddenly changes to 255.6, which results in the following parallels: 78b4–79a14 ≈ 255.6–257.8; 79a14–80b4 ≈ 275.15–279.6; 80b4–83b8 ≈ 267.5–275.15; from 83b8 onwards, the MS corresponds to 279.6 and what follows.

[240] For some textual loss which Arnzen assumes for the beginning of the passage, see his *apparatus criticus* in Aristotle's *DA*, p. 281. The loss also occurs in the Aya Sofya version.

[241] See Arnzen, *Aristoteles' DA*, pp. 412–13.

[242] Interestingly, the Aristotelian text underlying the first section excerpted from the *DA*-paraphrase does not mention touch at all but continues after the discussion of the

justified since, according to Aristotle, without touch no other sense can exist (*DA* 415a3–4, 435a12–13) and touch is the primary form of sense (413b4–5). A further common characteristic of the two sections taken from the *DA*-paraphrase is the idea of 'balance'. Yet, the first section talks about the balance of the sense[243] and how the sense perceives its corresponding sensibles as covering the range between a certain pair of contraries,[244] whereas the second passages discusses the balance of the four elements within a well-functioning sense organ.

The impairment or destruction of the sense organs by extreme sensibles is taken up three times in the *DA* (424a28–32, 426a27–b3, 435b7–14),[245] and twice (424a32–b2, 435a21–b3) linked to the topic of the missing sense perception of plants, the latter of which is paraphrased in the second part of our passage in the *PR*.

(156) 88b11–89a7 He says

Source: *DA*-paraphrase, ed. Arnzen with German tr., 289.13–291.4; MS Aya Sofya 4156, fols. 87b11–88a5.

This passage of the *DA*-paraphrase shows that, on account of the absence or presence of matter, an affection qualifies either as sensation or not. It is an elaboration of the topic raised in the previous passage, namely why plants do not have sense perception.

(157) 89a8–89b4 He says

Source: Nemesius, *De natura hominis* 6, ed. Morani, 56.12, 56.6–11, 56.14–16; tr. Sharples and van der Eijk in *Nemesius On the Nature*, p. 102; *Fī Ṭabīʿa*, ed. Haji-Athanasiou in 'Le traité de Némésius', 102.15–16, 10–15, 16–19.

corruption of the sense organs by extreme sensibles with the statement 'This explains also why plants cannot perceive ... The explanation is that they have no mean, and so no principle in them capable of taking on the forms of sensible objects but are affected together with their matter' (*DA*, 424a32–b2, tr. Smith). Cf. *Fī l-Nafs* (60.15–18):
و〈هذا يفسر〉 النامية لم لا تحسّ ... والعلّة في ذلك أنّها لا تحسّ أنّه ليس لها تقدير التوسّط والاعتدال ، وليس فيها إمكان لقبول الصور المحسوسة ، لكنّها تألم مع الهيولى .
Thus the *PR* seems to be closer to the original Aristotelian text than is its actual source, the paraphrase.

243 Sense, according to Aristotle, is a ratio (λόγος) – see *DA* 426b3.

244 See *DA* 422b23–5: 'Every sense seems to be concerned with a single pair of contraries, white and black for sight, sharp and flat for hearing, bitter and sweet for taste, (tr. Smith); and similarly in the paraphrase (ed. Arnzen, 277.16–279.2).

245 For further source texts of the *DA*-paraphrase, see Arnzen, *Aristoteles' DA*, pp. 414–15.

The idea that each sense organ is related to a different element has already been alluded to in a passage by Miskawayh quoted above (94). The topic is now expounded in the words of Nemesius, who in turn draws on Galen's account in the *De placitis Hippocratis et Platonis*.[246] The Galenic account disagrees with Aristotle's in *De sensu*, 438b16–439a5:[247] whereas Galen attributes fire to vision, water to taste and an intermediary level between air and water to smell, Aristotle correlates water with vision as the perceiving part of the eye is watery, smell with fire, and taste with earth, as it is just a particular form of touch. Aristotle's argument why vision cannot be related to fire occurs in *DS*, 437b10–15 where he refutes this doctrine, which he ascribes to Empedocles and Plato's *Timaeus*, the argument being that we would then see in the dark. The view which Aristotle himself adopts, i.e. that the eye is of water, is ascribed to Democritus (*DS* 438a5–6).

The readers of the *PR* are, however, not given any indication who might be the author of this passage and they would probably reach the conclusion that it is Aristotle, as he is the last author quoted by name.

(158) 89b4–5 He says

Source: Nemesius, *De natura hominis* 6, ed. Morani, 57.3–5; tr. Sharples and van der Eijk in *Nemesius On the Nature*, p. 103; *Fī Ṭabīʿa*, ed. Haji-Athanasiou in 'Le traité de Némésius', 103.11–13.

This short definition of sense and sense organ is also taken from Nemesius.

[246] The passage in Galen, *De placitis Hippocratis et Platonis* VII, 5, 42 – 6, 1, reads: 'We shall say, therefore, that it was needful that the organ of sight be luminous, the organ of hearing airy, that of smell vaporous, that of taste moist, that of touch earthy. It was not possible for them to be otherwise, for they required the alteration caused by similars; and this is what Empedocles meant when he said, "By earth we perceive earth, by water, water; air by shining air; and consuming fire by fire." For we do in fact perceive by the more earthy of the sense organs, touch, the earthy nature in objects of sense, and by the most luminous, sight, the luminous nature; and similarly the discernment of the properties peculiar to air is accomplished by the organ of hearing, which becomes air-like. And indeed we perceive tastes by the organ of taste, which is by nature moist and spongy. What remains is the organ of smell, ... This is a fifth sense organ, even though there are not five elements, since the class of odors is intermediate between air and water, as Plato wrote in this passage in the Timaeus: "As water changes to air and air to water, all odors have arisen in between."' (tr. De Lacy, p. 463).

[247] For the deviating account in *DA* 425a3–9, see passage (159).

(159) 89b6–7 He says

Source: unidentified; a lost part of the *DA*-paraphrase or some other account of *DA*, 425a4–6?

Cf.: *Fī l-Nafs*, ed. Badawī, 63.1–3.[248]

This passage must derive from some version of Aristotle's *DA* 425a4–6, maybe even the *DA*-paraphrase, although its two extant versions miss the rendering of the beginning of Book III.[249] It contradicts the previously quoted excerpt from Nemesius, (157), as it links vision to water and attributes no sense to fire.[250] For the reader the correct impression arises that this passage is by Aristotle who is still the last authority quoted. Why the compiler inserted the two partly contradicting passages by Nemesius is not obvious, and it is particularly surprising that he does not set them more clearly apart from the Aristotelian material to avoid the apparent contradiction between passages so close to each other.

(160) [blank]

Meant to be filled by: Miskawayh, alternative *Fawz*?

As in: *K. al-Ḥikma*, MS Esad Efendi 1933, fols. 68b5–69a5; 69a5–8; 69a8–69b5.[251]

Cf.: Miskawayh, *Fawz* I, 3; II, 4; III, 9, ed. ʿUḍayma, 42.14–44.2; 74.9–13; 141.11–142.10, with French tr. Arnaldez, pp. 21–2; 42; 90–1; tr. Sweetman, pp. 101–2; 128; 181–2.

This is the first of three deliberate gaps which the copyist (or even the compiler) of the *PR* has left in the Marsh manuscript. He resumes his text on fol. 92b8, starting in the middle of a passage attested in the *K. al-Ḥikma* and parallel to Miskawayh's *Fawz* III, 3 (see next passage). As at least one of

248 The passage reads (by brackets Badawī indicates additional explanatory material as in the extant Greek text, but probably not by Aristotle):
(وذلك أنّ الحدقة من ماء ، والسمع من هواء والشمّ من كليهما) ؛ ثم لا تصير النار حاسّة لشيء واحد ، بل تكون شائعة بينهما (لأنه ليس يكون شيء حاسّ بغير حرارة) .

249 Cf. Arnzen, *Aristoteles' DA*, p. 418.

250 With regard to that latter point Aristotle contradicts his own view expressed in *DS*, 438b24–5, where smell is related to fire. As no extant text of the Arabic *DS* is known (the MS edited by Hansberger is defective at the beginning and thus has no corresponding part to *DS*) we can only refer to Ibn Rushd's *Talkhīṣ* (ed. Badawī in *Fī l-Nafs*, 194.22–195.1):
ويخص آلات الثلاث قوى ، أعني السمع والبصر والشمّ ، أنّها منسوبة إلى البسائط : فالعين إلى الماء ، والسمع إلى الهواء ، والشمّ إلى الحارّ الناريّ الدخانيّ .

251 How much text must be missing is calculated assuming that each missing folio-page has thirteen lines and that seven lines are missing on fol. 92b where 6 lines appear on the lower half.

the other gaps appears within a lengthy quotation of Miskawayh, we may assume that all the gaps are caused by a defective manuscript of the alternative *Fawz* version and can thus be filled using the *K. al-Ḥikma*. The material that occurs in the *K. al-Ḥikma* before its parallel to the *PR* (from fol. 92b8 onwards) is mainly on sense perception and thus a very likely candidate to fill the gap in the Marsh manuscript.

The first section of this *K. al-Ḥikma*-material is paralleled in the standard *Fawz* I, 3, and refers to Aristotle's view expressed in the phrasing of the *DA*-paraphrase in passage (155): each sense organ is in a state of equilibrium and perceives its corresponding sensibles through their deviation from that equilibrium, the deviation constituting a certain perceptible quality.[252] The correspondence between each sense organ and its sensibles is based on their correlation to the same of the four elements. In correlating the senses to the elements Miskawayh follows Galen and Nemesius as quoted above in passage (157), and not Aristotle. The section then turns to a proof for God's existence based on motion following the tradition of Aristotle's 'First Unmoved Mover', but it seems doubtful whether the compiler of the *PR* would have introduced a proof for God's existence at this stage of his compilation.

The second section paralleled in the standard *Fawz* II, 4, is short and discusses the impairment or destruction of sense organs by sensibles which are extreme. It does not add anything that has not already been stated in the passage of the *DA*-paraphrase quoted above, in passage (155).

The third section presents a hierarchy of the senses, following Aristotle in ranking touch as the lowest, and vision and hearing as the highest, senses.[253] In the context of the standard *Fawz* version this classification occurs within a discussion of revelation. This might explain why noble notions are said to be only perceived by vision and hearing, as only these two senses qualify for receiving revelation.

(161) [blank] – 94a1

Meant to be filled by: Miskawayh, alternative *Fawz*?

As in: *K. al-Ḥikma*, MS Esad Efendi 1933, fols. 69b5–70b2.

[252] See, e.g., *DS* 445b23–31; Arnzen, *Aristoteles' DA*, pp. 406–7.

[253] Slightly different from the *DA*, the *DS* 436b12–437a5 lists touch and taste as the senses which all animals have, and smelling, hearing and seeing as the senses which are found in animals which have locomotion as they operate through external media. However, vision and hearing are further singled out, vision for supplying the primary wants of life and hearing for developing thought.

Cf.: Miskawayh, *Fawz* III, 3, ed. ʿUḍayma, 123.2–125.5, with French tr. Arnaldez, pp. 77–9; tr. Sweetman, pp. 167–9.

The compiler of the *PR* certainly intended to include this account of the common sense by Miskawayh into his compilation, as is apparent from the fact that the second half of it is preserved in the Marsh manuscript on fols. 92b8–94a1. The missing first half can be supplied by the parallel in the *K. al-Ḥikma* and the standard version of the *Fawz* III, 3. The concept of the common sense is Aristotelian and expressed in *DA* 426b21–427a9, and *DS* 449a5–20, but what particular source Miskawayh used must remain doubtful, especially as the discussion of the common sense is missing in the two extant versions of the *DA*-paraphrase and as the transmission of *DS* is dubious.[254] Miskawayh defines the function of the common sense as collecting the sense perceptions of all senses, discerning them as well as retaining them. The physiological account which is included in the discussion of the common sense follows Galen, Nemesius and Qusṭā ibn Lūqā, and is in line with the material the compiler of the *PR* excerpts in previous passages.[255] The common sense is thus said to be a psychic faculty related to the corporeality of (psychic) pneuma. Its ability to perceive 'numerically different objects simultaneously' is already stressed by Aristotle (*DS* 449a19, tr. Beare).

(162) 94a2–8 Imagination is better

Source: unidentified.

Although Miskawayh discusses imagination in the remaining part of the *Fawz* III, 3, which is paralleled in the *K. al-Ḥikma*, the compiler of the *PR* draws on some other source for this passage. In Aristotle's *DA* and its paraphrase the definition of imagination occurs in the context of distinguishing it from sense perception and intellect as in our passages, yet there are no striking parallels. However, the statement that sense perception functions only while one is awake, while imagination is also active during sleep, occurs in both texts.[256]

[254] Arnzen convincingly argues for the existence of such a discussion in the original version (*Aristoteles' DA*, p. 418). For the *DS* see Hansberger, 'The transmission of Aristotle's *Parva naturalia* in Arabic', p. 16, where she states that 'Of the extant part of the first *maqāla* of *K. al-Ḥiss wa-l-maḥsūs* practically nothing corresponds to the Greek *DS* closely enough to be classified as translation or paraphrase.'

[255] For Aristotelian parallels, see Sweetman, *Fawz*, p. 168.

[256] See *DA* 427a17–429a9 and esp. 428a6–7; *DA*-paraphrase, ed. Arnzen, pp. 293–303. Both

(163) 94a9–95a2 The Greek Sage says

Edition: ed. and tr. Rosenthal in 'Aš-Šayḫ', II, pp. 370–73: no. C 1–2; ed. Badawī in *Aflūṭīn*, 189.18–190.12: no. 10; tr. Lewis in 'Plotiniana', p. 153; tr. Hansberger in 'Mediating', p. 63: T1, pp. 63–4: T3, p. 64: T4.

Cf.: *Enn.* IV 5, 1, 3–27.

This passage attributed explicitly to the Greek Sage renders most of *Enn.* IV 5, 1, rather loosely. It does not mention the bodily sense organs which are similar to the sense objects and therefore a main reason for sense perception according to Plotinus. Instead the Arabic stresses the bodily intermediary, i.e. air for vision, and the Aristotelian concept that like can only be affected by like.[257]

(164) 95a3–97a2 Someone may say

Edition: ed. Badawī in *Aflūṭīn*, 190.13–192.2: no. 10; partial edn and tr. Rosenthal in 'Aš-Šayḫ', II, pp. 372–7: no. C 3–4; tr. Lewis in 'Plotiniana', pp. 157–9; partial tr. Hansberger in 'Mediating', pp. 65–6: T6, pp. 66–7: T8, p. 68: T11, p. 69: T12, p. 70: T15.

Cf.: *Enn.* IV 5, 2, 33 – 3, 10.

The Arabic rendering does not take up the first half of *Enn.* IV 5, 2, which is a doxographical account of theories of vision prior to Plotinus. The rest of the chapter is, however, discussed at great length and with reasonable faithfulness to the Greek. In accordance with Aristotle, light is shown to be crucial for the process of vision, as it makes the potentially transparent – for example air, water – actually transparent. However, Plotinus is far more interested in proving that air remains unaffected by the process of vision than in showing what is actually needed in that process, namely a moderate distance between the object and the perceiver and a lit transparent medium in between. The argument that the eye cannot see when a visible object is placed directly on its surface is already presented in passage (152), quoting from the *DA*-paraphrase.

versions of the *DA*-paraphrase use the term *wahm* and not *takhayyul* as our passage. The Arabic version of the *DA*, *Fī l-Nafs* (pp. 68–72) has *tawahhum*.

[257] For a more detailed discussion of this and the following two passages, see Hansberger, 'Mediating'.

(165) 97a3–6 He says

Edition: ed. and tr. Rosenthal in 'Aš-Šayḫ', II, pp. 376–7: no. C 4; ed. Badawī in *Aflūṭīn*, 192.3–5: no. 10; tr. Lewis in 'Plotiniana', p. 160; tr. Hansberger in 'Mediating', p. 71: T17.

Cf.: *Enn.* IV 5, 4, 2–4.

This short passage, drawing on the beginning of *Enn.* IV 5, 4, describes the relationship between light and air and their respective roles in the process of vision, still following the Aristotelian theory as expounded in *DA* II, 7.

(166) 97a7–9 Aristotle says

Source: unidentified; some commentary on *DS* 437a3–11?

This short statement may derive from some version of Aristotle's *DS* 437a3–11, where vision is said to be the superior sense when 'regarded as a supply for the primary wants of life', while hearing 'takes the precedence' 'for developing thought'. Further, 'we perceive the common sensibles, viz. figure, magnitude, motion, number' through vision, 'while hearing announces only the distinctive qualities of sound, and, to some few animals, those also of voice' (tr. Beare). This latter statement could be interpreted as referring to the composite audibles, i.e. the different qualities, in something simple, namely sound, whereas the simple visible may be said to be composite of different sensibles, namely common sensibles which are perceived along with colour. Yet neither the extant version of the *K. al-Ḥiss wa-l-maḥsūs* nor Ibn Rushd's *Talkhīṣ* contain a comparable passage.

(167) 97a10–b8 Someone says

Source: *Fī Nawādir al-falāsifa fī l-mūsīqā* as preserved in the *Rasāʾil Ikhwān al-Ṣafāʾ*, I, 236.15–20, 236.22–237.2?

The four short statements on the priority of hearing to vision are excerpted from a collection of sayings entitled *Fī Nawādir al-falāsifa fī l-mūsīqā* (*Rare Sayings of the Philosophers on Music*),[258] which is partly preserved at the end of the epistle 'On Music' in the *Rasāʾil Ikhwān al-Ṣafāʾ*. Another part of the collection constitutes the fourth chapter of the second treatise of

[258] Might it actually be a compilation produced by Ḥunayn, part of which is preserved in MS Koprülü 1608, fol. 52a, under the heading *Mimmā jammaʿahū Ḥunayn min alfāẓ al-falāsifa fī l-musīqā*?

al-Kindī's *R. fī Ajzāʾ khabarīya fī l-musīqā* (*On the Informative Parts of Music*). However, as the only manuscript of al-Kindī's *Risāla* known to be extant is defective at the end, the four statements in question are not included there.[259]

In the *PR* the four statements seem to be included in order to support the previously quoted Aristotelian excerpt on vision and hearing with its implicit claim for the priority of hearing.

The first statement apparently supports the extramission theory of vision, according to which the eye emits rays which reach the object and bring the sensation back to the eye. This theory can be used nicely to stress the priority of hearing which does not go out in order to search its sensibles as the eye allegedly does.[260]

The second statement on vision being exercised in straight lines and hearing from a circumference around the head is Aristotelian (*PA* 656b28–31) and, for example, also taken up in Nemesius' *De natura hominis*.[261]

The mention of the possible errors of vision in the fourth statement recalls passage (95).

(168) 97b9–98a4 Aristotle says

Source: a lost part of the *DA*-paraphrase or some other reworking of Aristotle's *DA* 419b13–25?

Cf. : *Fī l-Nafs*, ed. Badawī, 48.3–16.[262]

[259] The *Risāla* is edited by Zakarīyāʾ, *Muʾallafāt*, pp. 95–110, who supplies the *Ikhwān* version of the *Fī Nawādir al-falāsifa* to compensate the incomplete manuscript. However, as al-Kindī's extant material is, in part, different from the one in the *Ikhwān*, it is difficult to say whether his *Risāla* ever actually contained the four statements in question.

[260] The *PR* contains no further material on the topic of the extra- and intromission theories of vision and it is thus impossible to say which view our compiler himself held, if, in fact, he had any. For a general introduction to these theories, see Lindberg, *Theories of Vision*, and for al-Kindī in particular, who adhered to the extramission theory, see Adamson, 'Vision', esp. pp. 209–14.

[261] See *Fī Ṭabīʿa*, 108.1 and 118.2–4; also *Nemesius On the Nature*, tr. Sharples and van der Eijk, pp. 114 and 107, n. 533.

[262] The passage in the Arabic *DA* reads:
وإنما يكون القرع بالفعل إذا كان شيء يصدم شيئاً ، وذلك أن الضارب هو الذي يفعل القرع . من أجل ذلك لا يمكن القرع أن يكون شيئاً واحداً ، والضارب غير المضروب ، كذلك المضروب إنما يقرع بصوته شيئاً ، والقرع لا يكون إلا بحركة . وقد أخبرنا أن ليس كلّ ما اصطك من الأشياء حدث عنه قرع ، لأنّ الصوف إذا صُكّ أو صُكّ به لا يفعل قرعاً ألبتّة ، ويفعل ذلك النحاس وكلّ ما كان أملس مقعراً : أمّا النحاس فمن أجل ملوسته ، وأمّا المقعر من الأشياء فإنما يحدث الصوت عنه من أجل انطواء الجو فيه ، ويفعل خفقات كثيرة بعد الخفقة الأولى ، ويبقى مسموعاً طويلاً ، وذلك أن ما دفع القرع من الجو لا يمكنه الخروج سريعاً . وفي الماء قد يمكن القرع ، إلا أنه قرع ضعيف . . وليس الجو ولا الماء علّة للقرع ، لأن القرع يحتاج إلى أجسام كثيفة

This passage must be a reworking of the beginning of *DA* II, 8, where Aristotle mentions all five aspects on which the emergence of sound depends according to the *PR*: the bodies which can produce a sound are said to be solid (στερεός), smooth (λεῖος) and hollow (κοῖλος), and they must be struck by a sharp or vehement (σφοδρός) and sudden (ταχύς) blow. Contrary to the Aristotelian passage, where these terms are scattered, the *PR* gathers them in a neat list of criteria for the emergence of sound which indicates some later reworking or commentary. The most likely candidate, the *DA*-paraphrase, does not contain this or a similar passage in any of the two known versions, but may have in its original form. This seems even more plausible, as our passage is followed immediately, and without any indication of the beginning of a new excerpt, by passage (169), which is paralleled in the *DA*-paraphrase.

The aforementioned passage of the *DA* also states that, although sound is heard in both air and water, in the latter it is less distinct.

(169) 98a4–7 It is air

Source: *DA*-paraphrase, ed. Arnzen with German tr., 267.4, 267.1–3; MS Aya Sofya 4156, fol. 78b2–3, 78a14–78b2.

Cf.: passage (150).

This section appears as one unit with the previous one in the *PR*, but unlike passage (168) this one seems to draw on the *DA*-paraphrase, more precisely on the same material that is already taken up in passage (150). In the present passage the material is, however, quoted in a slightly rearranged order. So the possibility remains that passage (169) indeed formed a unit with passage (168) in the source text, which may have been a lost part of the *DA*-paraphrase. The repetition of the material in passage (150) would then have already occurred there.

(170) 98a8–13 He says

Source: unidentified; some reworking of Aristotle's *DA* 419b33–420a2?

Cf.: *Fī l-Nafs*, ed. Badawī, 49.3–8.[263]

تصدم الهواء فيصدم بعضها بعضاً . وإنما يكون ذلك إذا ثبت الجو عند ضرب الضارب فلم يتبدّد . لذلك إن ضرب ضارب فأسرع وشدد ضربه أجاب الهواء بخفقة ، لأنه ينبغي لحركة ضرب الضارب أن تسبق تبدّد الهواء ، كمن ضرت شيئاً من رمل .

[263] The passage reads in the Arabic version of the *DA*:
وأمّا الخلاء فنعم ما قيل إنه المستولي على السماع ، فقد يظنّ بالهواء أنّه خلاء وأنّه هو الذي

This and the following passages all seem to be some reworking of the first half of Aristotle's *DA* II, 8 on hearing, but none of them derives from any of the two extant versions of the *DA*-paraphrase.[264] The present passage stresses, again, the importance of the mediation of air and introduces the Aristotelian concept of a continuous mass of air which stretches from the source of the sound to the perceiving ears.

(171) 98b1–5 He says

Source: unidentified; some reworking of Aristotle's *DA* 420a3–5?

Cf.: *Fī l-Nafs*, ed. Badawī, 49.8–9.[265]

This paragraph elaborates on the idea of the air continuum introduced in the previous passage. The simile of waves is not found in the Aristotelian text of the *DA*.

(172) 98b6–10 He says

Source: unidentified; some reworking of Aristotle's *DA* 420a16–19?

Cf.: *Fī l-Nafs*, ed. Badawī, 49.18–21.[266]

The idea that the ear is filled with its own air which is in constant motion and produces a permanent hum or buzz in the ear is taken up here and in the passage after the next, i.e. (174), and most probably rephrases *DA* 420a16–19.

(173) 98b11–99a4 He says

Source: unidentified; some reworking of Aristotle's *DA* 420a11–15?

يفعل السماع إذا تحرّك باتّصال الكلّ . 〈و〉لسخافته وتخلخله ما كان ليكون سماع ، لولا أنّ المضرب أملس : وإذا كان المضروب أملس كان الهواء واحداً متّصلا . وكذلك حال السطح الأملس . فما كان محرّكاً هواءً واحداً متّصلاً إلى أن ينتهي إلى السمع ، فذاك فعّال للقرع .

264 As Arnzen (*Aristoteles' DA*, p. 406) remarks, *DA* 420a3–b4 is not accounted for in the paraphrase.

265 The passage of the Arabic version of the *DA* reads:
والهواء مجانس للسمع ، والقرع إنّما يكون في الهواء الخارج ، فمتى ما تحرّك الهواء الخارج فحرّك الهواء الداخل فينا ، كان سماع .

266 The Arabic version of the *DA* has the following passage:
وذاك أن الهواء أبداً محرّك في المسامع حركة خاصة 〈مثل القرن〉 . أمّا القرع فانّه 〈يظل〉 غريباً ليس بأهلي . وكذلك يزعمون أنّ السماع يكون في الخلاء الذي يحدث عنه وجبة لأنّا نسمع حيثما كان الهواء محدوداً .

Cf.: *Fī l-Nafs*, ed. Badawī, 49.14–17.[267]

The air in the ear is protected by the structure of the ears, i.e. the spirals, and the tympanic membrane, which is likened to the water in the eye and its protecting membrane. The phrasing of this sounds Galenic since it stresses the role of nature in designing the ears the way they are.

(174) 99a5–8 He says

Source: unidentified; some reworking of Aristotle's *DA* 420a16–19?

This statement repeats the idea expressed in passage (172), but makes the additional point that not only is the air in the ears always in motion, but air is in general constantly moving. This can be observed in the floating of dust particles.

(175) 99a9–12 He says

Source: unidentified; some reworking of Aristotle's *DA* 419b25–7?
Cf.: *Fī l-Nafs*, ed. Badawī, 48.17–18.[268]

This passage presents the Aristotelian explanation of an echo as given in the *DA*, namely that air is gathered, prevented from dissipation, and then rebounds from a hard surface like a ball from a wall. The wording differs completely from the account of the echo in the *DA*-paraphrase, even the Arabic term used for the phenomenon is different (*ṣadan* in the Arabic translation of the *DA* and the *PR*, and *ṭanīn* in the *DA*-paraphrase, ed. Arnzen, 269.1–4).

(176) 99a13–99b4 Plato says

Source: unidentified; some reworking of Plato's *Timaeus* 66e?

The ultimate source for this passage is most probably the *Timaeus* in which Plato defines the consistency of odours as intermediate between water and

[267] The Arabic version of the *DA* has the statement in reversed order:
ولذلك لا نسمع في الماء لأن الهواء لا يصل إلى الهواء المجانس له المركّب فينا ، ولا إلى السمع يصل من أجل السماخات . وإذا كان هذا ، لم يسمع ولا ⟨أيضاً إذا أ⟩ لم الشغاف ، بمثل ما تألم جلدة الحدقة .

[268] The Arabic version of the *DA* reads:
وأمّا الصدى فإنّما يكون من جو واحد من أجل الإناء الذي حجز ومنعه أن يتبدّد ، فيرجع مدفوعاً بمنزلة الكرة .

air, and gives the example of someone whose nose is blocked and who will not be able to smell, but will still be able to breathe.

(177) 99b5–100a2 Aristotle says

Source: unidentified.

This alleged Aristotelian account of the process of smell has no similarity to Aristotle's discussion of smell in the *DA*, *DS* or the *DA*-paraphrase.[269] It rather recalls the short Plotinian reference to smell when illustrating the process of emanation in passages (32) and (44). The process of smelling happens because the object smelt generates something perceptible in the air. It does so without being changed itself – i.e. it does not emit even the smallest particle of its own substance – and without changing the air.

(178) 100a3–4 Light is some activity

Source: *DA*-paraphrase, ed. Arnzen with German tr., 259.2–3; MS Aya Sofya 4156, fol. 76a4–7.

The elaborate discussion of light in passages (178–80) is excerpted from the *DA*-paraphrase. The beginning is marked by a version of the Aristotelian statement that light is the activity of the transparent (*DA* 418b9–10). The idea that light perfects the transparent body is, for example, expressed by Alexander of Aphrodisias.[270]

(179) 100a4–11 Some say

Source: *DA*-paraphrase, ed. Arnzen with German tr., 261.11–263.8; MS Aya Sofya 4156, fol. 77a1–13.

In contrast to the *DA*-paraphrase, the compiler of the *PR* starts the discussion of whether light is a body or not with arguments for the corporeality of light. He takes over all three arguments of the paraphrase which, in turn, derive from Philoponus' commentary on the Aristotelian text.[271]

[269] For Aristotle on smell, see Johansen, 'Aristotle on Smell'.

[270] In his *Maqāla fī kayfa yakūnu al-Ibṣār ʿalā madhhab Arisṭāṭālīs*, Alexander says that light is the perfection of the transparent (*istitmām al-mushiff*) (ed. Gätje in *Psychologie*, 155.13).

[271] See Arnzen, *Aristoteles' DA*, pp. 401–2.

(180) 100a12–101a4 Others say

Edition (partial): ed. Cottrell with French tr. in 'L'Anonyme', pp. 438–9.

Source: *DA*-paraphrase, ed. Arnzen with German tr., 259.4–14, 261.5–10, 263.9–11; MS Aya Sofya 4156, fols. 76a7–b6, 76b10–77a1, 77a13–b1.

The compiler of the *PR* quotes only three of the four arguments for light's incorporeality in the *DA*-paraphrase, but closes in the same way with an argument that is ascribed directly to Aristotle and refers to his *DA* 418b13–20. Most of the arguments are also taken up in (Ps.-) Ḥunayn, *Fī anna l-Ḍawʾ laysa bi-jism*,[272] and used by al-Kindī.[273]

The doctrine of the incorporeality of light is thus shown to be Aristotelian and therefore the one to follow.

(181) 101a5–10 The Greek Sage says

Edition: ed. and tr. Rosenthal in 'Aš-Šayḫ', II, pp. 378–9: no. D 1–2; ed. Badawī in *Aflūṭīn*, 192.6–13: no. 11; ed. Cottrell with French tr. in 'L'Anonyme', pp. 438–9; tr. Lewis in 'Plotiniana', p. 165; partial tr. Hansberger in 'Mediating', p. 72: T19.

Cf.: *Enn.* IV 5, 6, 11–13 and IV 5, 7, 13–17.

The doctrine of the incorporeality of light is also the basis for the hypothesis discussed by Plotinus that light may be a quality which needs a bodily substratum.[274] The compiler of the *PR* presents this Plotinian view on light by a close rendering of two passages in *Enn.* IV 5, 6 and 7.

(182) 101a11[275]-102b6 Abū Zayd says

Edition (partial): ed. Cottrell with French tr. in 'L'Anonyme', pp. 439–40.

The passage provides an account in the name of Abū Zayd, i.e. Abū Zayd Aḥmad b. Sahl al-Balkhī (~ 850–934), on colours, which follows the doctrine of al-Kindī closely.[276] According to it, colour must be caused by earth

[272] See Arnzen, *Aristoteles' DA*, pp. 63–71.

[273] See Adamson, 'Vision', esp. pp. 223–5, where he discusses a section of the *DA*-paraphrase quoted in the *PR*, passage (180).

[274] See Hansberger, 'Mediating', pp. 72–3.

[275] This is a rare case in which the compiler leaves a complete line blank so that there are only twelve lines on the folio page instead of the usual thirteen.

[276] Assuming that the entire passage is by Abū Zayd and not only the first section, as Cottrell seems to think since she understands it as a commentary on the previous *GS*-passage ('L'Anonyme', p. 440) and edits only its first half. For al-Kindī on colour, see Adamson, 'Vision', pp. 225–36.

as that is the only one of the four elements which is dense and thus blocks vision. Therefore, the perception of water, air and fire is actually due to earthy particles mixed with each of these elements, which are in themselves invisible.[277]

(183) 102b7–103b13 The Sage says

Edition: ed. Badawī in *Aflūṭīn*, 192.14–193.18: no. 12; partial edn and tr. Rosenthal in 'Aš-Šayḫ', II, pp. 378–81: nos. E 1–2; tr. Lewis in 'Plotiniana', p. 113.

Cf.: *Enn.* IV 4, 28, 10–18, 55–8, 38–43 and 30–2.

Having discussed the physiology of perceptions, the compiler of the *PR* turns to feelings. To do so, he picks the Arabic rendering of a section of the *Enneades* (IV 4, 28), in which Plotinus makes exactly the same transition. The text of the *GS*, to whom the compiler refers only as 'the Sage' this time, is a reasonably faithful version of the Greek model, though in rearranged order. Particularly remarkable is the rephrasing of the localisation of appetite in the liver. The Greek text does not formulate an objection to the claim that the Ancients placed the seat of the appetitive soul in the liver, but simply takes it for granted. The phrasing of the Arabic reminds us of passage (55) in which the Ancients were said to have placed the rational soul in the brain, but with all due caution. Passage (55) renders *Enn.* IV 3, 23, and the Greek text also mentions the liver and the heart. Thus the Arabic version of *Enn.* IV 4, 28 may actually incorporate aspects of the Greek text of *Enn.* IV 3, 23. Another difference between the Greek and the Arabic is that our text only mentions one sort of anger, namely the one which is caused by harm affecting the body: it stirs blood and bile and causes a feeling which informs the soul about the body being harmed. Plotinus further lists the opposite process starting from above, i.e. the rational soul is aroused by a wrong, which is not necessarily a bodily harm, and thus stirs anger.

The rearranged order of the Arabic allows for the assumption that not much text is missing at the end of our passage in the Marsh manuscript. The textual loss affects the rendering of the lines following line 32 of *Enn.*

[277] Similar aspects, namely the lack of colour in a transparent substance that has no defined boundaries, the non-transparency of earth, and the existence of colour through the mixture of fire and earth, are found in Alexander of Aphrodisias' treatise *Fī l-Lawn* (ed. Gätje in *Die arabische Übersetzung*, lines 58–62 and 91–3). However, Alexander's underlying theory is 'diametrically opposed' to al-Kindī's, as the former upholds 'a positive concept of the transparent', whereas the latter defines the transparent 'negatively' (see Adamson, 'Vision', pp. 229 and 228–32).

IV 4, 28, but the rendering of line 38 and what follows is found earlier in the passage. It seems plausible that only a few lines of the Plotinian text were written on fol. 104a, but that most of the page has been left blank, which may be the reason why it was – along with the other blank pages – finally removed from the manuscript. The fact that the Plotinian passage was followed – most probably directly – by a deliberate gap is apparent from fol. 107a which is extant, but only written on in the lower half. The gap may indicate another intended insertion of material by Miskawayh (as in the case of the other two gaps) which was not accessible to the compiler at the time of producing the *PR*.

(184) 107a7–107b1 Galen says

Parallel: al-Anṣārī, *Ādāb al-falāsifa*, ed. Badawī, 123.9–15; Muḥammad Ibn Dāwūd, *K. al-Zahra*, ed. Nykl, 18.14–19.2; Mubashshir, *Mukhtār al-Ḥikam*, ed. Badawī, 294.3–7; Ibn al-Jawzī, *Dhamm*, ed. ʿAṭā, 247.4–8; Ibn Abī Uṣaybiʿa, *ʿUyūn al-anbāʾ*, ed. Müller, I 88.27–32; tr., based on the same witnesses except for Mubashshir, in Biesterfeldt and Gutas 'Malady', p. 23, n. 22.

This passage attributed to Galen on love[278] is quite widespread in Arabic literature, but the earliest evidence of the text is so far al-Anṣārī's *Ādāb al-falāsifa*, which may or may not be the source for the compiler of the *PR*. The passage is especially interesting as it ascribes to Galen the doctrine of ventricular localisation. The extant Galenic texts contain a clear tripartition of the powers or faculties in the brain, but do not locate them spatially. However, Strohmaier suggests that the no longer extant early work *De demonstratione*, and in particular Book V, in which Galen dealt with the senses in general, may have suggested such a localisation. He further points out that, although Ḥunayn could no longer get hold of a complete copy of *De demonstratione*, he affirms that he translated Book V.[279] Be that as it may, the first extant Greek sources for a ventricular localisation are from the fourth century and by the physician Posidonius of Byzantium (according to the sixth-century author Aëtius of Amida) and Nemesius.[280]

[278] The Arabic term used is *ʿishq*, which implies an excessive quality of the love described (unlike *maḥabba*). See, e.g., Walzer, 'Some Aspects', p. 227.

[279] Strohmaier, 'Avicennas Lehre', pp. 335–8. Ḥunayn's attestation of his translation of Book V is preserved in his *R. fī Dhikr mā turjima min kutub Jālīnūs* (ed. Bergsträsser in *Ḥunain*, pp. 47–8: no. 115).

[280] See Rocca, *Galen*, pp. 245–7; *Nemesius On the Nature*, tr. Sharples and van der Eijk, pp. 25 and 118–23; Galen, *DPH* III, ed. De Lacy, p. 674.

(185) 107b2–11[281] He says

Source: Miskawayh, *Tahdhīb*, ed. Zurayk, 20.4–5, 22.11–13, 16–17, 21.7, 20.12–14; tr. Zurayk in *Refinement*, pp. 18–20; French tr. Arkoun in *Traité*, pp. 30–2.

Most of the definitions in this passage also appear in the same or very similar wording in Miskawayh's discussion of the different subordinate virtues belonging to each of the four cardinal virtues in his *Tahdhīb*.[282] Our passage only presents virtues which are linked to moderation and liberality which, although a virtue of moderation, also has its own sub-heading in the *Tahdhīb*.[283] Whether the list is compiled from the *Tahdhīb* and enhanced with some additional material[284] especially for the *PR* or whether there existed an exact model in some other, maybe no longer extant, writing of Miskawayh must remain doubtful. Miskawayh's source, in turn, may be the ethical treatise *Maqāla fī l-Mudkhal ilā ʿilm al-akhlāq*, which has a similar, although shorter list.[285]

(186) 108a1–4 Laughter is the balancing

Source: Galenic?

Cf.: al-Ṭabarī, *al-Muʿālajāt al-Buqrāṭīya*, Bodleian, MS Marsh 158, fol. 33a1–2; facsimile edn Sezgin, 45.5–6.[286]

This physiological explanation of laughter and weeping may be Galenic, as al-Ṭabarī has a similar definition of laughter which he ascribes to Galen.

(187) 108a5–11 Plato says

Source: ʿAlī ibn Rabban, *Firdaws*, ed. al-Ṣiddīqī, 88.9–11, 13–15, 19–21; ed. al-Jundī, 66.23–4, 26–7, 67.3–4.

281 This folio-page has only eleven lines.

282 For a discussion of Miskawayh's list of 'subordinate virtues' as well as his possible models, see Walzer, 'Some Aspects', pp. 222–3.

283 Zurayk (*Refinement*, p. 198, n. 14) suggests that this singling out of liberality is due to Aristotelian influence (see *EN* IV).

284 Additional are the second sentence dealing with shame and the entry on 'embarrassment'.

285 The list is found on pp. 407–9 of Badawī's edition in *al-Akhlāq*. The similarity to Miskawayh is particularly apparent in the definition of shame, which reads in the *M. fī l-Mudkhal* (407.21–2):
فأمّا الحياء فهو الإعراض عن القبائح بالطبع أو الخوف والحذر من السبّ الصادق .

286 See *al-Muʿālajāt al-Buqrāṭīya*, ch. 48 on laughter and weeping:
جالينوس في منافع الأعضاء وغيره من الكتب وأمّا الضحك فهو انبساط النفس بحركتها إلى السطح البدن .

The passage attributed to Plato is actually excerpted from ʿAlī ibn Rabban's *Firdaws*. The implicit disapproval of laughter reflects Plato's view on the matter, as, for example, expressed in *Res Publica* 388e and *Leges* 732c.

(188) 108a12 The innate disposition is a nature

Source: unidentified.

This short definition of the innate natural disposition may or may not be related to the next passage on smallpox. In any case, it seems a bit misplaced in the middle of this section on the physiology of feelings and should rather have been put at the beginning.

(189) 108a13–108b4 He says

Source: unidentified.

It is surprising that the compiler of the *PR* includes this account on what causes smallpox[287] in his compilation. One may speculate whether he has unintentionally copied it from his source along with either the previous or the following passage. However, al-Baladī reports that the 'modern physicians' (*muḥdathūn*) define smallpox as the transformation of the blood of a child into the blood of a youth.[288] This seems to indicate that smallpox must have been regarded as a common condition, which might be the reason for its discussion in the *PR*.

The link between smallpox and a surplus of food is already drawn by Galen, whose statement al-Rāzī summarises in his *K. al-Judarī wa-l-ḥaṣba* (*Treatise on the Small-pox and Measles*), saying 'the superfluous parts of the food that remain, which are not converted into blood, and remain in the members, putrefy, and become more acid, in process of time, until there are generated the erysipelas, *Small-Pox*, and spreading inflammation' (tr. Greenhill, p. 27).[289]

[287] If *judarī* indeed indicates 'smallpox' here. Another possible meaning of the term is 'measles'.

[288] The passage in his *Tadbīr al-ḥabālā* (321.10–11) reads:
فأمّا المحدثون فمنهم من قال إن الجدري انقلاب دم الطفوليّة إلى دم الشباب .

[289] The Galenic passage occurs, as al-Rāzī states, in *UP* IX (ed. Kühn, III, 686.1–5) which reads in the Arabic version (National Library of Medicine, MS A 30.1, fol. 130b14–17):
كذلك كلّ واحد من الأعضاء إذا فقد الغذاء المشاكل له اضطرته شهوته الغريزيّة الى أن يجتذب بعض الأخلاط الرديئة والخلّة الثانية أنّ هذه الفصول التي تجتمع في العضو تعفن على طول الزمان حتى أنّها تحتدّ وتسخن فتولّد الأورام الحارّة والحمرة والجدري والقروح الساعية والحمّيات وأمراض أخر كثرة شبيهة بما وصفت ...

In the same treatise al-Rāzī appears to be in line with the definition of small-pox by the 'Moderns' according to al-Baladī, claiming that hardly anyone escapes smallpox:

Now the Small-Pox arises when the blood putrefies and ferments, so that the superfluous vapours are thrown out of it, and it is changed from the blood of infants, which is like must, into the blood of young men, which is like wine perfectly ripened: and the Small-Pox itself may be compared to the fermentation and the hissing noise which take place in must at that time. And this is the reason why children, especially males, rarely escape being seized with this disease, because it is impossible to prevent the blood's changing from this state into its second state, just as it is impossible to prevent must (whose nature it is to make a hissing noise and to ferment,) from changing into the state which happens to it after its making a hissing noise and its fermentation. And the temperament of an infant or child is seldom such that it is possible for its blood to be changed from the first state into the second by little and little, and orderly, and slowly, so that this fermentation and hissing noise should not show itself in the blood: for a temperament, to change thus gradually, should be cold and dry; whereas that of children is just the contrary, as is also their diet, seeing that the food of infants consists of milk; and as for children, although their food does not consist of milk, yet it is nearer to it than is that of other ages. (tr. Greenhill, pp. 29–30)

The physiological explanation given for smallpox in the *PR* further resembles the style of the ones given for laughter and weeping before, and shivering and paleness of complexion in what follows.

(190) 108b5–13 The reason for the tremor

Source: unidentified.

The passage explains the symptoms of fear, namely shivering and paleness of complexion. That the explanation, i.e. the withdrawal of heat to the heart, must have been very common among Greek writers is apparent by its reoccurrence in the following two passages.[290]

[290] Domanski (*Psychologie*, p. 124) cites, e.g., Aristotle's *De motu animalium* 701b34–702a1; Plotinus, *Enn.* III 6, 3, 1–19; Galen, *De placitis Hippocratis et Platonis* II 7, 16–17 and *De sympt. caus.* II 5, pp.191–3.

(191) 109a1–9 He says

Source: Nemesius, *De natura hominis* 21, ed. Morani, 81.15–82.5; tr. Sharples and van der Eijk in *Nemesius On the Nature*, pp. 142–3; *Fī Ṭabīʿa*, ed. Haji-Athanasiou in 'Le traité de Némésius', 153.2–4, 153.7–9, 153.5–7 and 154.4–7.

The list of the different sorts of fear is excerpted, slightly rearranged and shortened, from Nemesius' *K. fī Ṭabīʿat al-insān*, which, in turn, may draw on the Stoic tradition.[291]

(192) 109a10–13 Plato says

Source: ʿAlī ibn Rabban, *Firdaws*, ed. al-Ṣiddīqī, 89.2–5, 88.22–89.1, 89.5–6; ed. al-Jundī, 67.7–9, 5–6, 9–10.

The physiological description of fear attributed to Plato and quoted in the words of the *Firdaws* resembles passage (190) above. The view that excessive fear or joy can cause death is also found in al-Majūsī's *K. Kāmil al-Ṣināʿa al-Ṭibbīya* I, 4, 7 (I, 141.23–6) in the chapter on death, where a more detailed account is given.

(193) 109b1–12 Apollonius was asked about passionate love

Parallel: long version of the 'Malady of Love', ed. Biesterfeldt and Gutas, pp. 37–9, and tr. pp. 43–4.

Cf.: dramatised version of the 'Malady of Love' (as above); and in al-Daylamī, *K. ʿAṭf al-alif al-maʾlūf*, 78.14–22.

The medico-philosophical *topos* of the 'malady of love' found its way from the Greek tradition into Arabic where it exists in at least four different versions, which have been termed 'short', 'hybrid', 'long' and 'dramatized' versions by Gutas and Biesterfeldt. The discussion of the opposite complexion of the lover and his beloved occurs in the two latter versions. Whereas in the dramatised version Aristotle is asked about love by Agathodaimon, the long version which is paralleled in the *PR* presents Apollonius as the main figure. The account takes up several aspects of the previous passages and thus fits well into the compilation.

[291] Sharples and van der Eijk in *Nemesius On the Nature*, p. 142, n. 719. The definition of shame (*ḥayāʾ*) is, however, Aristotelian – see *EN* 1128b11–12 in its Arabic version *K. al-Akhlāq* (ed. Akasoy and Fidora, 285.4): *al-khawf min al-danāʾa*.

(194) 109b13–110a2 Plato says

Source: ʿAlī ibn Rabban, *Firdaws*, ed. al-Ṣiddīqī, 52.11–15; ed. al-Jundī, 43.17–19; German tr. Siggel in *Propädeutische Kapitel*, p. 420.[292]

This account of the production of hair is excerpted from the *Firdaws*, which may ultimately go back to Plato's *Timaeus* 76b-d, although, apart from black and white, ʿAlī b. Rabban does not mention different hair colours as Plato does.

(195) 110a3–110b2 Hippocrates says

Parallel: al-Majūsī, *Kāmil* I, 2.16, 81.29–82.4.

This explanation for why men grow more hair than women and eunuchs is paralleled in al-Majūsī. Might this be an indication that this passage derives from the same source as passage (141) above which is also paralleled in al-Majūsī?

The idea that women have no beards because they stay mainly indoors and are thus not in need of protection is also found in Galen's *UP* XI, 14.

(196) 110b3–9 He says

Parallel: al-Majūsī, *Kāmil* I, 2.16, 82.6–10.

This account of where most hair grows must be taken from the same source as the previous passage as it is still paralleled in al-Majūsī.

(197) 110b10–111a4 Galen says

Source: Galen, *De temperamentis*, ed. Helmreich, II, 618, pp. 68–9; ed. Kühn, I, 627–8; *K. fī l-Mizāj*, MS Aya Sofya 3593, fol. 37b7–20; as a gloss in *K. fī l-Firāsa*, ed. Ghersetti with Italian tr., (14.18–15.3), 16.5–6, 8–10; 17.5–9; 16.13–17.3.

The ultimate source of this passage must be the Arabic version of Galen's *De temperamentis*, but the compiler of the *PR* probably quotes it from the *K. fī l-Firāsa* which is his source text for the following passage. The Arabic translator of the *K. fī l-Firāsa*, Ḥunayn, adds the Galenic excerpt into his translation as a gloss for which he indicates the author and the work quoted. Thus it is not surprising that the compiler of the *PR* rightly attrib-

[292] This *Firdaws* passage must also have inspired Saʿīd b. Hibat Allāh in his *Khalq al-Insān*, ch. 35 (150.17–18).

utes the passage to Galen. It also explains why he erroneously implies that even the next passage, taken from the Ps.-Aristotelian text, is Galenic as he introduces it with 'he says'.[293] The possibility that the compiler of the *PR* had access to a copy of the Galenic *K. al-Mizāj* can not, of course, be ruled out. However, as the *K. fī l-Firāsa* is undoubtedly a source text for our compiler in passages (113) and (198), and as it presents the Galenic text faithfully in the aforementioned gloss,[294] this hypothesis is not necessary and would only increase the assumed number of source texts to which our compiler had access.

(198) 111a5–9 He says

Source: Ps.-Aristotle, *Physiognomonica* 806b21–7; *K. fī l-Firāsa*, ed. Ghersetti with Italian tr., 18.1–3, 18.13–15.

These short statements on the indicative value of the flesh in a body, of the body's movements and of the voice derive from the Ps.-Aristotelian *K. fī l-Firāsa*. There, they occur shortly after the previous Galenic passage which the Arabic translator Ḥunayn adds in a gloss.

(199) 111a10–111b8 What comes to the body on the part of its form

Source: unidentified.

Every body is said to be defined by its form as well as by its matter. In the latter case the proximate and the remote matter must be distinguished. Depending on the form, the body has certain properties, whereas it has

293 It is possible that the compiler of the *PR* either had a copy of the *K. fī l-Firāsa* which did not clearly set Ḥunayn's glosses apart, or overlooked the end of the gloss (indicated with the words 'Aristotle says' in Ghersetti's edition, 17.16). The Arabic text of the *K. fī l-Firāsa* can be particularly misleading as there follows another gloss in which Ḥunayn criticises (Ps.-)Aristotle (18.3–11) shortly after the insertion of the Galenic excerpt.

294 A comparison of the gloss in the *K. fī l-Firāsa* with the Galenic work as preserved in Aya Sofya 3593 shows that Galen is quoted verbatim (except for minor deviations which may be explained by scribal errors).
The *K. fī l-Firāsa* may also have been Miskawayh's source for identifying the Turks as the most remote inhabitants in the north. Harvey ('Source', pp. 37–40) discusses Miskawayh's passage and indicates its Galenic flavour, but points out that neither Galen nor any classical writer could have made 'the identification of the people in the far north as Turks and of the people in the far south as Negroes' (p. 40). It seems, however, that Galen's translator, Ḥunayn, introduced the Turks (*al-turk*) into the Arabic version. The Black Africans (*al-zanj*) are already mentioned by Qusṭā in his *R. fī l-Farq* in a passage even quoted in the *PR*, passage (112).

accidents resulting from its matter. The example illustrating this is laughter and may actually be the reason why the compiler of the *PR* inserted the passage at this particular point. The activity of finding something funny and being surprised is linked to the body's form, the rational soul, whereas the bodily symptom of laughter, the uttering of sounds of laughter, depends on the body's matter and is therefore accidental.

(200) 111b9–13 Plato says

Source: Miskawayh, *Tahdhīb*, ed. Zurayk, 88.7–8; tr. Zurayk in *Refinement*, p. 79; French tr. Arkoun in *Traité*, p. 141; Aristotle, *NE* 1094a3[295] and 1097a19–21 in *K. al-Akhlāq*, ed. Akasoy and Fidora, 113.5 and 131.4–6.

The ascription of this passage to Plato is puzzling as all the material it contains is Aristotelian, or at least Ps.-Aristotelian. The first sentence is paralleled in the Ps.-Aristotelian treatise *Faḍāʾil al-nafs* which Miskawayh quotes verbatim in his *Tahdhīb* from the translation of Abū ʿUthmān al-Dimashqī.[296] The relative clause at the end of the first sentence and the entire second sentence occur at the beginning of Aristotle's *EN* 1094a3, which is also quoted at the beginning of passage (202), and 1097a19–21. Whether the compiler of the *PR* combined the statements from these two sources or whether he had an unidentified source text in which the passage occurred in its entirety must remain unknown.[297]

(201) 112a1–10 Aristotle says

Source: Miskawayh, *Tahdhīb*, ed. Zurayk, 76.17–77.2, 78.3–5; tr. Zurayk in *Refinement*, pp. 70–1; French tr. Arkoun in *Traité*, pp. 125 and 127.

Cf. also: Stobaeus, *Anthologium* II, 7, 19.14–29; Ps.-Aristotle, *Magna moralia*, 1183b19–37.

This passage, as well as its attribution to Aristotle, is taken over from Miskawayh's *Tahdhīb*, more precisely from the beginning of the third chapter, which introduces seven different divisions of the good with the

[295] This parallel is joined as a relative clause to the alleged Platonic quotation.

[296] On which, see Pines, 'Texte inconnu', pp. 16–43, whose conclusions Zurayk (*Refinement*, p. 202, n. 13) summarises as follows: 'this work is probably Neoplatonic, but preserves an authentic Aristotelian core'. Miskawayh refers to the author of the treatise only by the term 'sage (*ḥakīm*)', which may have been misunderstood to refer to Plato.

[297] Such a source could possibly have been the *Summa Alexandrinorum* of the *EN* which has, according to Harvey, been a source for Miskawayh's *Tahdhīb* to a much greater extent than so far assumed ('Greek Library', pp. 499–501).

words 'Aristotle's division of the good, as reported by Porphyry and others, is as follows' (tr. Zurayk). Pines and Walzer indicate the similarity between the first two divisions – and thus with the passage in the *PR* which only quotes the first division – and an account preserved in Stobaeus attributed there to 'the survey of Peripatetic Ethics by Arius Didymus'.[298] Zurayk (p. 202, n. 4) further mentions that the divisions in Miskawayh are closer to the Ps.-Aristotelian *Magna moralia* than to the *EN*, as the former lists four divisions occurring in Miskawayh's account (the one quoted in the *PR* is the first one).

(202) 112a11–113a5 He says in the *Ethics*

Source: *Ikhtiṣār al-Iskandarānīyīn* – *Summa Alexandrinorum*; cf. abbreviated fragments, ed. Dunlop in 'Taimur Pasha', 260.1–3, 6–8, 13–14, 14–15, no parallel to (e) and (f), 260.16–18;[299] ed. Badawī in, *al-Akhlāq*, 439.3–5, 440.1–3, 440.8–9, 440.9–10, no parallel to (e) and (f), 440.11–13.

Parallel (partial): Mubashshir, *Mukhtār al-Ḥikam*, ed. Badawī, 210.3–4 (to (c)) and 5–8 (to (g)); passage (208) below (to (f)); Ps.-ʿĀmirī, *as-Saʿādah*, facsimile edn Minovi, 30.15–18 (to (a)).

Cf.: Arabic-Latin translation *Summa* in two versions: ed. Marchesi, XLI.2–5, 17–22, XLIII.3–5, 5–8, 12–14, 32–4,[300] XLIII.34-XLIV.1; and ed. Fowler in 'Manuscript Admont', 195.2–196.1, 196.15–21, 198.16–18, 18–21, 26–9, 199.22–30, 200.8–12; ch. 1 of a Hebrew abridgement of the *Summa* in Shem Tov ibn Falaquera's *Shelemut ha-maʿasim*, ed. Jospe in *Torah und Sophia*, pp. 417–20;[301] Aristotle, *EN* 1094a1–3, 1094a6–18, 1096b13–14, 1097a34–b6, 1097b20–1, 1098a16–20 (cf. also 1101a14–6),[302] 1098b12–16; *K. al-Akhlāq*, ed. Akasoy and Fidora, 113.4–5 (quoted

[298] See Walzer, 'Some Aspects', p. 226; see also Pines, 'Texte inconnu', p. 6. For Arius Didymus' account, see Sharples, 'Classification'. However, Pines ('Texte inconnu', pp. 6–7) points out that the following divisions (five in Zurayk's edition, but only four in the Cairo edition which Pines used) have nothing in common with Stobaeus. He concludes that the entire section in Miskawayh must therefore derive from a different Greek account of the Peripatetic Ethics, which became linked to Porphyry in the Arabic tradition.

[299] Cf. also Dunlop, 'Introduction', pp. 79–82, where he compares some of the fragments of the *Summa* with the Latin translation.

[300] The two Latin versions differ considerably here, Marchesi's edition being much shorter. However, neither is particularly close to the Arabic.

[301] On which, see Harvey, 'Greek Library', pp. 497–501.

[302] The Arabic of the *PR*, which has no obvious parallel in the Latin of the *Summa*, sounds like a merging of the two passages *EN* 1098a16–20 and 1101a14–16. In the Arabic translation of the *EN* only the first passage is extant, the second is lost due to a missing folio (which Badawī replaces in his edition by his own Arabic translation of the Greek text). For 1098a16–20, see *K. al-Akhlāq* (ed. Akasoy and Fidora, 135.12–15):
وإذا كان الأمر هكذا فالخير الذي يخصّ الإنسان هو فعل للنفس على ما توجبه الفضيلة فإن

as possible source in the apparatus), 113.10–15, 127.9–10, 131.16–20, 133.9, 135.12–15, 137.16–139.5; Ps.-Aristotle, *Magna moralia*, 1184b1–10 (for (g)).

Although this alleged excerpt from Aristotle's *EN* starts with quoting the first two sentences of its Arabic translation, the source of the entire rest of the passage is the *Ikhtiṣār al-Iskandarānīyīn* of the Aristotelian work which is well known in its Latin translation as *Summa Alexandrinorum*.[303] This makes one wonder whether the first two sentences of the *Summa* were originally the same as in the Arabic translation of the *EN*[304]– of which the *PR* would then be a witness – but were abridged later on, or at least abridged in the version for which the only Arabic fragments of the *Summa* survive.

The excerpt in the *PR* is put together from several statements in the first chapter. Most of them are also preserved in the fragments of the *Summa* edited by Badawī, but the *PR* often has the longer and more accurate version, as a comparison with the Latin translation shows. This sits well with the result of Dunlop's comparison of the preserved fragments and Mubashshir's quotations with the Latin text, namely that Mubashshirs' quotations are closer to the Latin than the fragments which must thus have been abbreviated.[305] The fact that the *PR*'s definition of the happy man is not found in the Latin confirms another finding of Dunlop's comparison between the preserved fragments and the *Summa*,[306] namely that the Latin translation seems to be less complete than the lost original Arabic version.

كانت الفضائل كثيرة فهو فعل ما يوجبها أفضلها أو أكملها وإنّما يكون هذا الفعل في السيرة الكاملة لأنّ خطّافاً واحداً لا ينذر بالربيع ولا يوماً واحداً معتدل الهواء ينذر بذلك . كذلك أيضاً لا يجعل الإنسان سعيداً يوم واحد من حياته ولا زمان يسير منها .

'If the matter is so, then the good which specially characterises man is a function belonging to the soul according to what virtue necessitates, and if the virtues are many, it is a certain function which the best or most perfect of them necessitates. And this function is in a perfect life, because one swallow does not announce spring, nor does one day of temperate weather announce that' (tr. Dunlop).

And the translation of the Greek text of *EN* 1101a14–16, missing in the Arabic version as it is extant today: 'Why then should we not say that he is happy who is active in conformity with complete excellence and is sufficiently equipped with external goods, not for some chance period but throughout a complete life?' (tr. Ross and Urmson).

303 The same holds true for some alleged quotations from the *EN* in Miskawayh's *Tahdhīb*, as Harvey ('Greek Library', pp. 499–501) shows, which leads him to 'surmise that the version of the *Summa* that Miskawayh had before him was attributed to Aristotle and was thought by him to be Aristotle's *Nīqūmāḥiyya*' (*ibid*., p. 500).

304 There are more instances of similarities between the Arabic translation of the *EN* and the *Summa* which Dunlop tentatively explains by suggesting that the translator of the *Summa* had the Arabic *EN* before his eyes and added from it while translating (see 'Introduction', p. 71).

305 See Dunlop, 'Summa', p. 256.

306 See Dunlop, 'Taimur Pasha', p. 254.

The two divisions of the goods (in (c)) can be traced back to the *EN*, but appear also in the *Magna moralia* and in Miskawayh's *Tahdhīb*, although different in wording and detail. The threefold division (in (g)) is already used by Plato and Aristotle, but seems to have been established even before them. It also occurs, for example, in the *Maqāla fī Ithbāt faḍāʾil al-nafs* and in Ibn al-Ṭayyib's *Istithmār*.[307]

The definition of the happy man is taken up in the *PR* again, in passage (208). It stresses an important development of Aristotle's ethical theory, namely that happiness is not excellence or virtue alone, but also the 'activity in accordance with excellence. But it makes, perhaps, no small difference whether we place the chief good in possession or in use, in state or in activity' (*EN* 1098b31–3, tr. Ross and Urmson).

(203) 113a6–12 Plato says in the *Laws*

Edition: with tr. Rosenthal, 'Knowledge', p. 396.

Source: Plato, *Leges* I, 631b-c in an unidentified Arabic version.

Cf.: al-Fārābī, *Jawāmiʿ* (or *Talkhīṣ*) *K. al-Nawāmīs*, ed. Druart in 'Le sommaire', 126.17–21; ed. Gabrieli in Alfarabius, *Compendium*, 7.1–7; ed. Badawī in *Aflāṭūn*, 39.16–40.5.[308]

This division of goods into the divine and the human derives from Plato's *Leges*. Comparing the Greek text with the version in the *PR*, Rosenthal speaks of 'a radical alteration' in the Arabic, apparently referring to the reverse order of the presentation of the goods; in the Greek the human goods are presented first. The content of the Platonic passage is, however, rendered rather faithfully. As in the case of the material from the *Phaedo* in passages (27) and (28), where the compiler of the *PR* got this Platonic material from remains unknown.[309] The corresponding passage in al-Fārābī's *Jawāmiʿ* or 'Summary' of the *Leges* seems to be unrelated to the version in the *PR*. It is shorter, but retains the order of the Greek.

[307] See Ps.-Plato, 'Pseudoplatonischer Text', ed. Daiber, lines 49–54 and pp. 39–40; and Ibn al-Ṭayyib, *Proclus' Commentary*, 24.8.

[308] Druart's edition (126.17–21) reads :
وقسّم الفضائل وبيّن أن منها ما هي إنسيّة ومنها ما هي إلهيّة وأن الإلهيّة آثر من الإنسيّة وأن المقتني الإلهيّة لا يعدم الإنسيّة والمقتني الإنسيّة ربّما فاتته الإلهيّة . والإنسيّة كالقوّة والجمال واليسار والعلم وغير ذلك مما قد عدّوه في كتب الأخلاق . وذكر أن صاحب الناموس الحقّ هو الذي يرتّب هذه الفضائل ترتيباً موافقاً ليتأدّى ذلك إلى حصول الفضائل الإلهيّة لأنّ الفضيلة الإنسانيّة متى استعملها صاحبها على ما أوجبه الناموس كانت إلهيّة .

[309] See the *status quaestionis* of the knowledge of Plato in Arabic summarised by Arnzen, 'Arabisches Mittelalter', esp. pp. 440–1.

The division of goods into divine and human ones also occurs in the *Maqāla fī l-Mudkhal ilā ʿilm al-akhlāq*, where it is linked with the threefold division of human goods into the ones of the soul, the body and outside the body.[310]

(204) 113a13–114a11 The Sage says

Edition: ed. Badawī in *Aflūṭīn*, 194.1–21: no. 13; partial edn and tr. Rosenthal in 'Aš-Šayẖ', pp. 380–3: no. F; tr. Lewis in 'Plotiniana', p. 477.

Cf.: *Enn.* V 1, 2, 20–1;[311] I 8, 2, 2–7; and VI 7, 24, 24–5; VI 7, 21, 1–3; VI 7, 23, 10–15.

This lengthy section attributed to 'the Sage', and thus belonging to the *GS*, stresses that the First Cause is also the First Good which is in line with the discussion of the previous and the following passages. The first part of the passage expounds a common Neoplatonic doctrine, and it is therefore difficult to determine an exact model within the *Enneades*. The rhetorical questions may be inspired by *Enn.* VI 7, 21 and 24 and the final discussion of the evil not being opposed to the good renders VI 7, 23, 10–15 rather accurately.

(205) 114a12–114b2 Aristotle says

Source: unidentified.

The present passage recalls the second division of the goods mentioned in passage (202), although it discusses excellences or virtues rather than goods.[312] It explains why all three types of excellences are needed, even if, according to the previous passage, the goods and thus the excellences of the soul are the most desirable. This functions as an introduction to the following passage in which the Aristotelian view which emphasizes the importance of both body and soul for the attainment of happiness is contrasted with the (Neo)Platonic view which focuses on soul alone. In Miskawayh's *Tahdhīb* (94.6–9) from which the next passage derives, Aristotle is further credited with the following quotation in the same vein as the present passage: 'Divine happiness – even though it is as noble as we have stated and even though its life is more delightful and noble than any

[310] Cf. Badawī's edition of the passage in *al-Akhlāq*, pp. 400–1.

[311] Rosenthal ('Aš-Šayẖ', p. 381, n. 3) mentions this parallel which occurs in a different context though.

[312] The same threefold division of excellences or virtues also occurs in the *M. fī l-Mudkhal ilā ʿilm al-akhlāq*, ed. Badawī in *al-Akhlāq*, 401.8–9.

other life – is still in need of other external kinds of happiness in order that it may be manifested through them; otherwise, it will remain hidden and invisible' (tr. Zurayk, p. 84).

(206) 114b3–115a13 Aristotle says

Source: Miskawayh, *Tahdhīb*, ed. Zurayk, 79.7–80.12; tr. Zurayk in *Refinement*, pp. 72–3; French tr. Arkoun in *Traité*, pp. 130–1.

This presentation of Aristotle's account of happiness and its opposition to the doctrine of Pythagoras, Socrates and Plato is taken from Miskawayh's *Tahdhīb*, where a third view ascribed to the Stoics and the Naturalists is presented as well.[313] The fivefold Aristotelian division of happiness can be reduced to the threefold one which is used in the previous and the following passages and which is also alluded to in the accounts of the views ascribed to Pythagoras, Socrates and Plato, namely the division into happiness pertaining to the soul, to the body and to matters external to the body. The first kind of happiness clearly corresponds to the happiness belonging to the body and the second kind to the happiness pertaining to the external matters, whereas the last three kinds are linked to psychic activities like acting generously, deliberation and discernment and thus refer to the happiness of the soul.

Miskawayh clearly prefers Aristotle's approach which includes soul, body and even external factors outside the body,[314] to the purely otherworldly orientation of the Neoplatonic tradition which he presents in the name of Pythagoras, Socrates and Plato.[315]

(207) 115b1–11 Aristotle says somewhere else

Source: Miskawayh, *Tartīb al-saʿādāt*, ed. Awjabī in *Ganjīna*, 109.16–110.2.

Cf.: *Maqāla fī l-Mudkhal ilā ʿilm al-akhlāq*, ed. Badawī in *al-Akhlāq*, 399.4–10 and 411.1–412.14; passage (210).

This further division of happiness, as well as its attribution to Aristotle,

313 It reads: 'The Stoics and a group of the Naturalists, on the other hand, included the body as a part of man and did not regard it a mere tool as we have explained above. Thus, they were obliged to consider the happiness of the soul as incomplete unless it is coupled with the happiness of the body and of what is also external to the body, such as the things that exist as a result of fortune and good luck' (tr. Zurayk, p. 73).

314 See, e.g., *EN* I, 8.

315 See Walzer, 'Some Aspects', pp. 224–6, where he also suggests that Miskawayh may have been inspired by Porphyry.

derives from Miskawayh's *Tartīb*. It is threefold, as are the divisions of the goods and the excellences or virtues in the passages (202) and (205), and as is also implied in the account of the views of Pythagoras, Socrates and Plato in the previous passage.

In the *M. fī l-Mudkhal ilā ʿilm al-akhlāq* there are at least two similar passages: the first of them is also ascribed to Aristotle and lists three genera of goods which are required for happiness. They recur in the second passage linked to the Ancients, and in both instances they are illustrated with examples which are partly the same as in the present passage.

(208) 115b12–116a1 He says

Source: *Ikhtiṣār al-Iskandarānīyīn*?

Cf.: Passage (202) above; Aristotle, *EN* 1098a16–20 and 1101a14–16.

This definition of the happy man sounds like a merging of two passages from Aristotle's *EN* and may derive from the *Ikhtiṣār al-Iskandarānīyīn*, the *Summa* as it first appears in the *PR* in passage (202) among other excerpts from the *Summa*. It stresses that happiness is a lifelong activity in accordance with virtue.

(209) 116a2–7 Excellence is a state

Source: Aristotle, *EN* 1106a15–23; *K. al-Akhlāq*, ed. Akasoy and Fidora, 169.12–17? Or *Ikhtiṣār al-Iskandarānīyīn*?[316]

In this statement excellence or virtue (ἀρετή, *faḍīla*) is defined as something which renders something else faultless. Its ultimate source is clearly the *EN*, but as the passage deviates from the wording of the Arabic translation, it may also derive from the *Ikhtiṣār al-Iskandarānīyīn*, the *Summa*. However, the remains of the Arabic version of the *Summa* (the preserved fragments, the quotations in Mubashshir, and the Latin or Hebrew translation) do not contain a rendering of this particular section. In his *Tahdhīb* (pp. 12–13), Miskawayh discusses the meaning of 'the best (*al-afḍal*)' in the same vein as our passage and illustrates it with, among others, the examples of horse and man.

[316] A partial parallel version found in the *K. al-Saʿāda* (facsimile edn Minovi, 78.2–5) deviates considerably from the *EN* and the *PR*.

(210) 116a8–117b1 Abū ʿAlī Miskawayh says

Source: Miskawayh, alternative *Fawz.*

Parallel text: *K. al-Ḥikma*, MS Esad Efendi, fol. 75a15–b19.

Cf.: *Tartīb al-saʿādāt*, ed. Awjabī in *Ganjīna-yi*, 109.16–110.2, 109.11–14, 109.4–8, 110.2–7; passage (207).

As the *PR* attributes this passage explicitly to Miskawayh, and since it is paralleled in the *K. al-Ḥikma*, it seems almost certain that it derives from the alternative *Fawz*, even though it has no correspondence in the standard *Fawz*. It is, however, very similar in contents, if not in wording, to Miskawayh's *Tartīb al-saʿādāt*. In his account there Miskawayh refers once to Aristotle (for the threefold division of happiness) and once to Aristotle's *K. al-Akhlāq* (for the concept that two sorts of happiness are attained mainly by chance, whereas the happiness of the soul is due to effort and endeavour). Yet in the *PR* the entire passage is presented without any reference to Aristotle.

(211) 117b2–123b2[317] According to the division by the people of wisdom

Source: Miskawayh, alternative *Fawz.*

Parallel text: *K. al-Ḥikma*, MS Esad Efendi, fols. 76a1–78b15.

Cf.: al-Fārābī, *Iḥṣāʾ al-ʿulūm*, ed. Amīn, pp. 111–20, esp. 111.4–9, 112.1–4, 11–12, 113.2–4, 114.9–10, 15–17, 116.9–10, 116.14–117.3, 117.5–6, 10–13, 16–18, 118.5–119.2, 119.5–14.

This division of wisdom is taken from the same source as the previous passage, namely the alternative version of Miskawayh's *Fawz*. The link between the two excerpts is the sciences, which are defined as 'happiness of the soul' in passage (210). A similar combination of the two topics, happiness and science, is found in Miskawayh's *Tartīb al-saʿādāt*, where it is even announced in the title, *Tartīb al-saʿādāt wa-manāzil al-ʿulūm*, i.e. *Succession of the Kinds of Happiness and Steps of the Branches of Knowledge.*

[317] In this section the third lacuna in the Marsh manuscript is found. The last line of fol. 120b breaks off after the first word, and the catchword 'blank (*bayāḍ*)' clearly indicates that the copyist deliberately left some space. Whether the lacuna was initially as big as it is now or whether a folio with only some lines on it (i.e. the beginning after the gap) has been lost in the transmission process is hard to tell. The text as we have it today restarts on fol. 123a1 in the first line in the middle of a sentence. The entire missing material can be restored using the *K. al-Ḥikma*.

In the present passage Miskawayh divides philosophy into a theoretical and a practical part, which are further divided into metaphysics, mathematics (consisting of arithmetic, geometry, astrology/astronomy and music)[318] and natural sciences, and into ethics, oeconomics and politics. In this he follows the general classification of philosophy given in the *Prolegomena* of the Late Alexandrian curriculum.[319]

Particularly interesting is the subdivision of the natural sciences which is not part of the standardised classification of the Alexandrian *Prolegomena*, but has a striking parallel in al-Fārābī's *Iḥṣāʾ al-ʿulūm*. Although al-Fārābī's overall classification of the sciences is completely different from Miskawayh's,[320] he starts his account of the natural sciences with the same lengthy analogy between the natural and the artificial bodies and then employs the identical eightfold subdivision. The parallels between the two texts increase towards the end: the fourth to the eighth subdivisions are described with almost identical wording. One obvious difference between the two accounts is, however, the ascription of the division to Aristotle in Miskawayh and its lack in al-Fārābī. As it seems unlikely that Miskawayh took his material from al-Fārābī and added this ascription, the most probable assumption to explain the similarities between the two accounts is a common source. This assumption is further corroborated by the fact that al-Fārābī's section on logic in the *Iḥṣāʾ* clearly shares a common source with Miskawayh's *Tartīb*. This source is, as Gutas demonstrates, a no longer extant Arabic translation of a lost work by Paul the Persian to whom Miskawayh refers by name.[321] If al-Fārābī used this work in the part on logic of the *Iḥṣāʾ*, it is plausible that he also used it in the part on natural sciences; and if Miskawayh quoted from it in his *Tartīb* for a classification of the sciences according to Aristotle's works, it is probable that he also used it when dividing the sciences in a more general way in

[318] For the division of mathematics in Greek and early Arabic sources, see Hein, *Definition*, pp. 182–8. The quadrivium and some similarities in contents to Miskawayh's presentation can be found in Ibn al-Ṭayyib's *Commentary on Porphyry's* Eisagoge (ed. Gyekye, 25.18–26.8; tr. Dunlop, 'Existence', p. 92).

[319] The same classification is found, e.g., in Ammonius, Elias, David and Paul the Persian; see Hein, *Definition*, pp. 163–251, esp. pp. 163 and 226; Gutas, 'Paul', pp. 240–1, 261.

[320] At the beginning of the *Iḥṣāʾ* (43.6–10), al-Fārābī lists the following parts: linguistics (*ʿilm al-lisān*), logic, mathematics (including arithmetic, geometry, optics, astronomy, music, weights and mechanics), natural sciences, metaphysics, politics, *fiqh* and *kalām*. The intention behind this scheme is, according to Gutas ('Paul', p. 260), to show that 'Logic and grammar, philosophy and religion, are ... *complementary* parts of the *same* system'.

[321] Gutas, 'Paul', pp. 236–8. Gutas further suggests that Paul's work was written in Pehlevi and translated by Abū Bishr Mattā (pp. 244, 253–4).

his alternative *Fawz* version. Gutas shows that the material on logic common to the *Iḥṣāʾ* and the *Tartīb* is probably based on the fourth Alexandrian *Prolegomenon*, which provides 'a general introduction to the philosophy of Aristotle'.[322] The account of the natural sciences shared by the *Iḥṣāʾ* and the alternative *Fawz* may derive from an exposition of the first *Prolegomenon* which introduces philosophy in a general way. Thus Paul's work may have either discussed all the *Prolegomena* or concentrated on the classification of philosophy and presented different possible divisions.

(212) 123b3–124a10 He says

Source: Miskawayh, alternative *Fawz*.

Parallel: *K. al-Ḥikma*, MS Esad Efendi, fols. 73b6–74a4.

Cf.: Miskawayh, *Fawz* II, 9 and I, 2, ed. ʿUḍayma, 93.17–94.1 and 41.7–42.3, with French tr. Arnaldez, pp. 55 and 20; tr. Sweetman, pp. 144 and 100.

This is a particularly interesting passage of Miskawayh's alternative *Fawz*, as it presents in a unit what is split up into two parts and occurs in different contexts and chapters in the standard *Fawz*. The first part links the present passage nicely to the previous one in that the main division of knowledge into the theoretical and the practical is taken up and they are both said to be sent down to the people by God through His prophets. However, as his further elaboration in the standard *Fawz* demonstrates, Miskawayh does not imply that knowledge can only be achieved through revelation. For the truth of prophetic instruction may be verified by the findings of the sages.

In the second part of the present passage the prophets are said to be the physicians of the soul. The idea is probably inspired by the Late Antique definition of philosophy as medicine for the soul, which is attested in the Arabic gnomological literature in different guises.[323]

(213) 124a11–13 Abū l-Ḥasan al-ʿĀmirī says

Source: al-ʿĀmirī, *al-Itmām li-faḍāʾil al-anām*?

Cf.: al-ʿĀmirī, *Iʿlām bi-manāqib al-Islām*, ed. Ghurāb, p. 78.

[322] *Ibid.*, p. 240.

[323] See Gutas, *Literature*, pp. 66–7 ('[Pythagoras] said: Philosophy is the medicine of the souls'), pp. 164–5 ('[Aristotle] used to say: Drugs cure bodies and philosophy cures souls; so, if people are in the habit of praising doctors for healing bodies, I am more deserving of praise for healing souls') and commentary pp. 233–4.

The maxim that knowledge without acting accordingly is useless and that there can be no action without prior knowledge[324] introduces a long continuous quotation, actually the longest in the *PR*, in the name of Abū l-Ḥasan al-ᶜĀmirī, a contemporary of Miskawayh. Since not many of al-ᶜĀmirī's works survive, it is not particularly surprising that the quoted material is not found in any of the extant ones. Although the compiler of the *PR* gives no indication whether the entire excerpt in the name of al-ᶜĀmirī, i.e. passages (213–22) derive from one single work by that author, it seems highly probable. Therefore one may venture a guess as to which of the lost works, about which we have some information,[325] could have been the source of the material preserved in the *PR*. As the main topic addressed in the excerpts is the virtues, a likely candidate is *al-Itmām li-faḍāʾil al-anām*, i.e. *The Perfecting of the Virtues of Man*. Not only does the title seem fitting, but also al-ᶜĀmirī refers to this work in his *K. al-Iᶜlām bi-manāqib al-Islām*, where he first quotes the maxim of passage (213) in a slightly different wording and then continues saying that he discusses the topic in *al-Itmām*.[326]

In the present passage a further maxim, namely that man is soul united with body, is attached immediately to the first one, although the two statements have no apparent connection. The possible link may be that in man knowledge is the activity of the soul, whereas acting belongs mainly to the body.

(214) 124b1–125b8 The virtues are divided into two parts

Source: al-ᶜĀmirī, *al-Itmām li-faḍāʾil al-anām*?

Cf.: *Muntakhab Ṣiwān al-ḥikma*, ed. Dunlop, lines 2790–802; and ed. Badawī, pp. 306–7.

This passage not only presents a division of the virtues into physical and

[324] This recalls two different Aristotelian statements, namely that the first in thought is the last in action (e.g. *Metaphysica* 1032b; and see Stern, 'The First in Thought') and that the end of theoretical knowledge is truth, and of practical knowledge, action (e.g. *Metaphysica* 993b20–1, and see, e.g., Adamson, *Arabic Plotinus*, pp. 30–1).

[325] Al-ᶜĀmirī lists seventeen of his works at the beginning of the surviving *K. al-Amad* (pp. 53–5). Further information on al-ᶜĀmirī's lost works can be gained from references in his other extant writings (see Rowson, *Muslim Philosopher*, pp. 7–17; and *Rasāʾil Abī l-Hasan al-ᶜĀmirī*, ed. Khalīfāt, pp. 468–75).

[326] The passage in *al-Iᶜlām* (p. 78) reads:
فإن العلم مبدأ للعمل ، والعمل تمام العلم ، ولا يُرْغَبُ في العلوم الفاضلة إلا لأجل الأعمال الصالحة .
The reference to *al-Itmām* occurs in *ibid.*, p. 79.

psychic virtues, but also lists the features which make the psychic virtues nobler than the physical ones, i.e. perpetuity and continuity, power and increase, end and completion, and help and support. According to the *Muntakhab*, Shahīd ibn al-Ḥusayn al-Balkhī (d. around 940), who is known for refuting and being refuted by Abū Bakr al-Rāzī on the topic of pleasures, describes more or less the same features as the excellences by which the pleasures of the soul excel the pleasures of the body, namely perpetuity and continuity, the end and existence of an aim (*wujūd al-ghāya*), power and increase, and completion. Even some of the explanatory sentences are astonishingly similar.[327] Shahīd allegedly discusses the topic in a work called *Tafḍīl ladhdhāt al-nafs ʿalā ladhdhāt al-badan* (*The Excellence of the Pleasures of the Soul over the Pleasures of the Body*). Whether his source is al-ʿĀmirī or vice versa or whether they both draw on a common source, is impossible to say.

The division of virtues into virtues of the soul and of the body occurs frequently in gnomological sources, where it is attributed to either Plato or Aristotle.[328] As the four virtues of the soul are the cardinal virtues wis-

[327] For Shahīd, see Kraus, *Rasāʾil*, pp. 145–7, where he suggests (p. 147, n. 1) that the passage in the *Muntakhab* might be taken from Shahīd's refutation of al-Rāzī; see further Rowson, 'al-Tawḥīdī', p. 69. The passage in the *Muntakhab* (ed. Dunlop, lines 2790–802) reads:
أحد الفضائل التي يفضل بها لذّات الأنفس على لذّات الأبدان الدوام والاتصال . وذلك أنّ لذّة النفس بما تقتنيه من سرور بوجود مطلوبها من الحكمة والعلم ، وبتيقّن بفضلها على غيرها دائمة متصّلة لا نفاد لها ولا انقطاع . وأمّا لذّة البدن بوجود القوّة الحساسّة محسوسها فمنقضية زائلة سريعة التبدّل والاستحالة . والثاني الانتهاء ووجود الغاية : فإنّ النفس كلّما تحرّكت في وجود مطلوب لها فأدركته مرةً أنقضى سعيها وتمّ فعلها وفرغت من شغلها . وأمّا البدن فكلّما انقضى وطره من محسوس له يلتذّ به بعلل ما نال من اللذّة وعادت الحاجة إلى ما كانت فالحركة دائمة والحاجة إلى أبد الأزمنة . والانتهاء إلى غاية تكفي وتغني عن ذلك الشيء يعينه معدوم . والثالث القوّة والازدياد فإنّ النفس كلّما استفادت فضيلة من فضائلها واقتفت لذّة من لذّاتها ، قويت به على نيل مثلها والازدياد بما هو أفضل منها . فأمّا البدن فإنّه كلّما نال محسوسه الملتذّ به أكثر ، كانت قوّته على نيل مثله وما هو أفضل منه في جنسه أضعف . والرابع التمام فإنّ النفس كلّما تزّيدت في فضائلها وقنيتها صارت إلى تمام طبع الإنسانيّة . فأمّا البدن فإنّه كلّما أزداد استهتاراً باللذّات المحسوسة ، وانهماكاً فيها ، زادت لذّاته بالقوّة البهيميّة التي في الإنسان وبعدته من تمام طبعه وشرائط إنسانيّته .

[328] The ascription to Plato is based on *Res Publica IV*, 442b-d (see Gutas, *Literature*, p. 388), and is much more common. It occurs, e.g., twice in Istanbul, MS Köprülü 1608, first under the heading *wa-mimmā jummiʿa ayḍan min nawādir al-falāsifa mimmā tarjamahū Ḥunayn* (which is not contained in Badawī's edition of al-Anṣarī's *Ādāb*), fol. 33b10–13, and second under the heading *al-Alfāẓ li-Aflāṭūn al-ḥikmīya*, fol. 62b20–4; and in the *Multaqaṭāt Aflāṭūn* (ed. Badawī in *Aflāṭūn*, 253.13–15):
وقال : فضائل النفس أربع ، وفضائل الجسد أربع بإزائها : فالنفس لها الحلم وللجسد التمام والكمال ، وللنفس العدل وللجسد الجمال ، وللنفس الشجاعة وللجسد القوّة ، وللنفس العفّة وللجسد الصحّة .

The *Mukhtār min Kalām al-ḥukamāʾ* quotes the same passage in the name of Aristotle (Gutas, *Literature*, p. 166). For a discussion of the four cardinal virtues, i.e. the virtues of dom (*ḥilm* or *ḥikma*), justice (*ʿadl*), courage (*shajāʿa*) and moderation (*ʿiffa*), the attribution to Plato seems more accurate. The corresponding virtues of the body are completion (*tamām* or *tamām wa-kamāl*), perfection (*kamāl*) or beauty (*jamāl*), power (*quwwa*) and health (*ṣiḥḥa*). Although al-ʿĀmirī employs the same division, his lists of virtues are strikingly different. The virtues of the body are not even listed straight away, but only later on in passage (222). As for the cardinal virtues, he replaces 'moderation' and 'justice' by 'generosity' and 'governance' in the present passage, whereas he normally uses the standard terms, even later on in passage (221).[329] Interestingly, Miskawayh also replaces 'moderation' by 'generosity' in his *Tahdhīb* (17.5–6) when discussing how virtues must be extended towards one's fellow men to be virtues in the true sense of the word: 'a person is said to possess one of these [four cardinal] virtues and is praised for it only when it goes beyond him to others' (tr. Zurayk, p. 16). In this context, which is the same as in al-ʿĀmirī's quotation in the *PR*, a replacement makes sense, as it would be difficult to describe the extension of moderation towards others.

Although al-ʿĀmirī thus operates with a slightly different list of cardinal virtues, he still ascribes each of them to one particular part of the soul.[330] In doing so he follows the Platonic tradition, which was widespread in Arabic.[331]

In what follows, the compiler of the *PR* presents al-ʿĀmirī's detailed discussion of two of the four cardinal virtues, namely knowledge and governance, in passages (215–16) and (218–20), whereas generosity and courage are dealt with together in a rather short excerpt (217). Whether ʿĀmirī did not elaborate on these latter two or whether his discussion of them was simply not taken up by the compiler of the *PR* is impossible to say.

the soul, see, e.g., al-Kindī's *R. fī Ḥudūd al-ashyāʾ* under the entry *al-faḍāʾil al-insānīya* (*Rasāʾil al-Kindī*, ed. Abū Rīda, I, 177.4–178.1).

329 See also the quotation of al-ʿĀmirī preserved al-Tawḥīdī's *al-Baṣāʾir*, III, 94.1–4.

330 The same topic is also discussed in another fragment most probably deriving from another of al-ʿĀmirī's lost works preserved in Istanbul, MS Servili 179 (fol. 110b-111a) after his *K. al-Amad*. See my partial translation in Wakelnig, 'Metaphysics', pp. 49–50.

331 See, e.g., the following passage of the *Multaqaṭāt Aflāṭūn* (ed. Badawī in *Aflāṭūn*, 247.5–9):
وقال : من ضبط ثلاثة أشياء حتى يقوى على ضبط أعنّتها ، قوى على السياسة كلّها : على الخاصّ نحو ذاته ، وعلى العامّ نحو الكلّ ، وهي النطق والغضب والشهوة . وذلك أنّ من ضبط المنطق كان حكيماً ، ومن ضبط الغضب كان شجاعاً ، ومن ضبط الشهوة كان عفيفاً . ثم يستعمل العدل في هذه الثلاثة لأنّه في المنطق عدل ، وفي الغضب عدل ، وفي الشهوة عدل .
See also *M. fī Ithbāt faḍāʾil al-nafs* (Ps.-Plato, 'Pseudoplatonischer Text', ed. Daiber, lines 35–7).

(215) 125b9–126a11 Knowledge is divided into two parts

Source: al-ʿĀmirī, *al-Itmām li-faḍāʾil al-anām*?

Parallel: ʿAlī al-Riyāḍī, *K. Ṣaḥāʾif al-laṭāʾif*, MS Aya Sofya 4130, pp. 26–7.[332]

Cf.: al-ʿĀmirī, *Iʿlām bi-manāqib al-Islām*, ed. Ghurāb, pp. 84–5; tr. Rosenthal in *Classical Heritage*, p. 63; German tr. Rosenthal in *Fortleben*, pp. 91–2.

Like Miskawayh in passage (211), al-ʿĀmirī divides knowledge into two parts. Yet, instead of following the Late Antique division into theoretical and practical philosophy, he distinguishes between the philosophical and the religious sciences. As for the philosophical part, he lists only the theoretical sciences and contrasts them with the Islamic disciplines of *ḥadīth*, *kalām* and *fiqh*. It is not clear whether his intention is the replacement of practical philosophy by these disciplines or whether he is simply dividing knowledge along different lines from, for example, Miskawayh.

The quotation, except for the last paragraph on the relation of philology and logic to the other sciences, occurs also verbatim in ʿAlī al-Riyāḍī's *K. Ṣaḥāʾif al-laṭāʾif* in the name of Abū l-Ḥasan Muḥammad al-ʿĀmirī, but unfortunately without any indication of the source text. Since it is unlikely that ʿAlī al-Riyāḍī in the fifteenth century had access to the *PR*, his quotation is an additional witness of an independent treatise by al-ʿĀmirī containing our passage.[333] It therefore rules out the possibility that the compiler of the *PR* rephrased a section from al-ʿĀmirī's *Iʿlām* which also lists the two divisions of knowledge or the sciences, but does not discuss the relation of philology and logic to the other sciences. The passage in the *Iʿlām* further divides the three parts of the philosophical sciences and discusses the five branches of mathematics, namely arithmetic, geometry, astronomy/astrology, music and mechanics. As for the natural sciences, they are divided along the different existing substances, namely the heavenly bodies and the four elements and what is composed of the latter. This division resembles the categories of the simple and the composite substances in Miskawayh's discussion of natural sciences in passage (211) above. Among the sublunar existents, al-ʿĀmiri further distinguishes the things in the atmosphere, in the interior of the earth and in between these two strata, namely the animals and plants.

332 The manuscript is paginated, not foliated.

333 The composition date of the *K. Ṣaḥāʾif al-laṭāʾif* is, according to Minovi ('Az Khazāʾin', p. 74), 1478, whereas Gönül Alpay Tekin's entry 'al-Riyāḍī' in the *EI*², which mentions a *K. Ṣaḥāʾif al-laṭāʾif* among al-Riyāḍī's works, places him in the sixteenth and seventeenth century. At the end of his al-ʿĀmirī quotation, al-Riyāḍī continues with some explanatory remarks in Persian.

The last paragraph of the present passage reminds us of previous passages where we found the Aristotelian concept that some goods are only goods because they help to attain what is good in itself. Along the same lines, the existence of philology and logic is justified as they exist for the sake of the attainment of the other sciences, which in turn ensure happiness.

The question of whether logic is a part of philosophy or an instrument to it as al-ʿĀmirī claims, was already raised by Alexander of Aphrodisias and later by Ammonius and Philoponus.[334] The Greek discussion is, for example, taken up in the introductory part to Ibn al-Ṭayyib's *Commentary on Porphyry's* Eisagoge.[335]

(216) 126a12–127a1 Any of the things that can be truly known is reached

Source: al-ʿĀmirī, *al-Itmām li-faḍāʾil al-anām*?

The excerpt as presented by the compiler of the *PR* contains no clue as to how this threefold division of knowledge is to be harmonised with the twofold one in the previous passage. The passage in al-ʿĀmirī's *Iʿlām* which is partly parallel to passage (215) also mentions knowledge acquired by experience, but sets it aside from the true sciences, i.e. the philosophical and the religious. It claims that 'The common people sometimes use the word "science" for any trade whatsoever. Empiricists sometimes use it for ideas empirically acquired, for example, fortune-telling based on augury through observation of the flight of birds or based on shoulder blades and footprints' (tr. Rosenthal, *Classical Heritage*, p. 63). With this in mind, it is surprising that al-ʿĀmirī apparently grants knowledge attained through experience the same level of reality and truth as the knowledge gained through heavenly matters, i.e. revelation – probably corresponding to the religious sciences of the previous passage – and through deduction, which may correspond to the philosophical sciences.

(217) 127a1–11 Generosity

Source: al-ʿĀmirī, *al-Itmām li-faḍāʾil al-anām*?

In contrast to the detailed discussion of knowledge and leadership, gener-

[334] See Hein, *Definition*, pp. 153–162.

[335] See ed. Gyekye, pp. 26–7; tr. Dunlop, 'Existence', pp. 92–3.

osity and courage are dealt with sparsely, either by al-ʿĀmirī himself or due to a deliberate choice by the compiler of the *PR* in the present passage. An interesting aspect is the establishment of generosity as some kind of meta-virtue which becomes apparent in defining 'sharing the virtues knowledge and courage' as 'generous'.

(218) 127a12–128a11 Leadership is a craft

Source: al-ʿĀmirī, *al-Itmām li-faḍāʾil al-anām*?

Cf.: al-Balkhī's account in al-Tawḥīdī's *al-Baṣāʾir*, ed. al-Qāḍī, IX, pp. 146–7; tr. Rosenthal in 'al-Balkhī', pp. 289–90.[336]

If it can be assumed that this section still belongs to the quotation in the name of al-ʿĀmirī, it presents his rephrasing and rearrangement of a passage by his teacher, Abū Zayd al-Balkhī, which is lost in the original, but preserved in al-Tawḥīdī's *al-Baṣāʾir*. Al-Tawḥīdī indicates that he excerpted the passage from al-Balkhī's *K. al-Siyāsa*, although he does not specify whether his reference is to the large or the small book which Abū Zayd allegedly wrote on the subject.[337] As Rosenthal shows, the passage is firmly

[336] I re-arrange the text of *al-Baṣāʾir* according to its parallels to al-ʿĀmirī's passage in the *PR* to facilitate the comparison of the two passages (each dash indicates a switch within al-Tawḥīdī's account):
السياسة صناعة ، ثم هي من أجل الصناعات قدراً وأعلاها خطراً ، اذ كانت صناعة بها تتهيّأ عمارة البلاد وحِمَايَة من فيها من العباد وكلّ صانع من الناس فليس يستغني في إظهار مصنوعه عن خمسة أشياء تكون عللاً لها أحدها مادّة له آلة ومادّة يعمل بها . ومثال ذلك من صناعة البناء أنّ المادّة التي يعمل منها البناء هي التراب والطين والحجارة والخشب . ومثال ذلك من صناعة الطب أنّ المادّة التي يفعل بها الطبيب إنما هي أجساد الناس المحتملة الصحّة والسقم . والثاني صورة يَنْحُو بفعله نحوها . والصورة التي ينحوها بوهمه صورة البيت . والصورة التي ينحوها الطبيب بوهمه إنما هي الصحّة . والثالث حركة يستعين بها في توحيد تلك الصورة بالمادّة . والفاعل هو الطبيب المعالج . والفاعل هو البناء . والرابع غرض ينصبه في وهمه من أجله يفعل ما يفعل . والغرض الذي من أجله يفعل سكنى البيت وإحراز ما يحرز فيه . والغرض الذي بسببه يفعل الطبيب إنما هو بقاء جسم المعالج المادّة التي تتهيّأ له ان يبقاها . والخامس آلة يستعملها في تحريك المادّة . والآلة التي بها يعمل هي آلات النباء . والشيء الذي يتخذه الطبيب آلة في المعالجة وإفادة الصحّة هو كالقصد وسقي الادوية . فإذا نقل هذا المثال إلى صناعة السياسة قلنا إنّ المادّة فيها أمور الرعية التي يتولّى الملك القيام بها والصورة فيها إنما هي المصلحة التي ينحو نحوها وهي نظير الصحّة لأنّ المصلحة هي صحّة ما والصحّة مصلحة ما وكذلك المفسدة سقم ما والسقم مفسدة ما ، والفاعل هو عناية الملك بما يباشره من أمور الرعية وغرضه فيما يفعله هو يقاء المصلحة ودوامها والشيء الذي يقوم له مقام الآلة في صناعته إنما هو الترغيب والترهيب . وفعل السائس الذي هو نظير المعالجة من الطبيب ينقسم بكلّيّة إلى قسمين أحدهما التعهّد والآخر الاستصلاح ؛ أمّا التعهّد فحفظ المستقيم وأمور الرعية على استقامة وانتظام من الهدوء والسكون حتى لا يزول عن الصورة الفاضلة ؛ وأمّا الاستصلاح فردّ ما عارضه منها الفساد والاختلال إلى الصلاح والالتئام . ونظير هذا التعهّد والاستصلاح في صناعة السياسة من صناعة الطبّ التي هي سياسة الأجساد حفظ الصحّة وإعادة الصحّة ، وكما أنّ الطبّ كلّه مدرج في هذين البابين ، كذلك السياسة كلّها مدرجة في نظيريهما ، يعني التعهّد والاستصلاح .

[337] See Rosenthal, 'al-Balkhī', p. 287, and, for the following, pp. 290–5; see also Wakelnig, 'Fragments of al-ʿĀmirī', pp. 233–6.

rooted in the Aristotelian tradition, but enhanced with Persian and Islamic material. The idea that leadership is the craft of the king and that his subjects are the matter in this craft also occurs in Ps.-Māwardī's *Naṣīḥat al-mulūk*. Thus it fits well that one of its editors, Aḥmad, tentatively suggests Abū Zayd al-Balkhī to be the author of this work.[338]

(219) 128a12–130a3 In administering his affairs

Source: al-ʿĀmirī, *al-Itmām li-faḍāʾil al-anām*?

The idea that the king needs a number of people supporting him in exerting his power is, of course, ubiquitous, especially in the genre of Mirror of Princes of which this and the following passages are reminiscent.[339] The twofold division and the number of seven classes in each division are, however, peculiar and may well be al-ʿĀmirī's original contribution to the theme. Much of the other material he uses is commonplace – for example, the analogy between the secretaries and the sovereign's tongue.[340]

(220) 130a3–132a1 It is the king's duty

Source: al-ʿĀmirī, *al-Itmām li-faḍāʾil al-anām*?

This section describes the relation between the king and his subjects. Both of them have duties to the other, which al-ʿĀmirī lists in sets of six. As in the previous passage, the material may derive from the genre of the Mirror of Princes, but it also contains the single reference to the Koran in the entire *PR*. The verse, 9:120, is quoted to emphasise that the subjects are to

338 For the passage on leadership as craft and the subjects as matter, see Ps.-Māwardī's *Naṣīḥat al-mulūk*, ed. Khiḍr, 198.5–6. For Aḥmad's suggestion, which is presented in his edition of the *Naṣīḥat al-mulūk* to which I did not have access, see Crone, *Medieval Islamic Political Thought*, p. 439.

339 See, e.g., Ps.-Māwardī's *Naṣīḥat al-mulūk* (ed. Khiḍr, pp. 185–7).

340 Geert van Gelder (personal communication) was kind enough to indicate to me the following occurrences of this analogy: the earliest instance is probably the one in the *Risāla* by the Umayyad secretary ʿAbd al-Ḥamīd ibn Yaḥyā to a fellow secretary (*kātib*), quoted in Ibn Khaldūn's *Muqaddima*, which reads in Rosenthal's translation (II, pp. 29–30) as follows: 'Your position with regard to rulers is that (you are) the ears through which they hear, the eyes through which they see, the tongues through which they speak, and the hand through which they touch.' It is further used in the *Ādāb al-mulūk* attributed to al-Thaʿālibī (ed. al-ʿAṭiyya, 141.3–4):
ومعلوم أنّ الكتّاب ساسة الملك ، وعماد المملكة وألسنة الملوك وجهابذة الألفاظ وبأقلامهم تدبّر الأقاليم .

It is also attested in al-Tawḥīdī's *al-Imtāʿ* (ed. Amīn and al-Zayn, I, 102):
لأنه [الكاتب] لسانه الذي به ينطق .

submit entirely to their righteous sovereign, in which case God will reward them for any hardship they may encounter during their submission.

(221) 132a2–133a10 The doctrine has already been presented before

Source: al-ʿĀmirī, *al-Itmām li-faḍāʾil al-anām*?

Cf.: passages (109–10) and also passage (214); al-Tawḥīdī, *al-Baṣāʾir*, ed. al-Qāḍī, III, 94.1–4.

The reference to the previously mentioned doctrine is to passage (214) above,[341] where al-ʿĀmirī operates with a slightly revised list of the four cardinal virtues: generosity, courage, wisdom, leadership. However, the fact that he links these two passages indicates that he does not perceive them as different from each other. He then combines the Platonic cardinal virtues with the Aristotelian concept of virtue as the mean between two vices, which has already been taken up in the *PR* in the words of Miskawayh in passages (109–10). The ultimate source may in both cases be the Ps.-Aristotelian *De virtutibus et vitiis*, translated into Arabic twice.[342] The structure of the passage further recalls the account of the three actions, i.e. the two evil extremes and the one good mean, in the introductory part of Ibn al-Ṭayyib's *Commentary on Porphyry's* Eisagoge.[343]

In his *Baṣāʾir*, al-Tawḥīdī preserves a similar discussion of the virtues in the name of al-ʿĀmirī, but since the wording is, except for the main termini, completely different, it is unlikely that the *PR* and al-Tawḥīdī use the same source.[344]

(222) 133a11–133b13 The physical virtues

Source: al-ʿĀmirī, *al-Itmām li-faḍāʾil al-anām*?

[341] This is a further indication that the passages (213–22) are all excerpted from the same al-ʿĀmirī source.

[342] See Ps.-Aristotle, ed. Kellermann in 'Pseudoaristotelischer Traktat über die Tugend', pp. 39 and 59.

[343] The passage reads (ed. Gyekye, 25.7–13, and tr. Dunlop, 'Existence', pp. 91–2):
ثلاثة أفعال فعلان منهما قسمين شراً وهما الطرفان وواحد يسمّى خيراً ، وهو المتوسّط . ففعل
النفس الشهوانيّة المتوسّط يسمّى عفّة ، والطرفان أحدهما شرهاً والآخر يسمّى كلال الشهوة .
وفعل النفس الغضبيّة المتوسّط يسمّى شجاعة ، والطرفان يسمّى احدهما تهوّراً والآخر يسمّى
جبناً . وفعل النفس الناطقة المتوسّط حكمة والطرفان احدهما يسمّى خبّاً والآخر يسمّى بلادة .
وجملة الأفعال الثلاثة المتوسّطة المحمودة يسمّى عدالة ، فالإنسان يكون جميل الأخلاق بحسب ما
يقتضيه الإنسانيّة إذا ما كان عادلاً وهذا هو ان يكون حكيماً شجاعاً عفيفاً .

[344] Unlike to the compiler of the *PR*, al-Tawḥīdī may well have rephrased his source.

This is the complementary passage to passage (214) in that al-ᶜĀmirī finally lists the virtues of the body. They are completely different from the four bodily virtues which are found in, for example, gnomological sources in the name of Plato and Aristotle as counterparts to the cardinal virtues.[345]

The importance which is assigned to birth- and living place sounds Galenic and the doctrine was probably well known in Arabic.

[345] See comment on passage (214).

Appendices

Appendix 1

Excerpts from Galen, *Kitāb fī Manāfiʿ al-aʿḍāʾ* which the compiler of the *PR* changed and re-arranged considerably when quoting them in passages (118), (131) and (143).

Excerpt 1

K. fī Manāfiʿ al-aʿḍāʾ IX, 4 (Bibliothèque Nationale de France, MS Arabe 2853, fol. 156a3–10; John Rylands Library, MS Arabic 809, fol. 155b9–15; National Library of Medicine, MS A 30.1, fol. 134a11–17).

وإذ كان ذلك كذلك فالذي بيّنّاه في كتاب آراء بقراط وفلاطن من أنّ الروح النفسانيّ يتولّد من الروح الذي به تكون الحياة الذي يصعد في العروق الضوارب التي ترتقي من القلب إلى الدماغ وهو بمنزلة العنصر والمادّة المشاكلة صواب . والشيء الذي كتبت قبله في أوّل هذا الكتاب فأني أقوله هاهنا أيضاً وهو أنّه ليس يستطيع أحد أن يدرك منفعة شيء من الأعضاء الجزئيّة دون أن يعرف أوّلاً فعل جملة العضو المركّب من الجزئيّة وقد أتيت في كتاب آراء بقراط وفلاطن بالبرهان على أنّ النفس الناطقة تسكن الدماغ وإنما نتفكّر بالدماغ وأنّه يحوي روحاً نفسانيّاً كثيراً .

If it is like that then what we have explained in the *Book on the Opinions of Hippocrates and Plato* is correct, namely that the psychic spirit is brought forth by the spirit due to which life exists, which rises in the arteries ascending from the heart to the brain and serves as matter and proper material. The matter about which I have written before at the beginning of this book I will repeat here, i.e. that no one can perceive the benefit of any of the partial body parts without knowing the action of the entirety of the body part which is composed of the partial [parts] first. In the *Book on the Opinions of Hippocrates and Plato* I have brought forward the proof that the rational soul dwells in the brain, that we think with the brain and that it comprises abundant psychic spirit.

Excerpt 2

K. fī Manāfiʿ al-aʿḍāʾ V, 2–3 (Bibliothèque Nationale de France, MS Arabe

2853, fol. 79b16–80a6; John Rylands Library, MS Arabic 809, fol. 82a11–24; National Library of Medicine, MS A 30.1, fol. 52a3–15).

ولو لم تكن قد تقدّم في توسيع هذا الموضع واخلائه لما كان يمكن فعل شيء مما وصفت على ما ينبغي ولو كان المعاء الصائم لاصقاً بأسفل المعدة لكانت استدارته ستحدث في هذا الموضع من الزحمة والضيق امراً ليس باليسير ، فلسابق علم الفاعل بهذا لم يجعل للمعاء الأوّل الذي يلقى المعدة استدارات بل جعله ممتدّ⟨اً⟩ مستقيماً على الصلب بقدر الحاجة كانت إلى توسيع الموضع وإخلائه للأجرام التي ذكرتها . فإذا انتهى هذا إلى المعاء المتوالي التفّ ما يليه من الأمعاء وانثنى وانطوى وصارت له استدارات ، وهذا الجزء من الأمعاء يسمّى الصائم وإنما سمّى بذلك لأنّه يوجد فارغاً أبداً ليس فيه من الغذاء قليل ولا كثير . وأمّا ما فوق هذا المعاء الصائم بينه وبين أسفل المعدة وهو الذي عدم الاستدارات والتلفّف للسبب الذي ذكرناه فإنّ أصحاب التشريح سمّوه أوّل نبات الأمعاء . والآلات التي يصير إليها الغذاء بعد المعدة وهي الأمعاء منظومة هذا النظام أوّلها مما يلي أسفل المعدة المعاء المسمّى أوّل نبات الأمعاء والثاني المعاء الصائم والثالث المعاء الدقيق والرابع المعاء المعور والخامس المعاء المسمّى قولن والسادس المعاء المستقيم الذي في طرفه العضل الذي يضمّه ويمنع الثفل ويحبسه . فقد وصفت منافع جميع الأمعاء وذاك أنّي وصفت في هذه المقالة منفعة أوّل نبات الأمعاء ووصفت في المقالة السالفة منافع أصناف جميع الأمعاء الدقاق والغلاظ .

If there were no prearrangement for widening and emptying this space, the action of what we have described could not be as it should be, and if the fasting intestine stuck to the lower part of the stomach, its coil would cause some significant crowd and constraint in this space. Due to the Agent’s foreknowledge of this He did not make coils for the first intestine which meets the stomach, but made it extended and stretched along the spinal column as much as was necessary to widen and empty the space for the bodies which we have mentioned. When this ends in the successive intestine, the following part of the intestines winds, bends, folds and has coils. This part of the intestines is called fasting, which is called that because it is always found empty with no nutriment in it, neither little nor much. The part above this fasting intestine, [i.e.] between it and the lower part of the stomach, [the part] which has neither coils nor windings due to the reason mentioned is called ‘first outgrowth [into] the intestine’ by the anatomists. The instruments which the food reaches after [having passed] the stomach are the intestines organised in this order: the first of them belongs to what follows the lower part of the stomach. It is called the ‘first outgrowth [into] the intestine’, the second is the fasting intestine (*jejunum*), the third is the small intestine, the fourth is the blind gut (*caecum*), the fifth is the intestine called 'colon', and the sixth is the straight intestine (*rectum*) which has at its end the muscle which closes it, stops and confines the waste. I have already described the benefits of all intestines, i.e. in this chapter I have described the benefit of the 'first outgrowth [into] the intestine' and in the preceding chapter the benefits of all kinds of intestines, be they thin or thick.

Excerpt 3

K. fī Manāfiᶜ al-aᶜḍāʾ XIV, 2 and 4 (Bibliothèque Nationale de France, MS Arabe 2853, fols. 244b13–245a6, 246a21–b12; John Rylands Library, MS Arabic 809, fols. 246b18–247a9, 248b7–25; Escorial, MS Arab. 850, fols. 89b10–90a2, 91a17–b10; National Library of Medicine, MS A 30.1, fol. 203b6–19, 204b24–205a13).

ولكن لما كانت المادّة التي منها خُلقوا غير ممكن فيها ذلك إذ كان الشيء المخلوق المركّب من عروق ضوارب وغير ضوارب وعصب وعظام التي فيها ينفذ ولحم لا يمكن أن يكون غير قابل للفساد والتغيّر ، لطف بحكمته فأرفق خليقته بالشيء الذي به كان يمكن أن يصير إلى النجاة من التلف والموت . وذلك بمنزلة البناء الحكيم الحاذق الذي إذا بنى مدينة لم يجعل قصده فيها وعنايته ببنائها أن تسكن وتعمر في ذلك الزمان الذي يبنيها فيه فقط بل يحتاط في بنائها ويحتال أن تبقى له أبداً أو تمكث دهراً طويلاً ، إلا أنا إلى هذه الغاية لم نجد مدينة قد أتى عليها من الدهر ما درس فيه ذكر المجد لها حتى لا يعرف ولا يذكر . وأمّا أفعال الخالق وآثاره في الخلق فقد مضت عليها ربوات دهور كثيرة وستبقى كذلك بما لطف لها بحكمته من التناسُل الذي يقوّم به بدل الشخص الهالك من الحيوان شخص آخر . وقد قصدنا في كلامنا هذا إلى تعليمك آثار هذه الحكمة في الإنسان وغيره من الحيوان وتنبيهك على الحيلة التي بها يلطف الخالق لأن ينفي التلف عن أجناس الحيوان حتى لا يبيد شيء منها ويبقى على حالها سليمة غير مائتة . وأنا مبتدئ من هاهنا فأقول إنّه خلق لجميع الحيوان أعضاء يتناسل بها وجعل في تلك الأعضاء قوّة غريزيّة لها خاصّة بها تكوّن اللذّة وحبّبت إلى النفس المستعملة لتلك الأعضاء استعمالها وشوّقت إليه شوقاً عجيباً لا يوصف ، فصار الحيوان إذا بعثته هذه المحبّة وهذا الشوق المطبوع فيه وحرّكته الشهوة إلى مباطشة ذلك انبعث على طلب التناسل واستبقاء جنسه ألا يبيد …

والسبب في اختلاف عدد الأرحام أنّه إنما عنى بأمر الخلقة كيلا يبيد شيء من أجناس الحيوان ، تلطّف لما كان من الحيوان قليل البقاء قصير العمر إمّا لضعف بدنه وإمّا لأنّه طعمة بحيوان هو أقوى منه فجعل بتلافي تواتر فساده وتوالي تلفه بكثرة توالده . وذلك أيضاً من الأشياء العجيبة في الخلقة الا أنّ هاهنا شيء لا أشكّ أنّك تقرّ بأنّه أعجوبة تفوق كلّ عجب مساواة عدد تجويفات الأرحام لعدد

الأثداء ، وذلك أنّ هذا موضع ليس يمكن السوفسطائين أن يقولوا فيه أنّ العلّة التي لا تنقاس والبخت العديم الحكمة جعل للأرحام في الناس تجويفين وفي الخنازير تجويفات كثيرة ، وأمّا ما نجده من مساواة عدد الأثداء لعدد التجويفات فإنّه يدفع عن الفكرةِ التصديقَ بالكون الهذيانيّ الكائن من تلقاء نفسه ، مع أنا لو سلّمنا أنّ الذي نجده من ذلك في الإنسان والخنازير إنما يقع بالبخت فإنّ الذي يجده في سائر الحيوان من مساواة أولاده لعدد أثدائه مما لا يذهب عمن بلغ غاية القحة وصفاقة الوجه لو أنّه مما لا يكون عن غير تعمّد وعناية وقصد إلا يكون في غاية عدمان الحسّ . وكذلك ما يجده من كون اللبن في الثدى بعد تمام الجنين وقرب حاجته إلى الخروج بالولاد مما لا أحسب أحداً من الناس يظنّ أنّه من فعل البخت عن غير معرفة ولا تعمّد يفعل وينفي عنه أن يكون دليلاً على حكمة الصانع العجيبة . فإنّه لو لم يفّ شيء مما وصفت باقناعهم وردعهم وردّهم إلى التصديق بالرأي الحقّ والإقرار بأنّه فعل بحكمة لكانت هذه الواحدة ينبغي أن تفي بذلك وتصدفهم عن رأيهم الباطل ، وإنه لما كان جميع الحيوان في أوّل ميلاده ليّن البدن رخو وضعيف النحيزة صار لا يمكنه استمراء الغذاء الصلب فجعل له غذاء مهيئاً من والدته مثل الغذاء الذي كان يغتذي به منها وهو فى الرحم .

Since the matter from which [the animals] are created does not allow for that [i.e. permanence] – as it is impossible that something created and composite of arteries, veins, nerves, bones, in which they [the nerves] pass, and of flesh is incorruptible and unchangeable – He in His wisdom has been benevolent and kind with His creation by something through which deliverance from destruction and death has become possible. That is like the wise and skilled builder who when building a city does not aim at and exert providence that it may be inhabited and cultivated only at the time of being built, but in building it he makes provision and exercises a skill so that it always lasts for him or remains a long time. Yet, as to that extent, we do not find any city over which such a long time has passed that the mention of its founder has been wiped out so that he is neither known nor remembered. Over the Creator's acts and effects upon creation many ten thousands of lifetimes have passed and they will still remain the same due to procreation which He in His wisdom has benevolently provided for them and by means of which a deceased individual among the animals is replaced with another one. With this our speech we intend to inform you about the effects of this wisdom upon man and other animals and to call your attention to the skill by which the Creator is benevolent. For destruction is avoided for [all] kinds of animals so that none of them perishes, but remains safely and undyingly in its condition. I start from here saying that

He has created reproductive organs for all animals and has put into these organs an innate power specific to them, which produces pleasure, evokes in the soul, which uses these organs, a love for using them and awakens an amazing indescribable desire for doing so. So when this love and desire imprinted in it rouses the animal and passion moves it towards enforcing it, it is incited to desire procreation and to ensure the continuance of its kind so that it may not perish ...

The reason for the different numbers of uteri is to exert providence over creation so that none of the kinds of animals perishes. He has acted benevolently towards animals which are transient and short-lived because of either the weakness of their bodies or their being food for animals stronger than them. So He has prepared compensation for their constant corruption and continuous destruction by means of abundance in their reproduction. That also belongs to the astonishing things among creation, yet there is something which I do not doubt you acknowledge: that it is a wondrous thing exceeding all wonders, [namely] that the number of sinuses in the uteri equals the number of teats. Therefore this is not a topic about which the Sophists can say that an unreasonable cause and chance devoid of wisdom have put two sinuses in the human uteri and many in the pigs. In [logical] thinking the fact that we find the number of teats equal to the number of sinuses does away with the belief in an irrational creation generated spontaneously. Although even if we conceded that what we have found in humans and pigs has happened by chance, the equality of the number of offspring to the number of teats found in all other animals – if it belongs to that which cannot be caused without intention, providence and aim – could not escape the most impudent and insolent, unless he [also] completely lacks sensation. In the same way the fact that there is milk in the teats after the embryo has reached completion and is close to its necessary emergence through delivery belongs to that which, as far as I see, nobody believes to be caused by the action of chance without knowledge and intention to do it and which [nobody] rejects as proof for the amazing wisdom of the Creator. If nothing I have described suffices for convincing, repelling and bringing them back to believe in the true opinion and to consent to it being done wisely, this single [fact] must suffice for that and for turning them away from their corrupt opinion: when at the first time after its birth every animal is soft in body, fragile and weak in nature, it is not able to digest solid nutriment. So He has made nutriment for it, which is prepared by its mother like the nutriment which it has been fed by her while in the womb.

Appendix 2

A continuous excerpt from the *Kitāb al-Ḥaraka* (Ankara, Univ., MS Ismail Saib I 1696, 99b1–100b2; Istanbul, Süleymaniye Kütüphanesi, MS Hacı Mahmud 5683, fol. 128a5–129a17) with four parallels in the *PR*; due to shortage of space the text of MS Ismail Saib I 1696, 100a3–b1, and MS Hacı Mahmud 5683, fol. 128b13–129a15, which is more or less identical to *PR*, passages (12–13), is not reproduced (see the indicated passages and the corresponding apparatus).

فقال إنه ليس شيء من الأشياء الكائنة بعرض قبل الأشياء الذاتية فان لم تكن قبلها فليست اذن بعلل ، فإن كان هذا هكذا وكان الشيء الكائن بالبخت والشيء الكائن بالاتفاق ليسا عللاً لكون السماء فالعقل الأوّل والطبيعة هما علل لكون السماء . وقد تبيّن ذلك في كتاب سمع الكيان في المقالة الثانية حين وصف العلل وقسمها وميّز العلّة الفاعلة من سائر العلل . ثم وصف كم العلل التي ينبغي أن تضاف إلى كون العالم كلّه وأنّه علّة بمعنى وكيف ينبغي أن يميّز بعضها من بعض . فقال إنّ الأوائل المحرّكة بحركة طبيعيّة إثنان ، أعنى بالحركة الطبيعيّة الشيء المتحرّك بذاته لا بحيلة ولا بإرادة . ثم قال إنّ أحد الأوّلين المحرّكين ليس هو بطبيعيّ وذلك أنّه ليس فيه مبدأ حركة وهو الشيء المحرّك دائماً من غير أن يتحرّك ، مثل الشيء الذي هو أوّل الأشياء كلّها ولا يتحرّك البتّة . وإنما عنى الفيلسوف بقوله أنّ الشيء الأوّل هو مبدأ أنّيّة العالم كلّه وهو محرّك الأفلاك وناظم شرحها . وقد تبيّن هذا في المقالة الثانية من كتاب سمع الكيان . فقال إنّ الحركة تسلك في جميع الأشياء دائمة وأقام على ذلك دلالات مضطرّة وإنّ الحركة لا تعب فيها إنما يمكن أن تكون في الأجرام المتحرّكة حركة مستديرة وجعل قياسه في إثبات ذلك من قسمة الحركة وبيّن كيف يكون ذلك . ثم وصف بعد ذلك الأجرام المستوية والمستديرة فقال إنّه لا يمكن أن يكون عظم ما ولا شئ من الأجرام الطبيعيّة لا المستوية ولا الأجرام المستديرة لا نهاية له في العظم ، واحتج على ذلك في الكتاب الذي سميناه وفي كتاب السماء والعالم في المقالة الأولى واحتج على ذلك بالبرهان ، فلما بيّن أنّه لا يمكن أن يكون شيء من الأجرام لا نهاية له في العظم أتبع ذلك فقال إنّ الجرم المتناهي في العظم [see PR, passage (12)] ... فقال إنّ هاهنا غير هذه الجواهر اضطراراً [see PR, passage (13)]... وأنّه لم يكن ليمكن الشيء الذي لا يحتاج في كونه إلى العلّة الإلهيّة أن يقبل منها شرحاً ولا تدبيراً ولا ترتيباً ، وقد ذكر الفيلسوف في كتابه الذي يدعى ثلوجيا وهو كتاب الربوبيّة وأتى هنالك بمثل ضربه .

He says that none of the things which are accidental are prior to the essential things, so they are not prior to them and therefore no causes. If it is like that and if something which is by luck and something which is by chance are not causes for the being of the heavens, then the first intellect and nature are causes for the being of the heavens. He has already examined that critically in the second chapter of the *Physics* where he describes the causes, classifies them and distinguishes the efficient cause from the other causes. Then he has described how many causes [there are] which must be connected to the being of the whole world, what 'cause' means and how they must be distinguished from each other. He says that the principles which cause natural motion are two. By natural motion he means something moved by itself, not by skill or will. Then he says that one of the principles causing motion is not natural, that is to say there is no principle of motion in it. This is what moves perpetually without being moved, like the thing which is the First of all things and does not move at all. The philosopher means by his statement that the First Thing is the starting-point of the being of the whole world and is the mover of the spheres and the organiser of their order. He has already examined this critically in the second chapter of the *Physics*. He says that motion enters all things perpetually, establishes that by compelling proofs and [says that] effortless motion can only be in the moving bodies as circular motion. In order to show that, he starts his argumentation with a classification of motion and explains how it takes place. Afterwards he describes the linearly [moving] and the rotating bodies and says that it is not possible that there is any magnitude or anything among the natural linearly [moving] bodies or the rotating bodies which is infinite with regard to magnitude. He objects to that in the book which we have referred to and in the first chapter of the *Book on Heaven and Earth* and he objects to that with proof by the way of demonstration. After he has explained that no body can be infinite in magnitude, he investigates it saying that the body of finite magnitude ... He says there are necessarily other than these substances ... it could not possibly happen that something the generation of which has not required the Divine Cause could receive order, direction or arrangement from [this Divine Cause]. The philosopher has already mentioned [it] in his book which is called *Theology*, i.e. the *Book on Divinity* and there brought forward an example of its kind.

Bibliography

Primary Sources

ʿAbd al-Laṭīf al-Baghdādī, *ʿAbd al-Laṭīf al-Baġdādī's Bearbeitung von Buch* Lambda *der Aristotelischen* Metaphysik, ed. A. Neuwirth with German tr. (Wiesbaden: 1976).

al-Faṣl al-ʿishrūn min K. fī l-ʿIlm mā baʿd al-ṭabīʿa, ed. ʿA. Badawī in *al-Aflāṭūnīya al-muḥdatha ʿind al-ʿArab* , 2nd edn (Kuwait: 1977), 248–56.

Alexander of Aphrodisias, *De anima* in *Alexandri Aphrodisiensis Praeter commentaria scripta minora*, ed. I. Bruns, II 1 (Berlin: 1887), 1–100.

De intellectu in *Alexander Aphrodisiensis, De anima libri mantissa*, ed. R.W. Sharples (Berlin, New York: 2008), 42–52.

De intellectu in *Alexander of Aphrodisias. Supplement to 'On the Soul'*, tr. R.W. Sharples (London: 2004), 24–44.

Die arabische Übersetzung der Schrift des Alexander von Aphrodisias über die Farbe, ed. and tr. H. Gätje (Göttingen: 1967).

Maqālat al-Iskandar al-Afrūdīsī fī l-ʿAql, ed. J. Finnegan in 'Texte arabe du ΠΕΡΙ ΝΟΥ d'Alexandre d'Aphrodise', *Mélanges de l'Université Saint-Joseph* 33.2 (1956), 181–99.

Maqālat al-Iskandar al-Afrūdīsī fī l-ʿAql ʿalā raʾy Arisṭuṭālīs, ed. ʿA. Badawī in *Commentaires sur Aristote perdus en grec et autres épîtres* (Beirut: 1971), 31–42.

[Alexandrians], *Min K. al-Akhlāq: Ikhtiṣār al-Iskandarānīyīn*, ed. ʿA. Badawī in *al-Akhlāq* (Kuwait: 1979), 439–45.

Min K. al-Akhlāq Ikhtiṣār al-Iskandarānīyīn, ed. D.M. Dunlop in 'The Manuscript Taimur Pasha 290 aḫlāq and the Summa Alexandrinorum', *Arabica* 21.3 (1974), 260–3.

Summa Alexandrinorum, ed. G.B. Fowler in 'Manuscript Admont 608 and Engelbert of Admont (c. 1250–1331). Appendix 14: *Summa Alexandrinorum*', *Archives d'Histoire Doctrinale et Littéraire du Moyen Âge* 49 (1982), 195–252.

Summa Alexandrinorum, ed. C. Marchesi in *L'Etica nicomachea nella tradizione latina medievale* (Messina: 1904), XLI–LXXXVI; repr. in ʿA. Badawī, *al-Akhlāq* (Kuwait: 1979), 446–91.

ʿAlī ibn Rabban al-Ṭabarī, *Die propädeutischen Kapitel aus dem Paradies der Weisheit über die Medizin des ʿAlī b. Sahl Rabban aṭ-Ṭabarī*, German tr. A. Siggel (Mainz: 1953).

Firdaws al-ḥikma fī l-ṭibb, ed. M. al-Ṣiddīqī (Berlin: 1928).

Firdaws al-ḥikma fī l-ṭibb, ed. ʿA. al-Jundī (Beirut: 2002).

Mukhtaṣar Firdaws al-ḥikma, Istanbul, Süleymaniye Kütüphanesi, MS Aya Sofya 4857, fols. 5b-57b.

al-ʿĀmirī, Abū l-Ḥasan, *Faṣl*, Istanbul, Süleymaniye Kütüphanesi, MS Servili 179, fols. 110b-111a.

Kitāb al-Amad ʿalā l-abad, ed. and tr. E.K. Rowson in *A Muslim Philosopher on the Soul and its Fate: al-ʿĀmirī's Kitāb al-Amad ʿalā l-abad* (New Haven, Conn.: 1988), 52–177.

Kitāb al-Iʿlām bi-manāqib al-islām, ed. A. Ghurāb (Cairo: 1967).

Kitāb al-Fuṣūl fī l-Maʿālim al-ilāhīya, ed. E. Wakelnig with German tr. in *Feder, Tafel, Mensch. Al-ʿĀmirīs* Kitāb al-Fuṣūl fī l-Maʿālim al-ilāhīya *und die arabische Proklos-Rezeption im 10. Jh.* (Leiden, Boston: 2006).

Rasāʾil Abī l-Ḥasan al-ʿĀmirī wa-shadharātuhū al-falsafīya, ed. S. Khalīfāt (Amman: 1988).

Ps.-ʿĀmirī, *al-Fikr al-siyāsī wa-l-akhlāqī ʿinda al-ʿĀmirī. Dirāsat wa-taḥqīq Kitāb al-Saʿāda wa-l-isʿād fī l-sīra al-insānīya*, ed. A. ʿAṭīya (Cairo: 1991).

As-Saʿādah waʾl-isʿād (On Seeking and Causing Happiness) Written by Abūʾl-Ḥasan Muḥammad al-ʿĀmirī of Nēshābūr (992 A.D.), facsimile edn M. Minovi (Wiesbaden: 1957–8).

Anonymous, *Fuṣūl falsafīya majmūʿa min nawādir kathīra*, Istanbul, MS Köprülü 1608, fols. 38a-47b.

Kitāb al-Ḥikma, Istanbul, Süleymaniye Kütüphanesi, MS Esad Efendi 1933, fols. 56b-109b.

Mukhtār min Kalām al-ḥukamāʾ al-arbaʿa al-akābir, ed. and tr. D. Gutas in *Greek Wisdom Literature in Arabic Translation. A Study of the Graeco-Arabic Gnomologia* (New Haven: 1975), 62–213.

Risāla fī Ārāʾ al-ḥukamāʾ al-yūnānīyīn, ed. ʿA. Badawī in *Aflāṭūn fī l-Islām* (Tehran: 1974), 306–30.

Tamjīd al-falāsifa, Istanbul, MS Köprülü 1608, fols. 32a-33a.

al-Anṣārī, Muḥammad, *Ḥunayn ibn Isḥāq, Ādāb al-falāsifa ikhtaṣarahū Muḥammad ibn ʿAlī ibn Ibrāhīm ibn Aḥmad ibn Muḥammad al-Anṣārī*, ed. ʿA. Badawī (Kuwait: 1985).

Ps.-Apollonius, *Das 'Buch über das Geheimnis der Schöpfung' von Pseudo-Apollonios von Tyana*, ed. U. Weisser (Berlin, New York: 1980).

ʿArīb ibn Saʿīd al-Qurṭubī, *Kitāb Khalq al-janīn wa-tadbīr al-ḥabālā wa-l-mawlūdīn – Le livre de la génération du fœtus et le traitement des femmes enceintes et des nouveau-nés*, ed. H. Abdelkader and N. Jahier with French tr. (Algiers: 1956).

Aristotle, *De anima*, ed. W. D. Ross (Oxford: 1956).

De anima [Arabic]: *Fī l-nafs*, ed. ʿA. Badawī in *Arisṭūṭālīs Fī l-nafs*, 2nd edn (Kuwait, Beirut: 1980), 3–88.

De caelo, ed. D. J. Allan (Oxford: 1936, repr. 2005).

De caelo [Arabic]: *al-Samāʾ li-Arisṭūṭālīs*, ed. ʿA. Badawī in *Arisṭūṭālīs fī l-Samāʾ wa-l-āthār al-ʿulwīya* (Cairo: 1961), 125–87.

Ethica Nicomachea, ed. I. Bywater (Oxford: 1894, repr. 1962).

Ethica Nicomachea [Arabic]: *The Arabic Version of the Nicomachean Ethics*, ed. A. A. Akasoy and A. Fidora, tr. D. M. Dunlop (Leiden: 2005).

Ethica Nicomachea [Arabic]: *al-Akhlāq taʾlīf Arisṭūṭālīs* in ʿA. Badawī, *al-Akhlāq* (Kuwait: 1979), 51–387.

De generatione animalium, ed. H. J. Drossaart Lulofs (Oxford: 1965).

De generatione animalium [Arabic]: *Generation of Animals: The Arabic Translation Commonly Ascribed to Yaḥyā ibn al-Biṭrīq*, ed. J. Brugman and H. J. Drossaart Lulofs (Leiden: 1971).

Aristote. De la génération et de la corruption, ed. C. Mugler with French tr. (Paris: 1966).

Aristote. Histoire des animaux, ed. P. Louis with French tr., 3 vols. (Paris: 1964–9).

Historia animalium [Arabic]: *Arisṭūṭālīs, Ṭibāʿ al-Ḥayawān*, ed. ʿA. Badawī (Kuwait: 1977).

Metaphysica, ed. W. Jaeger (Oxford: 1957, repr. 1960, 1963, 1980).

Metaphysica [Arabic]: *Mā baʿd al-ṭabīʿa*, see Ibn Rushd.

Aristotelis meteorologicum libri quattuor, ed. F. H. Fobes (Cambridge, Mass.: 1919, repr. Hildesheim: 1967).

Aristote. Les parties des animaux, ed. P. Louis with French tr. (Paris: 1956).

De partibus animalium [Arabic]: *The Arabic Version of Aristotle's Parts of Animals, Books XI–XIV of the Kitāb al-Ḥayawān*, ed. R. Kruk (Amsterdam, Oxford: 1979).

Physica, ed. W. D. Ross (Oxford: 1956, repr. 1988).

Physica [Arabic]: *Arisṭūṭālīs, al-Ṭabīʿa tarjamat Isḥāq b. Ḥunayn*, ed. ʿA. Badawī, 2 vols. (Cairo: 1964–5, repr. vol. I, 1984).

De sensu and De memoria, ed. and tr. G. R. T. Ross (Cambridge: 1906).

The Complete Works of Aristotle: The Revised Oxford Translation, ed. J. Barnes, 2 vols. (Princeton: 1984).

(Aristotle), *De anima-paraphrase*, ed. R. Arnzen with German tr. in *Aristoteles'* De anima. *Eine verlorene spätantike Paraphrase in arabischer und persischer Überlieferung* (Leiden, New York, Cologne: 1998), 180–351.

[second version of the *DA*-paraphrase] *Kitāb fīhi Ṣifat al-nafs, ta᾿līf al-ḥakīm al-fāḍil Arisṭāṭālīs*, Istanbul, Süleymaniye Kütüphanesi, MS Aya Sofya 4156, fols. 54a-106a.

Ps.-Aristotle, *De virtutibus et vitiis libello*, ed. F. Susemihl in *Aristotelis Ethica Eudemia* (Leipzig: 1884, repr. Amsterdam: 1967), 181–94.

'Ein pseudoaristotelischer Traktat über die Tugend. Edition und Übersetzung der arabischen Fassungen des Abū Qurra und des Ibn aṭ-Ṭayyib', ed. M. Kellermann with German tr. (Diss. Universität Erlangen-Nürnberg: 1965).

Kitāb al-Ḥaraka, Ankara, Üniversitesi Dil ve Tarih-Coğrafya Fakültesi Kütüphanesi, MS Ismail Saib I 1696, fols. 93a-108b; Istanbul, Süleymaniye Kütüphanesi, MS Hacı Mahmud 5683, fols. 119b-140a.

Physiognomonica, ed. R. Foerster in *Scriptores physiognomonici*, I (Leipzig: 1893, repr. 1994), 1–90.

Physiognomonica [Arabic]: *Il Kitāb Arisṭāṭalīs al-faylasūf fī l-firāsa nella traduzione di Ḥunayn b. Isḥāq*, ed. A. Ghersetti with Italian tr. (Rome: 1999).

Risālat al-Siyāsa al-ʿāmmīya, ed. M. Grignaschi in 'La "Siyâsatu-l-ʿâmmiyya" et l'influence iranienne sur la pensée politique islamique', *Hommages et Opera Minora III. Monumentum H. S. Nyberg* (Tehran, Liège: 1975), 97–197.

Risālat al-Siyāsa al-ʿāmmīya, ed. M. Maróth in *The Correspondence between Aristotle and Alexander the Great: An Anonymous Greek Novel in Letters in Arabic Translation* (Piliscsaba: 2006), 25–84.

ThA, see Plotinus

Al-Baladī, Aḥmad b. Muḥammad b. Yaḥyā, *Tadbīr al-ḥabālā wa-l-aṭfāl wa-l-ṣibyān wa-ḥifẓ ṣiḥḥatihīm*, ed. M. Qāsim Muḥammad (Baghdad: 1987).

Al-Bīrūnī, Abū Rayhān Muḥammad b. Aḥmad, [*Taḥqīq mā li-l-Hind*] *Alberuni's India: An Account of the Religion, Philosophy, Literature, Chronology, Astronomy, Customs, Laws and Astrology of India about A.D. 1030*, ed. E. Sachau (London: 1887).

Alberuni's India: An Account of the Religion, Philosophy, Literature, Chronology, Astronomy, Customs, Laws and Astrology of India about A.D. 1030, tr. E. Sachau, 2 vols. (London: 1910).

Kitāb Taḥdīd nihāyāt al-amākin li-taṣḥīḥ masāfāt al-masākin, ed. P. Bulgakov (Cairo: 1962).

al-Daylami, Abū l-Ḥasan ʿAlī b. Muḥammad, *Kitāb ʿAṭf al-alif al-ma᾿lūf ʿalā al-lām al-maʿṭūf*, ed. J. C. Vadet (Cairo: 1962).

al-Fārābī, Abū Naṣr Muḥammad, *Al-Farabi on the Perfect State: Abū Nasr al-*

Fārābī's Mabādiʾ arāʾ ahl al-madīna al-fāḍila, ed. and tr. R. Walzer (Oxford: 1985).
Épître sur l'intellect, French tr. P. Vallat (Paris: 2012).
Iḥṣā al-ʿulūm, ed. ʿU. Amīn, 3rd edn (Cairo: 1968).
Jawāmiʿ Kitāb al-Nawāmīs, ed. ʿA. Badawī in *Aflāṭūn fī l-Islām* (Tehran: 1974), 34–83.
Jawāmiʿ Kitāb al-Nawāmīs: Alfarabius. Compendium legum Platonis, ed. F. Gabrieli (London: 1952).
Jawāmiʿ Kitāb al-Nawāmīs, ed. T.A. Druart in 'Le sommaire du livre des "Lois" de Platon (Ǧawāmiʿ Kitāb al-Nawāmīs li-Aflāṭūn) par Abū Naṣr al-Fārābī', *Bulletin d'Études Orientales* 50 (1998), 109–55.
Kitāb al-Siyāsa al-madanīya al-muqallab bi-Mabādiʾ al-mawjūdāt, ed. F.M. Najjār (Beirut, 1964).
Risāla fī l-ʿAql, ed. M. Bouyges, (Beirut: [1938]).
Galen, *Galeni Compendium Timaei Platonis*, ed. P. Kraus and R. Walzer (Nendeln, Liechtenstein: 1951, repr. 1973).
De foetum formatione [Greek]: *Galen. Über die Ausformung der Keimlinge*, ed. D. Nickel with German tr. (Berlin: 2001).
Maqāla fī l-Mawlūd li-sabʿa ashhar, ed. R. Walzer in 'Galens Schrift "Über die Siebenmonatskinder"', *Rivista degli Studi Orientali* 15 (1935), 323–57.
De methodo medendi, ed. C.G. Kühn in *Claudii Galeni opera omnia*, XI (Leipzig: 1826), 1–146.
Mukhtaṣar min Kitāb al-Akhlāq li-Jālīnūs, ed. ʿA. Badawī in *Dirāsāt wa-nuṣūṣ fī l-falsafa wa-l-ʿulūm ʿinda al-ʿArab* (Beirut: 1988), 190–211.
Mukhtaṣar min Kitāb al-Akhlāq li-Jālīnūs, ed. P. Kraus in 'Kitāb al-Akhlāq li-Jālīnūs', *Bulletin of the Faculty of Arts of the University of Egypt* 5.1 (1937), 1–51.
De nervorum dissectione [Arabic]: *Galen: 'Über die Anatomie der Nerven': Originalschrift und alexandrinisches Kompendium in arabischer Überlieferung*, ed. A. al-Dubayan with German tr. (Berlin: 2000).
De placitis Hippocratis et Platonis [Greek]: *Galen On the Doctrines of Hippocrates and Plato*, ed. and tr. P. De Lacy, 3 vols. (Berlin: 1978–84).
De propriis placitis, ed. V. Boudon-Millet and A. Pietrobelli in 'Galien Ressuscité: Édition princeps du texte grec du *De Propriis Placitis*', *Revue des Études Grecques* 118.1 (2005), 168–213.
De propriis placitis [partly Greek and Latin]: *Galen On My Own Opinions*, ed. and tr. V. Nutton (Berlin: 1999).
De semine [Greek]: *Galen On Semen*, ed. and tr. P. De Lacy (Berlin: 1992).

De symptomatum causis, ed. C.G. Kühn in *Claudii Galeni opera omnia*, VII (Leipzig: 1824), 85–272.

De temperamentis, ed. C.G. Kühn in *Claudii Galeni opera omnia*, I (Leipzig: 1824), 85–272.

Galeni De temperamentis libri III, ed. G. Helmreich (Leipzig : 1904).

De usu partium, ed. C.G. Kühn in *Claudii Galeni opera omnia*, III–IV (Leipzig: 1822).

Galeni De usu partium libri XVII, ed. G. Helmreich, 2 vols. (Leipzig: 1907–9).

De usu partium [English tr.]: *Galen On the Usefulness of the Parts of the Body*, tr. M.T. May, 2 vols. (Ithaca, NY: [1968]).

De usu partium [Arabic]: *Kitāb fī Manāfiʿ al-aʿḍāʾ*, Paris, Bibliothèque Nationale de France, MS Arabe 2853; Manchester, John Rylands Library, MS Arabic 809; Bethesda, Maryland, National Library of Medicine, MS A 30.1 [middle of Book I – middle of Book XV]; Madrid, Escorial, MS Arab. 850 [Books IX–XVII].

De usu partium [Book XVI, Arabic]: 'Galen On Nerves, Veins and Arteries', ed. and tr. E. Savage Smith, (Diss. University of Wisconsin: 1969).

Hippocrates, *Aphorismes*, ed. É. Littré with French tr. in *Œuvres complètes d'Hippocrate*, IV (Paris: 1844, repr. Amsterdam: 1962), 458–609.

Aphorismi [Arabic]: *Kitāb al-Fuṣūl*, Istanbul, Süleymaniye Kütüphanesi, MS Fatih 3565, fols. 1a-33a; Istanbul, Süleymaniye Kütüphanesi, MS Aya Sofya 3724, fols. 1b-17b; Istanbul, MS Raghib Pasha 1482, fols. 162b-177b.

De l'Aliment, ed. É. Littré with French tr. in *Œuvres complètes d'Hippocrate*, IX (Paris: 1861), 98–121.

De la Génération, de la nature de l'enfant et quatrième livre des maladies, ed. É. Littré with French tr. in *Œuvres complètes d'Hippocrate*, VII (Paris: 1851), 470–615.

Des Humeurs, ed. É. Littré with French tr. in *Œuvres complètes d'Hippocrate*, V (Paris: 1846), 476–503.

De humoribus et De alimento [Arabic]: *Kitāb Buqrāt fī l-akhlāṭ (Hippocrates: On Humours) and Kitāb al-Ghidhāʾ li-Buqrāṭ (Hippocrates: On Nutriment)*, ed. and tr. J.N. Mattock (Cambridge: 1971).

Kitāb al-Ajinna li-Buqrāṭ. On Embryos (On the Sperm and On the Nature of the Child), ed. and tr. M.C. Lyons and J.N. Mattock (Cambridge: 1978).

Ḥunayn ibn Isḥāq, *Adāb al-falāsifa*, see al-Anṣārī

Isagoge, ed. G. Maurach in 'Johannicius. Isagoge ad Techne Galieni', *Sudhoffs Archiv* 62 (1978), 148–74.

Isagoge, tr. E.T. Withington in 'The "Galenic" System of Medicine. Appendix IV', *Medical History from the Earliest Times* (London: 1894) 386–96.

al-Masāʾil ʿalā Ṭarīq al-taqsīm wa-l-tashjīr, Berlin, MS 6258 (= Sprenger 1885), fols. 1a-18b; Oxford, Bodleian, MS Huntington 602, fols. 48b-71a.

al-Masāʾil fī l-Ṭibb li-l-mutaʿallimīn, ed. M. ʿA. Abū Rayyān, M. M. ʿArab and J. M. Mūsā (Cairo, Beirut: 1978).

Mimmā jamaʿahū Ḥunayn min alfāẓ al-falāsifa fī l-musīqā, Istanbul, MS Köprülü 1608, fol. 52.

Risāla fī Dhikr mā turjima min kutub Jālīnūs, ed. G. Bergsträsser in *Ḥunain Ibn Isḥāq über die syrischen und arabischen Galen-Übersetzungen* (Leipzig: 1925).

Iamblichus, *Neuplatonische Pythagorica in arabischem Gewande. Der Kommentar des Iamblichus zu den Carmina aurea. Ein verlorener griechischer Text in arabischer Überlieferung*, ed. H. Daiber with German tr. (Amsterdam: 1995).

Ibn Abī Uṣaybiʿa, Aḥmad ibn a-Qāsim, *Kitāb ʿUyūn al-anbāʾ fī ṭabaqāt al-aṭibbāʾ*, ed. A. Müller, 2 vols. (Cairo, Königsberg: 1882–4).

Ibn al-ʿAmīd, Abū l-Faḍl, *Naturwissenschaft bei den Arabern im 10. Jahrhundert n. Chr. Briefe des Abū l-Faḍl ibn al-ʿAmīd (gest. 360/970) an ʿAḍudaddaula*, ed. H. Daiber with German tr. (Leiden, New York, Cologne: 1993).

Ibn Bakhtīshūʿ, Abū Saʿīd, *R. fī l-Ṭibb wa-l-aḥdāth al-nafsānīya – Über die Heilung der Krankheiten der Seele und des Körpers*, ed. F. Klein-Franke (Beirut: 1977).

Ibn Dāwūd al-Iṣfahānī, Abū Bakr Muḥammad, *Kitāb al-Zahrah (The Book of the Flower), The 1st Half*, ed. A. R. Nykl (Chicago: 1932).

Ibn Durayd, Muḥammad b. al-Ḥasan, *al-Mujtanā*, ed. ʿA. al-ʿAwwād ([Riyad]: 1996).

al-Mujtanā, ed. M. al-Dālī (Limassol: 1997).

Ibn Falaquera, Shem Tov, *Shelemut ha-maʿasim*, ed. R. Jospe in *Torah and Sophia: The Life and Thought of Shem Tov Ibn Falaquera* (Cincinnati: 1988), 417–38.

Ibn al-Jawzī, Abū l-Faraj ʿAbd al-Rahmān b. ʿAlī, *Dhamm l-hawā*, ed. A. ʿAṭā (Beirut: 1987).

Ibn Khaldūn, Abū Zayd ʿAbd al-Rahmān b. Muḥammad, *Ibn Khaldûn. The Muqaddimah. An Introduction to History*, tr. F. Rosenthal, 3 vols. (New York: 1958).

Ibn Māsawayh, *Maqāla fī l-Janīn wa-kawnihi fī l-raḥim*, ed. U. Weisser in 'Yūḥannā ibn Māsawaihs Abhandlung über die Leibesfrucht und ihre Entstehung im Mutterleib', *Zeitschrift für die Geschichte der Arabisch-Islamischen Wissenschaften* 8 (1993), 114–31.

Ibn al-Nadīm, Muḥammad ibn Isḥāq, *Kitāb al-Fihrist*, ed. G. Flügel, 2 vols. (Leipzig: 1871–2).

Ibn al-Qifṭī, Abū l-Ḥasan ʿAlī, *Ibn al-Qifṭī's Taʾrīḫ al-ḥukamāʾ*, ed. J. Lippert (Leipzig: 1903).

Ibn Rushd, Abū l-Walīd Muḥammad, *Averroès. Tafsīr Mā baʿd al-ṭabīʿat,* ed. M. Bouyges, 3 vols. (Beirut: 1938–52); 2nd edn (1967–73); 3rd edn (1990).

Ibn aṭ-Ṭayyib, *Ibn al-Ṭayyib's Commentary on Porphyry's* Eisagoge, ed. K. Gyekye (Beyrouth: 1986).

Proclus' Commentary on the Pythagorean Golden Verses, ed. and tr. N. Linley ([Buffalo, NY: 1984]).

Ikhwān al-Ṣafā', *Rasāʾil Ikhwān al-Ṣafāʾ*, 4 vols. (Beirut: 1983).

al-Isfizārī, Abū Ḥāmid, *Kitāb fī Masāʾil al-umūr al-ilāhīya*, ed. D. Gimaret in 'Un traité théologique du philosophe musulman Abū Ḥāmid al-Isfizārī (IVe*-Xe* S.)', *Mélanges de l'Université Saint-Joseph* 50.1 (1984), 207–52.

al-Kindī, Abū Yūsuf Yaʿqūb b. Isḥāq, *Risāla fī Ajzāʾ khabarīya fī l-mūsīqā*, ed. Y. Zakariyyāʾ in *Muʾallafāt al-Kindī al-musīqīya* (Baghdad: 1962), 95–110.

Rasāʾil al-Kindī al-falsafīya, ed. M. Abū Rīda, 2 vols. (Cairo: 1950–4).

'Trois épîtres d'al-Kindī', facsimile edn with French tr. in L. Veccia Vaglieri and G. Celentano, *Annali. Istituto Orientale di Napoli* 34.4 [N. S. 24] (1974), 536–62.

Koran, The Qur'an, tr. M. A. S. Abdel Haleem (Oxford: 2010).

al-Majūsī, ʿAlī ibn al-ʿAbbās, *Kitāb Kāmil al-Ṣinaʿa al-ṭibbīya*, 2 vols. (Cairo: [1877]).

al-Maqdisī, al-Muṭahhar b. Ṭāhir, *Kitāb al-Badʾ wa l-taʾrīḫ*, ed. C. Huart, 6 vols. (Paris: 1899).

Ps.-Māwardī, *Naṣīḥat al-mulūk*, ed. Kh. M. Khiḍr (Kuwait: 1983).

Miskawayh, Abū ʿAlī Aḥmad b. Muḥammad b. Yaʿqūb, *al-Fawz al-aṣghar. Le petit livre du salut*, ed. Ṣ. ʿUḍayma and French tr. R. Arnaldez ([Tunis]: 1987).

'al Fawz ul asghar. The Shorter Theology of Ibn Miskawayh', tr. J. W. Sweetman in *Islam and Christan Theology*, I.1 (London: 1945), 93–185.

al-Ḥikma al-khālida, ed. ʿA. Badawī (Cairo: 1952).

Risāla fī l-Nafs wa-l-ʿaql, ed. M. Arkoun in 'Deux épîtres de Miskawayh', *Bulletin d'Études Orientales* 17 (1961–2), 7–74.

Risāla fī l-Nafs wa-l-ʿaql, ed. ʿA. Badawī in *Dirāsāt wa-nuṣūṣ fī l-falsafa wa-l-ʿulūm ʿinda al-ʿArab* (Beirut: 1988), 57–97.

Risāla fī l-Nafs wa-l-ʿaql [French tr.]: G. Cuvelier in 'Les textes inédits attribués à Miskawayh: présentation et traduction', *Revue Philosophique de Louvain* 87 (1989), 215–34.

Tahdhīb al-akhlāq, ed. Q. Zurayk (Beirut: 1966).

Tahdhīb al-akhlāq [English tr.]: *The Refinement of Character*, tr. C. Zurayk (Beirut: 1968).

Tahdhīb al-akhlāq [French tr.]: *Traité d'éthique*, tr. M. Arkoun (Damascus: 1969).

Tajārib al-umam, ed. and tr. H.F. Amedroz and D.S. Margoliouth, 7 vols. (Oxford: 1920–1).

Tartīb al-Saʿādāt wa-manāzil al-ʿulūm, ed. ʿA. Awjabī in *Ganjīna-yi Bahāristān* (Tehran: 2000), 101–127.

'Textes inédits de Miskawayh (m. 421)', ed. M. Arkoun in *Annales Islamologiques* 5 (1963), 181–205.

Mubashshir ibn Fātik, Abū l-Wafāʾ, *Mukhtār al-Ḥikam wa-maḥāsin al-kalim*, ed. ʿA. Badawī, 2nd edn (Beirut: 1980).

Nemesius of Emesa, *Nemesii Emeseni De natura hominis*, ed. M. Morani (Leipzig: 1987).

De natura hominis [Arabic]: *Kitāb al-Abwāb fī ṭabīʿat al-insān*, ed. M. Haji-Athanasiou in 'Le traité de Némésius d'Émèse *De natura hominis* dans la tradition arabe', II (Diss. Université Panthéon – Sorbonne, Paris I : 1985).

Nemesius On the Nature of Man, tr. R.W. Sharples and P.J. van der Eijk (Liverpool: 2008).

Ps.-Nicolaus, *Maqāla fī l-Mudkhal ilā ʿilm al-akhlāq*, ed. ʿA. Badawī in *al-Akhlāq* (Kuwait: 1979), 394–431.

Plato, *Epistulae*, ed. I. Burnet in *Platonis Opera*, V (Oxford: 1907, repr. 1967).

Leges, ed. I. Burnet in *Platonis Opera*, V (Oxford: 1907, repr. 1967).

Phaedo, ed. I. Burnet in *Platonis Opera*, I (Oxford: 1900, repr. 1958).

Timaeus, ed. I. Burnet in *Platonis Opera*, IV (Oxford: 1902, repr. 1957).

Ps.-Plato, *Ḥujaj Aflāṭūn ʿalā baqāʾ al-nafs*, ed. ʿA. Badawī in *Arisṭū ʿind al-ʿArab* (Kuwait: ²1978), 73–4.

Maqāla fī Ithbāt al-nafs, ed. H. Daiber in 'Ein bisher unbekannter pseudoplatonischer Text über die Tugenden der Seele in arabischer Überlieferung', *Der Islam* 47 (1971), 29–34.

Risāla fī l-Radd ʿalā man qāla inna al-insān talāshā wa-fanā, ed. ʿA. Badawī in *Aflāṭūn fī l-Islām* (Tehran: 1974), 337–9.

Risāla fī l-Radd, ed. M. Türker with Turkish tr. in 'Un petit traité attribué à al-Fārābī', *Araştirma* 3 (1965), 51–63.

Risāla fī l-Radd, partial edn E. Cottrell with French tr. in 'L'Anonyme d'Oxford (Bodleian Or. Marsh 539): bibliothèque ou commentaire?', in C. D'Ancona (ed.), *The Libraries of the Neoplatonists*, (Leiden, Boston: 2007), 435–6.

Thamara laṭīfa min maqāyīs Aflāṭun fī anna al-nafs lā tafsud, ed. ᶜA. Badawī in *Aflāṭūn fī l-Islām* (Tehran: 1974), 331–2.

Plotinus, *Plotini opera*, ed. P. Henry and H.-R. Schwyzer, 3 vols. (Paris, Brussels: 1951–73).

Plotino. La Discesa dell'anima nei corpi (Enn. IV 8 [6]). Plotiniana Arabica (Pseudo-Teologia di Aristotele, Capitoli 1 E 7; 'Detti del sapiente greco'), ed. C. D'Ancona with Italian tr. (Padua: 2003).

Plotinus with an English translation, ed. and tr. A.H. Armstrong (Cambridge, Mass., London: 1966–88, repr. 2000–3).

'Plotiniana arabica', tr. G. Lewis in *Plotini Opera. Tomus II. Enneades IV–V*, ed. P. Henry and H.-R. Schwyzer (Paris, Brussels: 1959).

Risāla fī l-ᶜIlm al-ilāhī, ed. ᶜA. Badawī in *Aflūṭīn ᶜinda al-ᶜArab*, 3rd edn (Kuwait: 1977), 165–83.

Risāla fī l-ᶜIlm al-ilāhī, ed. P. Kraus with partial French tr. in G.C. Anawati, *Études de philosophie musulmane* (Paris: 1974), 182–221.

Texts of the Greek Sage [Arabic]: ed. and tr. F. Rosenthal in 'Aš-Šayḫ al-Yūnānī and the Arabic Plotinus Source', I–III, *Orientalia* NS 21 (1952), 461–492; 22 (1953), 370–400; 24 (1955), 42–66.

Texts of the Greek Sage [Arabic]: *Nuṣūṣ mutafarriqa li-Aflūṭīn waradat fī l-makhṭūṭ raqm 539 Mārsh sharqī bi-maktabat Būdliya bi-Uksfūrd*, ed. ᶜA. Badawī in *Aflūṭīn ᶜinda al-ᶜArab*, 3rd edn (Kuwait: 1977), 184–98.

Theologia Aristotelis [Arabic]: *Uthūlūjiyā Arisṭāṭālīs* in ᶜA. Badawī, *Aflūṭīn ᶜinda al-ᶜArab*, 3rd edn (Kuwait:1977), 1–164.

Theologia Aristotelis [Arabic]: *Kitāb Uthūlūjiyā Arisṭāṭālīs - Die sogenannte Theologie des Aristoteles*, ed. F. Dieterici (Leipzig: 1882).

Theologia Aristotelis [English tr. of the Arabic]: G.L. Lewis in 'A Re-examination of the so-called "Theology of Aristotle"' (Diss. University of Oxford: 1950).

Theologia Aristotelis [German tr. of the Arabic]: F. Dieterici in *Die sogenannte Theologie des Aristoteles* (Leipzig: 1883).

Porphyry, *Porphyrii Sententiae ad intelligibilia ducentes*, ed. E. Lamberz (Leipzig: 1975).

Proclus Diadochus, *The Elements of Theology*, ed. and tr. E.R. Dodds, 2nd edn (Oxford: 1963).

Proclus Arabus. Zwanzig Abschnitte aus der Institutio Theologica *in arabischer Übersetzung*, ed. G. Endreß with German tr. (Beirut: 1973).

Proclus Arabus [additions]: ed. and tr. F. Zimmermann in 'Proclus Arabus rides again', *Arabic Sciences and Philosophy* 4 (1994), 9–51.

'The Liber de causis (kalām fī maḥḍ al-ḫayr). A Study of Medieval Neoplatonism', ed. R.C. Taylor (Diss. University of Toronto: 1981), 136–277.

Liber de causis [Arabic]: *Kitāb al-Īḍāḥ fī l-Khayr al-maḥḍ*, ed. ʿA. Badawī in *ʿal-Aflāṭūnīya al-muḥdaṯa ʿinda al-ʿArab*, 2nd edn (Kuwait: 1977), 3–33.

Liber de causis [Arabic with German tr.]: *Die pseudo-aristotelische Schrift 'Über das reine Gute' bekannt unter dem Namen Liber de causis*, ed. O. Bardenhewer with German tr. (Freiburg im Breisgau: 1882, repr. Frankfurt: 2000).

Liber de causis II [Arabic]: ed. P. Thillet and S. Oudaimah with French tr., 'Proclus Arabe. Un nouveau *Liber de causis?*', *Bulletin d'Études Orientales* 53–4 (2001–2), 293–368.

Théologie platonicienne, ed. H.D. Saffrey and L.G. Westerink with French tr., 6 vols. (Paris: 1968–97).

Procli Diadochi *Tria opuscula* (*De providentia, libertate, malo*). Latine Guilelmo de Moerbeka vertente et Graece ex Isaacii Sebastocratoris aliorumque scriptis collecta, ed. H. Boese (Berlin: 1960).

Proclus (Procleius?), *Kalām li-Ubruqlus min kitābihī Usṭūkhūsīs al-ṣughrā ʿan al-makhṭūṭ raqm 539 Mārsh fī maktabat Būdliya bi-Uksfūrd*, ed. ʿA. Badawī in *al-Aflāṭūnīya al-muḥdaṯa ʿinda al-ʿArab*, 2nd edn (Kuwait: 1977), 257–8.

Ps.-Pythagoras, *Dicta aurea*, ed. E. Diehl and D. Young in *Theognis. Ps.-Pythagoras. Ps.-Phocylides. Chares. Anonymi aulodia*, 2nd edn (Leipzig: 1971), 86–94.

Waṣīyat Fīthāghūrus al-dhahabīya, ed. L. Cheikho, C. Eddé and L. Malouf in *Traités inédits d'anciens philosophes arabes musulmans et chrétiens avec des traductions de traités grecs d'Aristote, de Platon et de Pythagore par Isḥāq Ibn Honein* (Beirut: 1911, repr. Frankfurt: 1974), 59–63.

al-Qabīṣī, ʿAbd al-ʿAzīz b. ʿUthmān, *Al-Qabīsī (Alcabitius): The Introduction to Astrology*, ed. and tr. C. Burnett, K. Yamamoto and M. Yano (London: 2004).

Qusṭā ibn Lūqā, *Kitāb fī ʿAlal ikhtilāf al-nās fī akhlāqihim*, ed. P. Sbath in 'Le livre des caractères de Qosṭā ibn Loūqā grand savant et célèbre médecin chrétien au IXe siècle', *Bulletin de l'Institut d'Égypte* 23 (1941), 103–69.

Kitāb al-Farq bayn al-rūḥ wa-l-nafs, ed. H. Ülken in *Ibn Sina Risâleleri 2: Les opuscules d'Ibn Sina et Le livre de la différence entre l'esprit et l'âme par Qosta b. Luqa* (Ankara: 1953), 81–94.

Risāla fī l-Farq bayn al-rūḥ wa-l-nafs, ed. L. Cheikho in *al-Machriq* 14 (1911), 94–109.

Risāla fī l-Farq bayn al-rūḥ wa-l-nafs, ed. G. Gabrieli in 'La Risālah di Qusṭā

b. Lūqā "Sulla differenza tra lo spirit e l'anima"', *Rendiconti della Reale Accademia die Lincei, Classe di Scienze Morali, Storiche e filologiche, Serie Quinta* 19 (1910), 622–55.

Risāla fī l-Farq bayn al-rūḥ wa-l-nafs, ed. G. Troupeau, forthcoming.

Risāla fī l-Farq bayn al-rūḥ wa-l-nafs [English tr.]: J. W. Livingston in 'Qusṭā ibn Lūqā's Psycho-Physiological Treatise On the Difference between the Soul and the Spirit (Fī l-Farq bayn al-nafs wa-l-rūḥ)', *Scripta Mediterranea* 2 (1981), 63–74.

al-Rāzī, Abū Bakr, *A Treatise on the Small-pox and Measles by Abú Becr Mohammed ibn Zacaríyá ar-Rází (commonly called Rhazes)*, tr. W. A. Greenhill (London: 1848, repr. Birmingham, Ala.: 1987).

al-Riyāḍī, ʿAlī, *Kitāb Ṣaḥāʾif al-laṭāʾif*, Istanbul, Süleymaniye Kütüphanesi, MS Aya Sofya 4130.

Saʿīd ibn Hibat Allāh, Abū l-Ḥasan, *Khalq al-Insān*, ed. Y. Murād (Beirut: 2003).

al-Sāwī, ʿUmar ibn Sahlān, 'The "Mukhtaṣar Ṣiwān al-ḥikma" of ʿUmar b. Sahlān al-Sāwī', ed. R. M. Kartanegara (Diss. University of Chicago: 1996).

al-Shahrastānī, Muḥammad b. ʿAbd al-Karīm, *al-Milal wa-l-niḥal*, ed. ʿA. M. Al-ʿAbd, ([Cairo]: 1977).

Ps.-al-Sijistānī, [*Muntakhab*] *Ṣiwān al-ḥikmah wa-thalāth rasāʾil, taʾlīf Abū Sulaymān al-Manṭiqī al-Sijistānī*, ed. ʿA. Badawī (Tehran: 1974).

The Muntakhab Ṣiwān al-hikmah of Abū Sulaimān as-Sijistānī, ed. D. M. Dunlop (Hague, New York: 1979).

Stobaeus, *Ioannis Stobaei Anthologium*, ed. C. Wachsmuth and O. Hense, 5 vols. (Berlin: 1884–1912).

Ps.-Suidas, [*Suda*] *Suidae Lexicon*, ed. A. Adler, 5 vols. (Stuttgart: 1928–38).

al-Ṭabarī, Aḥmad b. Muḥammad, *al-Muʿālajāt al-Buqrāṭīya*, facsimile edn F. Sezgin, 2 vols. (Frankfurt: 1990).

al-Muʿālajāt al-Buqrāṭīya, Oxford, Bodleian Library, MS Marsh 158.

al-Tawḥīdī, Abū Ḥayyān, *Akhlāq al-wazīrayn*, ed. M. al-Ṭanjī (Damascus: 1965).

al-Baṣāʾir wa-l-dhakhāʾir, ed. W. al-Qāḍī, 10 vols. (Beirut: 1988).

Kitāb al-Imtāʿ wa-l-muʾānasa, ed. A. Amīn and A. al-Zayn, 3 vols. (Cairo: 1939–44); 2nd edn (1953).

al-Muqābasāt, ed. M. Ḥusayn (Baghdad: 1970).

al-Thaʿālibī, ʿAbd al-Malik b. Muḥammad, *Ādāb al-mulūk*, ed. J. al-ʿAṭiyya (Beirut: 1990).

Thābit ibn Qurra, *Maqāla fī Talkhīṣ mā atā bihi Arisṭūṭālīs fī kitābihi fīmā baʿd al-ṭabīʿa*, ed. and tr. A. Bertolacci and D. Reisman in 'Thābit ibn Qurra's

Concise Exposition' of Aristotle's Metaphysics', in R. Rashed (ed.), *Thābit ibn Qurra. Science and Philosophy in Ninth-Century Baghdad* (Berlin, New York: 2009), 715–76.

Themistius, *In Aristotelis metaphysicorum librum lambda paraphrasis hebraice et latine* in *Commentaria in Aristotelem Graeca V*, 5, ed. S. Landauer (Berlin: 1903).

Maqālat al-Lām, sharḥ Thāmisṭiyūs, tarjamat Isḥāq ibn Ḥunayn, ed. ʿA. Badawī in *Arisṭū ʿinda al-ʿArab*, 2nd edn (Kuwait: 1978), 329–33.

Min Sharḥ Thāmisṭiyūs li-ḥarf al-lām, ed. ʿA. Badawī in *Arisṭū ʿinda al-ʿArab*, 2nd edn (Kuwait: 1978), 12–21.

Thémistius. Paraphrase de la métaphysique d'Aristotle (livre lambda), tr. R. Brague (Paris: 1999).

Theophrastus, *On First Principles (Known as His Metaphysics): Greek Text and Medieval Arabic Translation*, ed. and tr. D. Gutas (Leiden, Boston: 2010).

al-Yaʿqūbī, Aḥmad b. Abī Yaʿqūb, *Ibn Wāḍiḥ qui dicitur al-Jaʿqūbī Historiae [Taʾrīkh]*, ed. M. Th. Houtsma (Leiden: 1883).

Yāqūt b. ʿAbd Allāh al-Hamawī, *Jacut's geographisches Wörterbuch*, ed. F. Wüstenfeld, 6 vols. (Leipzig: 1866–73).

Secondary Works

Adamson, P. *al-Kindī* (Oxford: 2007).

The Arabic Plotinus. A Philosophical Study of the Theology of Aristotle (London: 2002).

'The Kindian Tradition: The Structure of Philosophy in Arabic Neoplatonism', in C. D'Ancona (ed.), *The Libraries of the Neoplatonists* (Leiden, Boston: 2007), 351–70.

'Vision, Light and Color in al-Kindī, Ptolemy and the Ancient Commentators', *Arabic Sciences and Philosophy* 16 (2006), 207–36.

Afnan, S. M. *Philosophical Terminology in Arabic and Persian* (Leiden: 1964).

Altmann A., and S. M. Stern, *Isaac Israeli: A Neoplatonic Philosopher of the Early Tenth Century. His Works Translated with Comments and an Outline of His Philosophy* (Oxford: 1958, repr. Westport: 1979).

Anawati, G. C. 'Le néoplatonisme dans la pensée musulmane: État actuel des recherches', in his *Études de philosophie musulmane* (Paris: 1974), 155–221.

Aouad, M. '*La théologie d'Aristote et autres textes du* Plotinus Arabus', in R. Goulet (ed.), *Dictionnaire des philosophes antiques I* (Paris: 1989), 541–90.

Arberry, A.J. 'An Arabic Treatise on Politics', *Islamic Quarterly* 2.1 (1955), 9–22.

'Javidhan Khiradh', *Journal of Semitic Studies* 8.2 (1963), 145–58.

Arkoun, M. 'La conquête du bonheur selon Abū-l-Ḥasan al-ʿĀmirī', *Studia Islamica* 22 (1965), 55–90.

Arnzen, R. 'Arabisches Mittelalter', in C. Horn, J. Müller and J. Söder (eds.), *Platon Handbuch, Leben-Werk-Wirkung* (Stuttgart: 2009), 439–45.

Aristoteles' De Anima. *Eine verlorene spätantike Paraphrase in arabischer und persischer Überlieferung* (Leiden, New York, Cologne: 1998).

'On the Contents, Sources and Composition of Two Arabic Pseudo-Platonica: *Multaqaṭāt Aflāṭūn al-ilāhī* and *Fiqar ultuqiṭat wa-jumiʿat ʿan Aflāṭūn*', *Oriens* 37 (2009), 7–52.

Bergsträsser, G., *Neue Materialien zu Ḥunain ibn Isḥāq's Galen-Bibliographie* (Leipzig: 1932).

Bertolacci, A. 'On the Arabic translations of Aristotle's *Metaphysics*', *Arabic Sciences Philosophy* 15 (2005), 241–75.

Biesterfeldt, H.H., and D. Gutas, 'The Malady of Love', *Journal of the American Oriental Society* 104 (1984), 21–55.

Bürgel, J.C. 'A New Arabic Quotation from Plato's *Phaido* and its Relation to a Persian Version of the *Phaido*', in *Actas do IV Congresso de Estudos Árabes e Islamicos, Coimbra-Lisboa 1968* (Leiden: 1971), 281–90.

Cottrell, E. 'L'Anonyme d'Oxford (Bodleian Or. Marsh 539): bibliothèque ou commentaire?', in C. D'Ancona (ed.), *The Libraries of the Neoplatonists* (Leiden, Boston: 2007), 415–41.

Crone, P. *Medieval Islamic Political Thought* (Edinburgh: 2004).

Daiber, H. ʿAbū l-Farağ ibn aṭ-Ṭayyib on Plants: An Inquiry into his Sources' *Erdem* 9.26 (1996), 629–42.

'Hellenistisch-kaiserzeitliche Doxographie und philosophischer Synkretismus in islamischer Zeit', in W. Haase (ed.), *Aufstieg und Niedergang der Römischen Welt II: Principat 36.7* (Berlin, New York: 1986), 4974–92.

'New Manuscript Findings from Indian Libraries', *Manuscripts of the Middle East* 1 (1986), 26–48.

'Qosṭa ibn Lūqā (9. Jh.) über die Einteilung der Wissenschaften', *Zeitschrift für Geschichte der Arabisch-Islamischen Wissenschaften* (1990), 94–129.

review of N. Linley, 'Ibn aṭ-Ṭayyib. Proclus' Commentary on the Pythagorean Verses', *Der Islam* 66 (1988), 134–7.

D'Ancona, C. 'Divine and Human Knowledge in the Plotiniana Arabica', in

J.J. Cleary (ed.), *The Perennial Tradition of Neoplatonism* (Leuven: 1997), 419–42.

'La doctrine de la création "mediante intelligentia" dans le *Liber de causis* et dans ses sources', in C. D'Ancona, *Recherches sur le* Liber de causis (Paris: 1995), 73–95.

'Porphyry, Universal Soul and the Arabic Plotinus', *Arabic Sciences and Philosophy* 9 (1999), 47–88.

'Pseudo-Theology of Aristotle, Chapter I: Structure and Composition', *Oriens* 36 (2001), 78–112.

Recherches sur le Liber de causis (Paris : 1995).

'The Greek Sage, the Pseudo-*Theology of Aristotle* and the Arabic Plotinus', in R. Arnzen and J. Thielmann (eds.), *Words, Texts and Concepts Cruising the Mediterranean Sea* (Leuven, Paris, Dudley, Mass.: 2004), 159–76.

'Un Profilo filosofico dell'autore della "Teologia di Aristotele"', *Medioevo* 17 (1991), 83–130.

D'Ancona, C., and R.C. Taylor, 'Liber de causis', in R. Goulet (ed.), *Dictionnaire des philosophes antiques. Supplément* (Paris: 2003), 599–647.

Davidson, H.A. *Alfarabi, Avicenna, and Averroes on Intellect: Their Cosmologies, Theories of the Active Intellect, and Theories of Human Intellect* (New York, Oxford: 1992).

Delatte, A. 'La tétractys pythagoricienne', in A. Delatte, *Études sur la littérature pythagoricienne* (Paris: 1915), 247–68.

De Smet, D. *Empedocles Arabus. Une lecture néoplatonicienne tardive* (Brussels: 1998).

Diwald, S. *Arabische Philosophie und Wissenschaft in der Enzyklopädie. Kitāb Iḫwān aṣ-ṣafāʾ (III). Die Lehre von Seele und Intellekt* (Wiesbaden, 1975).

Domański, B. *Die Psychologie des Nemesius* (Münster: 1900).

Drossaart Lulofs, H.J. *Nicolaus Damascenus On the Philosophy of Aristotle. Fragments of the First Five Books Translated from the Syriac* (Leiden: 1965).

Drossaart Lulofs, H.J., and E.L.J. Poortman, *Nicolaus Damascenus De plantis* (Amsterdam: 1989).

Dunlop, D.M. 'Introduction', in A.A. Akasoy and A. Fidora (eds.), *The Arabic Version of the* Nicomachean Ethics (Leiden, Boston: 2005), 1–109.

'The Arabic Tradition of the Summa Alexandrinorum', *Archives d'Histoire Doctrinale et Littéraire du Moyen Âge* 49 (1982), 253–63.

'The Existence and Definition of Philosophy', *Iraq* 13 (1951), 76–94.

'The Manuscript Taimur Pasha 290 aḫlāq and the Summa Alexandrinorum', *Arabica* 21.3 (1974), 252–63.

*EI*²: *Encyclopaedia of Islam*, 2nd edn, 12 vols. (Leiden: 1960–2004).

Endreß, G. 'Mathematics and Philosophy in Medieval Islam', in J.P. Hogendijk and A.I. Sabra (eds.), *The Enterprise of Science in Islam: New Perspectives* (Cambridge, Mass., London: 2003), 121–76.

Proclus Arabus. Zwanzig Abschnitte aus der Institutio Theologica *in arabischer Übersetzung* (Beirut: 1973).

'The New and Improved Platonic Theology: Proclus Arabus and Arabic Islamic Philosophy', in A. Ph. Segonds and C. Steel (eds.), *Proclus et la Théologie Platonicienne* (Leuven, Paris: 2000), 553–70.

Fazzo, S. 'Nicolas, l'auteur du *Sommaire de la philosophie d'Aristote*. Doutes sur son identité, sa datation, son origine', *Revue des Études Grecques* 121 (2008), 99–126.

Ferrari, C. *Der Kategorienkommentar von Abū l-Farağ ʿAbdallāh ibn aṭ-Ṭayyib* (Leiden, Boston: 2006).

Finnegan, J. 'Texte arabe du ΠΕΡΙ ΝΟΥ d'Alexandre d'Aphrodise', *Mélanges de l'Université Saint-Joseph* 33.2 (1956), 159–202.

Freudenthal, J. 'Die durch Averroes erhaltenen Fragmente Alexanders zur Metaphysik des Aristoteles', *Abhandlungen der Königlichen Akademie der Wissenschaften zu Berlin, Philosophisch-Historische Klasse* (1885), 1–134.

Gätje, H. *Studien zur Überlieferung der aristotelischen Psychologie im Islam* (Heidelberg: 1971).

Geoffroy, M. 'La tradition arabe du Περὶ νοῦ d'Alexandre d'Aphrodise et les origines de la théorie farabienne des quatre degrés de l'intellect', in C. D'Ancona and G. Sera (eds.), *Aristotele e Alessandro di Afrodisia nella tradizione araba* (Padua: 2002), 191–231.

Goldziher, I. 'Säulenmänner im Arabischen', *Zeitschrift der Deutschen Morgenländischen Gesellschaft* 55 (1901), 503–8.

Grignaschi, M. 'La "Siyâsatu-l-ʿâmmiyya" et l'influence iranienne sur la pensée politique islamique', in *Hommages et Opera Minora III. Monumentum H.S. Nyberg* (Tehran, Liège: 1975), 33–286.

'La "Physiognomie" traduite par Ḥunayn ibn Isḥāq', *Arabica* 21.3 (1974), 285–91.

'Le roman épistolaire classique conservé dans la version arabe de Sâlim Abû-l-ʿAlâʾ', *Le Muséon* 80 (1967), 211–64.

'Les "Rasāʾil ʾArisṭāṭālīsa ʾilā-l-Iskandar" de Sālim Abū-l-ʿAlāʾ et l'activité culturelle à l'époque Omayyade', *Bulletin d'Études Orientales* 19 (1965–66), 7–83.

Gutas, D. *Greek Wisdom Literature in Arabic Translation: A Study of the Graeco-Arabic Gnomologia* (New Haven, Conn.: 1975).

'Paul the Persian on the Classification of the Parts of Aristotle's Philosophy: A Milestone between Alexandria and Baġdâd', *Der Islam* 60 (1983), 231–67.

'Review Article. On Graeco-Arabic Epistolary "Novels"', *Middle Eastern Literatures* 12.1 (2009), 59–70.

'The *Ṣiwān al-Ḥikma* Cycle of Texts [review of D.M. Dunlop, *The Muntakhab Siwân al-Ḥikmah of Abû Sulaimân as-Sijistânî*]', *Journal of the American Oriental Society* 102 (1982), 645–50.

Hansberger, R. 'Mediating the Medium: The Arabic Plotinus on Vision', in R. Hansberger, M. al-Akiti and C. Burnett (eds.), *Medieval Arabic Thought* (London, Turin: 2012), 61–76.

'Plotinus Arabus rides again', *Arabic Sciences and Philosophy* 21 (2011), 57–84.

'The Transmission of Aristotle's Parva Naturalia in Arabic' (Diss. University of Oxford: [2007]).

Harvey, S. 'A New Islamic Source of the Guide of the Perplexed', in A. Hyman (ed.), *Maimonidean Studies* 2 (New York: 1991), 31–60.

'The Greek Library of the Medieval Jewish Philosophers', in C. D'Ancona, *The Libraries of the Neoplatonists* (Leiden, Boston: 2007), 493–506.

Hasnawi, A. 'Deux textes en arabe sur les preuves platoniciennes de l'immortalité de l'âme', *Medioevo* 23 (1997), 395–408.

'Un élève d'Abū Bišr Mattā b. Yunus: Abū ʿAmr al-Ṭabarī', *Bulletin d'Études Orientales* 48 (1996), 35–55.

Hein, C. *Definition und Einteilung der Philosophie. Von der spätantiken Einleitungsliteratur zur arabischen Enzyklopädie* (Frankfurt am Main, Bern, New York: 1985).

Heinen, A. *Islamic Cosmology* (Beirut: 1982).

Johansen, T. 'Aristotle on the Sense of Smell', *Phronesis* 1996 (41), 1–19.

Jolivet, J. 'Pour le dossier du Proclus Arabus: Al-Kindī et la Théologie Platonicienne', *Studia Islamica* 49 (1979), 55–75.

Kraemer, J.L. *Humanism in the Renaissance of Islam: The Cultural Revival during the Buyid Age*, 2nd edn (Leiden: 1992).

Kraus, P. *Rasāʾil falsafiya li-Abī Bakr Muḥammad b. Zakariyāʾ al-Rāzī* (Cairo: 1939).

'Zu ibn al-Muqaffaʿ', *Rivista degli Studi Orientali* 14 (1934), 1–20.

Kruk, R. *The Arabic Version of Aristotle's Parts of Animals, Books XI–XIV of the Kitāb al-Ḥayawān* (Amsterdam, Oxford: 1979).

Kunitzsch, P. *Mittelalterliche astronomisch-astrologische Glossare mit arabischen Fachausdrücken*, (München: 1977).

Lane, E.W. *An Arabic-English Lexicon*, 8 vols. (London: 1863–93).

Leroux, G. 'Human Freedom in the Thought of Plotinus' in L.P. Gerson (ed.), *The Cambridge Companion to Plotinus* (Cambridge: 1996), 292–314.
Liddel, H.G., and R. Scott, *A Greek-English Lexicon* (Oxford: 1996).
Lindberg, D. *Theories of Vision from al-Kindī to Kepler* (Chicago: 1976).
Lyons, M.C. 'A Greek Ethical Treatise', *Oriens* 13–14 (1960–1), 35–57.
Maróth, M. *The Correspondence between Aristotle and Alexander the Great: An Anonymous Greek Novel in Letters in Arabic Translation* (Piliscsaba: 2006).
Meyerhof, M. 'ʿAlī Ibn Rabban Aṭ-Ṭabarī, ein persischer Arzt des 9. Jahrhunderts n. Chr.', *Zeitschrift der Deutschen Morgenländischen Gesellschaft* 85 (1931), 38–68.
Minovī, M. 'Az Khazāʾin-i Turkīya 2', *Majalla-yi Dānishshkada-yi Adabīyāt, Danišgāh-i Tehrān* 4.3 (1957), 53–89.
Nutton, V. *Ancient Medicine* (London, New York: 2004).
Galen On My Own Opinions (Berlin: 1999).
Peter, F.E. *Aristoteles Arabus: The Oriental Translations and Commentaries on the Aristotelian Corpus* (Leiden: 1968).
Pines, S. 'Hitherto Unknown Arabic Extracts from Proclus' Stoicheiōsis Theologikē and Stoicheiōsis Physikē', in S. Pines, *Collected Works of Shlomo Pines II: Studies in Arabic Versions of Greek Texts and in Mediaeval Science* (Jerusalem, Leiden: 1986), 287–93.
'Some Distinctive Metaphysical Conceptions in Themistius' Commentary on Book *Lambda* and their Place in the History of Philosophy', in J. Wiesner (ed.), *Aristoteles, Werk und Wirkung* II (Berlin, New York: 1987).
'Un texte inconnu d'Aristote en version arabe', *Archive d'Histoire Doctrinale et Littéraire du Moyen Âge* 23 (1956), 5–43.
al-Qāḍī, W. '*Kitāb Ṣiwān al-Ḥikma*: Structure, Composition, Authorship and Sources', *Der Islam* 58 (1981), 87–124.
Rocca, J. *Galen on the Brain: Anatomical Knowledge and Physiological Speculation in the Second Century A. D.* (Leiden, Boston: 2003).
Rosenthal, F. 'Abū Zayd al-Balkhī on Politics', in C.E. Bosworth, C. Issawi, R. Savory and A.L. Udovitch (eds.), *The Islamic World. From Classical to Modern Times. Essays in Honor of Berhard Lewis*, 3rd edn (Princeton, N.J.: 1991), 287–301.
'Al-Mubashshir: Ibn Fâtik. Prolegomena to an Abortive Edition', *Oriens* 13 (1960–1), 132–58.
'Arabische Nachrichten über Zenon den Eleaten', *Orientalia* 6 (1937), 21–67.
'Aš-Šayẖ al-Yûnânî and the Arabic Plotinus Source', *Orientalia* NS 21 (1952), 461–92; 22 (1953), 370–400; 24 (1955), 42–66.
Das Fortleben der Antike im Islam (Stuttgart: 1965).

'On the Knowledge of Plato's Philosophy in the Islamic World', *Islamic Culture* 14 (1940), 387–422.

'Sayings of the Ancients from Ibn Durayd's Kitāb al-Mujtanā', *Orientalia* 27 (1958), 29–54, 150–83.

'Some Pythagorean Documents transmitted in Arabic', *Orientalia* 10 (1941), 104–15, 383–95.

The Classical Heritage in Islam (London: 1992).

'The Symbolism of the Tabula Cebetis according to Abû l-Faraj ibn aṭ-Ṭayyib', in *Recherches d'islamologie: recueil d'articles offerts à Georges C. Anawati et Louis Gardet par leurs collègues et amis* (Louvain: 1977), 273–82.

'Two Graeco-Arabic Works on Music', *Proceedings of the American Philosophical Society* 110.4 (1966), 261–8.

Rowson, E. K. *A Muslim Philosopher on the Soul and its Fate: al-ʿĀmirī's Kitāb al-Amad ʿalā l-abad* (New Haven, Conn.: 1988).

'The Philosopher as Littérateur: Al-Tawḥīdī and his Predecessors', *Zeitschrift für Geschichte der arabisch-islamischen Wissenschaften* 6 (1990), 50–92.

Samir, SJ, K. 'Les versions arabes de Némésius de Ḥomṣ', in M. Pavan and U. Cozzoli (eds.), *L'eredità classica nelle lingue orientali.* Acta Encyclopaedica 5 (Rome: 1986), 99–151.

Savage-Smith, E. 'Galen's Account of the Cranial Nerves and the Autonomic Nervous System', *Clio Medica* 6 (1971), 77–98, 173–94.

Sebti, M. 'Une copie inconnue d'une paraphrase anonyme conservée en arabe du *De anima* d'Aristote: le MS Ayasofia 4156', in C. D'Ancona (ed.), *The Libraries of the Neoplatonists* (Leiden: 2007), 399–414.

Sharples, R. W. 'The Peripatetic Classification of Goods', in W. W. Fortenbaugh, *On Stoic and Peripatetic Ethics: The Work of Arius Didymus* (New Brunswick, London: 1983), 139–59.

Stern, S. M. 'Ibn al-Tayyib's Commentary on the "Isagoge"', *Bulletin of the School of Oriental and African Studies* 19.3 (1957), 419–25.

'New Information about the Authors of the "Epistles of the Sincere Brethren"', *Islamic Studies* 3 (1964), 405–28.

'The authorship of the Epistles of the Ikhwan-as-Safaʾ', *Islamic Culture* 20 (1946), 367–72.

'"The First in Thought is the Last in Action": The History of a Saying Attributed to Aristotle', *Journal of Semitic Studies* 7 (1962), 234–52.

Strohmaier, G. 'Avicennas Lehre von den "inneren Sinnen" und ihre Voraussetzungen bei Galen', in G. Strohmaier, *Von Demokrit bis Dante.*

Die Bewahrung antiken Erbes in der arabischen Kultur (Hildesheim, Zürich, New York: 1996), 330–41.

Swain, S. *Economy, Family, and Society from Rome to Islam: A Critical Edition, English Translation, and Study of Bryson's* Management of the Estate (Cambridge: 2013).

Taylor, R. 'Neoplatonic Texts in Turkey: Two Manuscripts containing Ibn Ṭufayl's "Ḥayy Ibn Yaqẓān", Ibn al-Sīd's "Kitāb al-Ḥadāʾiq", Ibn Bājja's "Ittiṣāl al-ʿAql bi-l-Insān", the "Liber de causis" and an Anonymous Neoplatonic Treatise on Motion', *Mélanges de l'Institut Dominicain d'Études Orientales* 15 (1982), 251–64.

Temkin, O. 'On Galen's Pneumatology', in O. Temkin, *The Double Face of Janus and Other Essays in the History of Medicine* (Baltimore: 1977), 154–61.

Thom, J. *The Pythagorean* Golden Verses (Leiden, New York, Cologne: 1995).

Türker, M. 'Un petit traité attribué à al-Fārābī', *Araştirma* 3 (1965), 1–63.

Ullmann, M. *Die Medizin im Islam* (Leiden, Cologne: 1970).

Die Nikomachische Ethik des Aristoteles in arabischer Übersetzung. Teil 2. Überlieferung. Textkritik. Grammatik (Wiesbaden: 2012).

Wörterbuch zu den griechisch-arabischen Übersetzungen des 9. Jahrhunderts (Wiesbaden: 2002).

Wörterbuch zu den griechisch-arabischen Übersetzungen des 9. Jahrhunderts. Supplement I–II (Wiesbaden: 2006–7).

Uri, J. and A. Nicoll, *Bibliothecæ Bodleianæ codicum manuscriptorum orientalium*, 2 vols. (Oxford: 1787–1821).

Van Bladel, K. *The Arabic Hermes - From Pagan Sage to Prophet of Science* (New York: 2009).

Veccia Vaglieri, L. and G. Celentano, 'Trois épîtres d'al-Kindī', *Annali. Istituto Orientale di Napoli* 34.4 [N. S. 24] (1974), 536–62.

Wakefield, C. 'Archbishop Marsh's Oriental Collections in the Bodleian Library. Raymond Gillespie Narcissus Marsh and his God', in M. McCarthy and A. Simmons (eds.), *The Making of Marsh's Library: Learning, Politics and Religion on Ireland, 1650–1750* (Dublin: 2004), 76–84.

Wakelnig, E. 'A New Version of Miskawayh's *Book of Triumph*: An Alternative Recension of *al-Fawz al-aṣghar* or the Lost *Fawz al-akbar*?', *Arabic Sciences and Philosophy* 19 (2009), 83–119.

Feder, Tafel, Mensch. Al-ʿĀmirīs Kitāb al-Fuṣūl fī l-Maʿālim al-ilāhīya *und die arabische Proklos-Rezeption im 10. Jh.* (Leiden, Boston: 2006).

'Fragments of the Hitherto Lost Arabic Translation of Galen's *On My Own Opinions* in the Philosophical Compilation Oxford, Bodleian Library, Oriental Collections, MS Marsh 539', in R. Hansberger, M. al-Akiti and

C. Burnett (eds.), *Medieval Arabic Thought: Essays in Honour of Fritz Zimmermann* (London, Turin: 2012), 221–38.

'Metaphysics in al-ʿĀmirī. The Hierarchy of Being and the Concept of Creation', *Medioevo*, 32 (2007), 39–59.

'Philosophical Fragments of al-ʿĀmirī Preserved Mainly in al-Tawḥīdī, Miskawayh and in the Texts of the *Ṣiwān al-Ḥikma* Tradition', in P. Adamson (ed.), *In the Age of al-Farabi: Arabic Philosophy in the Fourth/Tenth Century* (London: 2008), 215–38.

Walzer, R. 'A Diatribe of Galen', in R. Walzer, *Greek into Arabic: Essays on Islamic Philosophy* (Oxford: 1962), 164–74.

'Some Aspects of Miskawaih's *Tahdhīb al-Akhlāq*', in R. Walzer, *Greek into Arabic: Essays on Islamic Philosophy* (Oxford: 1962), 220–35.

Weisser, U. 'Noch einmal zur Isagoge des Johannicius: Die Herkunft des Lateinischen Lehrtextes', *Sudhoffs Archiv* 70 (1986), 229–35.

Zeugung, Vererbung und pränatale Entwicklung in der Medizin des arabisch-islamischen Mittelalters (Erlangen: 1983).

Werner, E., and I. Sonne, 'The Philosophy and Theory of Music in Judaeo-Arabic Literature', *Hebrew Union College Annual* 16 (1941), 251–319; 17 (1942–3), 511–73.

Westerink, L. G. 'Proclus commentateur des *Vers d'or*', in G. Boss and G. Seel (eds.), *Proclus et son influence* (Zürich: 1987), 61–78.

'Proclus on Plato's *Three Proofs of Immortality*', in *Zetesis. Album Amicorum Door Vrienden en Collega's aangeboden aan Prof. Dr. E. de Strycker* (Antwerpen, Utrecht: 1973), 296–306.

'The Alexandrian Commentators and the Introductions to their Commentaries', in R. Sorabji (ed.), *Aristotle Transformed: The Ancient Commentators and their Influence* (London: 1990) 325–48.

Witt, R. E. 'The Plotinian Logos and its Stoic Basis', *Classical Quarterly* 25.2 (1931), 103–11.

Zakeri, M. '*Ādāb al-falāsifa*: The Persian Content of an Arabic Collection of Aphorisms', *Mélanges de l'Université Saint-Joseph* 57 (2004), 173–90.

Persian Wisdom in Arabic Garb: ʿAlī b. ʿUbayda al-Rayḥānī (D. 219/834) and his Jawāhir al-kilam wa-farāʾid al-ḥikam, 2 vols. (Leiden, Boston: 2007).

Zimmermann, F. W. 'Proclus Arabus Rides Again', *Arabic Sciences and Philosophy* 4 (1994), 9–51.

'The Origins of the So-called *Theology of Aristotle*', in J. Kraye, W. F. Ryan and C. B. Schmitt (eds.), *Pseudo-Aristotle in the Middle Ages: The* Theology *and Other Texts* (London: 1986), 110–240.

Zonta, M. 'Aristote de Stagire. Les éthiques. Tradition syriaque et arabe', in

R. Goulet (ed.), *Dictionnaire des philosophes antiques, Supplément* (Paris: 2003), 191–8.

'Nicolas "de Laodicée" (Pseudo)' in R. Goulet (ed.), *Dictionnaire des philosophes antiques* IV (Paris: 2005), 680–3.

Un interprete ebreo della filosofia di Galeno. Gli scritti filosofici di Galeno nell'opera di Shem Tob ibn Falaquera (Turin: 1995).

Index

The numbers indicate page numbers in the introduction, the passages in brackets refer to the passages of the text, translation and commentary.

A person's name is in bold if there follow different works by this person.

For several works citations within these works are indicated (first are given the passages which refer to that work in general, then to chapters, then to lines).